Humanosphere Science School 2017 & 7th International Symposium for a Sustainable Humanosphere 2017

IOP Conference Series: Earth and Environmental Science
Volume 166

Bogor, Indonesia
1 - 2 November 2017

Editor:

Maya Ismayati

ISBN: 978-1-5108-6583-9
ISSN: 1755-1307

Printed from e-media with permission by:

Curran Associates, Inc.
57 Morehouse Lane
Red Hook, NY 12571

Some format issues inherent in the e-media version may also appear in this print version.

Printed by Curran Associates, Inc. (2018)

For permission requests, please contact the Institute of Physics at the address below.

Institute of Physics
Dirac House, Temple Back
Bristol BS1 6BE UK

Phone: 44 1 17 929 7481
Fax: 44 1 17 920 0979

techtracking@iop.org

Additional copies of this publication are available from:

Curran Associates, Inc.
57 Morehouse Lane
Red Hook, NY 12571 USA
Phone: 845-758-0400
Fax: 845-758-2633
Email: curran@proceedings.com
Web: www.proceedings.com

Table of contents

Volume 166

Humanosphere Science School 2017 & The 7th International Symposium for a Sustainable Humanosphere

1–2 November 2017, Bogor, Indonesia

Accepted papers received: 13 June 2018
Published online: 29 June 2018

Earth science

Community-based development and socioeconomic science

Applied science and technology

 doi:10.1088/1755-1315/166/1/011001

Foreword

Humanosphere Science School (HSS), since its initiation in 2006, has been dedicated to gather researchers, academicians, professionals and general public in sharing knowledge, disseminating research finding, exchanging success stories, and expanding both national and international collaboration. It has grown to becoming an international platform of science networking. This year forum also holds as the 360th Symposium on Sustainable Humanosphere for Kyoto University Japan and as the 7th International Symposium for Sustainable Humanosphere (ISSH) for Research Center for Biomaterials LIPI.

The organizing committee is gratefully acknowledging the support from many institutions and research programs in making this event successful. To name a few, the organizations are RISH and KURCA Kyoto University, JICA, JST, SATREPS, JASTIP, ARN, RC-Biomaterials LIPI, RC-Biology LIPI, ICIAR LIPI, and LAPAN. This year, HSS/ISSH had invited 16 distinguished Keynote Speakers who deeply care about the importance of our earth and environment sustainability. Our greatest appreciation to all of them for their substantial contributions.

We are also happy to welcome all presenters and participants, and would expect that with their various scientific backgrounds, i.e. earth science, bioscience, forest science, applied science and socioeconomic science, more ideas and innovation can be shared and collaborated, as an outcome of this platform. We are proud to announce that this year event had shown a significant increase on the number of presenters, with varied and exciting interdisciplinary topics, reflecting the forefront of global research, networking to the improvements of the humanosphere. We are excited to learn your presentation in both Oral and Poster Sessions. We wish you all the very best and an enjoyable meeting.

We look forward to seeing you all again in the future.

Dr. Riksfardini A. Ermawar
Chairperson of Organizing Committee 2017

Editorial Board

Organising Committee

General Coordinator	:	Prof. Dr. Sulaeman Yusuf, M.Agr. Prof. Hiroyuki Yano
Advisory Board	:	Prof. Takashi Watanabe Prof. Dr. Ir. Subyakto, M.Agr. Prof. Dr. Gadis Sri Haryani Dr. Wahyu Dwianto, M.Agr. Dr. Lisman Suryanegara, M.Agr. Dr. Ir. Euis Hermiati M.Sc. Dr. Dede Heri Yuli Yanto Ir. Halimurahman MT.
Chairperson	:	Dr. Riksfardini A. Ermawar, M.Bio.PB
Co- Chairperson	:	Yusup Amin, M.Si.
Secretary	:	Fahriya Puspita Sari, S.T. Eka Lestari, S. Hut. Febrina Dwiki Indriyani, S.I.P Linda Kriswati
Treasury and Meals	:	Triastuti, S.T. Dwi Ajias Pramasari, S.TP. Erlin Herlinawati, S.E
Program and Proceeding	:	Dr. Maya Ismayati, M. Eng Dr. Titik Kartika, M.Agr. Dr. Widya Fatriasari Dr. Firda Aulia Syamani Apriwi Zulfitri, M.Sc Ni Putu Ratna Ayu K, S.P., M.Si. Deni Zulfiana, M.Si. Anis Sri Lestari, M.S. Sita Heris Anita, M.Si Lilik Astari, M.For.Ecosys.Sc Wida Banar Kusumaningrum, M.T
Guest Editor for Proceeding	:	Prof. Dr. Ir. Eddy Hermawan M.Sc Prof. Dr. Henny Warsilah Dr. Puspita Lisdiyanti Irina Rafliana, M.Sc Amorisa Wiratri, MA Lidya Christin Sinaga, SIP, MA
Publication and Documentation	:	Syam Budi Iryanto, S. Kom. Agung Sumarno, M.T. Eko Widodo, S.T. Fazhar Akbar
Accommodation and Transportation	:	Herry Samsi, M.T. Adik Bahanawan, S.Hut.
Logistic	:	Teguh Darmawan, S.T. Aris Sigit Sarmanto Adi Muslia Muslim Sidki Awaludin
Sponsorship	:	Kurnia Wiji Prasetiyo, S.Hut., M.Si. Arief Heru Prianto, M.Si.

Reviewers

No	*Name*		*Affiliation*
1	Sayr Bahri, MT	:	Berlin Institute of Technology (TU Berlin), Germany
2	Dr. Tri Atmo Widi, M.Si.	:	Bogor Agricultural University, Indonesia
3	Ivan Permana Putra M.Si,	:	Bogor Agricultural University, Indonesia
4	Perdinan	:	Bogor Agricultural University, Indonesia
5	Prof. Aris Tri Wahyudi	:	Bogor Agricultural University, Indonesia
6	Ir. Sri Wahyuni	:	Center for Agro-socio-Economic and Policy Studies, Ministry of Agriculture, Indonesia
7	Prof. Dr. Eddy Hermawan, M.Sc	:	Center of Atmospheric Science and Technology, Indonesian National Institute of Aeronautics and Space (LAPAN), Indonesia
8	Dr. Arida Susilowati	:	Department of Forest Product, Faculty of Forestry, Universitas Sumatera Utara, Padang Bulan, Medan 20155, Indonesia
9	Dr. Ratih Damayanti	:	Forestry Engineering and Forest Product Processing R&D Center – FORDA -Ministry of Forestry, Indonesia
10	Dr. Ichsan Suwandhi	:	Institute Technology Bandung, Indonesia
11	Dr. Didi Tarmadi	:	Kyoto University, Japan
12	Dr. Sukma Surya Kusumah	:	Kyoto University, Japan
13	Dr. Nanang Masruchin	:	Kyungpook National University, Republic of Korea
14	Dr. Wheni Anastasia	:	Natural Materials Technology Research Institute, Indonesian Institute of Science (LIPI), Indonesia
15	Dr. Achmad Dinoto	:	Research Center for Biology, Indonesian Institute of Science (LIPI), Indonesia
16	Dr. Cahyo Rahmadi	:	Research Center for Biology, Indonesian Institute of Science (LIPI), Indonesia
17	Dr. Firda A.Syamani	:	Research Center for Biomaterial, Indonesian Institute of Science (LIPI), Indonesia
18	Dr. Khoirul HImmi Setiawan	:	Research Center for Biomaterial, Indonesian Institute of Science (LIPI), Indonesia
19	Dr. Maya Ismayati	:	Research Center for Biomaterial, Indonesian Institute of Science (LIPI), Indonesia
20	Dr. Wahyu Dwianto, M.Agr	:	Research Center for Biomaterial, Indonesian Institute of Science (LIPI), Indonesia
21	Prof Subyakto	:	Research Center for Biomaterial, Indonesian Institute of Science (LIPI), Indonesia
22	Prof. Dr. Sulaeman Yusuf, M.Agr	:	Research Center for Biomaterial, Indonesian Institute of Science (LIPI), Indonesia
23	Dr. Yantyati Widyastuti	:	Research Center for Biotechnology, Indonesian Institute of Science (LIPI), Indonesia
24	Dr. Yopi	:	Research Center for Biotechnology, Indonesian Institute of Science (LIPI), Indonesia
25	Dr.rer.nat. Fahrurrozi M.Si.	:	Research Center for Biotechnology, Indonesian Institute of Science (LIPI), Indonesia
26	Dra. Harmastini, M.Agr.Sc.	:	Research Center for Biotechnology, Indonesian Institute of Science (LIPI), Indonesia
27	Prof. Dr. Partomuan Simanjuntak	:	Research Center for Biotechnology, Indonesian Institute of Science (LIPI), Indonesia
28	Roni Ridwan, S.Pt, M.Si	:	Research Center for Biotechnology, Indonesian Institute of Science (LIPI), Indonesia
29	Esta Lestari	:	Research Center for Economy, Indonesian Institute of Sciences (LIPI), Indonesia
30	Dr. Rahmat Fajar Lubis	:	Research Center for Geotechnology, Indonesian Institute of

IOP Conf. Series: Earth and Environmental Science **166** (2018) 011004 doi:10.1088/1755-1315/166/1/011004

			Science (LIPI), Indonesia
31	Hilda Lestiana, S.Si, MT	:	Research Center for Geotechnology, Indonesian Institute of Science (LIPI), Indonesia
32	Dr. Herry Jogaswara, MA	:	Research Center for Population, Indonesian Institute of Sciences (LIPI),Indonesia
33	Gutomo Bayu Aji	:	Research Center for Population, Indonesian Institute of Sciences (LIPI), Indonesia
34	Dr. Cahyo Pamungkas	:	Research Center for Regional Resources, Indonesian Institute of Sciences (LIPI), Indonesia
35	Laksmi Rahmawati	:	Research Centre for Population, Indonesian Institute of Sciences (LIPI), Indonesia
36	Helbert, M.Sc	:	Tokyo University, Japan
37	Bobby Bachtiar	:	University of Birmingham, United Kingdom
38	Fikriyah Winata, M.S	:	University of Illinois at Urbana-Champaign, USA
39	Dr. Apri Heri Iswanto	:	University of North Sumatera, Indonesia
40	Dr. Samsuri	:	University of North Sumatera, Indonesia
41	Dr. Rida Oktorida Khastini	:	University of Sultan Ageng Tirtayasa, Indonesia

IOP Conf. Series: Earth and Environmental Science **166** (2018) 011005 doi:10.1088/1755-1315/166/1/011005

Peer review statement

All papers published in this volume of *IOP Conference Series: Earth and Environmental Science* have been peer reviewed through processes administered by the proceedings Editors. Reviews were conducted by expert referees to the professional and scientific standards expected of a proceedings journal published by IOP Publishing.

Morphological and Structural Changes in Microcrystalline Cellulose from OPEFB by Mechanical Grinding

T Herawan[1], F. R Panjaitan[1] and S Yamanaka[2]

[1]Research Group for Palm Oil Product Development & Quality
Indonesian Oil Palm Research Institute, IOPRI,
Brigjen Katamso 51-Medan 20158, Indonesia
[2]College of Environmental Technology, Muroran Institute of Technology
Mizumoto-cho 27-1, Muroran 050-8585, Japan

E-mail: panjaitanf@yahoo.com

Abstract. Microcrystalline cellulose derived from oil palm empty fruit bunch (OPEFB) was grinded in planetary ball mill with dry state (without solvent) and solvent-assisted (ethanol and acetone). The effect of dry state and solvent-assisted on morphological and structural changes of microcrystalline cellulose were investigated. The structure changes, including particle size, powder morphology, crystalline structure, and molecular structure during the mechanical grinding were investigated by Laser Diffraction Particle Size Analyzer, SEM, XRD and FT-IR, respectively. The original fibrous microcrystalline cellulose was changed into irregular shape with finer micronized particles by dry state and solvent-assisted. SEM results showed that solvent-assisted significantly prevented the agglomeration phenomena during the grinding process, compared to dry state. The crystallinity after 4h solvent-assisted grinding showed fairly low crystallinity, while amorphous characteristic was observed with dry state grinding. The solvent-assisted led the hydrophilic parts of microcrystalline cellulose become stiff during the grinding that might be less deformed, leading to a fairly retain in crystallinity. The finer micronized particles were obtained under acetone-assisted and its crystallinity was fairly kept. XRD results indicated that crystalline form of origin microcrystalline was not changed by mechanical grinding.

1. Introduction

Cellulose is an attractive material for sustainable substitute synthetic polymer, because it is renewable, abundant resources, biodegradable, and relatively cheap. Recently, microcrystalline cellulose (MCC) from cellulosic material has been recognised for its diverse applications such as biomedical and pharmaceutical, foods, sensor and detector, construction, packaging, cosmetics, oil absorbents, aerogels, and other industries [1, 2]. Alpha cellulose from MCC is a partially depolymerized cellulose and is composed of crystalline and amorphous domain. MCC biomass conversion to transportation fuels, chemical platforms, and others functionalized-cellulose has attracted extensive attention. In all these synthesis, the catalytic depolymerisation of MCC to glucose is an important step from which transportation fuels, chemical platforms, and others functionalized-cellulose are then produced [3]. However, utilization of MCC has several limitations. It has highly-ordered and recalcitrant structure which is difficult to hydrolyse and insoluble in common solvents and thus restricting its efficient applications. To improve the MCC reactivity towards catalytic depolymerisation prior to advance synthesis, physical pre-treatment-ball-milling is considered to be one of the facile techniques to its activation prior to the next process path [4]. Mechanical grinding, usually carried out by high-energy milling, refers to the use of friction, collision, impingement, shear or other mechanical actions to change the crystalline structure and physicochemical properties of solids [5]. Reduced the particle size, extent of specific surface area, amorphous cellulose, the crystalline structure is pre-requisite to make the MCC more accessible to degrading reagents. On these bases, the effect of dry state and solvent-assisted of mechanical grinding on morphological and structural of MCC was investigated. The significance of relate morphological and structural of MCC after grinding on transportation fuels, chemical platforms, and others functionalized-cellulose is our future research.

2. Material and Methods

2.1. Materials

α-Cellulose. α-Cellulose (raw material for OPEFB-MCC) was produced from oil palm empty fruit bunch (OPEFB). OPEFB cellulose has 88% of α-Cellulose content. The production of α-Cellulose was prepared from OPEFB by digested in sodium hydroxide to remove lignin and hemicellulose then bleached in sodium hypochlorite. Alkali hydrolysis was performed by soaking cellulose in sodium hydroxide for 12 hours. After soaking, cellulose was neutralized with water, dried and then ground to form powder cellulose. Detail methods described in the previous paper [6]. Acetone (≥99.5%), ethanol (99.5%), sodium hydroxide and sodium hypochlorite were purchased from Merck, Indonesia.

2.2. Preparation of OPEFB-MCC

OPEFB-MCC was obtained from α-Cellulose-OPEFB, as described in [7]. The chemical used for soaking α-Cellulose-OPEFB was sodium hydroxide (NaOH) 4%. The obtained slurry was filtered and washed for several times with distilled water. The hydrolysis of soaked α-Cellulose-OPEFB was carried out with sulfuric acid (H_2SO_4) 6% for 2 h. At the end of hydrolysis process, the suspension (called as OPEFB-MCC) was centrifuged to remove the excessive acid solution and washed with distilled water. Finally, OPEFB-MCC oven dried at 60°C.

2.3. Mechanical grinding of OPEFB-MCC

OPEFB-MCC was ground in a planetary ball mill (Fritsch Pulverisette-7 premium line, Germany). The planetary ball mill was equipped with two zirconia pots of 80 cm^3 volume, each containing 80 g of yttria-stabilized zirconia milling beads (3 mm diameter) in close system. According to the manufacturer (Nikkato Corporation, Japan) the grinding media has a density of 5.950 kg m^{-3} and chemical composition of 95% of zirconia and 5% yttria. Mechanical grinding treatments adapting the method as described in [8] based on original procedures described in [9]. Solvent-assisted used were ethanol and acetone. Mechanical grinding operation was stopped for 15 min after every 1 h of grinding to prevent the inside of the milling pot from overheating. The conditions used are listed in **Table 1**.

Table 1. Mechanical grinding of OPEFB-MCC conditions

Process conditions	
Zirconia pot volume, V	80 cm^3
Mass of milling beads, W_B	80 g
Bead-filling ratio relative to the pot volume, J^a	0.28
Bead diameter	3 mm
OPEFB-MCC weight charge in the pot, W_S	4 g
Sample loading ratio to void fraction of the beads, U^b	0.28
Revolution speed	450 and 900 rpm
Grinding time	30 min;4 h
Solvent-assisted weight charge (ethanol or acetone)	0.5 g

^{a} $J = [W_B/\{\rho_{ball} \times (1\text{-}0.4)\}]/V$
^{b} $U = (W_S/\rho_p)/(0.4 \times J \times V)$

2.4. Characterization

2.4.1. Particle size distribution determination

Particle size distribution of MCC was measured using a laser-diffraction analyser (MicroTrac MT3000EX, Nikkiso, Japan). All MCC samples were sonicated in an ultrasonic processor for 180 s before measuring.

2.4.2. Morphology

The morphology studies of the MCC were performed using a scanning electron microscope (SEM, JEOL JSM-6380A) with an acceleration voltage of 20 kV. Prior to analysis, the sample was covered a carbon film.

2.4.3. XRD analysis

Crystalline structures of MCC were investigated by using XRD on a Multi Flex-120 NP Rigaku (Japan) equipped with a Cu-Kα anode (λ = 1.5418 Å), operated at a tension and current of 40 kV and 20 mA, respectively. Measurements were recorded at room temperature over the 2θ range 3–50°, with a 0.02° step size. The crystallinity and the crystallite size were calculated using previous method in [10].

2.4.4. Functional groups

Surface functional groups of MCC were determined from their infrared spectra analyzed by FTIR (JASCO FTIR-460 PlusK spectrometer). The spectrum was recorded in the range 4000 to 400 cm^{-1} at a resolution of 4 cm^{-1} and 32 scans.

3. Results and Discussion

3.1. Morphological Changes

XRD analysis was performed in order to evaluate structural changes undergone by MCC as a function of solvent-assisted and revolution speed of the mechanical grinding. XRD pattern of the initial OPEFB-MCC was shown in **Figure 1** (a) - (b). The crystalline structure was cellulose II showing its reflections around 2θ= 12, 20 and 22 [10,11]. After 4 h mechanical grinding course, the original fibrous MCC was converted into the aggregated globular particles resulting in decrease of the crystalline index (CrI). As shown in **Figure 1** (a), the grinding with dry state (4 h and 450 rpm) caused the characteristic shape of cellulose II crystal to completely lost (amorphous). The CrI of MCC under solvent-assisted with acetone or ethanol with low revolution speed (450 rpm) was fairly kept, but with higher revolution speed (900 rpm) it resulted some amorphous crystals. According to Gong et al (2010), the CrI was calculated, and the CrI values of acetone-assisted, ethanol assisted and dry state were (20° = 74.38%; 22° = 64.39%), (20° = 55.68%; 22° = 56.62%), and amorphous, respectively [10]. The initial state of CrI was (20° = 84.53%; 22° = 82.97%). With further increase in revolution speed, 900 rpm, the crystal became amorphous in shorter time compared to revolution speed 450 rpm. These are attributed to the destruction of macromolecular chains and crystalline structure linked with ball impact and collision during the grinding course [12]. By increasing the milling speed improves the size reduction as shown in **Figure 1** (b). Milling speed seems to have positive effect on size reduction. It can be observed clearly that mechanical grinding caused decrease in peak intensity of MCC and broadening of diffraction peaks (**Figure 1** (a) and (b)). According to author in [13], the structures of MCC damaged during the grinding thus, resulting in the variation of morphologies such the increase of amorphous region ratio and hydrogen bond energy, and decrease crystallinity and crystalline size. The crystalline form of origin MCC was not changed by mechanical grinding.

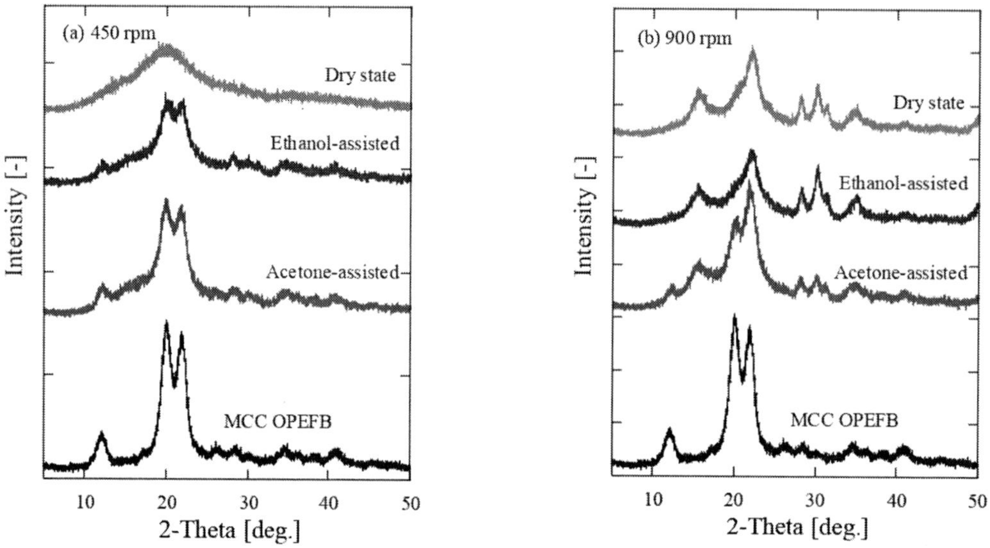

Figure 1. The XRD patterns (a) mechanical grinding with 450 rpm and 4 h; (b) 900 rpm and 30 min

3.2. Distribution of particle size

Mechanical grinding decreases particle size of MCC as results from impact, collision, friction and shear forces generated during the grinding course. This phenomenon changes the crystalline structure, induces the crystal defect and dislocation. At the same time, the grinding reduces the particle size and increases the specific surface area [4,8]. Using solvent-assisted method, coarse particles of MCC become finer (**Figure 2** and **3**). The D_{50} decreased from 435.38 μm to 6.54 μm, 7.51μm, and 61.41μm using acetone-assisted, ethanol-assisted and dry state methods at 450 rpm and 4 h, respectively. Increased in D_{50} was observed (**Figure 3**) when higher revolution speed of 900 rpm was applied. The D_{50} values are 11.27 μm, 21.148 μm, and 42.65 μm acetone-assisted, ethanol-assisted and dry state methods, respectively. This phenomenon was in agreement with Panjaitan et al (2016), by the SEM analysis, revealed agglomeration occurring during the grinding course [8]. The reason for distinguished D_{50} results on applying solvent-assisted was the interaction of solvents (ethanol or acetone) which partially coat the MCC during the grinding, thus prevents the agglomeration.

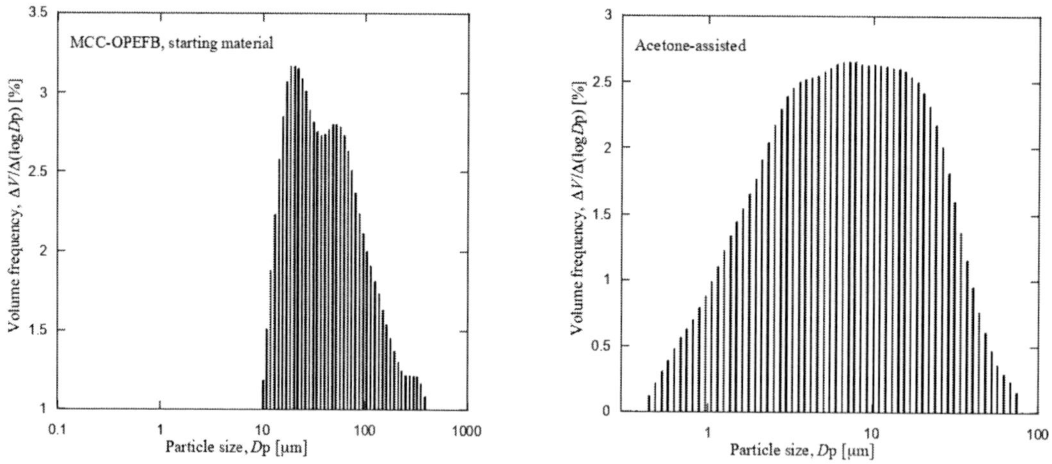

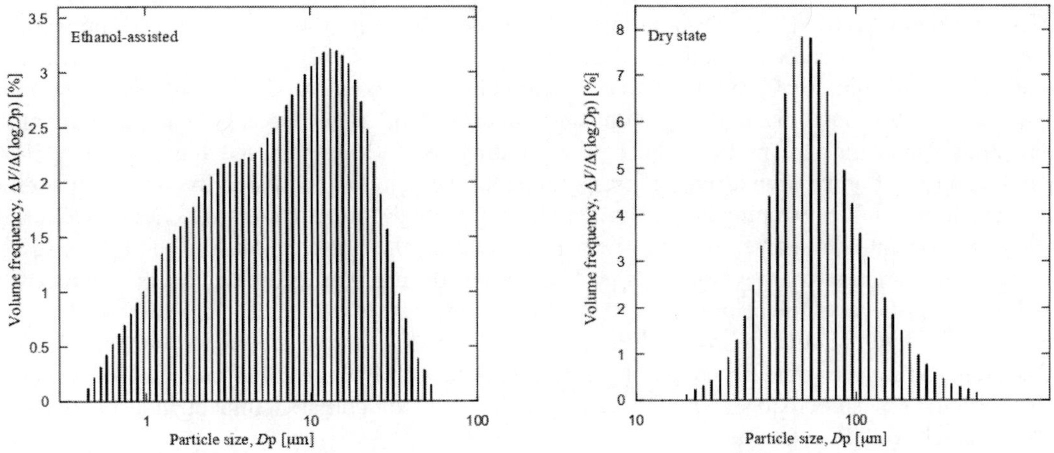

Figure 2. Distribution particle size of MCC at mechanical grinding with 450 rpm, 4 h

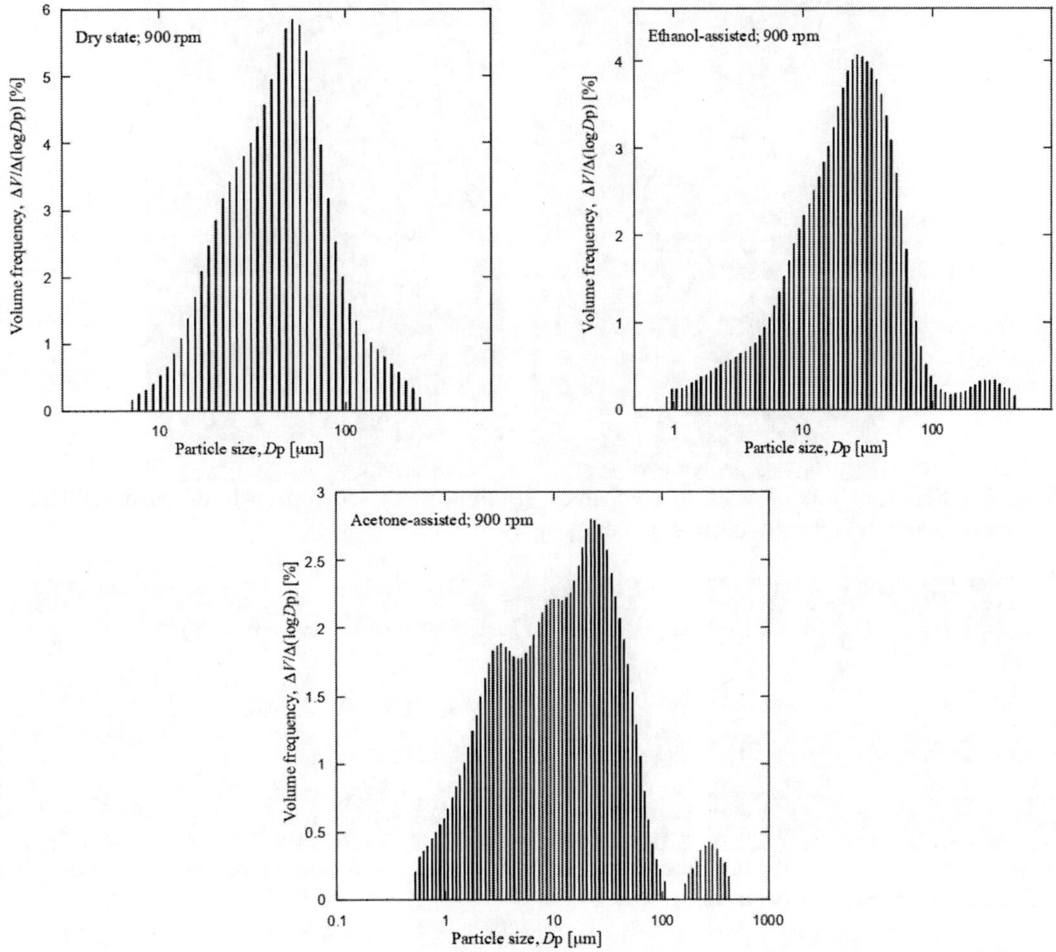

Figure 3. Distribution particle size of MCC at mechanical grinding with 900 rpm, 30 min

3.3. SEM Analysis

The morphologies of MCC after grinding course under different grinding conditions were shown in **Figure 4** and **5**. Mechanical grinding with solvent-assisted method generated better attrition and fracture in micronizing MCC compared to dry state method. Interaction of solvent during the grinding course is probably not only reacting with amorphous site of MCC leading to easier micronizing process but also coating the particles during the grinding course to prevent the agglomeration. Higher revolution speed (**Figure 5**) induced earlier agglomeration due to high collision frequency. In addition, it is possible that solvent coating concentration was not enough to reduce surface energy between the particles to prevent adhesiveness among the particles during the grinding course. The use of solvent-assisted and variation of revolution speed methods during the grinding course induced the transformation of crystalline cellulose of MCC. As a result, the MCC including crystalline, amorphous parts and their hydrophobic and hydrophilic planes [11] led the unique interaction between the MCC and solvent during the grinding course. Hydrophilic parts of MCC help the mobilization of solvent molecules to be adsorbed together with energy as a result of collision impact induced the cross-sectional cracks on MCC fibrous.

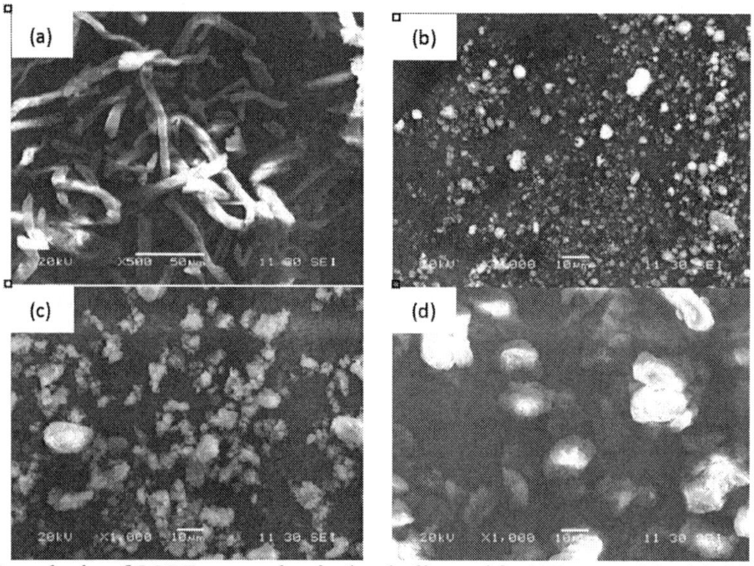

Figure 4. SEM analysis of MCC at mechanical grinding with 450 rpm, 4 h. (a) MCC OPEFB; (b) acetone-assisted; (c) ethanol-assisted; (d) dry state

Figure 5. SEM analysis of MCC at mechanical grinding with 900 rpm, 30 min. (a) acetone-assisted; (b) ethanol-assisted; (c) dry state

3.4. FT-IR Analysis

In order to understand the changes of MCC molecular groups during the mechanical grinding, FT-IR analysis was carried out. **Figure 6** presents the FT-IR spectra before and after mechanical grinding process. The difference spectra between the products and the original in the region O-H stretching from 3000 to 4000 cm^{-1} and C-O stretching vibrations from 800 to 1400 were compared among all products. The spectra in each region O-H stretching and C-O stretching was similar between the product with solvent-assisted and the original obtained. However, with dry state conditions, both in 450 and 900 rpm, each region O-H stretching and C-O stretching becomes wider in peak. According to XRD analysis (**Figure 1**), the crystals were amorphous. The O-H stretching bands of the products shifted toward lower wave numbers around 3350-3360 cm^{-1}. Possibly, during the grinding course, the MCC released free hydroxyl groups or ones weak hydrogen bonds causing the fracture in internal molecules of MCC [13].

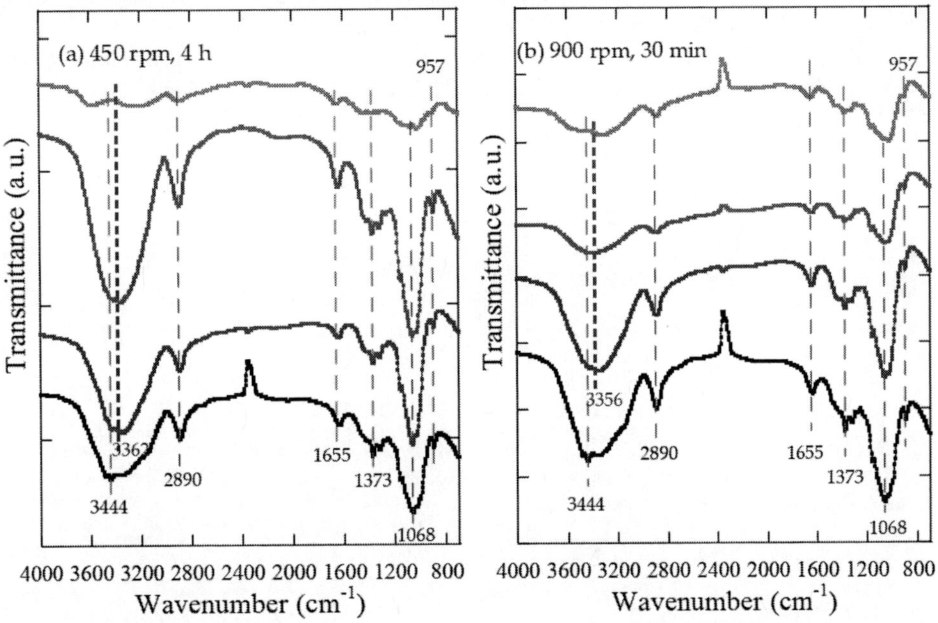

Figure 6. FT-IR spectra of MCC. (a) mechanical grinding with 450 rpm and 4 h; (b) 900 rpm and 30 min

4. Conclusion

The solvent-assisted method used in this work, i.e., acetone and ethanol affected the morphological and structural of MCC OPEFB under the mechanical grinding, giving the specific crystallinity, particle size distribution, and morphologies. The specific changes may be achieved by the solvent interaction and also the revolution speed of grinding process.

Acknowledgment
The authors would like to thank Oiso Takahiro and Yamanishi Tomoko in Surface and Chemical Engineering Lab., Muroran Institute of Technology, Hokkaido for their supporting in particle size distribution analyzing.

5. References

[1] Korhonen Juuso T, Kettunen M, H A Ras Robin and Ikkala O 2011 *ACS Appl. Mater. Interfaces* **3** 1813
[2] Haafiz M K Mohamad, Hassan A, Zakaria Z and Inuwa I M 2014 *Carbohydrate Polymers* **103** 119
[3] Zhang Q, Benoit M, De Oliveira Vigier K, Barrault J, Jegou G, Philippe M and Jerome F 2013

Green Chem. **15** 963
[4] Yuan X, Liu S, Feng G, Liu Y, Li Y, Lu H and Liang B 2016 *Korean Journal of Chemical Engineering* **33** 2134
[5] Ago M, Endo T and Hirotsu T 2004 *Cellulose* **11** 163
[6] Herawan T and Rivani M 2015 *Advanced Materials Research* **1123** 173
[7] Herawan T and Rivani M 2013 *J. Pen. Kelapa Sawit* **21** 31
[8] Panjaitan F R, Yamanaka S and Kuga Y 2016 *Advanced Powder Technology* **28** 1627
[9] Yamanaka S, Suzuma A, Fujimoto T and Kuga Y 2013 *J. of Nanoparticle Research* **1573** 1
[10] Gong J, Li J, Xu J, Xiang Z and Mo L 2017 *RSC Adv.* **7** 33486
[11] Ago M, T. Endo and K. Okajima, 2007 *Polymer Journal* **39**(5) 435
[12] Huang Z, Liang X, Hu H, Gao L, Chen Y and Tong Z, 2009 *Polym Degrad Stab* **94** 1737
[13] Hu, H, Y Zhang, X Liu, Z Huang, Y Chen, M Yang, X Qin and Z Feng, 2014 *Polym. Bull* **71** 453

Characterization of Polypropylene Composite Reinforced with Wood Flour or Cellulose Fiber

I Risnasari[1], E Herawati[1] and E. N Sirait[1]

[1]Faculty of Forestry, University of Sumatera Utara, Jl. Tridharma Ujung No.1 Kampus USU, Medan, Indonesia

E-mail: iwanrisnasari@gmail.com

Abstract. The use of lignocellulosic materials as reinforced polymer composite has been in great demand during the last decade due to some of its advantages. The lignocellulosic material may be in the form of wood flour or cellulose fiber, both of which have different properties and structures due to effects of the resulting composite product. This study aims was to compare lignocellulosic materials as reinforcing agent in polypropylene composite, i.e in the form of wood flour or cellulose fibers. The characterization conducted were the physical and mechanical properties of the composite. The characteristics of lignocellulose materials were investigated using FTIR and XRD analysis. Cellulose fibers were obtained by pulping with an alkaline solution, followed by bleaching treatment. The results showed that the chemical treatments could increased the crystallinity of the fibers, therefore resulted in an increase of physical and mechanical properties of polypropylene composites.

1. Introduction

The use of natural fibers in the plastic composite products is in great demand now days. Natural fiber has several advantages, such as low cost, easy processing and biodegradable [1,2]. The problem in making plastic composites so far is the compatibility between hydrophilic cellulose and the hydrophobic polymer matrix. The content of wood materials, consisting of cellulose (40-50%), hemicellulose (20-30%), lignin (20-30%) and small amounts of inorganic and extractive materials [1,3]. Wood has hygroscopic properties (can absorb and release water from the surrounding environment) due to the presence of hydroxyl groups in the wood component. Therefore wood has hydrophilic properties (strong water absorbing ability), rigidity and can be biologically degraded. This property causes cause wood is not easy to be combined with non-organic materials such as hydrophobic (water-resistant) plastics. Combination of these two different properties materials will result in weak bond between fiber and matrix, and make it incompatible with a hydrophobic plastic matrix, due to the low wettability and weakness of the interface [4, 5, 6]. A weak bond between a hydrophilic fiber and a hydrophobic matrix can affect the physical and mechanical properties of the composite product. Faruk *et.al* [7] explains that in order to increase the interfacial adhesion between fibers and matrices, modifications must be made to one of the components. Modification of fibers can be done by physical and chemical methods. Corona / plasma and mercerisation are used to influence mechanical bonding with the polymer matrix. While the chemical fiber modification is used to form reaction of the hydroxyl group on the fiber with the appropriate reactive compound to form a covalent bond. The newly formed group will act as an interface between the fibers and matrix resulted in better

IOP Conf. Series: Earth and Environmental Science 166 (2018) 012002 doi:10.1088/1755-1315/166/1/012002

properties composites. Another chemical modification is the use of additional component to matrix and fiber known as compatibilizer or coupling agents. The coupling agent serves to modify one or both composite components in order to improve the adhesion [8]. In this study, plastic composites were made using fillers of wood powder and purified fibers (cellulose fibers), then evaluate the characteristics of the plastic composite.

2. Materials and methods

The material used in this research was *Eucalyptus pellita* wood fiber obtained from PT. TPL sector AekNauli. Polypropylene impact copolymer (block copolymer) BI9.0GA was obtained from PT. Chandra Asri Petrochemical, Cilegon. The most important characteristic of the BI9.0GA impact copolymer (block copolymer) is the impact resistance(not easily broken), low temperature resistance (reaching -30°C), melt flow rate (MFR) 9.0 g / 10 min and density 0.9 g / cm^3. The coupling agent used was Licocene PPMA 6452 TP (Clariant Chemical, Ltd. German). Characteristics of Licocene PPMA 6452 TP are acid number: 41 mg KOH / g, softening point: 143°C, density: 0.93 g / cm^3 and 7% grafting rate of maleic anhydride. Chemicals used in this study (pro analysis grade) include sulfuric acid (95%), ethanol, benzene, sodium chlorite ($NaClO_2$), acetic acid, sodium hydroxide, distilled water and hydrochloric acid.

Fibers were produced by pulping process using 25% NaOH solution at 160-170°C, 8 atm. Then the fibers were washed with water until neutral pH. The bleaching process conducted with hypochlorite solution in 4 cycles.

2.1. Composite Manufacture

Preparation of composites was carried out through the following stages: the blending of wood flour/cellulose fibers, polypropylene (PP) matrix and maleated polypropylene (MAPP) coupling agent using HaakePolydrive lab plastomill at 175°C, 60 rpm for 20 min. Three different compositions (percentage) between fiber / PP / MAPP of the material used were 30/65/5, 40/55/5 and 50/45/5. Furthermore, the mixture was formed into a pellet, then made into a sheet/composite size 11 x 11 cm by using hot presses at a temperature of 180°C for 20 min. After that it was cooled for 10 minutes. Furthermore, the resulting composite was conditioned at room temperature.

2.2. Fiber and Composite Characterization

Determination of the degree of crystallinity was conducted by measuring the structure and Miller index of fiber using X-Ray Diffraction at 40kV and 30 mA, an angle of 10° -40° with CuKα radiation Ni filter (λ = 1.54060 Å) as well as JCPDS analysis. The molecular functional group bonding analysis and the absorption band change were performed using ABB MB 3000 (Reliable FTIR Laboratory Analyzer, Canada) at wave numbers 4450-500 cm^{-1} with a resolution of 4 cm-^{1}. The sample analysis was performed by placing 2 mg of sample in KBr powder (0.4 g) to obtain KBr-IR pellet. The entire spectra will be recorded at room temperature condition. In particular this procedure produced spectral data of the relationship between absorbance and wave number.

Physical properties of composite films analysis including water absorption and thickness swelling used sample size of 2x2 cm. Before the testings were conducted, the weight and thickness of each sample were measured. Subsequently the sample was immersed in aquades at room temperature for two days. After two days each sample was wiped with tissue paper to remove residual water and then weighed and measured its thickness. The mechanical properties testing include tensile strength, modulus young and break elongation of composite film were conducted based on ASTM D882-75b (Standard Test Method for Tensile Properties of Thin Plastic Sheeting) at 50 mm / min crosshead speed.

3. Results and Discussion

3.1 X-ray Diffraction Analysis

X-ray diffraction patterns of the lignocellulose materials showed improvement of cellulose structure after bleaching treatment. Bleached pulp showed the highest crystallinity value of 61.16% followed by pulp (50.49%) and wood flour (31.31%) as shown in **Figure 1**. Fiber composition was modified by removing the amorphous portion of the fiber increased the crystallinity of the fiber.

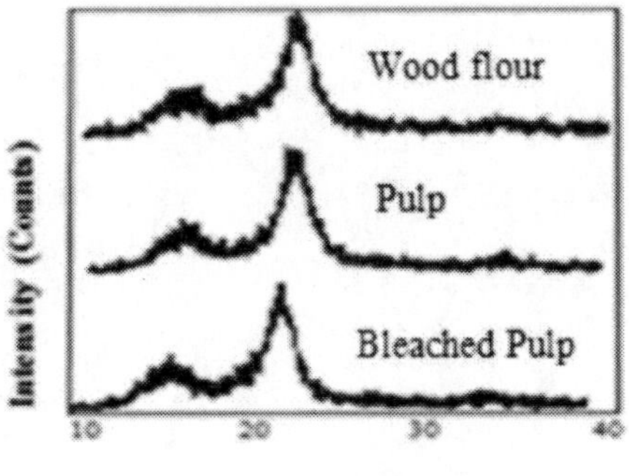

Figure 1. Diffraction Results of X-ray Wood Flour and Fibers

3.2 FTIR analysis

FTIR spectra indicated the formation of new chemical functional groups after pulping and bleaching treatment. In general, the entire sample spectrum confirmed the presence of absorption bands of xylan, glucomannan and cellulose as pulp-making chemical components. The presence of glucomannan was indicated by peak in an area of 800-900 cm^{-1}. The presence of absorption at peak 1700, 1600 and 1200 cm^{-1} showed the presence of xylan. The presence of cellulose was indicated by peak 1100, 1300 and 1400 cm^{-1} as well as changes in CH and OH waves in regions 2900 and 3300 cm^{-1}. The entire infrared spectrum of the sample still showed the presence of lignin in the uptake range of 1500 cm^{-1} [9,10, 11, 12].

In peak 952 cm^{-1} (**Figure 2**) there was a peak change between fiber without treatment and fiber with pulping and bleaching treatment. Absorbance of clusters (C = O) at peak 952 cm^{-1} were 0.65, 0.68 and 0.60. These changes occured due to the degradation of lignin, cellulose and hemicellulose content in the composites with each treatment.

Before the pulping and bleaching treatment, the absorbance of carbon group (C-H) at the point 2788 cm^{-1} was 0.12 and after the pulping process were 0.30 and 0.29. The addition of carbon clusters showed that composite films were degraded due to the influence of pulping and bleaching processes. The change of every 2788 cm^{-1} peak was influenced by changes that occur in the content of lignin, cellulose and hemicellulose. As stated by [13,14] that peak changes in each fiber treatment are due to stretching of cellulose, hemicellulose and lignin vibrations and the occurrence of C-H deformation of cellulose and lignin.

OH group with absorbance 0,31 was found in composite film at point 3456 cm^{-1}. This is due to the matrix and the wood powder separation created a cavity that allows the presence of OH groups attached to the composite film. After pulping and purifying process, the absorbance of OH groups were 0.41 and 0.31. This was possible at the time of pulping due to the presence of water bound to OH. The absorbance decreased in the bleaching process because of the bond between the fibers and the matrix is better. The process of crystallinity of fiber composite film can occur by estimating the strain ratio. A decrease in the strain ratio will cause a decrease in crystallinity. According to FTIR, pulp bleaching results had a higher crystallinity value than those of all treatments [13,15,16]. The higher the crystallinity value means the better the physical and mechanical properties of the composite.

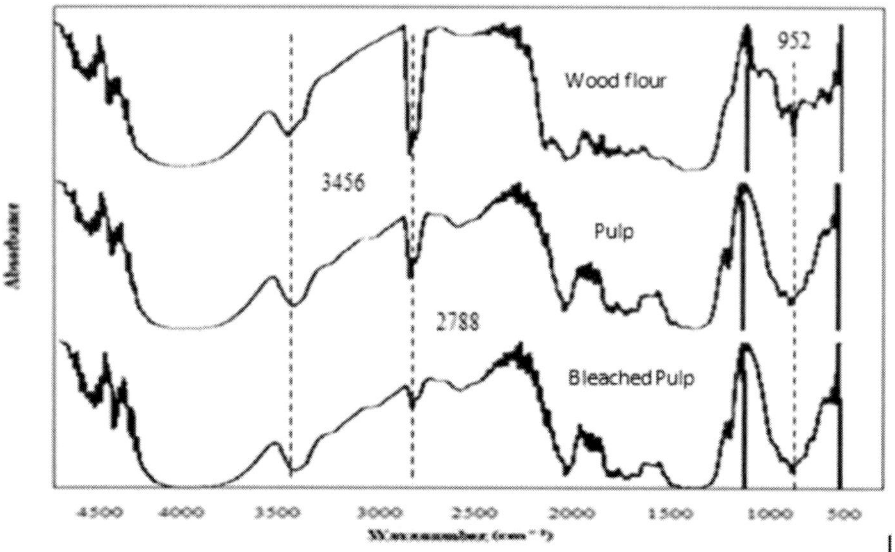

Figure 2. Characteristics of Infrared Fibers Spectra

3.3 Physical and Mechanical Characteristics of Plastic Composites

Figure 3 shows the composite film density values in the range 0.40 - 0.83 g / cm^3. Density is one of the physical properties that is very influential on the quality of composite film. A relatively high density indicates that the resilience of the composite film is high [17]. This is in accordance with the statement of [17] that basically, composites can be made from various cellulose fibers and the higher the density, the higher the mechanical properties.

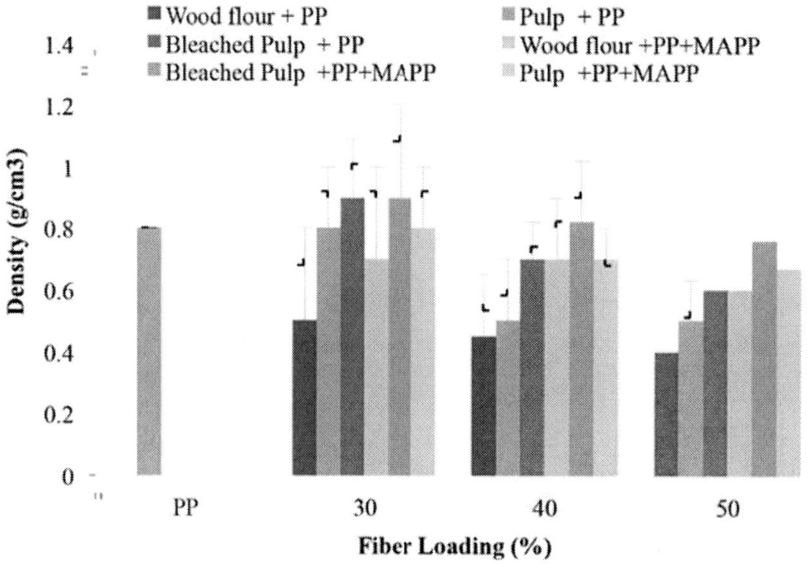

Figure 3. Density of plastic composite

Figure 4 shows the average water absorption rate of a composite film immersed for 24 hours. The water absorption of film composite soaked for 24 hours ranged from 2% - 10%. Although the PP used as an adhesive is a hydrophobic material, but due to different fiber and matrix compositions, it was likely that not all fibers were covered by adhesives. Thus the water can still enter either through the film cavities or the pores of the fibers. According to [1], although the general perception states that wood fibers in wood plastic products (WPC) are encased by resin plastics, but in fact the water absorption still occurs in commercial WPC products. Water absorption occurs on wood particles for example, on the part near the surface of the film rather than the plastic phase.

The water absorption of composite film without MAPP tended to be higher than of the composite film with MAPP. This was happening because when the composite was soaked, the compatibilizer successfully glue the dust matrix well so that the water absorption becomes low.

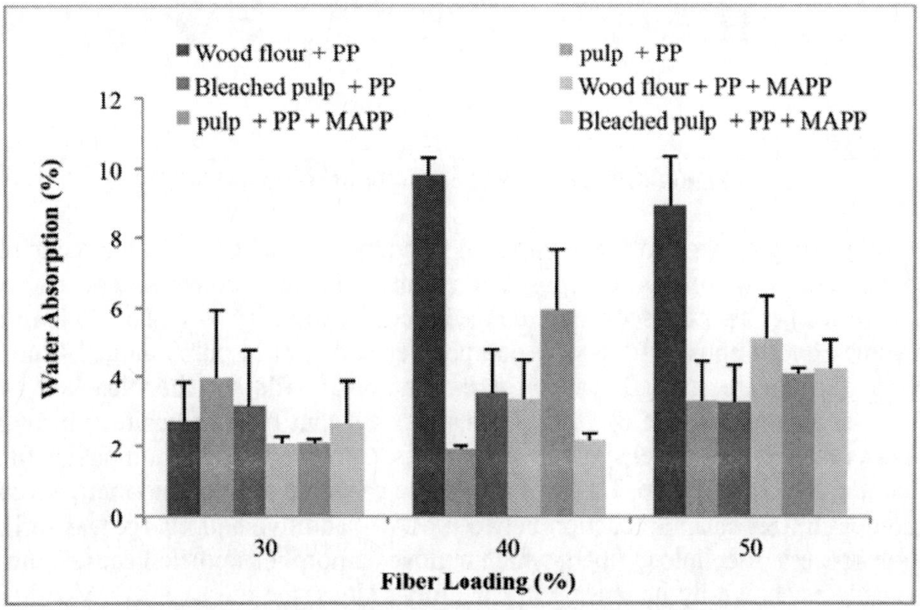

Figure 4. Water Absorption of the plastic composite

Averagely, film thickness was swelling after soaking for 24 hours in water. The increment of composite film thickness after 24 hours varied between 7% - 30% (**Figure 5**). The high value of the composite film thickness was in line with Manning, et.al [12], the varies fibers and matix composition of fibers and matrix varies so that the fibers are not completely covered by matrix, compatibilizer and showed high water absorption.

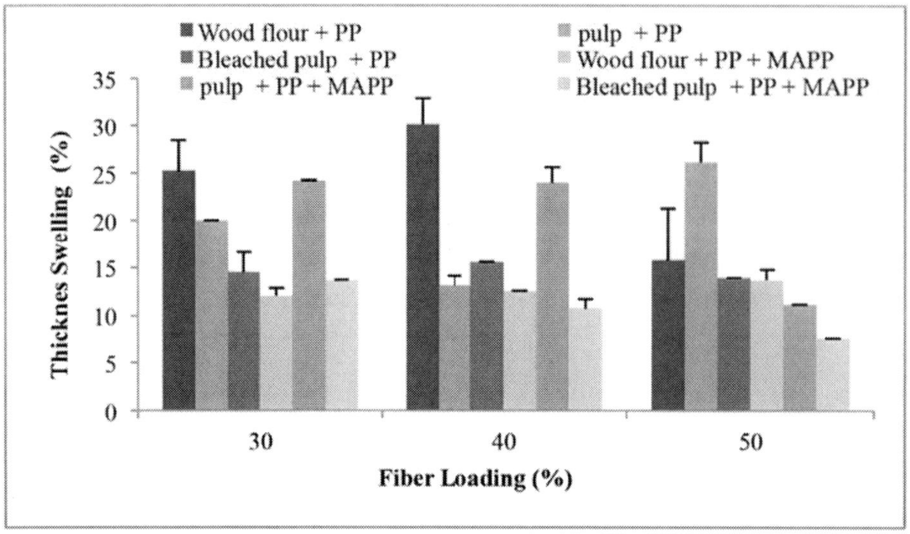

Figure 5.Thicknes Swelling of Plastic Composites

The tensile strength values of the composite film tend to decrease with the reduced number of matrices. The low matrix composition resulted in low cohesiveness and interface interaction with fiber as a consequence reduced material capacity to withstand loads. Tensile strength values for composite films of wood powder and control pulps ranged from 13 to 36 MPa. While the tensile strength value for the composite film of the bleached pulp were between 17 - 35 MPa (**Figure 6**). The results showed that the composite film of bleached pulp had a greater range of tensile strength values compared to the composite film of the wood flour and the control pulp. This was due to the variation in fiber and matrix composition and the compatibilizer causing reaction between MAPP additive and matrix was well. Beside, purification process of cellulose fibers which removes amorphous portion causes the fibers to be more stable as shown by increasing crystallinity values. the increase of crystallinity value has an impact on the increase of compatibility between fiber and matrix Furthermore the adhesion between matrix and fiber increased, resulting in the compatibility of composite film.

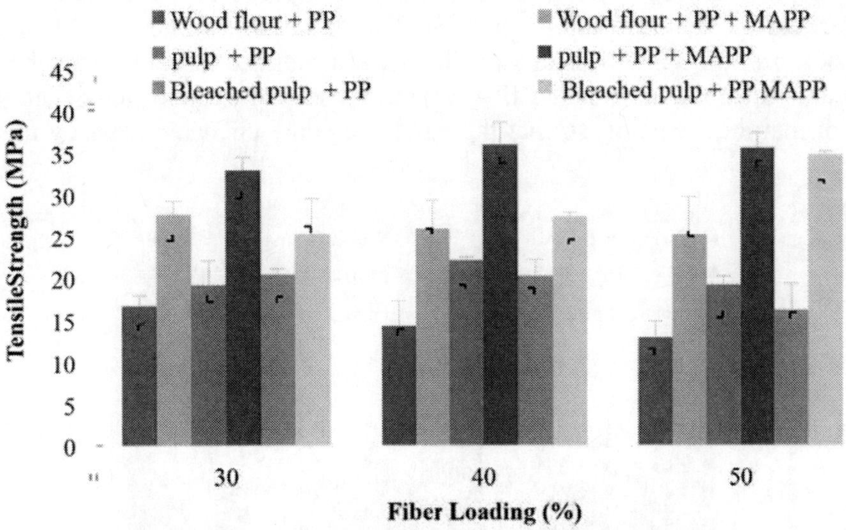

Figure 6. Tensile Strength of Plastic Composites

The modulus young values for composite film of wood flour and pulp ranged from 1 - 2.2 GPa. While the modulus young values for the composite film of the bleached pulp were between 1.5 - 3.3 GPa. The result showed that the composite film of the bleached pulp had a larger range of modulus young values compared to the composite film reinforced with wood flour and the pulp (**Figure 7**). This is because the fibers and matrix are mixed evenly and well bonded by MAPP so that the compatibility of the resulting composite film was achieved.

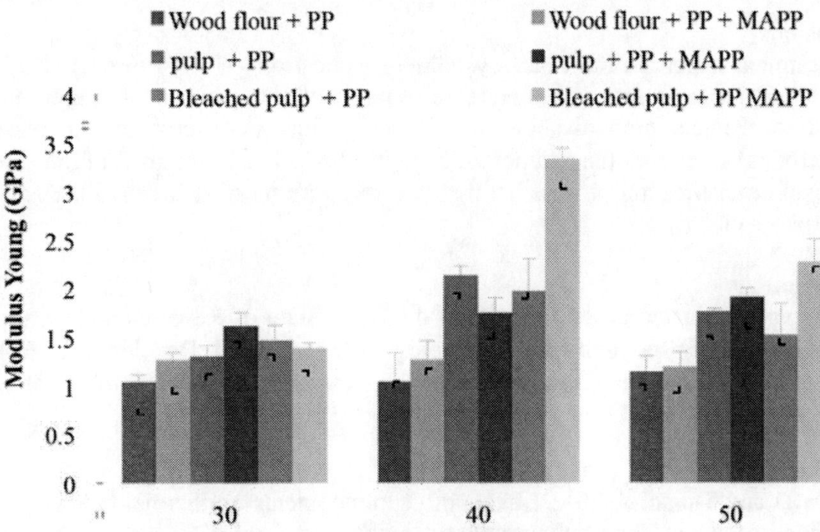

Figure 7. Modulus Young of Plastic Composites

The values of break elongation for film composite of wood powder and control pulps ranged from 3.3 to 5.1 (%). While the value of fracture element for the composite film of the bleaching pulp ranged from 2.8 to 4.5 (%). The result shows that the composite film of the bleaching pulp has a smaller fault elongation range compared to the composite film of the wood powder and the control pulp (**Figure 8**). The higher the number of fibers so that the value of fission breakdown will decrease. The highest elongation value is in the number of 70% matrix. It is assumed that smaller fiber size increases the surface area of the filler while the number of matrices remains, so that there may be some uncoated filler by the PP matrix optimally.

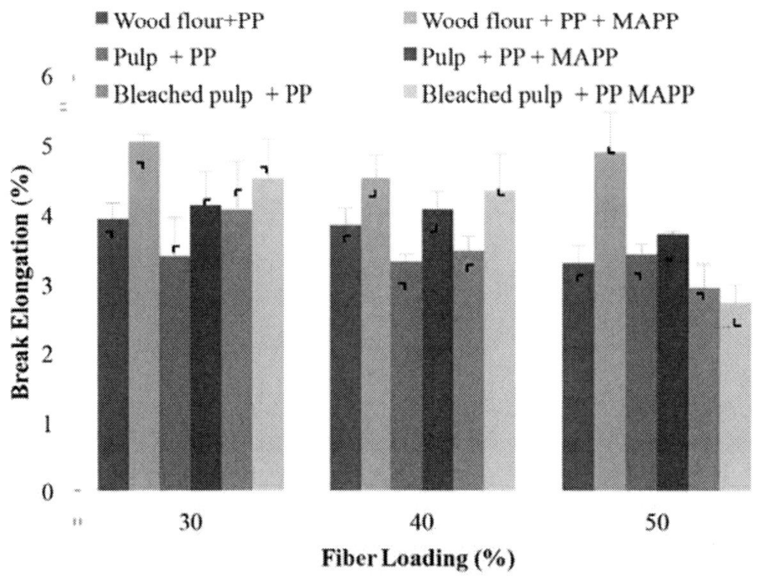

Figure 8. Break Elongation of Plastic Composite

4. Conclusions

Chemical treatment increased the fiber crystallinity value from 31.31% to 61.16%. The increase of crystallinity value has an effect on the increase of physical and mechanical properties. The optimum physical and mechanical properties of plastic composites are shown by the plastic composites reinforced cellulose fiber with the addition of MAPP. This is due to removal of the amorphous region (hemicellulose) at purification process, so that the fibers are more stable, and the addition of MAPP further improves its interfaces.

Acknowledgements

The research was conducted at the expense of the Directorate of Research and Community Service, Directorate General of Research and Technology Research and Development Higher Research, Technology and Higher Research in accordance with Research Contract No. 003 / SP2H / LT / DRPM / IV / 2017 dated April 20, 2017.

5. Reference

[1] Ndiaye D and Tidjani A 2012 Effects of coupling agents on thermal behavior and mechanical properties of wood flour/polypropylene composites *Journal of Composite Materials* **46(24)** 3067–3075

[2] Lu T, Jiang M, Jiang Z, Hui D, Wang Z and Zou Z 2013 Effect of Surface Modification of Bamboo Cellulose Fibers on Mechanical Prcperties of Cellulose/Epoxy Composite *Composites: Part B* **51** 28-34

[3] Ye X, Wang H, Zheng K, Wu Z, Zhou H, Tian K, Su Z and Tian X 2016 The interface designing and reinforced features of wood fiber/polypropylene composites: Wood fiber adopting nano-zinc-oxidecoating via ion assembly *Composites Science and Technology* **124** 1-9

[4] Li Z, Zhou X and Pei C 2011 Research Article Effect of Sisal Fiber Surface Treatment on Properties of Sisal Fiber Reinforced Polylactide Composites *International Journal of Polymer Science* DOI 10.1155/2011/803428

[5] Lima PRL, Santos RJ, Ferreira SR and Filho RDT 2014 Characterization And Treatment Of Sisal Fiber Residues For Cement-Based Composite Application *Eng. Agric., Jaboticabal.* **34(5)** 812-825

[6] Taramian A, Doosthoseini K, Mirshokraii SA and Faezipour M 2007 Particleboard manufacturing: an innovative way to recycle paper sludge *Waste management* **27(12)** 1739-1746 DOI 10.1016/j.wasman.2006.09.009

[7] Faruk O, Bledzki AK, Fink HP and Sain M 2012 Biocomposites reinforced with natural fibers: 2000–2010. *Prog. Polym. Sci.* **37(11)** 1552-1596 DOI 10.1016/j.progpolymsci.2012.04.003

[8] Lu JZ, Wu Q, Harold S and McNabb J 2000 Chemical Coupling in Wood Fiber and Polymer Composites: A Review of Coupling Agents and Treatments *Wood Fiber Sci.* **32(1)** 88-104

[9] Bivi RM, Siti NFMD, Khairulmazmi A, Idris A, Ahmed OH, Zamri R and Sariah M 2012 *In vitro* effects of salicylic acid, calcium, and copper ions on growth and sporulation of *GanodermaboninenseAfr. J.Biotecnol.* **11(70)** 13477- 13489

[10] Arbelaiz A, Fernandez B and Cantero G 2005 Mechanical properties of flax fiber/polypropylene composites. Influence of fiber/matrix modification and glass fiber hybridization *Composites Part A: Applied Science and Manufacturing* **36** 1637-1644

[11] Casey M 1980 Effect of Fiber Pretreatmen Conditioning on Interfacial Strenght and Mechanical Properties of Wood Fiber/PP Composites *Journal of Applied Polymer Science* **41** 1245-1252

[12] Manning MJ, Ascherl FM and Mankowski ME 2006 Wood-plastic composite durability and the compelling case for field testing In: *Yusoff MNM et al.* editor. Advance and Challenges in Biocomposites Symposium. *Proceedings of the 8th Pacific Rim Bio-Based Composites*; Kuala Lumpur **92** 19-27

[13] Fahma F, Iwamoto S, Hori N, Iwata T and Takemura A 2010 Isolation, Preparation, and Characterization of nanofibers from oil palm-empty-fruti-bunch (OPEFB) *Cellulose* **17** 977-985

[14] Popescu CM, Popescu MC, Singurel G, Vasile C, Argyropoulos DS and Willfor S 2007 Spectral Characterization of Eucalyptus Wood *Applied Spectroscopy* **61(11)** 1168-1177

[15] Chen Z, Hu TQ, Jang HF and Grant E 2016 Multivariate Analysis of Hemicelluloses in Bleached Kraft Pulp Using Infrared Spectroscopy *Applied Spectroscopy* **70(12)** 1981-1993

[16] Rosa MF, Medeiros ES, Malmonge JA, Gregorski KS, Wood DF, Mattoso LHC, Glenn G, Orts WJ and Imam SH 2010 Cellulose nanowhiskers from coconut husk fibers: Effect of preparation conditions on their thermal and morphological behavior *Carbohydr. Polym.* **81** 83–92

[17] Bjarnestad S and Dahlman O 2002 Chemical compositions of hardwood and softwood pulps employing photoacoustic fourier transform infrared spectroscopy in combination with partial least-squares analysis *Analytical Chemistry* **74(22)** 5851-5858

IOP Conf. Series: Earth and Environmental Science **166** (2018) 012003 doi:10.1088/1755-1315/166/1/012003

Asssesment of Initial Colony Founding by Swarming Reproductives of the Western Drywood Termite, *Incisitermes minor*

S. K Himmi[1], T Yoshimura[2], Y Yanase[3], T Torigoe[4], and S Imazu[5]

[1]Research Center for Biomaterials, Indonesian Institute of Sciences (LIPI), Indonesia
[2]Research Institute for Sustainable Humanosphere, Kyoto University, Japan
[3]Graduate School of Agriculture, Kyoto University, Japan
[4]Nara National Museum, Japan
[5]Nara University, Japan

E-mail: khoirul_himmi@biomaterial.lipi.go.id

Abstract. Colony founding activities in drywood termite is difficult to assess, since the colony forage and nest inside the wood. Non-destructive observation using X-ray computed tomography was used to evaluate nest-gallery establishment and colony breeding of the incipient colonies. The infested timbers were marked by year of infestation and were left on its original position to allow termite colonies develop in natural environment. Selected timbers were subjected for X-ray scanning analysis in Kyushu National Museum, Japan, in which colony founding from 5 royal pairs in the first year development was monitored. The CT data of the first six months colony founding indicated two different outcome in colony founding; two colonies distributed the energy to breed and slowed down in foraging simultaneously and other three concentrated in foraging. The CT scan after one year showed that only one royal pair showed no breeding activity, while the other four had been observed to have 2 –5 new colony members. Colony extraction also confirmed that colony size in the first and second year varied from 0 – 5 new colony members. The new colony members were mostly in larval instars, sometimes with the presence of one small soldier. Colony founding through nuptial flight is a risky process, as 59% of the observed chambers were failed colony founding, resulted from the dead of a royal pair or being abandoned (empty chamber).

1. Introduction

The western drywood termite, *Incisitermes minor* (Hagen) (Kalotermitidae) is considered to be the most invasive drywood termite in the world [1], and one of the five most economically important termites in the USA [2]. The colonies live entirely within a piece of sound and drywood [3,4], and thus, they can be easily transported around the world as a result of human activities. This invasive species was reported originally from the southwestern USA and northern Mexico [3], and spread all over Canada [5], China [6], and Hawaii [7], and more than half of the prefectures in Japan are listed as infested areas [8-11].

Although it is an economically important pest, there is very little scientific literature on the nesting biology of *I. minor*. Himmi et al. [12] has reported in situ nest-gallery development by foraging groups of *I. minor*, but in situ development of nest gallery by swarming reproductives is not yet reported. The establishment of a new colony following the dispersal of alates during swarming has been described as "vulnerable solitary founding" [13]. The level of failure is high, as the queen must brood and forage at the same time. Harvey [14] observed that the royal pair of *I. minor* had a period of inactivity after spending tremendous energy in swarming, shedding their wings, mating and excavating the chamber. The hibernation period was estimated to last for nine months, and many pairs do not survive through it. Successful colony founding depends on the royal pair's ability to maintain their energy while producing a workforce and to maintain their foraging efficiency. The colony has to distribute the time and energy spent in excavating the nest-gallery and foraging in such a way as to maintain its fitness [15].

The present report aimed to monitor the first-year development of royal chambers and incipient colonies following the nuptial flight of *I. minor* reproductive as monitored by X-ray CT. The results provide important information regarding the nesting biology of the royal pairs in the very early stage of colony founding.

2. Materials and methods

2.1 Timbers set up

Evaluation of nest-founding activity of *I. minor* reproductives was carried out at four infested houses in Wakayama Prefecture, Japan. In total, 60 pieces of Sitka spruce (*Picea sitchensis* Bong. Carriere) timbers and 40 pieces sugi (*Cryptomeria japonica* D. Don) timbers were laid with gaps of ~10 mm between them in the highly infested attic area of the houses (**Figure** 1). All timbers were made up of a combination of sapwood and heartwood with dimensions of 50 (R) x 50 (T) x 1000 mm (L). The timbers were arranged in random positions, without considering whether tangential sections, radial sections, sapwood parts, and heartwood parts were oriented in any particular direction. The experimental set-up was conducted on August 3, 2012, one month before *I. minor* swarming season in the Wakayama area, which was reported to be in September [11].

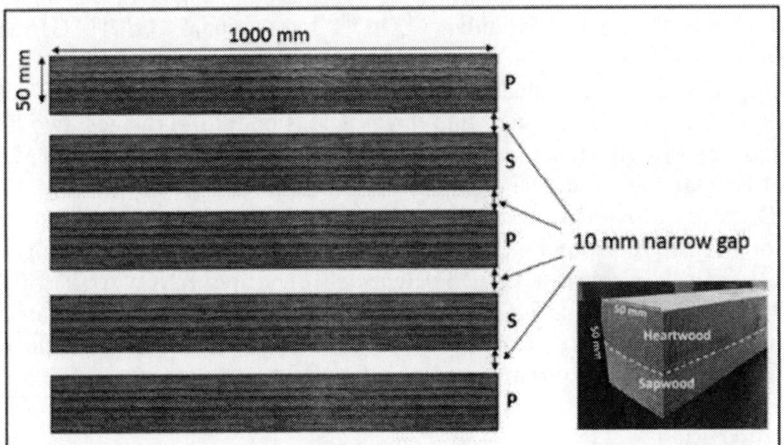

Figure 1. The arrangement of timbers with "P-S-P-S-P" arrangement and narrow gap (~10 mm) between timbers in each of test area, which consists of three Spruce timbers (P) and two Sugi timbers (S). In total, 20 test areas were set up in August 2012 at four highly infested houses.

The first evaluation was conducted two months after *I. minor* swarming season at Wakayama, on November 15, 2012. Due to low result of new infestations, we evaluated the arrangement of timber positions. In the first set up, Sitka spruce and sugi timbers were adjacently laid each other in straight position with narrow gap (~10 mm) between two timbers (Figure 1). We decided to modify the lay out in the following spring (March 2013), prior to a swarming season of *I. minor* in Wakayama. Figure 2 shows the modification in lay out of timbers. The modification was made by arranging "a close gap" in the edge of two timbers and "an open gap" in the other edge. Close gap is a position in which two edge of timbers are touch each other. Open gap is a position in which two edge of timbers are away from each other by ~10 mm gap.

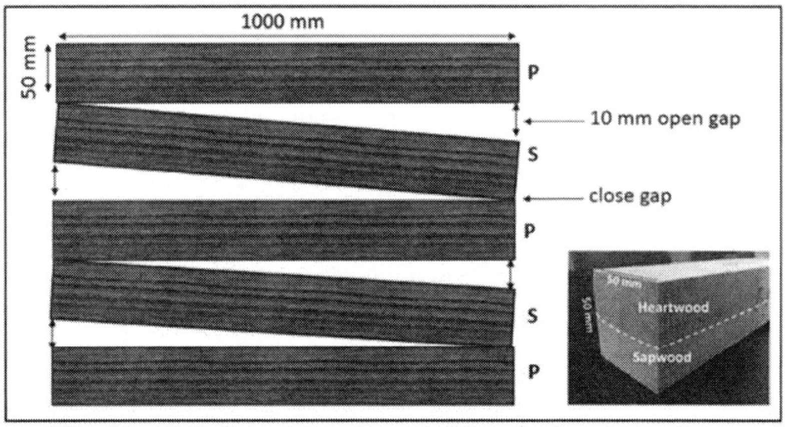

Figure 2. The modification of timbers set up by arranging the lay out to have a closed gap (CG) between two timbers at one end and an open gap (OG) between the two timbers at the other end. The CG is the area where the edges of the two timbers are touching each other, while the OG is the area where edges of the two timbers are separated by a ~10 mm gap. S: Sugi timber, P: Spruce timber.

2.2 Data analysis

The monitoring was conducted once a year, in November, around two months after swarming season. The infested timbers were brought back to the laboratory and were prepared for Computed Tomography (CT)-Scan analysis. The timbers were transported to Fukuoka, Japan, and were scanned by a large-scale X-ray CT (Y.CT Modular, YXLON International GmBH, Germany) at Kyushu National Museum.

The monitoring of 5 newly excavated nests was proceeded in biannual X-ray CT scanning analysis. The digital X-ray data were reconstructed into three-dimensional (3D) images, two dimensional (2D) section image and series of virtual cuts using volume graphics software (VGStudio MAX 2.1, Volume Graphics GmbH Germany), with digital data thickness 1 mm, 1024 x 1024 pixels, 3.17 2D image slice per mm. The 3D images of *I. minor* nest were reconstructed from digital image data and displayed as discrete values of a density function F (x, y, z), which was representing physical property.

The statistical analysis was carried out by Repeated Measures ANOVA ($P < 0.05$) using SPSS PASW (Ver. 18.0) (SPSS Inc., U.S.A). We relied on the volume of five scanned nests as the dependent variable to measure the development over time stages (the initial stage, the half year and the one year stage), and generated the differences under those three time stages.

3. Results and discussion

3.1 Nest founding activities

The first monitoring on November 15, 2012 found many alate wings were found on the test areas, indicating the high swarming activities. Two types of infestation were found in three spruce timbers: one timber was infested by a group of pseudergates, and two timbers were infested by primary reproductives. No attack was found in sugi timbers. The second monitoring was conducted in November 2013, indicated that lay out modification accomplished better results as the number of infestations were increased significantly. Six Sitka spruce timbers and four sugi timbers were found to be infested by *I. minor* reproductives. All the infestations were found in the close gap areas in sapwood part of the timbers. The details of infestations are presented on Table 1.

Table 1. Nest founding record by swarming repoductives of *I. minor* on the timbers

No.	Timber specimen	Year of collection	Details of Infestation									
			Type of Infestation		Infestation spot at the timber							
			Rp	W	Og	Cg	Sw	Hw	Bo	Sd	Up	Bm
1	Sitka spruce	2012		1	O*		O					O
2	Sitka spruce	2012	4		O*		O			O		
3	Sitka spruce	2012	1		O*		O					O
4	Sitka spruce	2013	1			O	O			O		
5	Sugi	2013	1			O	O					O
6	Sitka spruce	2013	1			O	O					O
7	Sitka spruce	2013	1			O	O			O		
8	Sitka spruce	2013	2			O	O			O		O
9	Sugi	2013	1			O	O			O		
10	Sugi	2013	1			O	O			O		
11	Sitka spruce	2013	1			O	O			O		
12	Sugi	2013	1			O	O			O		
13	Sitka spruce	2013	1			O	O					O

*Timber lay out in 2012 do not have a close gap (Cg) area.

Note:

Rp = Pairing Reproductive
W = Pseudergates (false worker)
Sw = Sapwood
Hw = Heartwood
Bo = Border
Sd = Side surface of timber
Up = Upper surface of timber
Bm = Bottom surface of timber
Og = Open gap
Cg = Close gap
O = Infestation event

Figure 3 shows nest founding actvities of a pairing reproductive of *I. minor*. The pair were assessing a spruce timber by chewing several spots before selecting a certain site and working together to excavate a royal chamber on that site. The assessment of nest site location may related to the feeding biology of drywood termite, which is suggested to be influenced by both the physical and chemical properties of the wood [16-18]. Himmi et al. [19] suggested that *I. minor* has preferences on nest-site selection, i.e., on the softer part of the timber [18] which related to physical properties of the timber [20,21]. The results supported previous reports, as *I. minor* reproductives showed preference in establishing the nest on the sapwood (Table 2). With respect to the timber arrangement, *I. minor* reproductives preferred the Cg area (Table 1). Previous studies have reported that *I. minor* reproductive love wood cracks, crevices, and holes as locations in which to establish the royal chamber [22,23]. When designing the timber lay out in the present study, the Cg area was situated to resemble wood cracks and crevices, and the results confirmed that royal pairs do exhibit selectivity in determining their nest location.

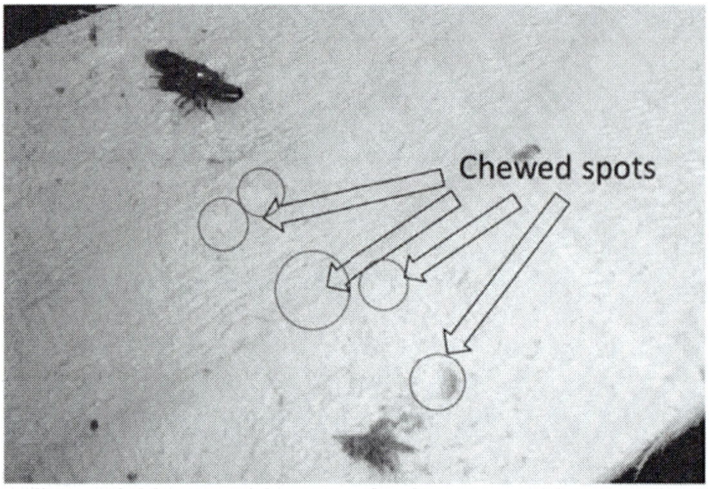

Figure 3. Nest founding activities: a pairing reproductive are working together to search a spot to establish the royal chamber in a Sitka spruce Timber.

3.2 Initial development of the nest-gallery

The first year development of drywood termite nest-gallery is recorded using X-ray CT scan. After assessing the preferred spot on timber for establishing a nest, a pairing reproductive of *I. minor* excavated a tiny hole and sealed it from inside using cement pellet. This tiny sealed-hole is often referred as a royal chamber. In current report, we presents CT data of a royal chamber that was established by a pairing reproductive from nuptial flight in a Sitka spruce timber. Figure 4 and Figure 5 illustrate the first year development of drywood termite nest-gallery in 3D and 2D CT images, respectively. The initial royal chamber (Figure 4.a) was excavated in an entrance slope (Figure 5.a – c). The chamber was extended in sloping vertical direction in springwood part of particular growth ring where the entrance hole was excavated (Figure 5.d – f) in the first six months, and further developed in vertical direction, parallel to vertical axis of that of springwood ring (Figure 5.g – i).

The initial cashew-nut shape of royal chamber (Figure 5.b) with ~ 4.6 mm and 41.4 mm^3 in diameter and volume (Figure 6), respectively, had been extended to be a new spacious chamber (Figure 5.e) with the maximum diameter of ~10.6 mm and the volume of 209.8 mm^3 (Figure 6) in the first six months. The chamber was excavated perpendicular to the longitudinal axis in the first six months, and parallel to the longitudinal axis of the timber through extension of the tunnel gallery in the one-year CT images. The royal pair excavated the nest-gallery on the springwood of a particular growth ring in which the entrance hole was established. The extension was indicated by the accretion in depth and length value (Figure 6). In the first year development, the fecal pellets were stored in the excavated chamber, and not yet disposed to outside of the nest. The first new colony member (body length ~ 2.1 mm) was observed in the first six months, which indicated that the royal pair distributed their energy in foraging and breeding.

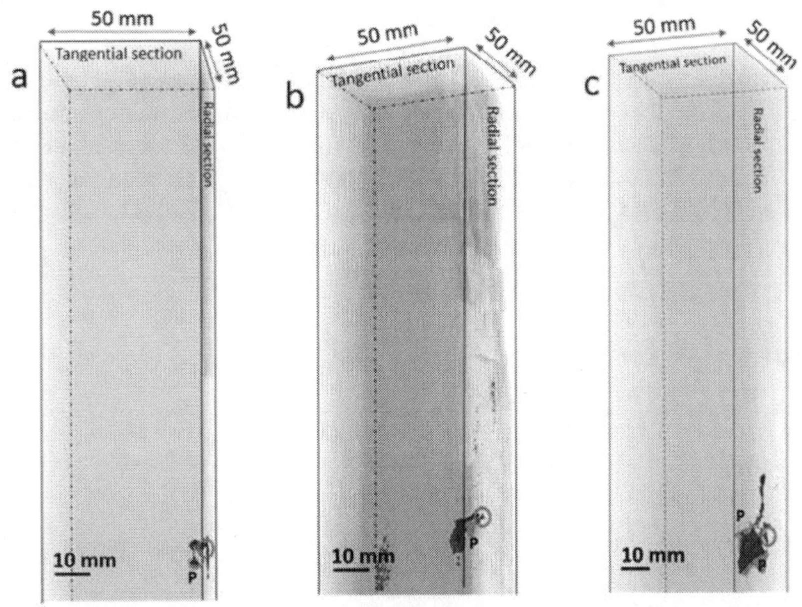

Figure 4. Three-dimensional CT images of nest-gallery development of *I. minor* in a Sitka spruce timber: (a) initial breeding chamber; (b) half year; (c) one year. P: fecal pellets. Entrance hole is marked by circular blue color. Orange color represents space inside the chamber. Termite presence is indicated by uncolored area inside the chamber.

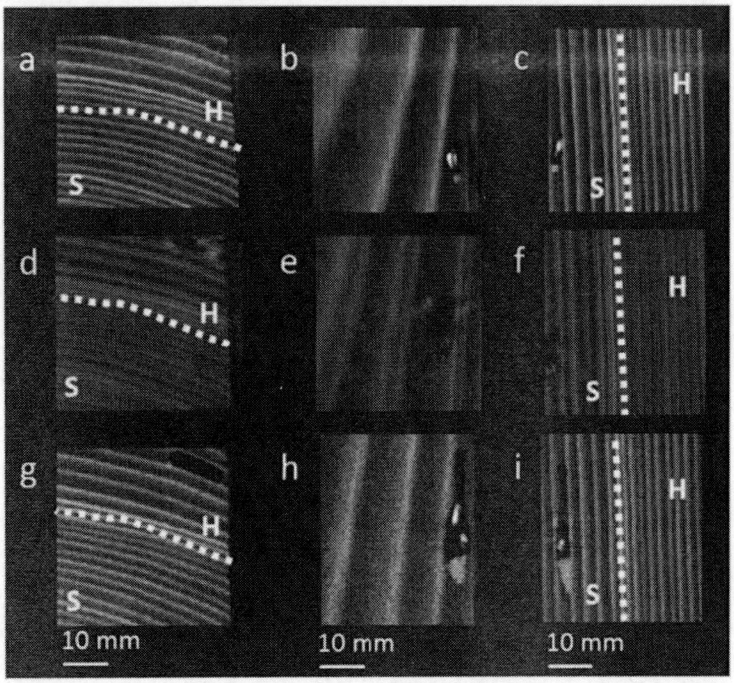

Figure 5. Two-dimensional CT images of the breeding chamber development in Timber C: (a – c): Initial breeding chamber; (d – f): half year; (g – i): one year. S: sapwood; H: heartwood. Cross section: a, d, g; Tangential section: b, e, h; Radial section: c, f, i. Springwood is indicated by darker color in CT images. Summerwood is indicated by lighter color in CT images.

 IOP Publishing
doi:10.1088/1755-1315/166/1/012003

After one year, more spacious chamber (17.5 mm and 445.8 mm^3 in diameter and volume, respectively) was excavated. No further development was observed in parallel direction to the axial system of the timber, which was shown by the stagnant value of the depth (Figure 6.b) as the chamber has reached the edge of timber (Figure 2.20.g). One exploratory tunnel was excavated in the vertical axis of spring growth ring. The queen and the king grew from initial 6.2 mm and 5.8 mm body length, to 6.7 mm and 5.9 mm after one year, respectively. The royal pair bred a new member, spotted with body length ~2.2 mm. The colony had a royal pair and two labor forces after one year development.

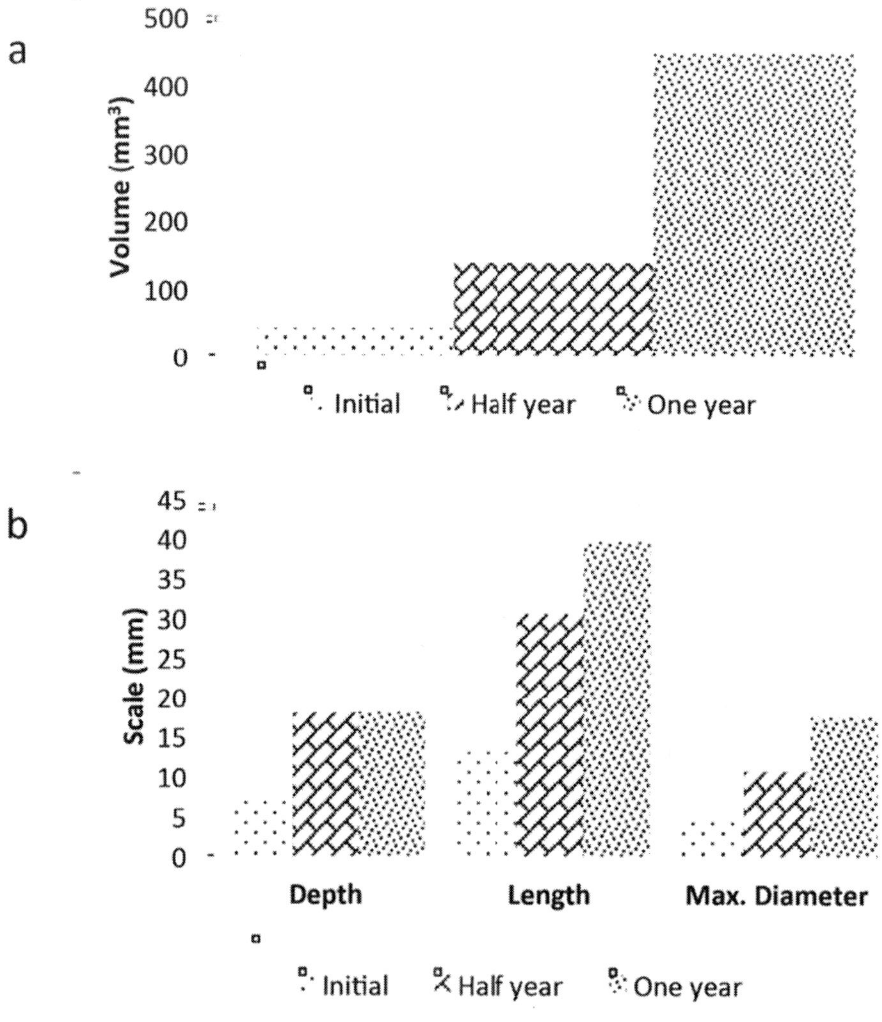

Figure 6. The properties of first year development of the nest gallery. (a) accretion in volume; (b) depth, length and maximum diameter.

Qualitative generalization of stage development of the nest-galleries was carried out using a repeated measures ANOVA ($P < 0.05$), by relying on volume properties of 5 scanned nests as a dependent variable. A repeated measures ANOVA with a Greenhouse-Geisser correction determined that the volume differed significantly ($p < 0.0005$) between time stages (initial, half year and one year). The average accretion in volume between time stages: initial, half year and one year development are 31.9 mm3, 201.7 mm3, and 447.6 mm3, respectively. Post hoc tests using the Bonferroni correction also revealed that biannual accretion in volume from initial establishment to half year and one year later were significantly different ($p = 0.001$ and $p < 0.0005$, respectively).

Moreover, the volume of breeding chambers after one year was also significantly different from the previous six months ($p = 0.002$).

The monitoring of five scanned nest-galleries provides important biological feedback of nesting biology of drywood termite. Harvey [14] argued that a nine-months inactivity period was a necessity in early stage of colony establishment, generated by tremendous fund of energy spent by royal pair in swarming, divesting wings, mating and excavating royal chamber. On the present study, the pairing reproductives in all chambers showed no hibernation period in the first six months and continuously excavated the galleries, in contrast with that of suggested by Harvey. One year CT analyses also confirmed that the reproductives continued foraging, which were indicated by significant accretion in volume and structure of the chambers. The chambers were developed in a certain pattern: cavernously excavated in a particular annual growth ring where the breeding chamber was established.

The results of present study are too early to draw any conclusive argumentation related to certain period of inactivity phase. All the naturally-infested timbers were kept in conditioned room (28 ± 2^0C; RHs $80 \pm 10\%$), thus the pairing reproductives never experienced any seasonal changing. Even though either temperature or humidity was reported as essential factors to influence feeding activities of *I. minor* [24-26], we argue that hibernation period of primary reproductive in the early stage of colony establishment is a biological fact that is need to be carefully evaluated.

In the second field monitoring in November 2013, we encountered a spruce timber infested by a pairing reproductive, which had volume properties approximately equal to that of a half-year chamber (Himmi, unpublished data), indicating that timber must be infested earlier than the common nuptial flight in September [11]. Thus the pairing reproductive were experiencing natural seasonal changing. If an inactivity period lasted for nine months as described by Harvey [14], the chamber should result in the same profile with that of initial royal chamber, and remain undeveloped.

There are no adequate literature about the actual time interval of primary reproductive in starting laying eggs after establishing the first royal chamber, neither in natural seasonal changing nor in laboratory assessment. The only comparative report was provided by Atkinson [27] who recorded seasonal changes in caste composition of 38 *I. minor* colonies dissected from their natal nest. However, he did not provide the adequate information whether the observations were carried out in actual seasonal changing, or in a conditioned temperature and humidity. He estimated that eggs were not encountered in colonies from October through May, with the peak production range from August to September. However, our data suggested a different life cycle. The first half-year CT evaluation conducted in May, 2013, showed breeding activities, which indicated that the oviposition occurred between November 2012 and April 2013.

4. Conclusion

The results showed that the royal pairs of *I. minor* can start breeding new colony members in the first six months since successful nest founding activities, and by the end of the first year, an incipient colony can have 0 – 5 new members. All the nests were established by pairing reproductives on the sapwood. The excavation of nest-galleries in the first year showed a preference for the springwood part of the particular growth rings where the entrance holes were excavated, which may related to foraging efficiency based on physical properties of the timbers.

Acknowledgments
The authors would like to thank Mr. Akio Adachi for preparing the wood samples used in this study and for his technical assistance in the experiment, and to Mr. Masao Oya for his important discussion and contribution in the set-up of timbers.

5. References

1. Evans TA, Forschler BT, Grace JK (2013) Biology of invasive termites: a worldwide review. Annu Rev Entomol 58:455-474. doi:10.1146/annurev-ento-120811-153554
2. Su NY, Scheffrahn RH (1990) Economically important termites in the United States and their control. Sociobiol 17:77-94
3. Light SF (1934) The distribution and biology of the common drywood termite, *Kalotermes minor*: Distribution and means of identification. In: Kofoid CA (ed) Termites and termite control. University of California Press, Berkeley, pp 210 - 216
4. Evans TA, Inta R, Lai JCS (2010) Foraging choice and replacement reproductives facilitate invasiveness in drywood termites. Biol Invasions 13 (7):1579-1587. doi:10.1007/s10530-010-9915-4
5. Grace JK, Cutten GM, Scheffrahn RH, Kevan DKM (1991) First Infestation by *Incisitermes minor* of a Canadian Building (Isoptera: Kalotermitidae). Sociobiol 18 (3):299-304
6. Xie Y, Li Z, Gregg WP, Li D (2001) Invasive species in China: an overview. Biodeversity and conservation 10:1317-1341
7. Haverty MI, Woodrow RJ, Nelson LJ, Grace JK (2000) Cuticular hydrocarbons of termite of the Hawaiian islands. J Chem Ecol 26 (5):1167-1191
8. Yasuda I, Kinjo K, Yaga S (2003) A new record of *Incisitermes minor* (Hagen) from Okinawa island (in Japanese). Jpn J Entomol 6 (2):103-104
9. Harunari M, Tomioka Y (2004) Occurrence and control of exotic termite, *Incisitermes minor* (Hagen), in Yokohama (in Japanese). House Househ Insect Pests 26 (2):107-113
10. Indrayani Y, Yoshimura T, Fujii Y, Yanase Y, Okahisa Y, Imamura Y (2004) Survey on the infestation of houses by *Incisitermes minor* (Hagen) in Kansai and Hokuriku areas. Jpn J Appl Entomol Zool 15 (4):261-268
11. Yoshimura T (2011) Strategies towards the integrated management of the invasive dry-wood termite, *Incisitermes minor* (in Japanese). Mokuzai Gakkaishi 57 (6):329-339
12. Himmi SK, Yoshimura T, Yanase Y, Oya M, Torigoe T, Akada M, Imadzu S (2016) Nest-gallery development and caste composition of isolated foraging groups of the drywood termite, *Incisitermes minor* (Isoptera: Kalotermitidae). Insects 7 (3). doi:10.3390/insects7030038
13. Cronin AL, Molet M, Doums C, Monnin T, Peeters C (2013) Recurrent evolution of dependent colony foundation across eusocial insects. Annu Rev Entomol 58:37-55. doi:10.1146/annurev-ento-120811-153643
14. Harvey PA (1934) The distribution and biology of the common drywood termite, *Kalotermes minor*: Life history of *Kalotermes minor*. In: Kofoid CA (ed) Termites and Termite Control. University of California Press, Berkeley, pp 217-233
15. Lee S-H, Su N-Y (2010) Simulation study on the tunnel networks of subterranean termites and the foraging behavior. J Asia-Pac Entomol 13 (2):83-90. doi:10.1016/j.aspen.2010.01.003
16. Kasseney BD, Deng T, Mo J (2011) Effect of wood hardness and secondary compounds on feeding preference of *Odontotermes formosanus* (Isoptera: Termitidae). J Econ Entomol 104 (3):862-867. doi:10.1603/ec10216
17. Sajap AS, Sahri MH (1983) Responses to wood and wood extractives of *Neobalanocarpus heimii* and *Shorea ovalis* by the drywood termite, *Cryptotermes cynocephalus* (Isoptera: Kalotermitidae). Pertanika 6 (3):28-31
18. Himmi SK, Yoshimura T, Yanase Y, Mori T, Torigoe T, Imazu S (2016) Wood anatomical selectivity of drywood termite in the nest-gallery establishment revealed by X-ray tomography. Wood Science and Technology 50 (3):631-643. doi:10.1007/s00226-016-0800-x
19. Himmi SK, Yoshimura T, Yanase Y, Oya M, Torigoe T, Imazu S (2014) X-ray tomographic analysis of the initial structure of the royal chamber and the nest-founding behavior of the drywood termite *Incisitermes minor*. J Wood Sci 60 (6):453-460. doi:10.1007/s10086-014-1427-x
20. Himmi SK, Yoshimura T, Yanase Y, Torigoe T, Akada M, Ikeda M, Imazu S (2018) Volume visualization of hidden gallery system of drywood termite using computed tomography: a new

approach on monitoring of termite infestation. In: McLellan B (ed) Sustainable future for human security: Environment and resources. vol 2. Springer Nature Singapore, Singapore, pp 61-68. doi:10.1007/978-981-10-5430-3_6
21. Choi B, Himmi SK, Yoshimura T (2017) Quantitative observation of the foraging tunnels in Sitka spruce and Japanese cypress caused by the drywood termite Incisitermes minor (Hagen) by 2D and 3D X-ray computer tomography (CT). Holzforsch 71 (6):535. doi:10.1515/hf-2016-0140
22. Harvey PA (1934) Colonization of the common dry-wood termite in wooden structures. In: Kofoid CA (ed) Termites and Termite Control. University of California Press, Berkeley, pp 239-265
23. Cabrera BJ, Scheffrahn RH (2001) Western drywood termite, *Incisitermes minor* (Hagen) (Insecta: Isoptera: Kalotermitidae). Univ Fla IFAS Ext (EENY248):1-7
24. Cabrera BJ, Rust MK (1994) The effect of temperature and relative humidity on the survival and wood consumption of the western drywood termite, Incisitermes minor (Isoptera, Kalotermitidae). Sociobiol 24 (2):95-114
25. Cabrera BJ, Rust MK (1996) Behavioral responses to light and thermal gradients by the western drywood termite (Isoptera: Kalotermitidae). Environ Entomol 25 (2):436-445
26. Indrayani Y, Yoshimura T, Yanase Y, Fujii Y, Imamura Y (2006) Evaluation of the temperature and relative humidity preferences of the western dry-wood termite *Incisitermes minor* (Hagen) using acoustic emission (AE) monitoring. J Wood Sci 53 (1):76-79. doi:10.1007/s10086-006-0817-0
27. Atkinson TH (1994) Seasonal changes in composition of colonies of the western drywood termite, *Incisitermes minor* (Hagen) (Isoptera: Kalotermitidae). Ann Entomol Soc Am

IOP Conf. Series: Earth and Environmental Science **166** (2018) 012004 doi:10.1088/1755-1315/166/1/012004

Properties of Particleboard Made from Wood Waste with Various Size

L Astari[1], K. W Prasetiyo[1] and L Suryanegara[1]

[1]Research Center for Biomaterial, Indonesian Institute of Sciences (LIPI)
Jl. Raya Bogor Km. 46, Cibinong, Bogor, 16911, Indonesia

E-mail: lilik.astari@biomaterial.lipi.go.id

Abstract. Particleboard is recognized as the most important products in forest products industry. Recently, utilization of industrial by-products such as waste wood from traditional sawmill is gaining attention. This research was aimed to investigate the physical and mechanical properties of particleboard made from wood waste with several size: coarse (3.360 mm), fine (2.00 mm) and mixed. Particle were oven dried prior to board production, mixing of glue and particle was conducted using drum-mixer then hot pressed with at temperature 150°C for UF and 130°C for PF. Mechanical testing was carried out using universal testing machine. Adhesive used in this study was urea formaldehyde and phenol formaldehyde with several concentration: 8%, 10% and 12%. The produced particleboards were tested for its mechanical and physical properties. Standard that using in this study was Japan Industrial Standard (JIS A 5908:2003). Research results indicated that more amount of adhesive increase both mechanical and physical properties. Boards density was meet the target as well as the thickness. It is also reported that fine size particle with 12% adhesive both UF and PF give the superior properties among others.

1. Introduction

In 2017, production of particleboard is forecasted remain increase. Several publications [17-20] also mentioned that particleboard still be an excellent products of engineered wood based product. The utilisation of wood waste into particleboard give more advantages due to it does not directly result in carbon emissions. Wood waste in from traditional sawmill in Indonesia is considered abundance. The utilization of this waste commonly as boiler or as landfill. Yet, the wood waste can be as an alternative material in particleboard industry. Iwakiri et al. [1] mentioned that wood waste from sawmill industry is a viable alternative in chipboard industry.

In particleboard development, there are several factor that affecting the properties of the board. Particle size and adhesive percentage are among those factors. Sekaluvu et al. [2] studied the effect of resin content and particle size in particleboard production using maize cobs. It is reported that particle size and adhesive affecting the board density, MOE and MOR. In addition, Cheng et al. [3] also investigated the effect of particle geometry and adhesive percentage in mechanical and physical properties of particleboard from peanut hulls. Similar to [3], it is reported that particle geometry, adhesive type and adhesive mass percentage directly affecting the mechanical performance of the board. It is also reported that the type of adhesive consequence to the suitable of particle origin (wood or non- wood) with the adhesive which further affecting the properties of the board.

This research was conducted to figure out the suitability of wood waste with differentiated size from home industry sawmill as raw material for particleboard. The research also aimed to investigate both physical and mechanical properties of particleboard made of different size of wood waste and different type of adhesive.

2. Materials and Method

2.1. Preparation

Wood waste were collected from home industry sawmill in Dramaga district, Bogor. The wood species of the wood were mentioned as kamper, sengon and meranti. After collection, wood waste were sieved and categorized into coarse size which pass 6 mesh or around 3.360 mm and fine size which pass 10 mesh or about 2.00 mm. Wood particle was oven dried for 24 hours at temperature 103±2°C.

Adhesive used for this study were commercial urea-formaldehyde (UF) and phenol-formaldehyde (PF). Percentage for both adhesive were 8wt%, 10wt% and 12wt% of board weight. Solid content of the adhesive were 50% for PF and 45% for UF.

2.2. Board Production

Each type of board were produced in duplo, the matrix are shown in **Table 1** below:

Table 1. Board Production Table

Adhesive type & Percentage	Particle size		
	Coarse (3.360 mm)	Fine (2.00 mm)	Mixture (coarse and fine)
UF 8%	2	2	2
UF 10%	2	2	2
UF 12%	2	2	2
PF 8%	2	2	2
PF 10%	2	2	2
PF 12%	2	2	2

Particle and adhesive were weighted according to the formula, afterward mixture were conducted using drum mixer and spray gun. The mixture were moulded in 25x25 mould before hot pressing process. The hot press was set into 150°C for particleboard with PF and 130°C for particleboard with UF. The pressing time was 10 minutes at 25 kgf/cm^2. After pressing, the board was conditioned for 7 days at room temperature before testing. Targeted density of the board was 0.8 g/cm^3 and targeted thickness was 10 mm.

2.3. Board Testing

The testing was conducted both for physical and mechanical. Board density, water absorption and thickness swelling were parameters measured for physical properties. Board mechanical properties were measured for its flexural strength, internal bond and screw withdrawal. Mechanical properties were tested using Shimadzu Universal Testing Machine. Standard occupied in this research was Japan Industrial Standard (JIS) A 5908-2003 [4]

3. Results and Discussions

3.1. Physical Properties

Thickness swelling is the main parameter to describe the stability of dimension for particleboards. The thickness swelling of particleboards in this research were decreased with the increased percentage of the adhesive, both for PF and UF (**Figure 1**).

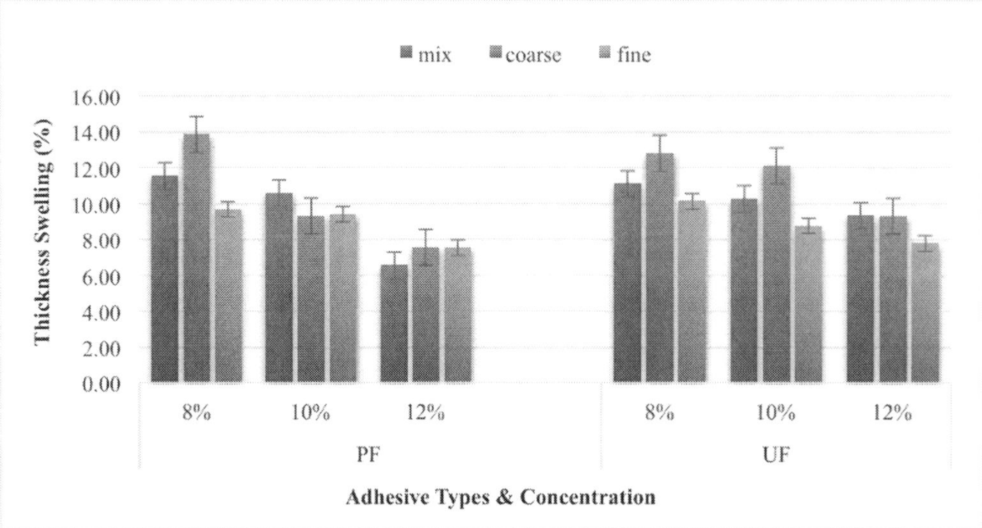

Figure 1. Thickness swelling of the particleboards

The similar results also reported by Li *et al.* [5], it is mentioned that thickness swelling have better value with the increase amount of adhesive. The significance decrease of TS was shown by coarse size particle with PF. It reduced from 13.85% to 7.56%, therefore the decrease was approximately 45.4%. It is also reported that three boards type were not meet the JIS A 5908:2003 standard for TS on particleboard. All those boards are from coarse size particle with 8% PF, 8% and 10% UF. According to Nemli*et al.* [6], the high percentage of thickness swelling can be enhanced by increasing the percentage of adhesive or coating application using hydrophobic materials on the board. On the other hand Amini [7] argue that thickness swelling of particleboard were more depends on the board density rather than type of adhesive.

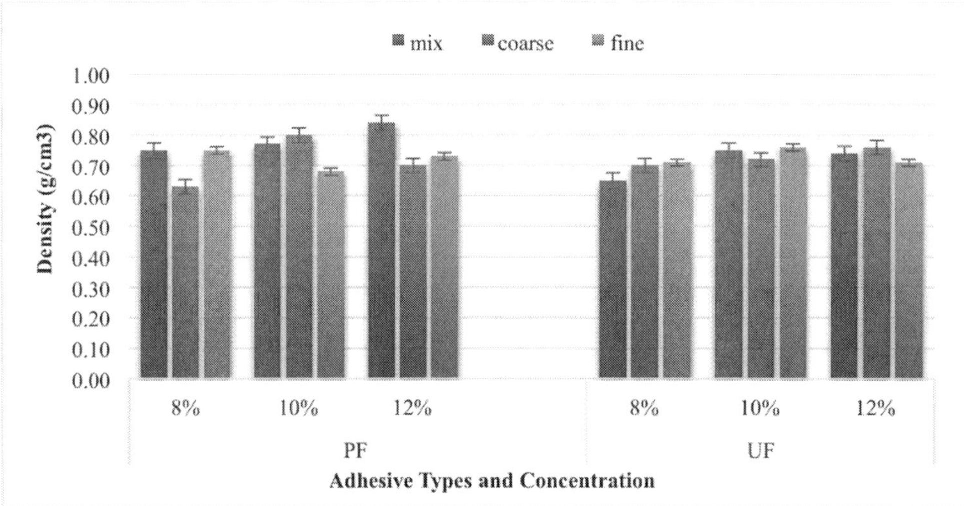

Figure 2. Density of the particleboards.

Targeted density of boards for this study was 0.8 g/cm^3. The range density of boards on this study was 0.63-0.8 g/cm^3 (**Figure 2**). The standard according to JIS was on the range 0.4-0.9 g/cm^3. Several studies report that low density particleboard correspondence to the high mechanical properties (Sellers et al. 1993[8]; Xu et al. 2004 [9]. Yet different result reported by Liao et al [16], it is argued that higher density board has a larger amount of particles per unit volume than that of lower density board; thus it shows higher MOR due to the superior bonding among the tightly packed particles. This explains that density of board has a role on mechanical properties.

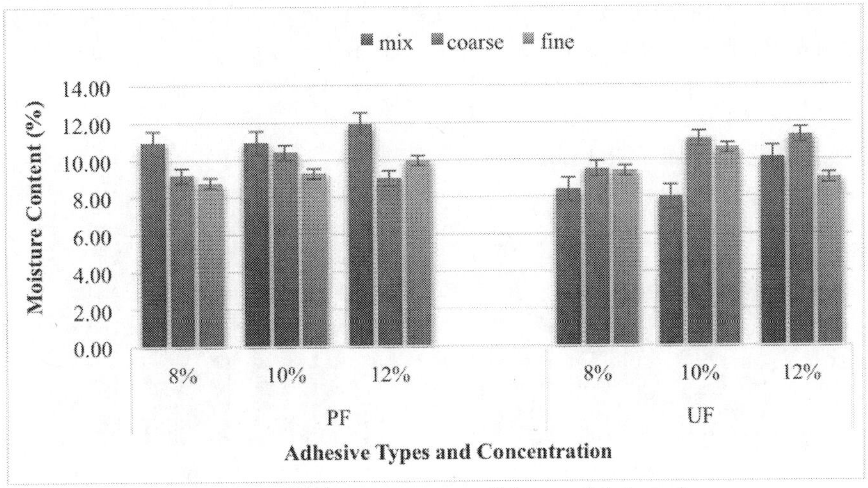

Figure 3. Moisture content of the particleboards.

Figure 3 shows that moisture content value of the board ranged between 8 – 11.9%. The highest value of water content was performed by mixed particle with 12% PF. All the boards are meet the requirement of JIS for water content value. It is observed that particleboard with PF adhesive have higher percentage of water content compare to those with UF.

3.2. Mechanical Properties

There are three particleboard types based on JIS A 5908:2003: type 8, type 13 and type 18. The higher number of type the better mechanical properties of the board. Result of flexural strength both MOR and MOE are shown in **Figure 4** and **5**.

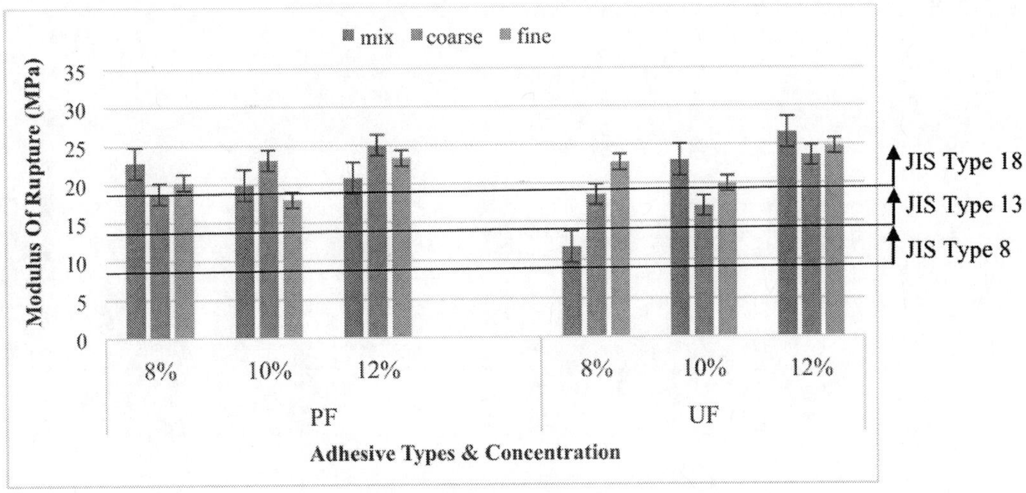

Figure 4. MOR value of the boards.

The value of MOR of all boards were meet the requirement of JIS type 8. The value were range from 11.69 – 26.54 MPa. The highest MOR is performed by mixed size raw material and 12% UF and the lowest MOR is given by board with mixed size raw material and UF 8%.It is observed that the higher percentage of adhesive, the higher value of MOR. **Figure 4** also shows that both PF and UF result in nearly similar value of MOR. Particle size also affects the MOR value, fine particle size gives higher value of MOR. This result correspondence to the research by Mohsen [10] and Samson [11] that reported MOE and MOR are influenced by panel density, percentage of adhesive and particle geometry. In the context of adhesive effect, high percentage of adhesive increases the bond contact between individual particles [12]. Slenderness ratio (chip length to thickness, SL) was a factor that influenced the MOR properties. The increased SL ratio gave high MOR value [13]. Different MOR values in different studies were due to different SL ratios.

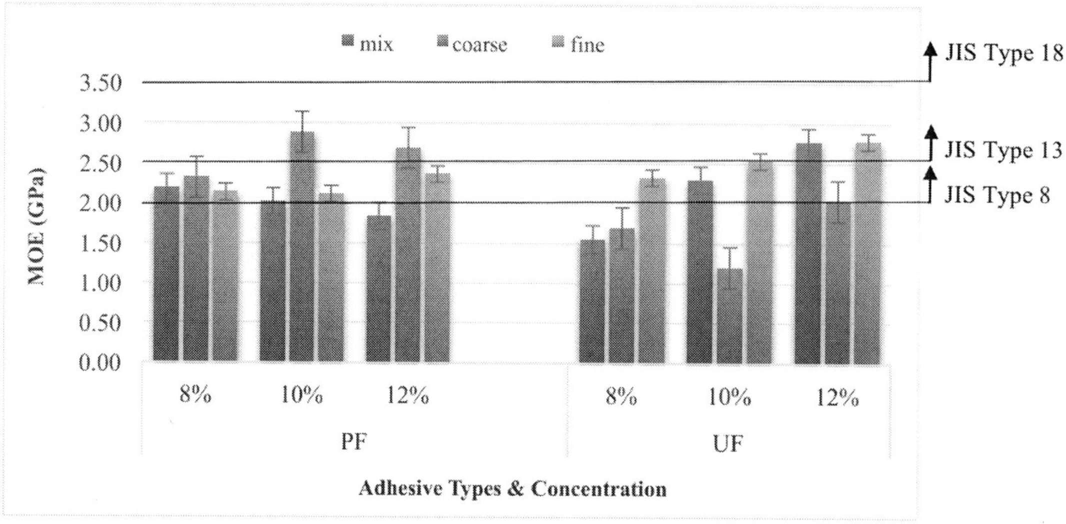

Figure 5. MOE value of the boards

Figure 5 depicts that only several boards meet the requirement of JIS A 5908 for board type 8 and 13. There are no board meet the standard for board type 18. The highest MOE value is performed by coarse size particle with 10% PF, followed by mix and fine particle with 12% UF. The graph also shows that PF adhesive gives better average on MOE value. Maloney [14] mentioned that MOE was affected by several factors such as adhesive content, adhesive type, bonding performance and fiber length.

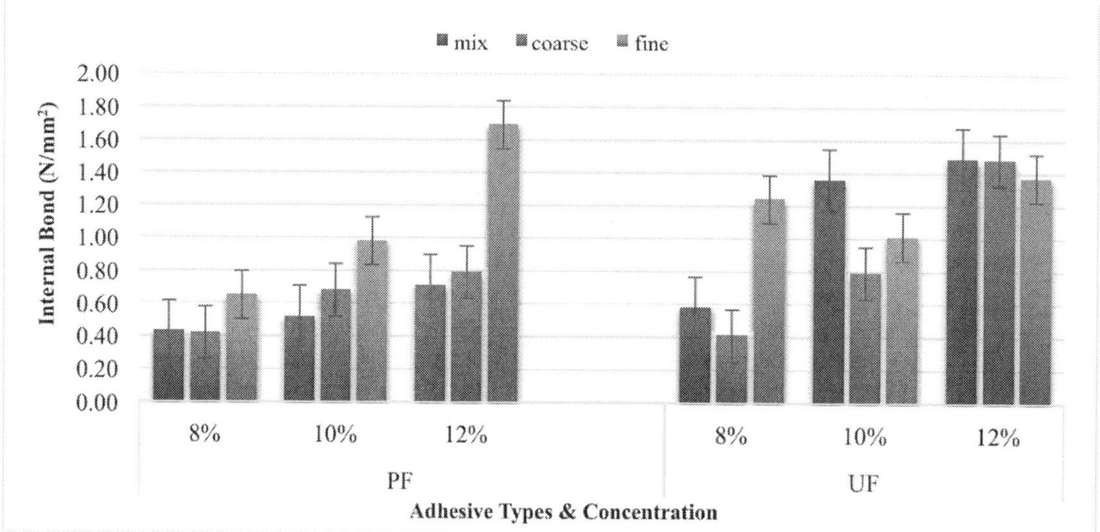

Figure 6. Internal bond value of the boards

Internal bond (IB) values of the boards in this study were range between 0.41-1.69 N/mm2. The highest IB was performed by fine particle with 12% PF whilst the lowest was from coarse particle with 8% UF. The graph illustrate that IB value of board with PF and fine size particle gives better result in every adhesive percentage. This condition difference with IB value from board with UF adhesive. It is showed that for UF adhesive, mix particle perform better IB value. Umemura et al. [15] mentioned that the adhesiveness on the inner layer of particleboard influence the internal bond value. In addition, adhesiveness of the surface layer of particleboard affecting the bending properties. Higher bonding strength properties of the composites made from uniform particle can be related to better glue line development due to their geometry. Uniformity of particle enhanced MOR, MOE and IB values of the composites.

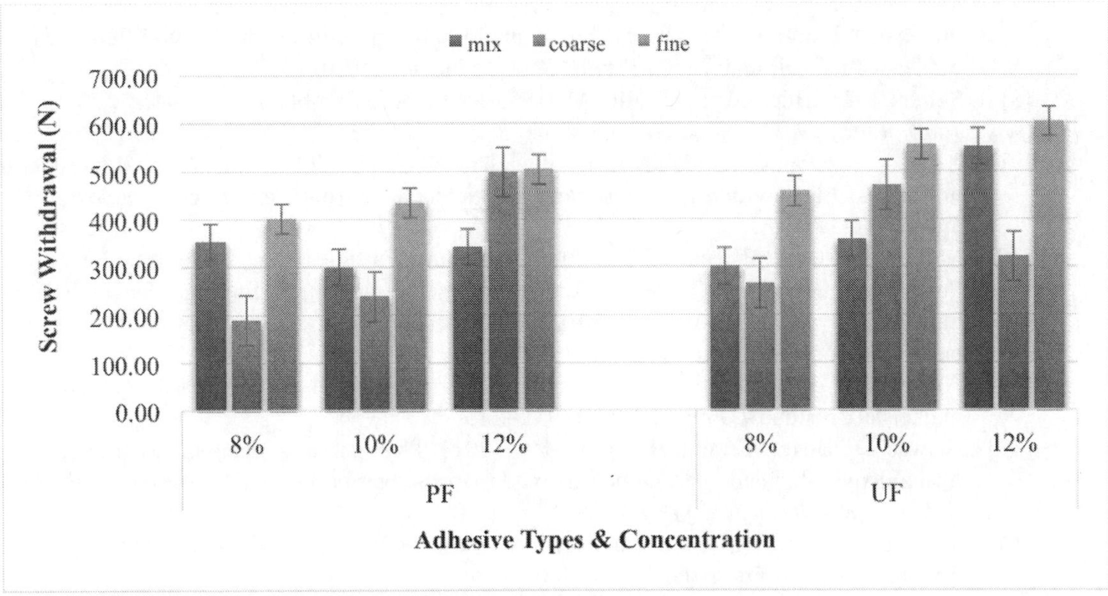

Figure 7. Screw withdrawal values of the board.

Another properties that investigated on this study was screw withdrawal values. The values were range between 189.06-650.13 N. There were three boards that did not meet the JIS Standard for particleboard type 8: board with coarse particle size and 8%PF, coarse with 10%PF and coarse with 8% UF. The highest screw withdrawal was given by board with fine particle and 12%UF. Poor screw withdrawal on coarse size particleboard due to the wide surface area of particle lead to low adhesive absorption.

4. Conclusion

In conclusion, wood waste from home industry sawmill can be utilized as raw material for particleboard. Dimensional stability of the boards increases with the increase of adhesive concentration. Physical properties of the boards mostly meet the requirement of standard (JIS A5908-2003). The amount of adhesive affects the mechanical properties of the boards. 12% adhesive performs higher properties both physical and mechanical. Urea formaldehyde (UF) adhesive was more suitable with wood waste since it gives better testing result. Fine size particles (6 mesh or 2.00 mm) with either UF or PF 12% give superior board properties.

5. References

[1] Iwakiri S 2012 Evaluation of particleboard panels properties produced from sawmill waste of nine tropical Amazonian wood species. In: Amazon Acta (1ND), Manaus 59-64
[2] Sekaluvu L, Tumutegyereize P and Kiggundu N 2014 Investigation of factors affecting the production and properties of maize cob-particleboards. *Waste Biomass Valor.* **5** 27–32
[3] Cheng X, He X, Xie J, Quan P, Xu K, Li X and Cai Z 2016 Effect of the Particle Geometry and Adhesive Mass Percentage on the Physical and Mechanical Properties of Particleboard made from Peanut Hull *BioResources* **11(3)** 7271-7281
[4] Japan Industrial Standard (JIS) A 5908-2003: Particleboards
[5] Li X, Cai Z, Winandy JE and Basta AH 2010 Selected properties of particleboard panels manufactured from rice straws of different geometries *Bioresource Technology* **101(12)** 4662-4666
[6] Nemli G, Örs Y and Kalaycıoğlu H 2005 The choosing of suitable decorative surface coating

material types for interior end use applications of particleboard *Construction and Building Materials* **19(4)** 307-312
[7] Amini MH, Hashim R, Sulaiman NS, Hiziroglu S, Sulaiman O, Mohamed M, Sarmin SN, Masri MN, Sulaiman MA and Abu Bakar MB 2016 Effect of urea formaldehyde addition to the dimensional stability of particle board made using glutardialdehyde modified corn starch as binder with FT-IR analysis *Materials Science Forum* **840** 108-111
[8] Sellers T, Jr Miller GD and Fuller MJ 1993 Kenaf core as a board raw material *Forest Products Journal* **(43)** 69-71
[9] Xu J, Sugawara R, Widyorini R, Han G and Kawai S 2004 Manufacture and properties of low-density binderless particleboard from kenaf core *Journal of Wood Science* **50(1)** 62-67.
[10] Mohsen S 2011 Effects of Hardener Type and Particles Size on Formaldehyde Emission Pollution. International Conference on Environment Science and Engineering IPCBEE **8**
[11] Samson AB and Hans-Wolf R 2010 Fibers of coffee husk and hulls for the production of particleboard. Mater. Struct. **(43)** 1049–1060
[12] Ajaye B 2011 Durability characteristics of cement-bonded particleboards manufactured from maize stalk residue *J. For. Res.* **22(1)** 111–115
[13] Biswas D, Bose SK and Hossain MM 2011 Physical and mechanical properties of urea formaldehyde-bonded particleboard made from bamboo waste *International Journal of Adhesion & Adhesives* **(31)** 84–87
[14] Maloney TM 1993 Modern Particleboard and Dry Process Fiberboard Manufacturing *Miller Freeman Inc.* San Fransisco
[15] Umemura K, Sugihara O and Kawai S 2015 Investigation of a new natural adhesive composed of citric acid and sucrose for particleboard II: effects of board density and pressing temperature *J Wood Sci* **(61)** 40-44
[16] Liao R, Xu J and Umemura K 2016 Low density sugarcane bagasse particleboard with citric acid and sucrose: Effect of board density and additive content *BioResources* **11(1)** 2174-2185
[17] Nurhazwani O, Jawaid M, Paridah MT, Juliana AH and Hamid SA 2016 Hybrid Particleboard made from bamboo (*Dendrocalamus asper*) veneer waste and rubberwood (*Hevea brasilienses*). *BioResources* **11(1)** 306-323
[18] Hazrat-Benhagh M, Zarea-Hosseinabadi H, Daliri-Sosefil M, Abginechi Z and Hemmati A 2016 Mechanical and insulating performances of ultralight thick particleboard from sugarcane residues and woods planer shaving. *Eur. J. Wood Prod.* **74** 161–168
[19] Candan Z and Akbulut T 2015 Physical and mechanical properties of nanoreinforced particleboards composites *Maderas Ciencia y Tecnología* **17(2)** 319 – 334
[20] Ganne-Chedeville C and Diederichs S 2015 Potential environmental benefits of ultralight particleboards with biobased foam cores *International Journal of Polymer Science* DOI 10.1155/2015/383279

IOP Conf. Series: Earth and Environmental Science **166** (2018) 012005 doi:10.1088/1755-1315/166/1/012005

Species Composition and Structure of Forests in the Muara Kendawangan Nature Reserve, West-Kalimantan, Indonesia

Purwaningsih[1], and K Kartawinata[2]

[1]Botany Division, Research Center for Biology, Indonesian Institute of Sciences (LIPI)
[2]Integrative Research Center, The Field Museum
1400 Lake Shore Drive, Chicago, IL 60605. USA

E-mail: purazali@yahoo.co.id

Abstract. Indonesia's forests are globally known for their high species diversity but many regions remain little known and so are floristic compositions of many conservation areas. This study investigated the tree communities in the heath, peat swamp, dryland dipterocarp and seasonal peat swamp forests at the Muara Kendawangan Nature Reserve, West Kalimantan. We established a plot size of (50mx60 m) in each forest type, thus the total area sampled was 1.2 ha. We enumerated trees with stem diameter ≥7cm on dryland dipterocarp and swamp forests as well as individuals with stem diameter ≥2 cm in heath forest. In the plot of 1.2 ha, we recorded 1.416 trees, representing 87 spesies, 57 genera and 28 families. Floristically the entire sampled forets consisted of *Schima wallichii-Shorea balangeran* association and *Baeckea frutescens-Rhodomyrtus tomentsa* association. The *Schima wallichii-Shorea balangeran* association consists of *Rhodamnia cinerea-Diospyros hermaphroditica* subassociation, *Baccaurea lanceolata-Calophyllum inophyllum* subassociation and *Syzygium fusticuliferum-Adinandra dumosa* subassociation. The forest communities investigated along the habitat gradient have low species richness. Further intensive studies in the above forest types as well as in the other community types present in the Muara Kendawangan Nature Reserve are necessary to provide data for better management of the reserve.

1. Introduction

The lowland forest of West Kalimantan covers mangrove, freshwater swamp, peat swamp, dryland dipterocarp and heath forests [1]. Much of these forests have been exploited for timber and converted into oil palm plantation. The remaining good undisturbed natural forests occur in national parks and nature reserves. One of these reserves is the Muara Kendawangan Nature Reserve (MKNR) in the Ketapang Regency, West Kalimantan.

Unlike the nearby Gunung Palung National Park, the species composition of forests within MKNR is less known. So far, the earlier floristic and vegetation studies have been carried out by [2-3]. Several surveys on ecological potentials have been carried out around the reserve, including in the Panggok, Bintan, Nata Lembawang and Bukit Marau forest regions [4-5]. MKNR comprised several ecosystems, including coastal, peat swamp, heath and lowland forests [6].

The MKNR was established on 12 October 1982 with a total area of ± 175,000 ha by the Decree of the Minister of Agriculture (*Surat Keputusan Menteri Pertanian Nomor 575/Kpts/Um/10/1982*). The area was further reduced to 149,079 ha by the Decree of the Minister of Forestry (*Surat Keputusan Menteri Kehutanan Nomor 174/Kpts-II/1993 tanggal 4 Nopember 1993*). MKNR is the largest among six nature reserves in West Kalimantan [6-7]. MNKR comprised coastal forest, peat swamp forest, seasonal peat swamp forest, heath forest, grassland, and dryland dipterocarp forest. Economically important species can be found in MNKR, including *meranti* (*Shorea* spp.), *ulin* (*Eusideroxylon zwageri*) and *jelutung* (*Dyera lowii*) [2-3].

It was indicated that various ecosystems in MNKR were the habitat and breeding ground of primates, turtles and birds [6-7]. The coastal forest constitutes the hatching ground of a turtle species known locally as *penyu belimbing* (*Dermochellelys coriaceae*) and *kuragading* (*Orlitia borneensis*) as well as breeding site for various bird species. The peat swamp forest was the habitat of several primate species, including *bekantan* (*Nasalis larvatus*).

The presence of the above plant and animal species made MNKR unique, but the uniqueness of MNKR has been steadily disappearing due to intensive human disturbances, including uncontrolled illegal logging, conversion to other uses, and forest fire. During the period of January-October 2001, hotspots were recorded in 11 districts (*kecamatan*) with the highest (100 spots) occurred in Kendawangan district. The highest occurrence of hotspots was recorded in August 2001, especially in Dusun Hilir district.

The objective of the present study was to perform quantitative measurements of species composition and structure of the peat swamp, seasonal peat swamp, dryland and heath forests. The scope of this paper is limited to the description of the forest in terms of the main structural characteristic, species richness, pattern of relative abundance and family composition. Such data are important for measuring the suitability and the priority of conservation [8] as well as for ecological restoration of disturbed forests.

2. Study Site

The study site was located on the side of the Membuluh River at 2°32.418' South and 110°20.992' East in the Tiang Balai subdistrict (*desa*) with the altitude of 0-30 m asl (above sea level). An areal photo with a scale of 1:50.000 showed that the grassland covered larger area than the forests. Several extensive grasslands dominated by *Hymenachme* sp. *Scleria* sp. and *Fimbristylis* sp. could be found in the Tiang Balai region, including the area of Kakora Lake and Arang-Arang Lake. MNKR consisted mainly of grasslands and a few islands covered by peat swamp and heath forests on sandy soils. Stagnant water occurred on many sites. The seasonal swamp forest was located in the Kampas Menali forest area, where the Bengkuang River tributary was flowing and was bordering with the Kakora Lake. It was flooded during wet season and dry during dry season. The Kampas Menali forest is situated at 2°34.776' South and 110°22.402' East with the altitude of 20 m asl. The dryland forest at Tiang Balai was smaller in area compared to the peat swamp forest and the species composition differed from that of other vegetation types. It was an old secondary forest, consisting of young trees with sparse herbaceous undergrowth. The plot on the dryland forest is located at 2°34.656' South and 110°22.184' East, at the elevation of 30 m asl, where the topography was relatively flat and the soil was dry. The heath forests were small patches of less than one ha and occurred sparingly. The soil consisted of sand on the top overlying black dry layer. The plot is located at 2°35.085' South and 110°22.029' East with flat topography and the altitude of 10 m asl, about ±5.5 km from the Tiang Balai subdistrict.

The study site was located in the rainfall type A according to [9]. The climate diagram (Figure 1) of the nearest rainfall station at Pangkalanbun (19 m asl) shows the mean annual rainfall of 2,679 mm and wet throughout the year, where the mean monthly rainfall was > 100 mm with the highest mean in January (425 mm) (LMG 1969).

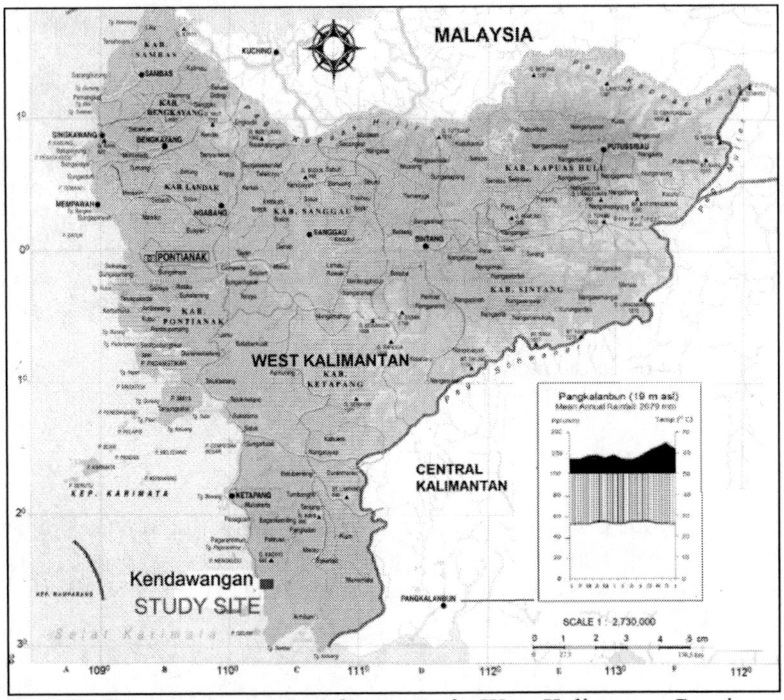

Figure 1. Map of the study site at Muara Kendawangan in West Kalimantan Province. The Walter's climate diagram was constructed for the closest meteorological station with about the same elevation at Pangkalanbun, Central Kalimantan.
Map modified from http://madhan33.blogdetik.com/2014/03/24/interpretasi-peta- kalimantan-barat.
Source of climate data: https://id.climate-data.org/location/1014061/

3. Methods

Field sampling was carried out in the Muara Kendawangan forests, using the quadrat method [10-11]. Forest investigation was done by establishing plots of 0.3 ha (50x60 m) each in dryland forest (Plot 1), seasonal swamp forest (Plot 2), peat swamp forest (Plot 3) and heath forest (Plot 4). Their orientations were fitted into the topographic conditions of each forest type. Thus the total sampled area of forest was 1.2 ha. Each plot was divided into 30 quadrats of 10 m x10 m each. Within each quadrat, DBH (Diameter at Breast Height) of each tree with stem diameter ≥ 7 cm (≥ 2 cm for heath forest) was measured, its bole and total tree heights were estimated, its identity was identified, and its position was recorded. Woody lianas with DBH $\geq$ 10 cm were treated as trees. A forest profile diagram for each plot was constructed from a strip of 10 m x 50 m nested within each plot.

We used the standard method [10-12] in defining density, frequency and dominance. Density is defined as the number of individuals per unit area. The number of individuals per species was later calculated for the total area of the plot, which was 0.30 hectare. The density in the plot is the sum of the individuals of all species and is expressed in terms of the number of individuals per hectare. The Relative Density (RD) for each species was calculated as follows:

$$RD = \frac{number\ of\ individuals\ of\ a\ species}{total\ number\ of\ individuals}\ x100\ \% \qquad (1)$$

Frequency was the number of times a species occurred in quadrats in the plot and was expressed as the percentage of the total number of quadrats. Relative Frequency (RF) was obtained as follows.

$$RF = \frac{frequency\ of\ a\ species}{sum\ frequency\ of\ all\ species}\ x100\ \% \ \ldots\ldots\ldots\ldots\ldots\ldots (2)$$

Tree dominance was usually determined by the stem cover, which was expressed as basal area. The basal area (BA) was calculated with the formula:

$$BA = (½d)^2 \pi \quad (3)$$

Where *d* stands for diameter. The dominance of a species was measured by summing the BA values for all individuals in the species. The Relative Dominance (RDo) was calculated as follows

$$RDo = \frac{dominance\ of\ a\ species}{dominance\ of\ all species}\ x100\ \% \quad (4)$$

The importance of a species in the plot was indicated by the sum of RD, RF and RDo. The Importance Value (IV) is then calculated as follows:

$$IV = RD + RF + Rdo \quad (5)$$

The Family Important Value (FIV) was calculated by summing up Importance Values of all species in a family [13] .
Shannon's index of diversity was calculated using the formula [14]:

$$H = -\sum_{i=1}^{s} \dot{P}_i \ lnP_i \quad (6)$$

In each subplot percentage of canopy gaps and the projected tree crown coverage were estimated, and topography and soil type were noted. Herbarium specimens for each species within the plot were collected and identified at the Herbarium Bogoriense, Research Center for Biology–LIPI, Cibinong. The identity and nomenclature of each tree species are conducted by follow the plant list.

4. Results

4.1. Species composition

The total number of tree species in four plots of 1.2 ha were 1180 individuals representing 87 species, 57 genera and 28 families (**Table 1**). The highest number of species (55) occurred in the seasonal swamp forest with the Shannon's diversity index of 3.64. It was followed by the dryland forest (38 species), peat swamp forest (17) and heath forest (8 species). Table 2 shows the species with IV >10 in four plots in the dryland, peat swamp, seasonal swamp and heath forests, showing that *Shorea balangeran* and *Syzygium griffithii* were the prevalent species with IV of 29.39 and 24.74, respectively.

Density and basal area in the dryland forest plot was higher than those in the three other plots, but the Shannon's diversity index (2.7) ws lower than plot seasonal swamp forest (**Table 1**). The dominant species in each plot differed. In the dryland forest, the dominant species as indicated by the IVs (**Table 3**) were *Schima wallichii*, *Vatica pauciflora*, *Calophyllum pulcherrimum* and *Rhodamnia cinerea*; in heath forest they were *Baeckea frutescens*, *Calophyllum pulcherrimum*, *Tristaniopsis whiteana* and species of Myrtaceae dan Clusiaceae; in the peat swamp forest it was *Shorea balangeran;* and in the seasonal forest they were *Madhuca motleyana* and *Pternandra azurea*.

Table 1. Vegetation characteristics of forests in MKNR

Vegetation characteristics	All plots (plot1-4)	Dryland forest (Plot 1)	Seasonal peat swamp forest (Plot 2)	Peat swamp forest (Plot 3)	Heath forest (Plot 4)
Plot size (ha)	1.2	0.3	0.3	0.3	0.3
Number of species	87	38	55	21	8
Number of genera	57	33	39	17	6
Number of families	28	19	23	11	2
Number of trees per hectare	1180	1073	516	786	2343
Number of basal Area (m^2/ha)	20.85	29.01	26.13	25.59	2.69
Shannon Index of diversity (H')	3.21	2.7	3.64	2.39	1.58

Table 2. Important species with Importance Value (IV > 10), Density (D) and Basal Area (BA) at MKNR, based on all four plots

Species	Family	D	BA	IV
Shorea balangeran	Dipterocarpaceae	63	5.16	29.39
Syzygium griffithii	Myrtaceae	249	0.31	24.74
Baeckea frutescens	Myrtaceae	185	0.37	19.03
Schima wallichii	Theaceae	46	1.93	15.28
Tristaniopsis whiteana	Myrtaceae	115	0.31	14.39
Combretocarpus rotundatus	Rhizophoraceae	46	1.39	14.38
Calophyllum pulcherrimum	Clusiaceae	109	0.12	13.04
Vatica pauciflora	Dipterocarpaceae	48	1.41	12.26
Pternandra azurea	Melastomattaceae	45	0.76	11.24
Rhodamnia cinerea	Myrtaceae	65	0.67	11.04
Gordonia ovalis	Theaceae	43	0.91	10.79

The synthesis table (**Table 3**) shows the grouping of species that can be designated as associations based on the 100% constancy [11] (of species in dryland, peat swamp, seasonal swamp and heath forest plots. The synthesis table method was used to designate the association and subassociations, which was purely based on floristic distribution [11]. Two associations can be identified:

(i) ***Schima wallichii-Shorea balangeran* association**, or for short ***Schima-Shorea* association.** Six species in Group 1 (**Table 3**) characterized the association. Two prevalent species, *Schima wallichii* and *Shorea balangeran,* with high IVs in dryland and swamp forests, respectively, can be used to name the association. Within the *Schima-Shorea* association three subassociations could be recognized, including:

(1) ***Rhodamnia cinerea-Diospyros hermaphroditica* subassociation** or ***Rhodamnia-Diospyros* subassociation,** whose character species are listed in Group 3 (**Table 3).** This subassociation constituted the forest on the dryland, where the soil was dry. It differed from other forest by the 15 species that form the character species whose distribution were restricted to the dry soil habitat. Other component species were listed in Group 1 (character species of the ***Schima-Shorea association*),** 6, 7 and 8, which occurred also in other subassociatios. Overall, we recorded 38 species, 33 genera and 19 families with density of 1073 trees/ha and basal area of 29.01 m^2/ha. The density and basal area were higher than those in the other subassociations. The dominant species included *Schima wallichii* (IV=46.22), *Vatica pauciflora* (IV=40.14), *Calophyllum pulcherrimum* (IV=34.22) and *Rhodamnia cinerea.* (IV=32.93). Undergrowth species were common and shrub species rarely present. Regeneration was dominated by primary forest species, including those of Lauraceae, Ebenaceae, Burseraceae and Sapindaceae.

(2) ***Baccaurea lanceolata-Calophyllum inophyllum* subassociation** or ***Baccaurea-Calophyllum* subassociation**. Thirty species constitute the character species as listed in Group 4 (**Table 3**). The subassociation formed the seasonally flooded swamp forest, whose character species were restricted in distribution to theis habitat. Other species composing the subassociation are listed in

Groups 1, 7 and 9. We recorded 55 species with density of 516 trees/ha and basal area of 26.13 m²/ha. The species richness was the highest compared to the other subassociation. The prevalent species with relatively high IV included *Madhuca motleyana* (IV=25.3) and *Pternandra rostrata* (IV=21.62). In this forest, *Shorea rugosa* (IV=3.15) and *Aquilaria malaccensis* (IV= 3.12) were recorded but with low density. The dry section of the forest was occupied by species of Burseraceae, Ebenaceae, Sapotaceae and Euphorbiaceae while the swamp section was dominated by species of Myrtaceae.

(3) ***Syzygium fusticuliferum-Adinandra dumosa*** **subassociation** or ***Syzygium-Adinandra*** **subassociation,** whose character species are listed in Group 5 (Table 3). Other component species are listed in Groups 1, 6, 8, 9 and 10. This subassociation formed the peat swamp forest, which could be found on the edge of Arang-arang Lake. Within the plot of 0.3 ha we recorded 21 species, 17 genera and 11 families, with density of 590 trees/ha and BA of 24.43 m²/ha (**Table 1**). The species richness is small. The dominant species was *Shorea balangeran* (IV=101.6), with mean height of 15 m and diameter of 15-20 cm. Other prevalent species included *Haemocharis ovalis* (**IV**= 33.89) *Combretocarpus rotundus* (IV=30.12) *Eugenia fusticuliferum* IV=32.66) and *Pternandra rostrata* (IV=24.94). Climbing species were quite common in the undergrowth layer in the swamp forest. It appears that it was related to the gaps resulted by tree removal. The regeneration in the swamp forest was dominated by species of Myrtaceae and Clusiaceae

(ii) The ***Baeckea frutescens-Rhodomyrtus tomentosa*** **association or** ***Baeckea frutescens-Rhodomyrtus*** **association.** The character species whose presence was restricted to this community were *Baeckea frutescens, Rhodomyrtus tomentosa, Syzygium bankense* and *Eugenia* sp. as listed in Group 2 (**Table 3**). Other component species are listed in Groups 6 and 10. This association formed the heath forest over the sandy soil. The heath forest plot is the poorest, containing only 8 tree species with the species richness index of 0.30. It was dominated by small trees with diameters of 2-5 cm with height of 3-5 m occurring on sandy soil. The BA was only 2.69 m²/ha, which was the lowest in comparison to those of the other three plots. The density was 2343 individuals/ha, which made the vegetation looked more like a scrub rather than a forest. The dominant species belong to Myrtaceae (**Table 3**). The most prominent species was) *Syzygium griffithii* (IV = 96.82) and *Baeckea frutescens* (IV=91.83). The leaves of *B. frutescens* are sclerophyllous emitting the fragrance of cayuput oil when crushed and reported to be good for curing stomach disorder. Other species with high IVs but present also in other forests were *Syzygium griffithii* (IV=86.82), *Calophyllum pulcherrimum* (IV=45.62) and *Tristaniopsis whiteana* (IV=42.01). Undergrowth community was relatively sparse and is mostly of the seedlings and saplings of the tree species which were mostly species of Myrtaceae. They were also common in the peat swamp forest and seasonal peat swamp forest.

Tabel 3. Grouping of species based on their presence in the plots at MKNR; D = Density (trees/ha), BA = Basal Area /ha, and IV = Importance Value.

No.	Species	Dryland Forest Plot			Seasonal Peat Swamp Forest Plot			Peat Swamp Forest Plot			Heath Forest Plot		
		D	**BA**	**IV**	**D**	**BA**	**IV**	**D**	**BA**	**IV**	**D**	**BA**	**IV**
	GROUP 1												
1	*Baccaurea polyneura*	3	0.02	0.96	7	0.15	3.48	3	0.26	2.19			
2	*Gordonia ovalis*	37	1.08	11.2	3	0.01	1.51	103	1.93	31.2			
3	*Pternandra azurea*	3	0.02	0.95	50	1.01	21.62	97	1.49	30.1			
4	*Schima wallichii*	130	5.79	43	7	0.2	3.66	17	0.44	6.1			
5	*Shorea balangeran*	3	0.34	2.07	7	1.59	8.2	200	15.3	102			
6	*Syzygium chloranthum*	17	0.17	5.03	17	1.05	11.29	20	0.41	7.14			
	GROUP 2												
7	*Baeckea frutescens*										617	1.22	91.8
8	*Syzygium bankense*										30	0.01	7.31
9	*Eugenia* sp.										3	0	0.98
10	*Rhodomyrtus tomentosa*										137	0.06	20.9
	GROUP 3												
11	*Antidesma cuspidatum*	10	0.57	4.61									
12	*Artocarpus nitidus*	13	0.37	4.23									
13	*Cratoxylum formosum*	10	0.13	3.1									

14	*Cryptocarya ferrea*	3	0.63	3.05									
15	*Diospyros hermaphroditica*	30	0.76	10									
16	*Garcinia dioica*	23	0.68	7.95									
17	*Guioa diplopetala*	3	0.01	0.93									
18	*Helicia moluccana*	7	0.38	3.07									
19	*Horsfieldia wallichii*	7	0.06	1.99									
20	*Licania splendens*	3	0.1	1.21									
21	*Litsea umbelliflora*	3	0.02	0.94									
22	*Nephelium cuspidatum*	3	0.03	0.98									
23	*Rhodamnia cinerea*	217	2.23	39.9									
24	*Gymnanthes borneensis*	3	0.11	1.26									
25	*Xylopia ferruginea*	10	0.23	3.45									
	GROUP 4												
26	*Agathis alba*				7	0.38	4.35						
27	*Alseodaphne umbeliflora*				10	0.38	4.99						
28	*Aporosa frutescens*				17	0.26	7.44						
29	*Aquilaria malaccensis*				7	6.67	3.12						
30	*Artocarpus lanceifolius*				3	0.01	1.5						
31	*Baccaurea macrocarpa*				3	0.24	2.37						
32	*Baccaurea javanica*				7	6.67	3.13						
33	*Baccaurea lanceolata*				13	2.14	13.19						
34	*Barringtonia reticulata*				3	0.15	2.02						
35	*Bouea oppositifolia*				7	0.08	3.19						
36	*Calophyllum inophyllum*				13	1.6	11.92						
37	*Canarium hirsutum*				10	0.44	6.03						
38	*Chisocheton patens*				3	0.04	1.59						
39	*Cryptocarya lucida*				17	0.1	6.84						
40	*Dacryodes rostrata*				10	0.64	6.79						
41	*Diospyros cf.confertiflora*				3	0.02	1.53						
42	*Elaeocarpus stipularis* var.*brevipes*				7	0.05	3.08						
43	*Elaeocarpus glaber*				3	0.03	1.58						
44	*Cryptocarya agathophylla*				7	0.03	3						
45	*Ficus* sp.				3	2.55	11.21						
46	*Gomphia serrata*				3	0.13	1.93						
47	*Haplolobus floribundus*				3	0.06	1.67						
48	*Hopea nigra*				7	0.08	2.4						
49	*Koordersiodendron pinnatum*				3	0.06	1.68						
50	*Litsea firma*				3	0.05	1.63						
51	*Litsea costalis* var.*nidularis*				3	0.02	1.53						
52	*Litsea resinosa*				3	0.06	1.68						
53	*Litsea robusta*				3	0.95	5.08						
54	*Macaranga depressa*				20	0.16	7.71						
55	*Macaranga triloba*				40	0.38	15.65						
56	*Ochanostachys amentacea*				10	0.5	6.28						
57	*Polyalthia lateriflora*				3	0.06	1.69						
58	*Sandoricum borneense*				3	0.68	4.05						
59	*Scorodocarpus borneensis*				13	1.26	10.64						
60	*Shorea rugosa*				7	6.67	3.15						
61	*Syzygium claviflorum*				7	0.06	2.33						
62	*Syzygium valdevenosum*				27	0.75	11.25						
	GROUP 5												
63	*Adinandra dumosa*							23	0.34	7.32			
64	*Parastemon urophyllus*							7	0.05	1.79			
65	*Symplocos fasciculata*							10	0.07	3.79			
66	*Syzygium fusticuliferum*							107	1.43	32.7			
	GROUP 6												
67	*Calophyllum pulcherrimum*	140	3.68	33.2				10	0.1	3.92	353	0.3	45.6
68	*Tristaniopsis whiteana*	30	0.97	9.6				13	0.24	4.89	350	0.28	42
	GROUP 7												
69	*Aglaia edulis*	7	0.06	1.98	3	0.67	4.02						
70	*Artocarpus tamaran*	17	0.56	6.36	3	0.45	3.17						
71	*Barringtonia samoensis*	3	0.01	0.93	7	0.08	3.23						
72	*Dacryodes laxa*	3	0.07	1.14	7	0.22	3.76						
73	*Dillenia serrata*	3	0.15	1.4	7	0.02	2.19						
74	*Endospermum peltatum*	3	0.51	2.63	20	0.15	7.67						
75	*Nephelium uncinatum*	3	0.07	1.11	3	0.09	1.78						
76	*Tetramerista glabra*	10	0.36	3.33	3	0.19	2.19						
77	*Vatica pauciflora*	157	4.58	40.1	3	0.12	1.92						
	GROUP 8												
78	*Calophyllum rigidum*	30	0.86	9.23				3	0.01	1.23			
79	*Combretocarpus rotundatus*	80	2.64	25.7				73	2	28.4			
80	*Eugenia lineata*	47	0.77	13.3				13	0.19	4.69			
	GROUP 9												
81	*Diospyros maingayi*				17	1.4	12.61	3	0.02	1.24			
82	*Garcinia tetandra*				3	0.07	1.71	7	0.12	2.83			
83	*Madhuca motleyana*				33	4.08	25.3	7	0.12	2.81			

84	*Syzygium sp*	3	0.02	1.52	23	0.12	8.68			
	GROUP 10									
85	*Cratoxylum glaucum*				33	0.75	11.7	37	0.04	4.5
86	*Syzygium griffithii*				13	0.24	5.66	817	0.77	86.8

Based on the calculation of the FIV (Family Importance Value) in all plots, Myrtaceae was recorded to be the family with highest number of species (14 species), followed by Euphorbiaceae (9 species). Dipterocarpaceae was not important in terms of the number of species (4), but had relatively high IV (104.38) mainly in the peat swamp forest (FIV=104.38). In the heath forest, only Myrtaceae (FIV=213.82) and Clusiaceae (FIV=86.18) were present. Myrtaceae practically dominated the community, including *Baeckea frutescens, Syzygium griffithii* and *Tristaniopsis whiteana,* amounted to 83%.

4.2 Structure

The forest structure refers to the size, density and horizontal as well as vertical distribution of trees (Kershaw 1964[15]). Graphs (**Figures 2** and **3**) and profile diagrams (**Figures 4a-e**) show the structure of forests in the study sites. Diameter classes show the dominance of trees with small diameter (> 30 cm) and only a few trees having diameters > 30 cm (**Figure 2**). In the heath forest plot, all trees in the plot were small with the range of 10-14 cm consisting of *Baeckea frutescens.* In the peat swamp forest plot, the 90 % of the trees having diameters of ≤ 30 cm and 13 trees had diameters of 40-50 cm consisting of *Shorea balangeran.* In the seasonal peat swamp forest plot, a total of 16 trees (10 %) had diameters of ≥ 40 cm and 83 % of the trees had diameters of ≤ 30 cm. Trees with diameters > 50 cm included *Ficus* sp. and *Madhuca motleyana.* In the dryland forest 92 % of the trees had small diameters and only 7% had diameters of 45-55 cm including *Combretocarpus rotundatus, Cryptocarya ferrea* and *Schima wallichii* (**Table 4**).

Figures 4a-d show the profile diagrams of the four forest types. **Figure 4d** is the profile diagram of the heath forest where the tree height did not exceed 8 m and the mean height was only 3.5 m, with very high tree density. In the peat swamp forest the tree height reached 25 m with the mean height of 12. In the dryland forest and the seasonal swamp forest the tree height was almost the same with the highest was recorded to be 35 m but the mean height was 15 m.

High population of trees with small diameters indicates the regeneration status of the forest. Tree species with small diameters included *Combretocarpus rotuodatus, Syzygium* spp., *Gordonia ovalis. Pternandra azurea, Macaranga triloba, Aporosa frutescens, Madhuca motleyana, Schima wallichii, Rhodamnia cinerea, Calophyllum pulcherrimum, Diospyros hermaphroditica* was fitly regenerating. In the heath forest, the species with good regeneration included *Baeckea frutescens, Calophyllum pulcherrimum, Cratoxyllum glaucum, Eugenia* spp., *Rhodomyrtus tomentosa* and *Tristaniopsis whiteana.* The pattern of diameter class distribution of the other species is as follows:

(1) Species that were represented in diameter classes up to 40 cm but with low density were *Calophyllum pulcherrimum, Schima wallichii, Pternandra azurea, Syzygium chlotanthum, Combretocapus rotuodatus, Shorea balangeran* and *Vatica pauciflora.*
(2) Species with diameters of 10-40 cm but its distributional pattern was irregular and low density (1-14 trees/ha) were *Calophyllum rigidum, Diospyros maingayi, Diospyros hermaphroditica, Xylopia ferruginea, Baccaurea lanceolata, Calophyllum inophyllum, Cratoxylon formosum, Artocarpus tamaran,* and *Ochanostachys amentacea.*
(3) Spesies that were represented in high diameter classes (>50 cm) but its occurrence in lower diameter classes was low were *Ficus* sp. (80.0-100 cm), *Litsea robusta* (50-70 cm). *Sandoricum borneense* (50-70 cm) and *Aglaia edulis* (50-70 cm).
(4) Vertical distribution in the plots except the heath foret plot shows three strata, the lowest stratum consisted of trees with height of 8-15 m, the height of next stratum was 15–25 m and the height of the uppermost stratum was 25–35 m. Some tree species had height of more than 30 m, including *Madhuca motleyana* (in seasonal peat swamp forest), *Shorea balangeran* (in

peat swamp forest) and *Schima walichii* (in dryland forest). In the heath forest the tree height was less than 10 m with the average height of 6 m.

Table 4. Species with large diameters in each plot in MKNR

Species	Family	Diameter (cm)			
		Seasonal peat swamp forest	Peat swamp forest	Dryland forest	Heath forest
Aglaia edulis	Meliaceae	51	-	0	0
Artocarpus tamaran	Moraceae	41	0	0	0
Baccaurea lanceolata	Euphorbiaceae	75	0	0	0
Baeckea frutescens	Myrtaceae	0	0	0	10-14
Calophyllum inophyllum	Clusiaceae	45-60	0	0	0
Combretocarpus rotundatus	Rhizophoraceae	0	0	55	0
Cryptocarya ferrea	Lauraceae	0	0	49	0
Diospyros hermaphroditica	Ebenaceae	0	0	45	0
Diospyros maingayi	Ebenaceae	42	0	0	0
Endospermum peltatum	Euphorbiaceae	0	0	44	0
Ficus sp.	Moraceae	99	0	0	0
Litsea robusta	Lauraceae	60	0	0	0
Madhuca motleyana	Sapotaceae	43-66	0	0	0
Sandoricum borneense	Meliaceae	51	0	0	0
Schima wallichii	Theaceae	0	0	45-48	0
Scorodocarpus borneensis	Olacaceae	42	0	0	0
Shorea balangeran	Dipterocarpaceae	76	40-52	0	0
Syzygium chloranthum	Myrtaceae	42	0	0	0

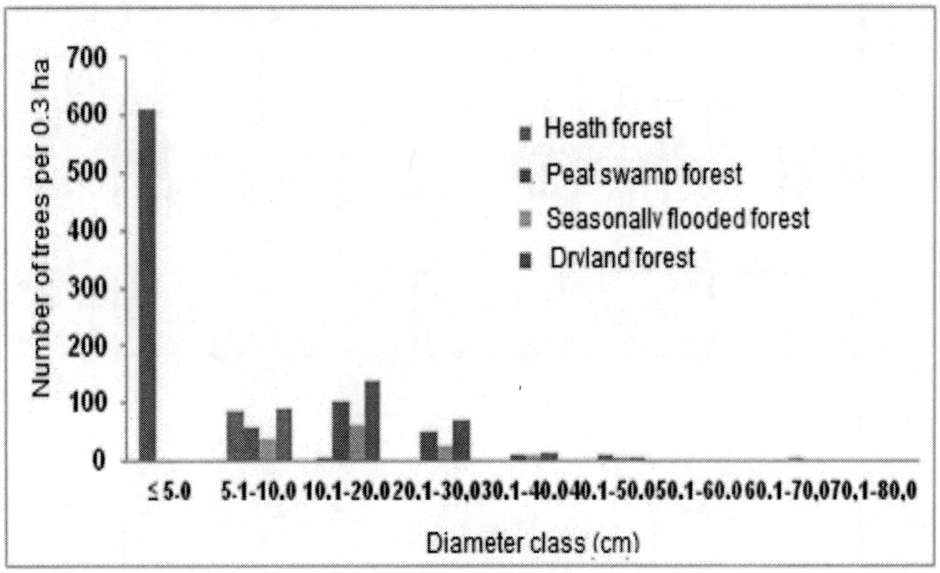

Figure 2. Number of trees according to the three diameter classes in the four plots in Muara Kendawangan Nature Reserve

IOP Conf. Series: Earth and Environmental Science **166** (2018) 012005 doi:10.1088/1755-1315/166/1/012005

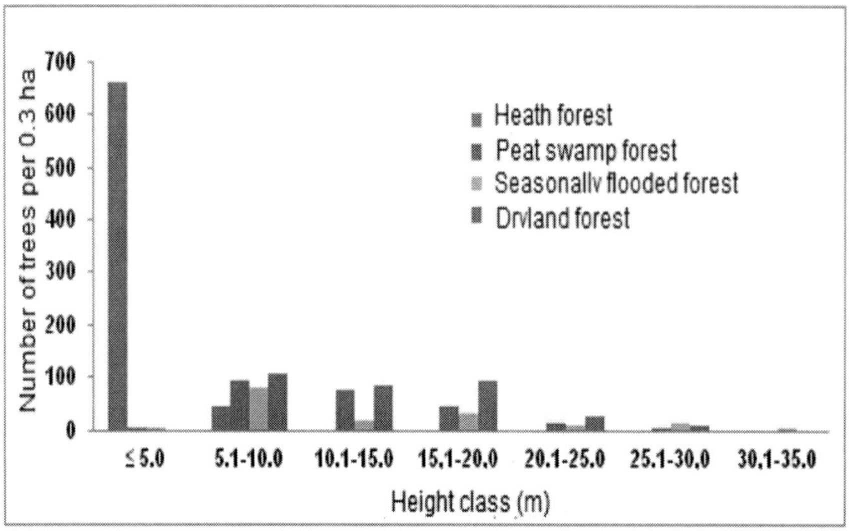

Figure 3. Number of trees according to the height classes in the four plots in Muara Kendawangan Nature Reserve

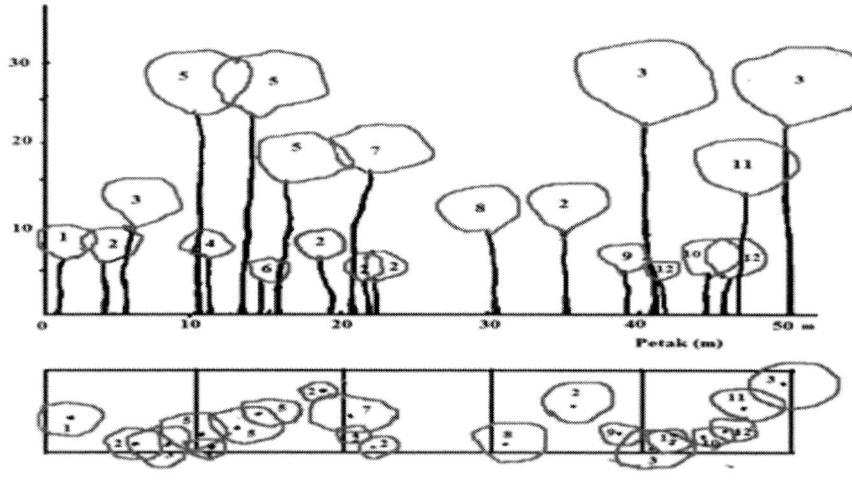

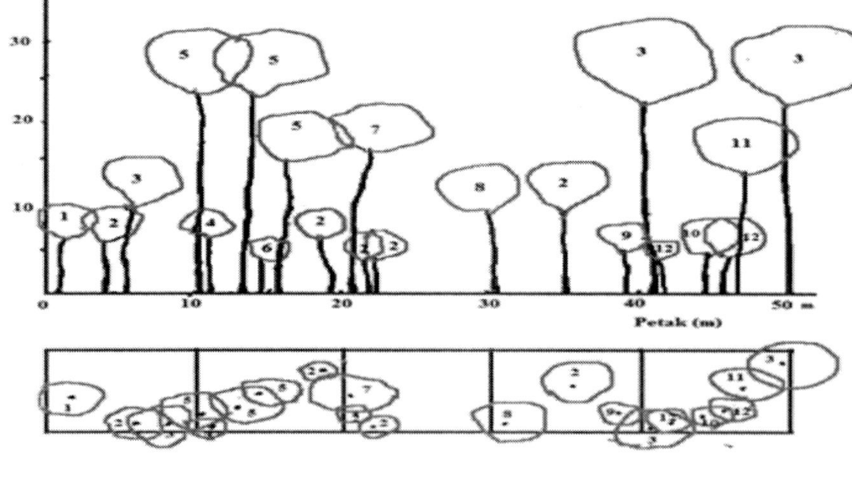

Figure 4a. The profile diagram of dryland forest (plot 1.): 1. *Artocarpus nitidus*; 2. *Rhodamnia cinerea*; 3. *Schima wallichii*; 4. *Combretocarpus rotundatus*; 5. *Vatica pauciflora*; 6. *Barringtonia samoensis*; 7. *Cryptocarya ferrea*; 8. *Gymnanthes borneensis*; 9. *Tristaniopsis whiteana*; 10. *Aglaia edulis*; 11. *Licania splendens*; 12. *Litsea umbelliflora*.

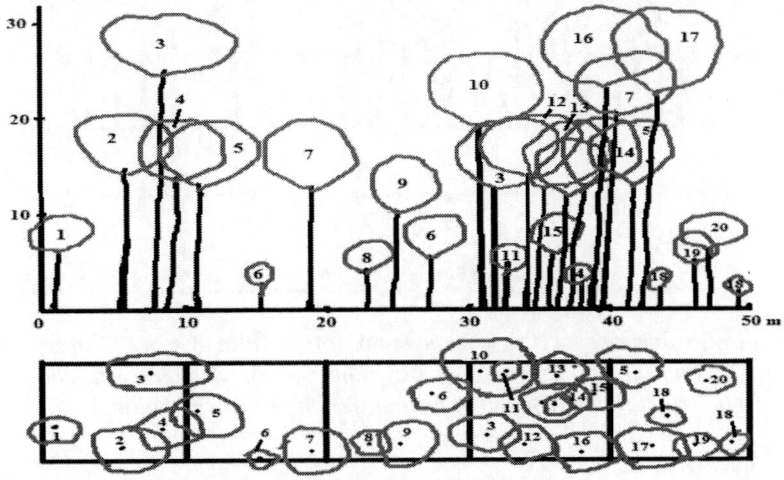

Figure 4b. The profile diagram of seasonal peat swamp forest (plot 2.): 1. *Cryptocarya lucida*; 2. *Gomphia serrata*; 3. *Ochanostachys amentacea*; 4. *Dacryodes rostrata*; 5. *Scorodocarpus borneensis*; 6. *Macaranga triloba*; 7. *Canarium hirsutum*; 8. *Schima wallichii*; 9. *Litsea firma*; 10. *Baccaurea macrocarpa*; 11. *Cryptocarya agathophylla*; 12. *Baccaurea polyneura*; 13. *Shorea rugosa*; 14. *Hopea nigra*; 15. *Aporosa frutescens*; 16. *Agathis alba*; 17. *Diospyros maingayi*; 18. *Alseodaphne umbelliflorum*; 19. *Baccaurea javanica*; 20. *Syzygium chloranthum*

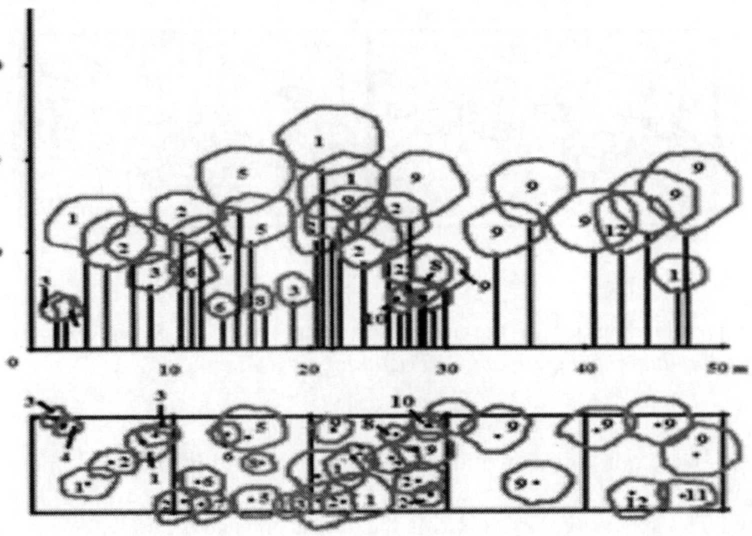

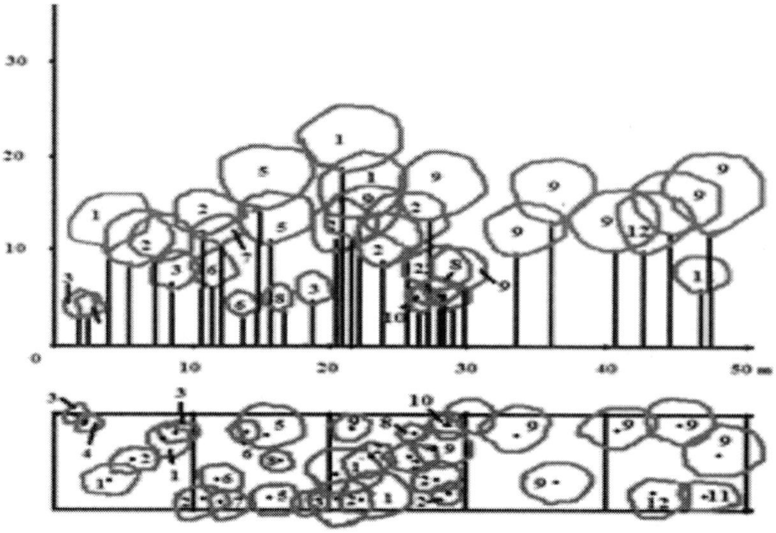

Figure 4c. The profile diagram of the peat swamp forest (Plot 3.): 1. *Cratoxylum glaucum*; 2. *Gordonia ovalis*; 3. *Pternandra azurea*; 4. *Syzygium* sp; 5. *Schima wallichii*; 6. *Parastemon urophyllus*; 7. *Syzygium griffithii*; 8. *Syzygium fusticuliferum*; 9. *Shorea balangeran*; 10.

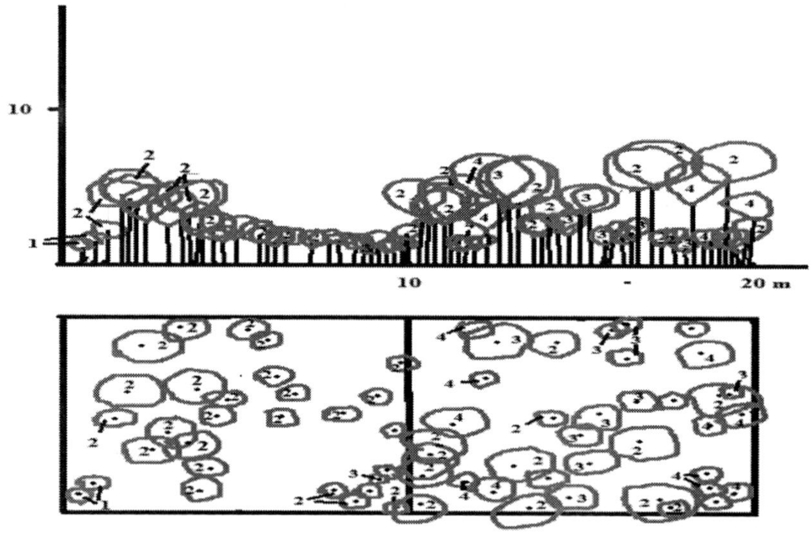

Figure 4d. The profile diagram of the heath forest (plot 4.): 1. *Syzygium griffithii*; 2. *Baeckea frutescens*; 3. *Calophyllum pulcherrimum*; 4. *Tristaniopsis whiteana.*

5. Discussion

The MNKR forests were rain forests with high rainfall and could be included in the rainfall type A of Schmidt & Ferguson (1951) [9] Forest with type A is characterized by everwet conditions, evergreen and thick canopy. The species composition of the forest consisted mainly of primary forest species (80%) including species of Myrtaceae (14 species), Lauraceae (8), Burseraceae (4), Dipterocarpaceae (4), Theaceae (4), Ebenaceae (3), Meliaceae (3), Sapindaceae (3). *Shorea balangeran,* which is a member of the character species of the *Shorea-Schima* association gained dominance in the swampy habitat, which could be designated as swamp forest indicator. Meanwhile *Schima wallichii* distinguished the dryland forest and *Baeckea frutescens* typified heath forest.

The density (trees/ha) in the dryland forest was higher than in the peat swamp forest. Similarly with the BA, the dryland forest was the highest (29.0 m^2/ha). In the seaonal swamp forest the mean diameter was 20.1 cm, while in the dryland forest it was 16.37 cm (**Table 1**). The trees with big diameters were more common in the seasonal swamp forest, including a *Ficus* sp. which could reach a diameter of > 80 cm and in the peat swamp forest *Shorea balangeran* which could have diameter of up to 75 cm. Myrtaceae and Dipterocarpaceae contributed the trees with large diameters including *Shorea balangeran* (Dipterocarpaceae) with a total BA=5.16) from 63 trees and *Syzygium grifithii* (Myrtaceae) with a total BA= 0.31 contributed by 249 trees. Other secies with with high BA included *Schima wallichii, Vatica pauciflora, Combretocarpus rotundus, Madhuca motleyana* dan *Calophyllum pulcherimum*.

In heath forest, 98.7% of trees were small with mean diameter of 3.51 cm but we observed *Syzygium bankense* having diameter of 42 cm. Small trees were also dominant in the other forest types. A total of 73 tree species had diameters <20 cm (Figure 2). Such pattern characterizes disturbed tropical rain forests [16-17].

The diversity of tree species in seasonal swamp forest according to diversity index shows the highest value (3.64). The high number of tree species in seasonal swamp forest area is suspected to be more influenced by land factor. The forest experienced longer dry period compared to swamp forest and when with dryland forest can still hold water during dry season. The area of the swamp forest in MKNR has been decreasing due to illegal logging, Species that have been disappearing because of illegal logging included *belangir (Shorea balangeran), gembor wood* (*Nothaphoebe umbelliflora), ramin wood* (*Gonystylus bancanus*) and *gaharu* (*Aquilaria malaccensis*). Currently, the population of these species were very small and even *ramin* and *gembor* were no longer present in the research sites. On the other hand, *Shorea balangeran* was one of the most abundant species of the Dipterocarpaceae occurring in the peat swamp plot (D = 200 trees/ha) but not so in seasonal peat swamp forest and dryland where there were only seven trees in each plot. The abundance and high occurrence of this species (Frequency =100%) led to lower species heterogeneity compared to the seasonal swamp forest and dryland forest plots. This species is endemic to Borneo and has been listed in the critical category in the IUCN [18-20].

The density of *Schima wallichii* was relatively low in seasonal swamp and peat swamp forests but it was high and dominant in hill forest plots. It was one of the five big species in the dryland forest research site. In Indonesia, it is widely distributed from lowland to montane forests [21-23]

The peat swamp forest was mostly inhabited by trees with small diameters of 10-20 cm, with a density of 590 trees/ha and BA of 24.43 m^2/ha. The common species included *Shorea balangneran, Syzygium fusticuliferum, Gordonia ovalis, Garcinia dioica* and *Calophyllum pulcherrimum*. The forest at Kempas Menali has been much disturbed as it is located adjacent to native agricultural fields. The disturbance led to the formation of canopy gaps, which were usually occupied by scrubs overgrown by rattans and vines. In the parts of the forest where there were still many large trees, the undergrowth was dominated by the species of Myrtaceae, Clussiaceae, Ebenaceae, and Sapotaceae.

Along with the rapid rate of regional development and population growth, disruption to wetland ecosystems has been also increasing. It has been threatening the integrity of the wetlands. The degradation of the area has been partly due to forest conversion to non-forest uses, including oil palm plantations, rice fields, settlements, and transmigration area.

In the heath forest plots, 98.7% of trees were small in size with an average diameter of 3.51 cm and average height of 6 m. But elsewhere, *Syzygium bankense* could reach a diameter of 42 cm [24]. Myrtaceae was one of the families that had the highest number of species in the heath forest. The forest in the plot was almost entirely dominated (83%) by *Baeckea frutescens, Syzygium subrufum*, and *Tristaniopsis whiteana*. In the Malesian forests [16] [25-27], including in riparian and swamp forests [28-29], Myrtaceae is known to be rich in species. Heath forest is one of the important forest types in Indonesia developed on podsol and quartz sandy soils, which are poor in nutrients and low pH

[16-17, 30]. The heath forest at MKNR occurred as scattered mosaic of small areas, sand in places they were planted to pineapple.

The density of trees in the three plots of research was quite high. Dryland forests were noted to have a higher density compared to the peat swamp and seasonal swamp forests. However, large diameter trees were more prevalent in seasonal swamp forests. In the seasonal swamp forest, *Baccaurea lanceolata* and *Shorea balangeran* stem diameters reached > 70 cm, but if in the peat swamp forest *Shorea balangeran* mostly had diameters of 40-60 cm. Myrtaceae and Dipterocarpaceae were the largest contributors to the Basal area. The trees with largest basal area were *Shorea balangeran* (BA = 5.16 m^2; density n = 63 trees/ha) and *Syzygium griffithii* (BA = 0.52 m^2; density = 262 trees/ha) (**Table 2**). Other species with relatively high BAs included *Calophyllum pulcherrimum*, *Combretocarpus rotundatus, Madhuca motleyana, Schima wallichii* and *Vatica pauciflora*.

6. Conclusion

The forests at MNKR could be included as the disturbed primary forest where 80 % of the species were primary forest species. The small statured heath forest was almost entirely dominated by species of Myrtaceae. High population of trees with small diameters indicates the regeneration status of the forest. The forest communities had low species richness and were designated as the *Schima wallichii-Shorea balangeran* association, composed of three subassociations, and the *Baeckea frutescens-Rhodomyrtus tomentosa*. The associations and subassociations were related to dryland, peat swamp, seasonal peat swamp and sandy dryland. It is predicted that the species composition of the associations and subassociations will not change in the coming years. The recovery of the disturbed sites and degraded lands to conditions similar to the original forests prior to disturbances would take a very long time. The recovery could be enhanced by applying ecological restoration through planting tree species native to the sites, complemented with actions to prevent tree poaching and forest encroachment. . The ecological restoration will help in increasing species diversity, improving forest conditions and expanding the coverage of forest to satisfy the requirements stipulated in Forestry Law (*UU no.41 Tahun 1999 tentang Kehutanan Pasal 18 Ayat 2*) and Spatial Arrangement Law (*UU26/2007 tentang Tata Ruang*) to preserve 30% of the total land area of a region as forests in order to maintain various ecosystem functions [12]. Species that will survive and maintain themselves in the forest in the future are currently represented in almost all diameter classes, although with low density. They included the *Schima wallichii, Shorea balangeran, Baeckea frutescens* and *Rhodomyrtus tomentosa*. Further intensive studies in the above forest types as well as in the other community types present in MNKR are needed to provide data for better management of the reserve.

7. Reference

[1] Mackinon K, Hatta G, Halim H & Mangalik A 1996 *The Ecology of Kalimantan*
[2] Uji T 2003 *Biodiversitas* 4(1): 112-117
[3] Sambas E N, Susiarti S and Suhardjonoi 1994 Struktur dan Komposisi hutan gambut di Kabupaten Sanggau, Kalimantan Barat *In*: Pros. Sem. Hasil Litbang SDH Puslitbang Biologi-LIPI, Bogor pp.344-350
[4] Siregar M & Ruskandi A 2000 *In*: Uji T *Eksplorasi dan Inventarisasi Flora dan Fauna Ikan di Cagar Alam Muara Kendawangan, Kalimantan Barat* Bogor: Puslitbang Biologi-LIPI
[5] Noor Y R, Hanafia E W 1995 A preliminary survey on the ecological potential of the Cagar Alam Muara Kendawangan, West Kalimantan. Directorate General of Forest Protection and Nature Conservation; Bogor: Asian Wetland Bureau (AWB), Indonesia Programme; [S.l.]: Overseas Development Administration (ODA), 1995
[6] Anonimous 1998 Informasi Dawasan Donservasi Dalimantan Barat Pontianak D: Sub Balai Konservasi Sumber Daya Alam Kalimantan Barat
[7] Anonimous 2017 Cagar Alam Muara Kendawangan Seksi konservasi wilayah I ketapang. http://skw1ketapang.blogspot.co.id/p/blog-page.html
[8] Keel S, Gentry A H & Spinzi L 1993 Using vegetation analysis to facilitate the selection of

conservation sites in Eastern Paraguay *Conservation Biology* 7(1):66–75
[9] Schmidt F H & Ferguson J H A, 1951. Rainfall types based on wet and period ratios for Indonesia with Western New Guinea. *Verhandelingen 42.* Jawatan Meteorologi dan Geofifisika, Jakarta
[10] Cox G W 1992 Laboratory Manual of General Ecology. Econd Ed. Dubuque, (Iowa: Wm. C. Brown Company Publisher)
[11] Mueller-Dombois D & Ellenberg H 1974 *Aims and methods of vegetation ecology* (John Wiley & Sons, New York)
[12] Purwaningsih, Polosakan R, Yusuf R and Kartawinata K 2017 *Reinwardtia* 16 (1): 31-45
[13] Kartawinata K, Samsoedin I, Heriyanto M & Afriastini J J 2004 A tree species inventory in a one-hectare plot at the Batang Gadis National Park, North Sumatra, Indonesia *Reinwardtia* 12: 145 – 157
[14] Kent M and Paddy C 1992 Vegetation Description and Analysis a Practical Approach (London: Belhaven Press)
[15] Kershaw K A 1964 *Quantitative and dynamic ecology* (Edward Arnold Publishing Co. Ltd London)
[16] Whitmore T C 1984 *Tropical rain forest of the Far East* 2nd edition Clarendron Press. Oxford
[17] Richards P W 1996 *The tropical rain forest, an ecological study* Second edition Cambridge University Press, Cambridge
[18] Ashton P S 1982 Dipterocarpaceae. *In*: Van Steenis C G G J (ed.) *Flora Malesiana* 9: 237-552
[19] Ashton P S 1989 Species richness in Tropical forest. *In:* G. Meurant (Ed.) Tropical Forests: Botanical Dynamics, Speciation & Diversity. 240-251
[20] Ashton P S 1998 *Shorea balangeran* The IUCN Red List of Threatened Species 1998: e.T33103A9756028
[21] Simbolon H & Mirmanto E 1997 Vegetation analysis of Citorek area, Gunung Halimun National Park. *In* Simbolon H, Yoneda M & Sugardjito J (Eds). *Gunung Halimun: The last submontane tropical forest in West Java*. LIPI, JICA, PHPA, Bogor
[22] Schmidt L H & Lieu N X 2004 *Schima wallichii* Choisy. Seed Leaflet, (99)
[23] Keng H 1978 Theaceae. *In*: F.S.P. Ng (ed.) *Tree Flora of Malaya. A Mannual for Foresters* 3: 291-292.
[24] Onrizal, Kusmana C, Sahardjo B H, Handayani I P, Kato T 2005 *Biodiversitas* 6 (4): 263-265
[25] Lyn A C, Sunarti S, Mudiana D, Yulistyarini T, Wardani M. 2003 *Foribunda* 2 (4): 89-94
[26] Ridley H N 1925 *The Flora of Malay Peninsula*. (L. Reeve and Co. London)
[27] Ng F S P 1978 *Tree flora of Malaya* Vol.3 & 4 Longman Malaysia, Kuala Lumpur
[28] Waldes N J L & Page S E 2002 Foresy structure and tree diversity of a peat swamp forest in Central Kalimantan, Indonesia. agris.fao.org
[29] Riswan S 1982 Ecological studies on primary, secondary and experimentally cleared mixed dipterocarp forest and kerangas forest in East Kalimantan, Indonesia. *Thesis (Ph.D.)*
[30] Hilwan I1996 Ekologi dan Diversity Ekosistem Hutan Tropika Indonesia Bogor: Pusat Pengkajian Keanekaragaman Hayati, Institut Pertanian Bogor

IOP Conf. Series: Earth and Environmental Science **166** (2018) 012006 doi:10.1088/1755-1315/166/1/012006

Effect of Particle Pre-treatment on Physical and Mechanical Properties of Particleboard Made from Oil Palm Trunk

R Hartono[1], A. H Iswanto[1], T Sucipto[1], and K. M Lubis[1]

[1]Faculty of Forestry, Universitas Sumatera Utara Jl. Tridharma Ujung No 1 Kampus USU Medan, Indonesia 20155

E-mail: rudihartono@usu.ac.id

Abstract. Oil palm trunk can be used as raw material for particleboard. On the other hand, oil palm trunk contains many extractive substances, which can disturb the adhesion of particleboard. The objective of this research was to the effect of particle-pretreatment on particleboard properties made from oil palm trunk. The particle had been pre-treated with cold water (24, 48, and 72 hours) and hot water (1, 2 and 3 hours). The particles were dried to 3% in the oven. Board were made with size 30 (length) x 30 (width) x 1 cm (thick) with density target of 0.7 g/cm^3. Particleboard was made bonded urea formaldehyde with pressing temperature of 130^{0}C and pressure of 25 kg/cm^2 for 10 minutes. Control (particleboard without pre-treatment) was also produced. The board properties were tested in accordance with SNI 03-2105-2006. The results showed that particle pre-treatment improved physical and mechanical properties of particleboard. Pre-treatment (immersion in cold and hot water) decreased the moisture content (MC) and thickness swelling (TS) and increased modulus of rupture (MOR), modulus of elasticity (MOE) and Internal Bond (IB) value. The value of density, MC, MOR and IB fulfilled the SNI 03-2105-2006 standard, while the all MOE and TS value have not fulfilled the standard. The best pre-treatment condition was particle immersion into hot water for 2 hours.

1. Introduction

The trend of natural forests in supplying timber for industry has decreased in the last decade. To overcome the problem, an alternative material to substitute wood is necessary. One of the non-timber materials that contain lignocellulose and the amount of biomass is oil palm trunk (OPT). Based on study [1], oil palm plantations can produce OPT of about 183 m^3/hectare during replanting and continue to increase in the next replanting period. For each hectare, the oil palm plantation can produce 50.1 m^3 of sawn timber from the outer part of OPT, which is generally only one-third of OPT [2].

The utilization of OPT has weakness properties such as high density variations. The OPT density of 40 years old tree was 0.23-0.74 g/cm^3 [3], while OPT density of 25 years old tree was 0.14-0.60 g/cm^3 [4]. In addition, OPT has high moisture content and high shrinkage. Therefore, the improvement of OPT quality by compressing inner part of OPT [5] and compregnating phenol formaldehyde into the OPT [6] are indispensable. The results of the Technical Feasibility Study and Economical Study revealed that the OPT was suitable to substitute wood for Biocomposite Industry. It is recommended that OPT was suitable for particleboard [7]

Biocomposite technology is a suitable technology of utilizing all materials containing wood and or other lignocellulosic materials, such as OPT. Biocomposites are panels made of wood or other small lignocellulosic material and then rearranged into new products such as particleboard, fiberboard, strandboard, structural composite wood, mineral plank, and composite plastic board [8].

The OPT contains extractive substances. The extractive content of OPT after being dissolved in cold water, hot water, alcohol-benzene and 1% NaOH were 11.2-17.82%; 12.51-18.22%; 9.90-10.84%; and 27.64-32.26%, respectively [9]. The extractive content is increasing from the bark to the center and from the bottom to height of trunk [10]. Wood extractive can effect on the adhesive consumption and its rate of maturation [11], wood wettability and adhesive spread to particles [12]. The presence of extractives in the OPT can disturb the adhesion during the compression of particleboard.

Based on the above information, it is necessary to carry out the study of pre-treatment techniques on OPT particle. The objective of this study was to analyze the particle pre-treatment on physical and mechanical properties of OPT particleboard.

2. Materials and Method

2.1. Material

The material used in this study was OPT from University of North Sumatera. The OPT was converted into 30 mesh particles. The other material was urea formaldehyde (UA-140) from Palmolite Adhesive Industry, Probolinggo with resin content of 63%.

2.2 Methods

2.2.1. Preparation of materials

Pre-treatment of OPT particle was carried out by soaking in cold water (24, 48, and 72 hours) and hot water (1, 2 and 3 hours). The code of pre-treatment for soaking 24, 48 dan 72 hours and dipping into hot water for 1, 2 and 3 hours were CW-24, CW-48, CW-72, HW-1, HW-2, HW-3, respectively. Then, the OPT particle was dried inside an oven until reaching a moisture content around 3%. Soaking technique in cold water referred to pre-treatment study of Sentang wood [13] as well as soaking technique in hot water [13-15]

2.2.2. Particleboard manufacturing

After pre-treatment of particle in cold and hot waters, the OPT particle was dried inside an oven with temperature of 90 ^{0}C for 24 hours to reach a moisture content about 3%. Then, the OPT particle was mixed with UF adhesive with concentration of 10% and placed in a forming box of 30 cm x 30 cm size. The board thickness was 1 cm and made with target density of 0.7 g/cm^3. Furthermore, the OPT particle was pressed using hot press at the temperature of 130^{0}C and pressure of 25 kg/cm^2 for 10 minutes, and then conditioning for 2 weeks. The next step was to cut the particleboard for several test referring to SNI standard for evaluating the physical properties such as density, moisture content, thickness swelling, mechanical properties including modulus of elasticity and modulus of rupture [16].

2.2.3. Physical and mechanical test

The measurement of physical properties was density (ρ) and moisture content (MC) with a sample size of 10 x 10 x 1 cm, thickness swelling (TS) with a sample size of 5 x 5 x 1 cm. Meanwhile, the mechanical properties measured were Modulus of Elasticity (MOE) and Modulus of Rupture (MOR) with a sample size of 5 cm x 20 cm x 1 cm, and internal bond with a sample size of 5 cm x 5 cm x 1 cm. The experiment was conducted in triplicate. The results were then compared with the SNI [16].

2.3. Data Analysis

The data of physical (density, moisture content and thickness swelling) and mechanical properties (MOE and MOR) were compared with SNI standard particleboard type 8 [16] as shown at **Table 1**.

Table 1. The SNI 03-2105-2006

Properties	Standard
Density	0.4-0.9
Moisture Content	Max 14
Thickness Swelling	Max 12
MOR	≥ 82 kg/cm^2
MOE	≥20400 kg/cm^2
Internal Bond	≥ 1,5 kg/cm^2

3. Results and Discussion

3.1. Physical properties

The physical properties in this study were density, MC and TS. The density, MC and TS of particleboard with several pretreatment were shown in **Figure 1, 2 and 3**, respectively. Based on **Figure 1**, the density of pre-treatment types of particle board ranged from 0.66-0.70 g/cm^3. The final density of the particleboard is influenced by several factors such as wood type (wood density), the level of pressing pressure, the amount of wood particles in layer, the adhesive content and other additives [17].

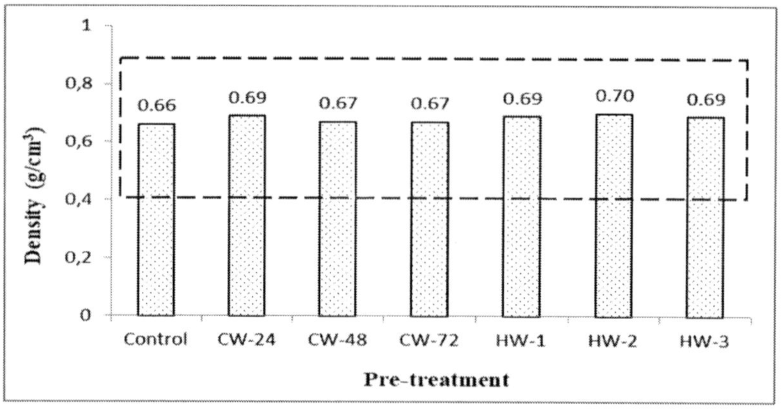

Figure 1. Density value of particleboard

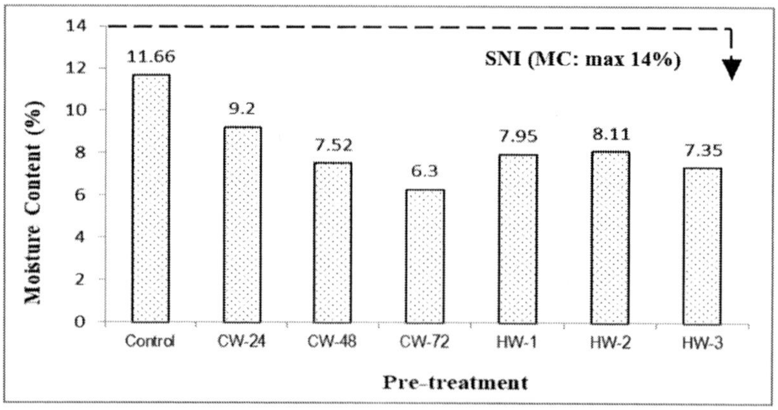

Figure 2. Moisture content value of particleboard

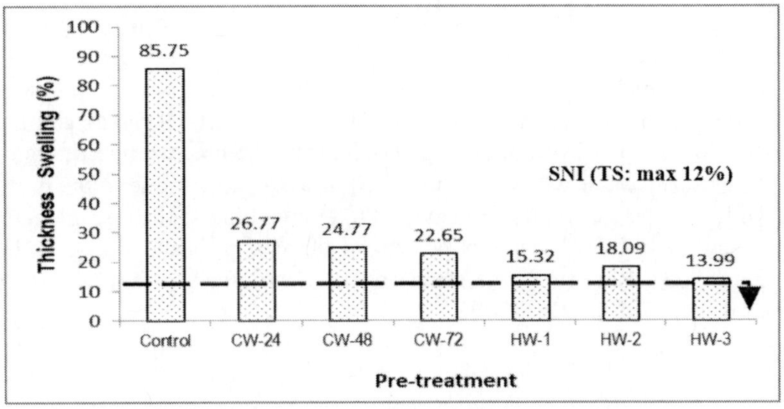

Figure 3. Thickness swelling value of particleboard

Figure 1 showed that the pre-treatment did not cause changes in particleboard density. In manufacturing particleboard, high density is not the main target but how to produce panel wood with low density and the properties fulfil the standard requirements [18]. The particleboards produced in this study are categorized into medium density boards. Medium density particleboard has a density between 0.59-0.80 g/cm^3. Based on the SNI standards, the density particleboard between 0.4 - 0.9 g/cm^3, and all density of particleboards have fulfilled standard [16].

The MC of pre-treatment type of particleboard ranged from 6.30-11.66 %. **Figure 2** showed that MC value of particleboard tends to decrease. Pre-treatment particle by soaking in cold and hot waters was essentially to remove the presence of extractive substances. This pre-treatment will decrease the extractive content in the wood (tannins, gums, sugars and pigments). The corelation extractive content with glue bond strength was poor. The increase of extractive content can reduce glue bond strength. Also, the higher extractive content had a significantly greater degree of adhesive failure and made the lower mechanical strength [19].

The particle pre-treatment will increase the adhesive ability to penetrate the cell wall and resulting in a good bonding process. Therefore, water vapour accessibility can be reduced [20]. Dissolution of extractive substances can increase the bonding performance of wood particles with their binder. In addition, the effect in hot pressing process using temperature of 130 ^{0}C for 10 min leads to the decrease in the hydrophilic properties of the lignocellulose materials. It will result low MC and TS. Based on SNI standard, the maximum MC is 14 % [16]. All MC of pre-treatment fulfils the SNI standard.

Figure 3 shows the TS of particleboards ranged from 15.32 to 85.75 %. The pre-treatment caused a decrease of TS of the particleboard. Particle immersion with cold water (24, 48, and 72 hours) and hot water (1, 2, and 3 hours) decreased the extractive content. The same result was obtained that the immertion of Jatropha particle using cold and hot water can removed some extractive component and caused the TS value to decrease compared untreatment [13]. The extractive component has influence in decreasing hygroscopic, wood permeability and also increasing wood durability. Although its few number, extractive influential in wettability of wood, affect of pH, consumtion, and penetration of glue [21]. The extractive components can be dissolved by immersing in hot water and cold water [18]. The dilution of extractive substances can increase the connectivity between wood particles with the binder. The adhesive ability to penetrate into the wood takes place well, whereby the presence of free hydroxyl groups on wood is filled by the adhesive. Consequently, the hygroscopic nature of the board can be minimized [20].

The value of thickness swelling with UF adhesive is still extremely high. This is due to UF adhesive was for interior purposes and this adhesive was not water resistant. The weakness of UF adhesives is that they were not resistant to moisture or high humidity causing this adhesive not suitable for exterior usage. The thickness swelling of particleboard is influenced by the quality of bonding and the adhesive properties [22]. The UF adhesive bond is not as strong as the isocyanate adhesive bond. This is because the particleboard that uses UF adhesives absorbs more water and significantly affects the value of its higher thickness swelling [23]. The absence of paraffin in the manufacture of particleboards is also one of the factors of high TS. The reduction of TS can be conducted through the addition of paraffin/wax or other hydrophobic material during the manufacture of the board [24].

The sample still absorbs water when it is immersed in water; this is due to the presence of free and bonded water. The free water lies in the cell cavity, the intercellular space and the gap on the adhesive bond within the wood. The bound water is present in the cell wall and may also be present in the adhesive-wood tissue [25]. Based on SNI standard, the maximum TS is 12% [16]. All TS of pre-treatment did not fulfilled SNI 03-2105-2006.

3.2. Mechanical properties

3.2.1. Modulus of elasticity (MOE), modulus of rupture (MOR) and internal bond (IB)

The MOE, MOR and IB value with several pretreatment were shown in Figure 4, 5 and 6.

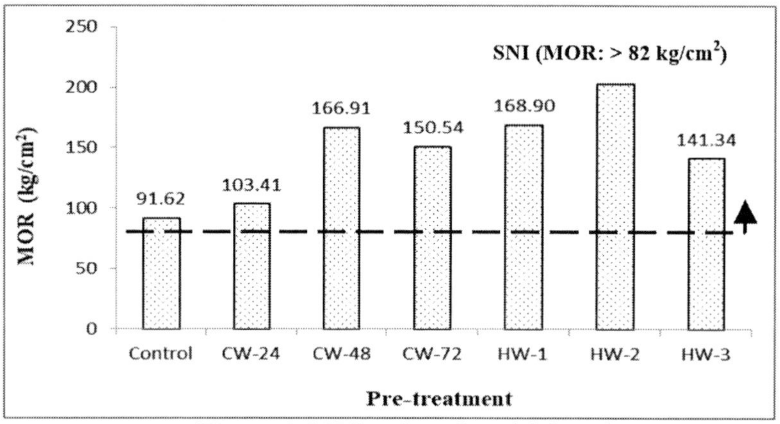

Figure 4. MOR value of particleboard

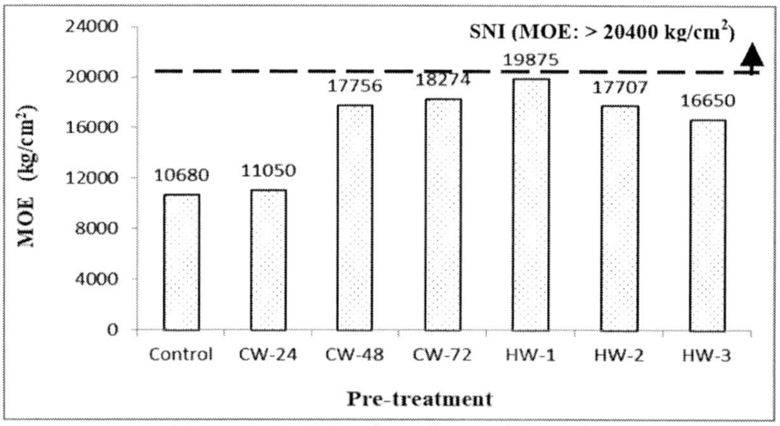

Figure 5. MOE value of particleboard

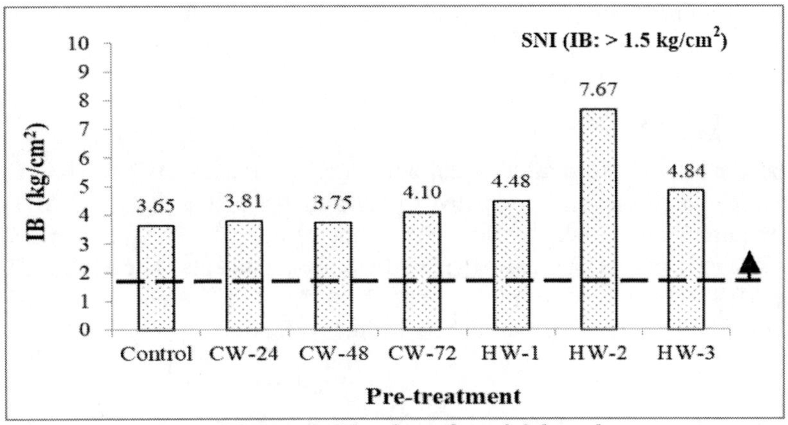

Figure 6. IB value of particleboard

Based on **Figure 4, 5 and 6**, the MOR, MOE, and IB values of the particleboard were ranged from 91.62-203.23 kg/cm^2, 10680-19875 kg/cm^2 and 3.65-7.67 kg/cm^2, respectively. The values of MOR, MOE and IB after pre-treatment tend to increase. The MOR, MOE and IB values of pre-treatment samples were higher than that of control. All of MOR and IB values of particleboard fulfilled the SNI standards. By contrast, MOE values of particleboard with several pre-treatments did not reach the SNI standards [13].

Figure 4 showed that MOR control was 91.62 kg/cm^2. After pre-treatment of particle using cold water (24, 48, 72 hours) and hot water (1, 2 and 3 hours), the value of MOR tend to increase from 103.41 kg/cm^2 until 203.23 kg/cm^2, respectively. The increase of MOR value using pre-treatment was 12.86-121.82%. The highest of MOR value was 203.23 kg/cm^2 achieved with the hot water immersion for 2 hours.

Figure 5 showed that MOE control was 10680 kg/cm^2. After pre-treatment of particle by using cold and hot waters, the MOE value tend to increase from 11050 kg/cm^2 to 19875 kg/cm^2. It means that the pre-treatment improved the MOE value around 3.46-85.16%. However, the MOE did not fulfilled SNI 03-2105-2006.

Figure 6 showed that IB value was tend to increase from 3.75 kg/cm^2 to 7.67 kg/cm^2 compared to control (3.65 kg/cm^2). The pre-treatment improved the IB value around 2.7%-110.13%. The highest IB value was 7.67 kg/cm^2 and achieved with the hot water treatment for 2 hours. The MOR value is influenced by the content and type of adhesive materials used, adhesive strength and fibre length. Factors affecting the MOR value of the panels are wood density, particle geometry, particle orientation, adhesive content, moisture content and hot press process. Meanwhile, the value of MOE is influenced by the content and adhesives types, adhesive strength and fibre length [11]. The pre-treatment by soaking in hot water will reduce alkaline extractives, decrease the pH value and increase the absolute acid buffer capacity of the particles [26].

The extractive substances are not a part of structural wood. The presence of extractives can inhibit penetration of adhesives into particles/strands, and resulting in lower mechanical properties of composite [16]. Soaking the strands in hot water can improve the surface tension of strands, strands adhesion and adhesives [27]. Thus, it can improve the mechanical properties of composite. It was reported that immersing red meranti wood flakes in hot water for 2 hours can improve the strength of OSB [28-29]. Also, immersing Sentang wood flakes in hot water for 2 hours and steam for 1 hour can increase MOR, MOE and IB value [13].

The other research reported that extractives have correlation with the wettability of wood surfaces to form good bonding of adhesives [30]. Wettability can be used as an indication of its gluing properties. One way to measured the wettability was by using cosine-contact angle method. The previous report that OPT have poor wettability or great of OPT contact angle compared Meranti and Sengon wood. The liquid resin did not flow easier and formed a higher contact angle [31].The improved of gluing properties will affect on particleboard quality. It is necessary to pre-treatment in the form of submersion in hot water or in cold water. It showed that good wettability will improve gluing properties and particleboard quality.

4. Conclusion
Particle pre-treatments by soaking in cold and hot waters decreased MC and TS. Thus, the physical properties of particleboard increase compared to control. All of densities and MC of particleboard fulfil the SNI requirements. However, the TS has not fulfilled SNI requirements. The mechanical properties (MOR, MOE and IB) increase after particle pre-treatments. The MOR, MOE and IB value increase around 18.87-121.82%; 3.46-85.16% and 2.7%-110.13%, respectively. All of MOR and IB values fulfilled the SNI requirement, but the MOE values were not.

5. Reference
[1] Bakar E S 2003 *Forum Komunikasi Teknologi dan Industri Kayu* 2 5-6
[2] Febrianto F and Bakar E S 2004 *Kajian Potensi, Sifat-sifat Dasar dan Kemungkinan Pemanfaatan Kayu Karet dan Biomassa Sawit di Kabupaten Banyuasin* (Indonesia: Lembaga Manajemen Agribisnis dan Agroindustri. Institut Pertanian Bogor. Bogor)
[3] Hartono R, Wahyudi I, Febrianto F, Dwianto W 2011 *J. Ilmu dan Teknologi Kayu Tropis* 9(1): 73-83
[4] Erwinsyah 2008 Improvement of Oil Palm Wood Properties Using Bioresin Dissertation of graduated school of Universität Dresden.
[5] Hartono R, Wahyudi I, Febrianto F, Dwianto W, Hidayat W, Jang J H, Lee S H, Park S H, Kim N H 2016 *J. Korean Wood Sci. Technol* 44(2): 172-183.
[6] Hartono R, Hidayat W, Wahyudi I, Febrianto F, Dwianto W, Jang J H, Kim N H 2016 *J. Korean Wood Sci. Technol* 44(6): 842-851
[7] Iswanto A H, Sucipto T, Azhar I, Effendi K 2009 *Studi Kelayakan Teknis dan Ekonomis Penggunaan Batang Kelapa Sawit Sebagai Bahan Baku Alternatif Pengganti Kayu Pada Industri Biokomposit* (Indonesia: Laporan Penelitian Hibah Unggulan Strategis Nasional Batch I. Medan).
[8] Marra A A 1992 *Technology of Wood Bonding: Principles in Practise* (New York: Van Nostrand Reinhold).
[9] Bakar ES, Rachman O, Hermawan D, Karlinasari L, Rosdiana 1998 *Jurnal Teknologi Hasil Hutan* XI(1): 1-11.
[10] Omar N S, Bakar ES, Jalil N M, Tahir P M, Yunus W M Z W 2011 *Wood Research Journal* 2(2) 73-77
[11] Maloney T M 1993 *Modern Particleboard and Dry-Process Fiberboard Manufacturing* (USA: Miller Freeman, San Francisco).
[12] Sernek M 2002 Comparative Analysis of Inactivated Wood Surface. Dissertation of Polytechnic Institute and State University, Virginia.
[13] Iswanto A H, Febrianto F, Wahyudi I, Hwang W J, Lee S H, Kwon J H, Kwon S M, Kim N H, and Kondo T 2010 *J. Fac. Agr., Kyushu Univ* 55(2): 371–377.
[14] Claude M, Yemele N, Koubaa A, Diouf P N, Blanchet P, Cloutier A, Stevanovic T 2008 *Wood and Fiber science Journal* 40(3): 339-351.
[15] Paredes J J, Jara R, Shaler S M, van Heiningen A 2008 *Forest Products Journal* 58(12): 56-62.
[16] SNI 03-2105-2006 Papan Partikel Badan Standarisasi Nasional
[17] Kelly M W 1977 *Critical Literature Review of Relationship Between Processing Parameter and Physical Properties of Particleboard* (General Technical Report FPL-10, Wisconsin).
[18] Bowyer J L, Shmulsky R, Haygreen J G 2003 *Forest Products and Wood Science: An Introduction* Fourth edition (USA: Lowa State University Pr.).
[19] Nussbaum R M and Sterley M 2002 *Wood and Fiber Science* 34(1) 57-71

[20] Pari G, Roliadi H, Setiawan D, Saepuloh 2006 *J Penelitian Hasil Hutan* 24 89-97.
[21] Ruhendi S, Sucipto T 2007 *Wetabilitas Tandan Kosong Sawit* In: Prosiding Seminar Nasional Mapeki X, Pontianak, pp 17-23
[22] Boquillon N, Elbez G, Schonfeld U 2004 *J. Wood Sci.* 50 230–235.
[23] Pan Z, Yi Zheng, Zhang R, Jenkins B M 2007 *Industrial Crops and Products* 26 185-194.
[24] Nemli G, Hiziroglu S, Usta M, Serin Z, Ozdemir T, Kalaycioglu H 2004 *For.Prod.J.* 54 36–40
[25] Boonstra M J, Pizzi A, Zomers F, Ohlmeyer M, Paul W 2006 *J Holz als Roh-und Werkstoff* 64 157-164.
[26] Xing C, Zhang SY, Deng J 2004 *Holzforschung* 58 408–412.
[27] Alamsyah E M, Yamada M and Taki K 2008 *J Wood Sci 54 208–213*
[28] Hadi Y S 1988 *J Teknologi Hasil Hutan* 2 16–24
[29] Febrianto F, Royama L I, Hidayat W, Bakar E S, Kwon J H and Kim N H 2009 *J Wood Sci and Tech* 37 121–127
[30] Rafael E 2016 *Appl Microbiol Biotechnol* 100:1589–1596
[31] Sucipto T, Hartono R, Dwianto W 2018 *IOP Conf. Ser.: Earth Environ. Sci.* 122 012141

IOP Conf. Series: Earth and Environmental Science **166** (2018) 012007 doi:10.1088/1755-1315/166/1/012007

Analysis of Forest Landscape Fragmentation in Samosir Island, North Sumatra

Samsuri[1], A Zaitunah[2] and J. F Manurung[3]

[1]Forestry Faculty – University of Sumatera Utara,
Kampus USU Padang Bulan, Medan 20155, Indonesia
[2]Forestry Faculty – University of Sumatera Utara,
Kampus USU Padang Bulan, Medan 20155, Indonesia
[3]Leuser Ecocystem Institute
Medan 20155, Indonesia

E-mail: samsuri@usu.ac.id

Abstract. Forest coverage has changed regarding to structure and size, which further causes forest fragmentation. Forest cover change of Samosir Island disrupts the forest ecosystem. This research were aimed to determine the land cover changes in Samosir Island of year 2005, 2011 to 2015 and to determine fragmentation degree in Samosir Island's forest. This research was carried out in four step such as delineation of watershed, image analysis, land cover changes analysis, and fragmentation analysis. The four sub-watershed of Samosir Island were divided into four aspect maps. Research found that during the period of 2005 to 2015, Samosir Island had 10 land cover classes; there were water body, forest, mixed garden, dry land farming, bare land, grassland, settlements, swamp, paddy field, and shrubs. Within the period of 2005 to 2011, the largest land cover change was 3.912,02 Ha of converted forest into mixed farm. During the period of 2011 to 2015, the change from forest to shrubs was 3.334,671 Ha and during period 2005 to 2015, the change from forest to shrubs was 4.137,277 Ha. The Samosir island's forest in 2005 – 2015 was classified as the lowest class fragmentation of the west sub-watershed, moderate class fragmentation of the South sub-watershed, and highest-class fragmentation of the north sub-watershed.

1. Introduction

Deforestation and forest degradation lead to forest fragmentation. Forest fragmentation occurs because large interconnected forests are separated into smaller blocks due to road, agriculture, urbanization or other development aspects. Consequently, the function of the forest as a home range of plant and animal species is reduced. Fragmentation can be caused by the removal of vegetation on a large area and the construction of a separated path habitat [1-3].

The changes of shape and forest extent that occur in the forests of Samosir Island lead to disruption of forest ecosystems. Reduced forest area resulted in decreased of land capability to sustainane absorb run off. Reduced forest is triggered by forestlands conversion. Economics benefit is one of the reasons for the community to land forest conversion. To prevent and reduce the damage caused by forest fragmentation, it requires efforts to control the level of forest fragmentation. The degree of the fragmentation rate is required to determine the strategy of forest fragmentation control. Therefore, it is necessary to analyse the degree of fragmentation [4].

IOP Conf. Series: Earth and Environmental Science **166** (2018) 012007 doi:10.1088/1755-1315/166/1/012007

Fragmentation occurs through loss of habitat. Otherwise, habitat loss may be viewed as the result of fragmentation. However, fragmentation may be accompanied by habitat loss as large habitat pockets divided into smaller and more isolated habitat pockets. If habitat loss and fragmentation are separately observed, the loss of habitat has a more significant impact on species survival than fragmentation. However, since fragmentation and habitat loss occur together, it is extremely difficult to determine which is more important for habitat change [5-7].

Measuring fragmentation is an interesting action. Many sizes can be used to view habitat fragmentation. For measuring aspects of landscape structure, three indices such as (1) landscape composition (2) landscape configuration and (3) patch form within the landscape can be used. The composition shows the number of different closure types found in the landscape. Compatible patch analysis program with ArcGIS is reliable enough to calculate fragmentation statistics, as it is a modification of Fragstat and can be used to calculate spatial statistics, either polygon or raster files [5, 8].

The analysis of land cover and degree of forest fragmentation on Samosir Island was carried out using remote sensing method and Geographic Information System. This study aimed to determine the change of land cover in Samosir Island from 2005, 2011 to 2015 and to examine the degree of fragmentation in watershed of Samosir Island.

2. **Method**

The research was conducted on April - August 2016 in Samosir Island, Samosir district, North Sumatra Province. Processing and data analysis was completed at Forest Management Laboratory, Faculty of Forestry, and University of Sumatra Utara, Indonesia.

2.1.Research site

Samosir Island is a peninsula connected by a 200-meter length of lands with west side area of Lake Toba. In 1906, the Dutch government built a canal on this land so that the land of Samosir was separated into an island. Samosir Island is located in Samosir District. Geographically, Samosir Island is located within 2°35' North and 98°49' East. Samosir district consists of nine sub districts where six of them are in Samosir Island and three in the outer ring of Lake Toba (**Figure 1**). The island has an area of 63,000 Ha. The 6-sub districts of Samosir Island are Simanindo, Pangururan, Ronggurnihuta, Palipi, Nainggolan and Onan Runggu.

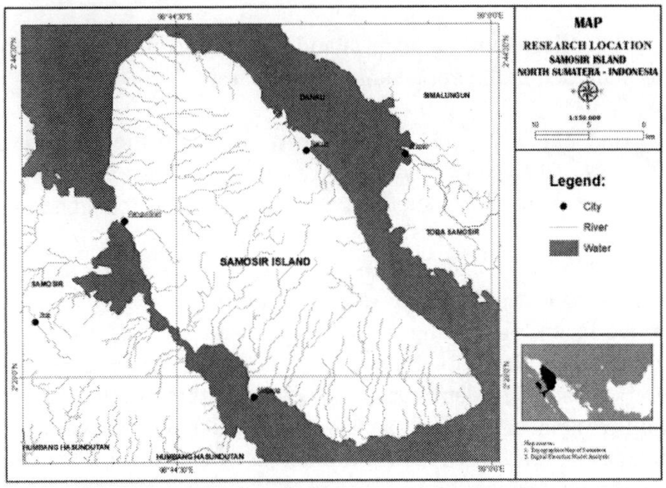

Figure 1. Map of Samosir island in Samosir district

IOP Conf. Series: Earth and Environmental Science **166** (2018) 012007 doi:10.1088/1755-1315/166/1/012007

2.2 .Tools and Research Data

The data used in this research were Landsat TM 5 (path/row 129/58) for 2005 and 2011 and Landsat 8 OLI imagery for 2015 (path/row 129/58). Image were acquired from the site (Http:\\\www.USGS.EarthExpoler.com), administrative map of Samosir regency 2015 with scale 1: 250.000 (Ministry of Environment and Forestry), forest area map of Samosir district issued by Ministry of Environment and Forestry by scale 1: 250.000 (Ministry of Environment and Forestry). The tools used in this research are Personal Computer, Arc GIS, ENVI, and ERDAS IMAGINE, Fragstat, Global Mapper, Geographics Positioning System (GPS) and camera.

2.3 .Research Procedure

The study was carried out in four step those were (1) watershed delineation, (2) land cover analysis, (3) land cover change analysis, and (4) fragmentation degree analysis.

2.3.1. Watershed delineation

It comprised of hill shade creating and fill sink. The hill shade functions to produce a clearer picture of topographic conditions. This function describes the Digital Elevation Model (DEM) with its topographic effects. The fill-sink function eliminates depression or sinks where conditions vary greatly with elevated area coverage.

2.3.2. Analysis of land cover

2.3.2.1. Image correction

Before further processed, correction was carried out to obtain a clear object on the image. It also facilitated visual image interpretation activities [9]. This correction was completed with EDAS version14 software, which consists of radiometric, and geometrics correction. Radiometric correction is a process for improving the visual quality of images and improving pixel values that do not match the actual reflectance values or spectral objects. One of the corrections used in the image is the histogram correction. Geometric correction was performed because there are geometric errors occurring at the time of recording. Geometric correction aims to rectify or justify the image coordinates to match the geographical coordinates. The raw data from the Landsat image consistently has geometric errors. Correction was carried out by defining projection and Geo referencing function. The selection of this projection was adjusted to the area division of the Universal Transverse Mercator (UTM) system. Samosir Island belongs to North Sumatra, which is located at UTM 47 zone and datum World Geographic System 84 (WGS 84).

2.3.2.2. Field checking

The field sample plot was collected on Samosir Island with a Garmin 62S GPS to assist in rectification of image classification included in the image analysis process.

2.3.2.3. Subset image

Subset Image was carried out using ERDAS 14 software since the Landsat image acquired from *website*http\\: *www.USGS.EarthExpoler.com* not only covers Samosir Island but also area surrounding Samosir Island. Subset areas were constructed based on the Samosir Island administration map.

2.3.2.4. Supervised classification

The images from years 2005, 2011 and 2015 were interpreted using a supervised classification. The method used in this research is visual image interpretation method [9, 10]. In this method, there are considerations of various factors, including the opportunity of a pixel or the appearance of the image to be explained into a particular class or category.

Image classification of Samosir Island was conducted for three time series of years at 2005, 2011 and 2015.Classification was performed for Landsat images such as Landsat TM 5 year 2005 and 2011 and Landsat 8 OLI image for year 2015. The combination of bands used for true colour composite on Landsat TM 5 was a combination of 321 as opposed to Landsat 8 OLI combination of 432 with RGB format (Red, Green, Blue) and using supervised classification. Digitization on screen was performed to assist image analysis in ERDAS 2014 to obtain land cover identification in the study area.

2.3.2.5. Accuracy analysis

Accuracy is analysed using a contingency matrix, which is a square matrix containing the number of pixels classified. This matrix is called "error metric" or "confusion metric". The error matrix compares the information from the reference area with information from the classified image in selected areas. The square error matrix with elements in the matrix row represents the area in the classified image, whereas the elements in the matrix columns represent the area of the reference data.

The reference data comes from a number of image pixels that have been identified through field checking or photo interpretation and are assumed correct. The error matrix is very effective at knowing the accuracy level of the classification result and the error occurring in the classification step. This accuracy is usually measured by the division of the point correctly explained by the total point used (the number of points within the diagonal of the matrix by the number of points used). Accuracy calculations by using this contingency matrix can also calculate the magnitude of the producer's accuracy and the user's accuracy. The systematically accurate calculation scheme [10] are as follows:

a. *User Accuracy*

$$Z = \frac{100\ \%}{N} \tag{1}$$

Note:

N= the number of validation coordinate
Z= the number of validated coordinate

b. *Procedure Accuracy*

$$\frac{X}{N}\ 100\% \tag{2}$$

Note:

N= the number of validation coordinate
X = the number of validated coordinate

c. *Overall Accuracy*

$$\frac{M}{N}\ 100\% \tag{3}$$

Note:

N= the number of validation coordinate
m = the number of validated coordinate

d. *Kappa Accuracy*

$$\frac{N\sum_{1}^{r} X_{ii} - \sum_{1}^{r} X_{i+}X_{+i}}{N^2 - \sum_{r}^{r} X_{i+}X_{+i}} 100\% \tag{4}$$

Note:

X_{ii} = diagonal value of contingency matrix row-i and column-i
X_{+i} = the number of point in column-i
X_{i+} = the number of point in row-i
N = the number of point

2.3.3. Analysis of land cover change.

Analysis of land cover change was carried out using ArcGIS 10.1. The analysis was carried out with land cover overlay for 2005, 2011 and 2015 in order to obtain three periods of land change data. It resulted in changes of extent and shape of the land cover class as it shifted its function to other area of use.

Fragmentation analysis. Data obtained from the classification, then processed using Fragstat 4.2 to obtain forest landscape metrics. It met with Janet et al [11] that stated that habitat fragmentation and forest loss could be examined by landscape metrics analysis [11]. Landscape metric as calculated was based on wildlife habitat such as macaca species [12- 15]. Landscape matrix used to determine the

IOP Conf. Series: Earth and Environmental Science **166** (2018) 012007 doi:10.1088/1755-1315/166/1/012007

level fragmentation is the Area Patch Area. (AREA), Patch Density (PD), Proximity Index (PROX), and Contiguity Index (CONTIG) [1, 5]. Each landscape metric is formulated as follows:

Area *patch*

Patch area was formulated:

$$\text{Area} = a_{ij}\frac{1}{10000} \tag{5}$$

Note:

AREA	= Area patch (m^2), divided by 10.000 (convert to be hectare);
a_{ij}	= area (m^2) patch ij

Patch density

Patch density is the number of patch unit on area have 100 Ha as a landscape unit.

$$\text{PD} = \frac{N}{A}(10000)(100) \tag{6}$$

Note:

PD	= the number of patch per 100 Ha;
N	= the number of forest patch, and
A	= area of forest landscape

Contiguity index

Contiguity index (CONTIG) is spatial size of forest patch connected as single patch with other forest patch. The higher contiguity is more connected the forest patch.

$$\text{CONTIG} = \frac{\frac{\sum_{r=1}^{z} c_{ij}}{a_{ij}} - 1}{V-1} \tag{7}$$

Note:

C_{ij}	= contiguity value of pixel "r" in patch-ij;
V	= the number of pixel value sized 3 x 3;
a_{ij}	= area patch i ij in a number of cell

Proximity index (PROX)

Proximity index is a tendency of patch to be relatively isolated (e.g. distance) from other patches on adjacent or similar ecological classes.

$$\text{PROX} = \sum_{g=1}^{n} \frac{a_{ijs}}{b_{ijs}^2} \tag{8}$$

Noted:

a_{ijs}	= area (m^2) patch ijsin surrounding of cell (m) patch ij;
b_{ijs}	= radius (m) between patch ijs and patch ijs, base on edge distance to edge patch. It calculated from cell centre to other cell centre

Rescaling score.

To define the standard fragmentation level then the total score of fragmentation transformed (rescaling) using the equation [16-17].

$$ind_{FLF} = \frac{score_{input-}\ score_{min}}{score_{tot\_max} - score_{tot\_min}} x(ind_{FLF} - ind_{FLFmin}) \tag{9}$$

Noted:

Ind_{FLF}	= fragmentation index
$\text{Score}_{\text{input}}$	= total score as input
$\text{Score}_{\text{tot- min}}$	= total score minimum
$\text{Score}_{\text{tot-max}}$	= total score maximum
$\text{Ind}_{\text{FLF max}}$	= maximum fragmentation index
$\text{Ind}_{\text{FLF min}}$	= minimum fragmentation index

IOP Conf. Series: Earth and Environmental Science **166** (2018) 012007 doi:10.1088/1755-1315/166/1/012007

3. Results and Discussion

3.1. Division of the Watershed

The division of the watershed was carried out with hydrology function of DEM data. DEM is digital data describing the geometry of the surface shape earth. It depicts the value of ground surface without pay attention to the object on it. The DEM used for Samosir Island was sourced from http:\\www.USGS.EarthExpoler.com and produced 4 (four) parts of the Samosir Island sub-watersheds.

The DEM data of Samosir Island was analysed using Arc Map 10.1. Furthermore, depressions or sinks were eliminated in pixels with different elevations with very striking area coverage using flow direction function generated flow direction at each pixel. Furthermore, the flow accumulation function using the flow direction produces an accumulated amount of water flow at each pixel in coverage of a particular region. The same accumulated value of each pixel having the same tendency flow direction produces a network of rivers or steam links. After the establishment of the network, the river Samosir Island was determined by the order of the river with Stehler and Shreve method to observe the connectivity of each river network in the different sub-watershed.

Generally, the pattern of river network in sub-watershed on Samosir Island in general was a dendritic pattern where tributaries tended to parallel the bloodstock. This model is like a tree having a branch structure and twigs as branches and its tributaries. Determination of watershed boundary was performed by watershed function [18-19]. This function is the most important in the delineation of watershed boundaries. This process works by searching for pixels that have a direction to the mouth of the river and separating other pixels.

The function of elimination polygon was done in sub region of watershed, which has area under 10 Ha. This function produces four sub watersheds in Samosir Island (**Figure 2**). To observe the difference of forest fragmentation level from each sub watershed in Samosir Island, each of these sub watersheds was divided into four directions namely North sub-watershed, Eastern sub-watershed, Southern sub-watershed and Western sub-watershed.

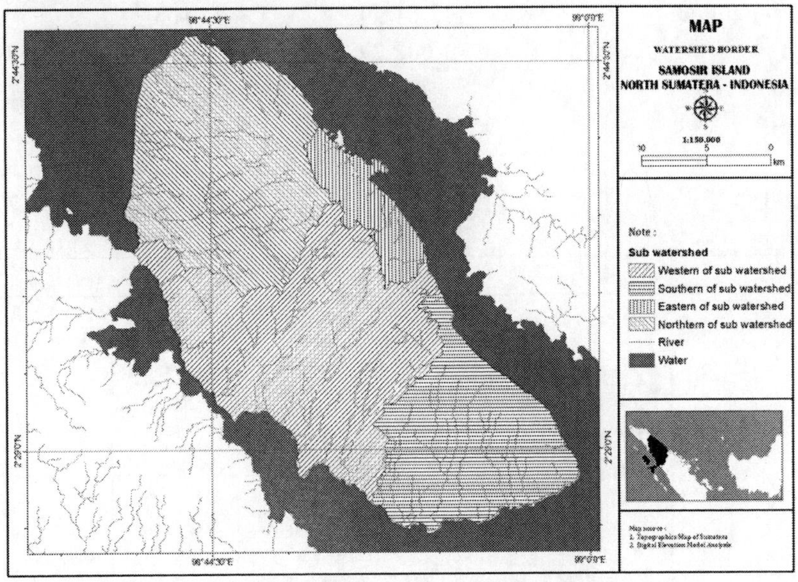

Figure 2. Sub watershed bordered

3.2. Classification of Land Cover

Based on classification results of Landsat imagery Samosir Island, 10 classes of land cover were founded. Land cover class conforms to the technical guidelines for the classification of low-resolution imagery by the Forestry Planning Board using a monogram of Sumatran landscape imagery. The

IOP Conf. Series: Earth and Environmental Science **166** (2018) 012007 doi:10.1088/1755-1315/166/1/012007

channel combination (Band) used was 345. The channel is appropriate to describe land cover in the field. Band 5 can differentiate between snow and cloud; band 4 can distinguish plant species, activities and to limit the body of water and band 3 is made to assist in viewing areas that absorb chlorophyll [9]. Band 3 can be used to assist in separating plant species, as well as for the observation of agricultural cultivation.

Land cover on Samosir Island in 2005, 2011 and 2015 was generally divided into 10 classes namely water agency, forest, mixed garden, fields, grasslands, settlements, swamps, rice fields, shrubs and clouds. Value Kappa accuracy was obtained from supervised classification for image of the year 2005 equal to 93,29%, for image of the year 2011 equal to 92,79% and for image of the year 2015 equal to 97,71%. It met with [10,20] that state that accuracy test using maximum likelihood aims to test the truth of the interpretation results obtained by checking in the field and measurements of several points selected from each form of land measurement. The level of accuracy is tested from the results of accuracy. The higher the accuracy of the results will be interpretation that is more accurate. Changes in shape and extent of land cover occur throughout the year. Additions and reductions occur on each land cover. The spatial distribution of the land cover was illustrated in **Figure 3** for year 2005, **Figure 4** for 2011 and **Figure 5** for 2015 and the extent of land cover was presented in **Table 1**.

Table 1. Land cover of Samosir island at 2005, 2011 and 2015.

Land cover types	Area (Ha) Year 2005	Area (Ha) Year 2011	Area (Ha) Year 2015
Cloud and cloud shadow	2.617,27	-	-
Water body	600,32	541,70	404,25
Forest	26.366,32	23.665,76	25.147,84
Mixed garden	1.466,04	11.615,81	11.912,51
Dry land agriculture	11.161,99	4.016,03	93,83
Bare land	1.082,89	1.246,50	915,43
Grass land	1.797,05	6.080,02	7.234,32
Settlement	1.309,40	4.325,00	6.373,00
Swamp	1.144,75	974,00	509,04
Paddy field	16.026,13	7.409,04	4.799,72
Shrubs	841,60	4.539,90	7.023,82
Total	64.413,76	64.413,76	64.413,76

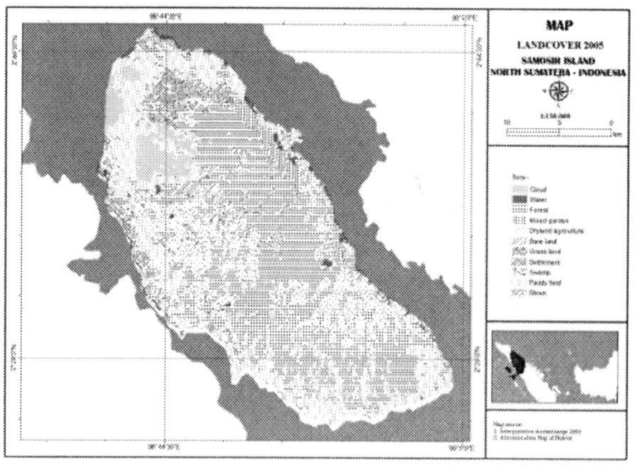

Figure 3. Land cover of Samosir islands at 2005

IOP Conf. Series: Earth and Environmental Science **166** (2018) 012007 doi:10.1088/1755-1315/166/1/012007

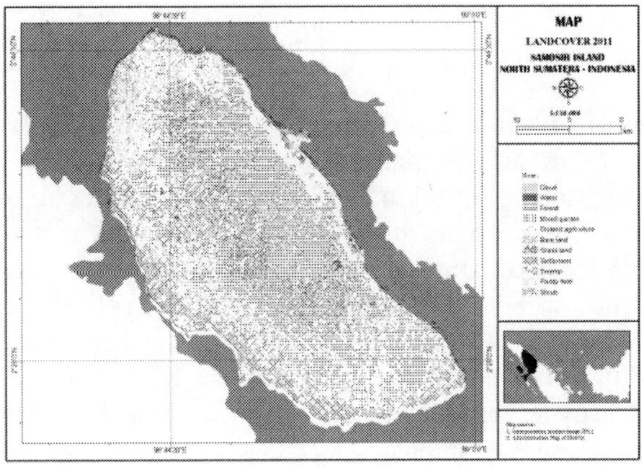

Figure 4. Land cover of Samosir islands at 2011

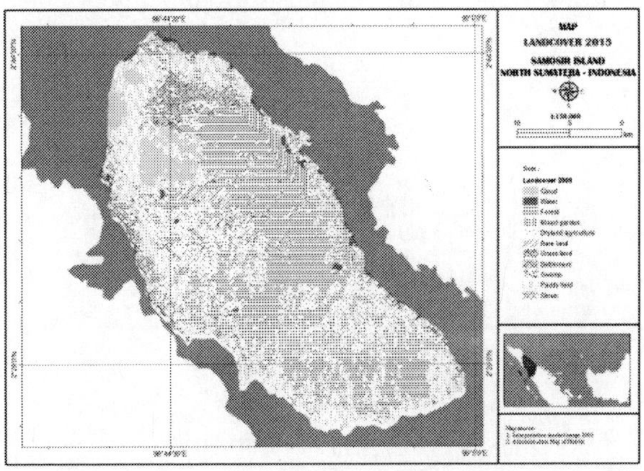

Figure 5. Land cover of Samosir islands at 2015

3.3. Forest Fragmentation
Fragmentation of forests generally occurs through loss of habitat. Otherwise, habitat loss can be viewed because of fragmentation [12, 30]. However, fragmentation may be accompanied by habitat depletion as resolving or distributing large habitat pockets into small and more isolated habitat enclaves [6, 22]. The degree of fragmentation can be expected by knowing the patch area, patch density, contiguity index, and proximity index [5, 7, 23]. The level of fragmentation can be grouped into 5 (five) classes namely lowest, lower, moderate, higher and highest.

3.4. Area
The area indicates the area of each forest patch. The forest is the largest area of land cover in Samosir Island. Within the period of 2005 to 2011, forest landscape of each sub-watershed decreased (Figure 7). The decreased occurred in western of sub-watershed have 2,341.91 Ha, southern part of sub-watershed of 259.32 Ha, eastern part of sub-watershed of 95.41 Ha and north sub-watershed of 14.44 Ha. The decreased was caused by the conversion of forestland and illegal logging.

Changes of forest patches area in 2011 - 2015 indicated the high increase in each sub watershed (**Figure 6**). The largest forest increase occurred in eastern part of sub watershed comprising a surface of 1,170 Ha followed by western part of watershed comprising a surface of 349.85 Ha, southern part of sub watershed comprising an area of 43.11 Ha, and northern part of watershed that has decreased

IOP Conf. Series: Earth and Environmental Science **166** (2018) 012007 doi:10.1088/1755-1315/166/1/012007

for an area of 71.06 Ha. People chose to abandon the less productive land and select plants that are tolerant to the dry season such as pine because of the increased forest area.

3.5. Patch density

Patch density is the number of forest patches per 100 Ha viewed from the landscape metric values. Patch Density of eastern part of sub-watershed, Southern part of sub-watershed and northern part of sub-watershed have increased in the period of 2005 - 2011 because the forest was divided into smaller parts and the smaller compactness. Meanwhile, the sub watershed West has decreased value of patch density from 2005-2011 (**Figure 6**). During the period of 2011 - 2015, degradation of patch density has occurred in 3 (three) sub watersheds in Samosir Island namely North sub-watershed, Western sub-watershed and Southern part of sub watershed which show compactness and compact forests. Meanwhile, the increased occurred in Eastern part of sub-watershed.

Decreased of patch density value indicates that the compactness of forests is consistently high, while the increased of patch density indicates that the forest area has been fragmented into small patches and improves habitat isolation. It met with [3, 16, 24] that stated tropical forest fragmentation triggers a decline in biodiversity conservation function. Breaking the patches becomes smaller resulting in reduced number of species, smaller populations and higher rate of extinction [25].

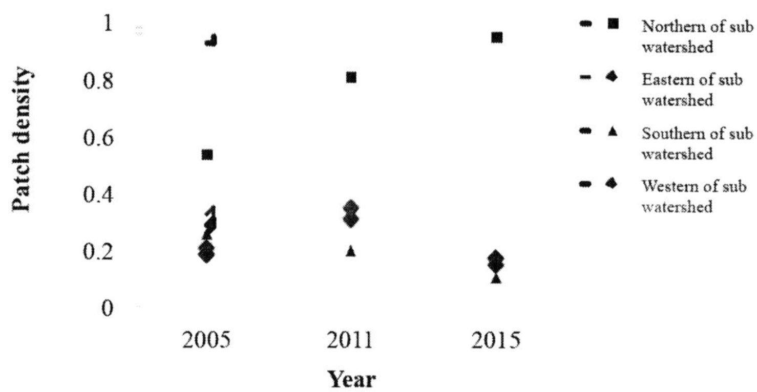

Figure 6. Patch density year 2005, 2011, and 2015

3.6. Contiguity index

Contiguity index is a spatial measure of the connectedness of individual forest patches with other forest patches. During 2005 – 2011, the Northern part of watershed, Eastern Sub-Watershed, and Southern Sub-watershed had decreased value of contiguity index (**Figure 7**), while the Western part has experienced increased value of contiguity index. In 2011 - 2015, three sub-watersheds in Samosir Island namely Eastern part of sub-Watershed, Western part of sub-watershed and Northern part sub-watershed have experienced an increased value of contiguity index, while Southern sub-watershed had decreased value of contiguity index.

IOP Conf. Series: Earth and Environmental Science **166** (2018) 012007 doi:10.1088/1755-1315/166/1/012007

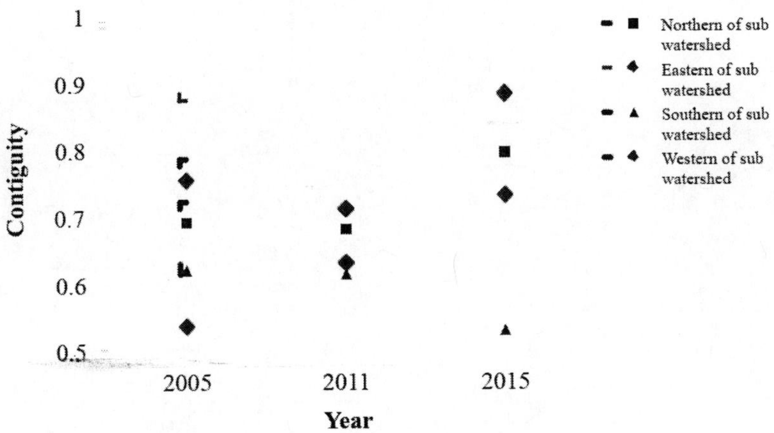

Figure 7. Contiguity index year 2005, 2011, and 2015

3.7. *Proximity index*

Proximity index is a tendency of patch to be relatively isolated from other patches. Fragmentation may be accompanied by habitat loss along with splitting or dividing pockets of habitats into smaller and more isolated sizes [6]. In 2005 – 2011, the Northern part sub - watershed, Eastern part sub - watershed and sub - watershed had an increased value of the proximity index, as opposed to the Southern part sub - watershed (**Figure 8**). In 2011-2015, four-sub watersheds have a decreased value of proximity index. Increasing the value of the proximity index indicates that the forest area has been reduced and the isolation occurred, whereas if the proximity index value decreased the forest area increased and the insulation decreased.

Figure 8. Proximity index in 2005, 2011, and 2015

IOP Conf. Series: Earth and Environmental Science **166** (2018) 012007 doi:10.1088/1755-1315/166/1/012007

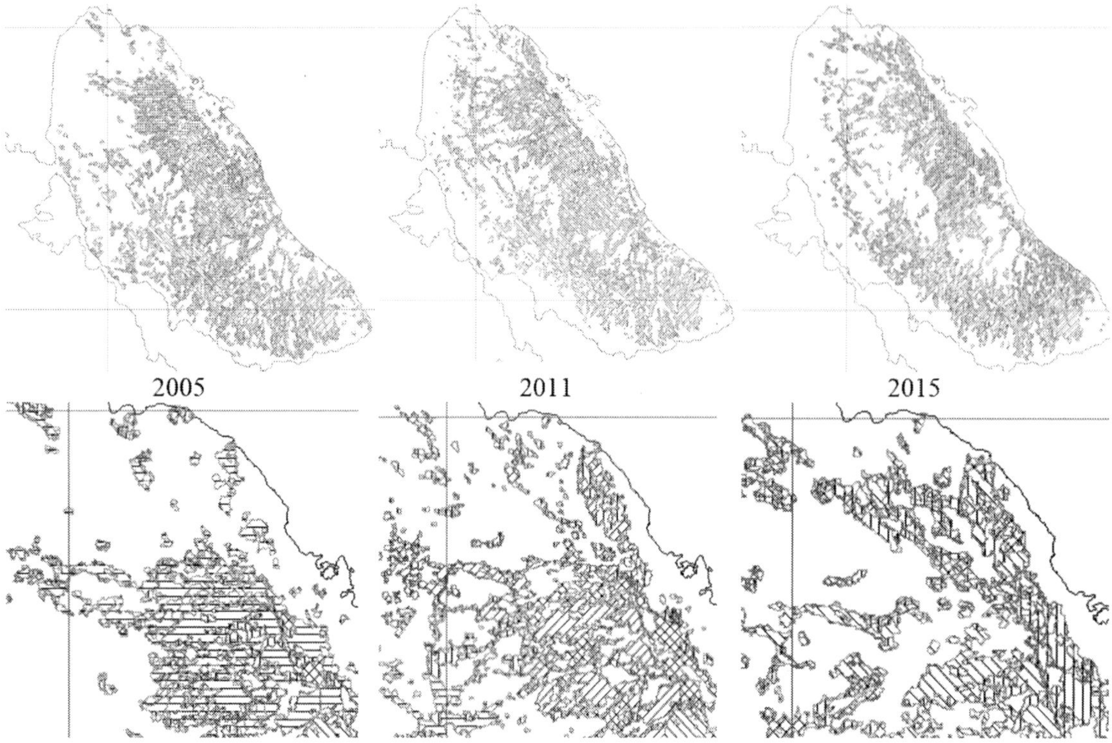

Figure 9. Change of forest fragmentation

Comparisons of forests changing during 10 years were illustrated in **Figure 9**. Compact forests in 2005 was separated into forest patches in 2011, while in 2015 forest areas became compact even though it is not similar with year 2005. The rescaling scores of fragmentation are grouped into five classes. Rescaling landscape fragmentation results of Samosir island forests in 2005 (**Figure 10**) showed the Eastern part of sub-watershed was dominated by highest class. Highest class of fragmentation degree in the Eastern part sub-watersheds was caused by the connection between decreased patches. The lack of compact forests leads to highest forest fragmentation.

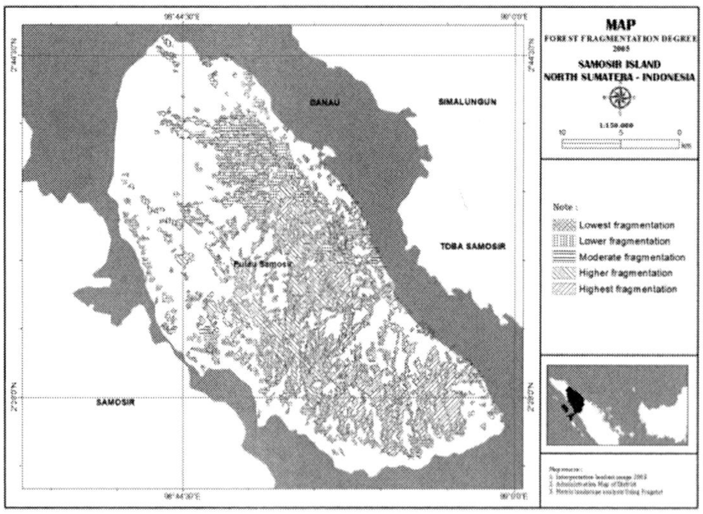

Figure 10. Map of the Samosir island forest landscape fragmentation degree year 2005

IOP Conf. Series: Earth and Environmental Science **166** (2018) 012007 doi:10.1088/1755-1315/166/1/012007

Moderate class dominates the degree of fragmentation of Northern part watershed, while low class dominates the fragmentation degree of Western part sub-watershed and Southern part sub-watershed. It can be concluded that the three sub-watersheds have compact forests and high connectivity. The degree of fragmentation in Eastern part sub-watersheds is dominated by highest class, which indicates the occurrence of land conversion resulting in uncompact forests.

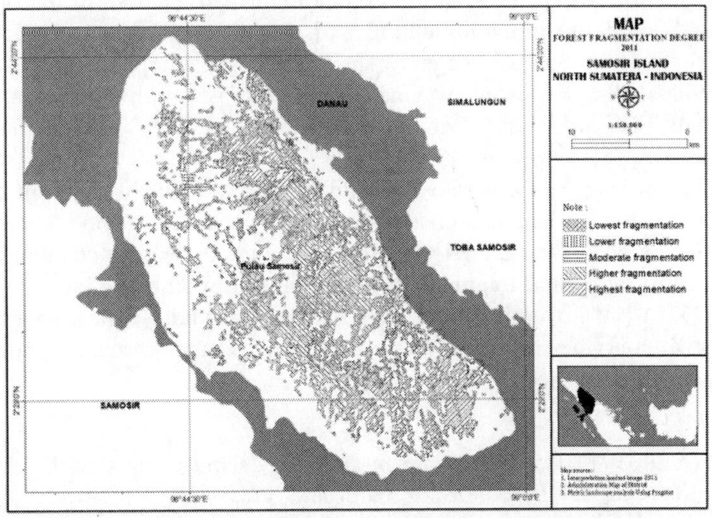

Figure 11. Map of the Samosir island forest landscape fragmentation degree at 2011

The rescaling score of forest metrics landscape of 2011 (**Figure 11**) showed that the fragmentation degree of Eastern sub-watershed and Northern sub-watershed were dominated by highest class. The highest degree of fragmentation was caused by the conversion of forest functions into rice fields and mixed gardens, resulting fragmented and separated forests. The sub-watersheds and Southern of watershed was dominated by low degree of fragmentations. Its shows a compact forest with high connectivity between each forest patches.

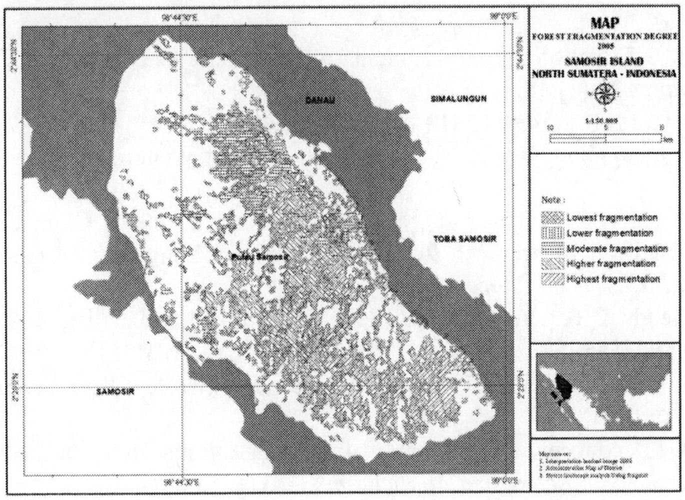

Figure 12. Map of the Samosir island forest landscape fragmentation degree year 2015

IOP Conf. Series: Earth and Environmental Science **166** (2018) 012007 doi:10.1088/1755-1315/166/1/012007

The rescaling degree of forest fragmentation of 2015 (**Figure 12**) showed that the Eastern sub-watershed is dominated by highest-class fragmentation degree. In Eastern sub-watershed, highest fragmentation degree showed isolated and uncompact forests that are linked by low linkages between forest and reduced forest area.

The degree of fragmentation in Northern watershed was dominated by low value, while extremely low class dominated the Western sub-watershed and Southern sub-watershed. The low to very low fragmentation degree of three watersheds indicated a compact and connected forest. This is due to increased forest area and low habitat isolation. Population density and kinds of community livelihood determine the natural conditions in the region [13, 26]. Social statistic data of 2015 collected from the Central Bureau of Statistics of Samosir Regency shows the highest population density in Pangururan is about 250,91 individuals/km^2 followed by Onan Runggu 175,51 individuals/km^2 and Rongurnihuta 90,99 individuals/km^2. The main employment of Samosir Regency is 49,359 farmers, manufacture 1815 persons and service 18,199 persons [27].

Farmers are the most abundant in Samosir Island leading the increased of agricultural land. Local people would left unproductive land and cleared the land by burning method to convert land use from forest into rice fields, bare land and mixed gardens to meet the increasing economic needs [28, 26]. It proves that increased of landscape fragmentation was caused by human practice related to providing cultivation lands [29, 30]. If government and NGOs does not increase the education of local people, the community will continue to destroy and convert natural resources continuously without considering the future consequences.

4. Conclusion

During the period of 2005 – 2015, land cover on Samosir Island comprised 10 classes namely water body, forest, mixed garden, dry land farming, bare land, grassland, settlements, swamp, paddy field, and shrubs. The largest land cover area are forests with 26,366.32 Ha in 2005; 23,665.76 Ha in 2011 and 25,147.84 Ha in 2015. The largest land cover change in 2005 - 2011 was forests into mixed gardens of 3,917.02 Ha, in 2011 - 2015 forest into shrubs covering an area of 3,334.671 Ha, and in 2005 - 2015 from forests into scrubs covering an area of 4,134.277 Ha. The degree of forest fragmentation in Samosir Island from 2005 - 2015 was considered as low fragmentation class dominating the Western of sub-watershed. The Southern of sub-watershed was dominated by moderate fragmentation class while the Eastern of sub watershed and Northern of sub-watershed were dominated by highest fragmentation class.

5. References

[1] Fahrig L 2003 Effect of habitat fragmentation on biodiversity *An Rev. of Ecology, Evolution, and Systematics* 34 (1):487-515.

[2] Gunawan P and L B Prasetyo 2013 *Fragmentasi Hutan Teori yang mendasari penataan ruang menuju pembangunan berkelanjutan* (Bogor: Pusat Penelitian dan Pengembangan dan Rehabilitas)

[3] Samsuri, Jaya I N S, Kusmana C, and Murtilaksono K. 2014 Connectivity and ecological indicators analysis of tropical forest landscape in Langkat watershed – *Indonesia J. Agriculture Fisheries and Forestry* 3 (3): 147-154

[4] Sonali S, Reddy C S, Vazeed P, Kalloli D, Saranya, and Satish K V. 2017 *Ecological Engineering* 99. 543–551

[5] Forman R T T and M Gordon 1986 *Landscape Ecology* (New York : John Wiley and Sons) p 45

[6] Haila Y 2002 A conceptual genealogy of fragmentation research: from island biogeography to landscape ecology *J. Ecological Applications* 12:321–334

[7] McGarigal K and Marks B J 1995 *FRAGSTATS: spatial pattern analysis program for quantifying landscape structure* (Gen.Tech. Rep PNW-GTR-351)

IOP Conf. Series: Earth and Environmental Science **166** (2018) 012007 doi:10.1088/1755-1315/166/1/012007

[8] Budiyanto E 2002 *Sistem Informasi Geografis Menggunakan ARC VIEW GIS* (Yogyakarta: Andi Yogyakarta)

[9] Lillesand T M and Kiefer R W 1990 *Penginderaan Jauh dan Interpretasi Citra* (Yogyakarta: Gadjah Mada University Press) p 88-132

[10] Jaya I N S 2007 *Analisis Citra Digital: Perspektif Penginderaan Jauh Untuk Pengelolaan Sumberdaya Alam* (Bogor: Fakultas Kehutanan IPB) p 75

[11] Janet E N, Sawaid A, and Gunter A F 2017. *Global Ecology and Conservation* 11 134-145

[12] Harvey PH and AM Lyles 1986 *Inbreeding in natural population of birds and mammals*ed ME Soule Conservation Biology : The Science of Scarcity and Diversity (Sunderland : Sinauer Assoc. Inc)

[13] Morrison M L Marcot B G and RW Mannan 1992 *Wildlife-Habitat relationships* (Wisconsin : The University of Wisconsin)

[14] Roonwal M L and Mohnot S M 1977 *Primates of South Asia* (Cambrige : Harvard University Press)

[15] Supriatna J and H E Wahyono 2000 *Primata Indonesia, Panduan Lapangan* (Jakarta: Yayasan Obor Indonesia, Jakarta) p 122

[16] Samsuri, Jaya I N S, Kusmana C, dan Murtilaksono K. 2014 Fragmentation of tropical forest landscape on Langkat Watershed - North Sumatra *J. Tropical Forest Management* 20 (2):77-85

[17] Syarifi M A Herwijnen M and van Toorn W H 2007 *Spatial decision support system* (Enschede : International Institut for Aerospace Survey and Earth Science (ITC))pp 123-127

[18] Purwanto T H 2013 *Ekstraksi Morfologi Daerah Aliran Sungai Dari Data Digital Surface Model (Studi Kasus DAS Opak)* (Yogyakarta: UGM Press)

[19] Rauf A Lubis A K and Jamilah 2011 *Dasar-Dasar Pengelolaan Daerah Aliran Sungai* (Medan: USU Press) p 12-24

[20] Sukojo B M and Diah S 2003. Penerapan Metode Penginderaan Jauh dan Sistem Informasi Geografis Untuk Analisis Perubahan Penggunaan Lahan *J Teknologi* 7 (1)

[21] Meret J 2007 *Habitat Fragmentation and Wildlife Corridors* http://www.science. mcmaster.ca.htm. Access on 02-04-2017.

[22] Saura S, Vogt P, Velázquezc J, Hernandoa A, and Tejeraa R 2011 Key structural forest connectors can be identified by combining landscape spatial pattern and network analyses. *J. Forest Ecology and* Management 262 : 150–160.

[23] Krishna P A, Prakash K P, Shant RJ, Prem R N and Michael K 2017 *Journal of Ecological Indicator.* 80 74–83

[24] Samsuri, Jaya I N S, Kusmana Cand Murtilaksono K 2014 Development index of restoration priority of degraded tropical forest landscape in Batang Toru watesheds, North Sumatera, Indonesia*J. Biotropia*. 21 (2):110-123

[25] Taylor P D Fahrig L Henein K and Merriam N G 1993*Connectivity is a vital element of landscape structure* (Oikos 68) pp. 571-573.

[26] Barnes T G 2000 *Landscape Ecology and Ecosystems Management* Cooperative Extension Services, University of Kentucky, College of Agriculture. UK. http://www.ca.uky.edu. Access on February 24, 2007.

[27] Badan Pusat Statistik 2015 *Kabupaten Samosir Dalam Angka Tahun 2015* (Samosir : Badan Pusat Statistik)

[28] Nagendra H Southworth J and Tucker C 2003 Accessibility as a determinan of Landscape Transformation in Westeren Honduras. *J. Landscape and Urban Planning* 83 154 – 167

[29] Izaya N, Sonaira S S, Mark A C, and Marcus V 2017. *Forest Ecology and Management* 401. 135–146

[30] Ana IR, Cabral, and Fernando LC 2017 *Applied Geography* 82 115-128

IOP Conf. Series: Earth and Environmental Science **166** (2018) 012008 doi:10.1088/1755-1315/166/1/012008

Biodiversity of Termites and Fungi in Two Botanical Gardens in Batam, Riau Island Province and Kuningan, West Java Province

T Kartika[1], S. K Himmi[1], B Wikantyoso[1], A. S Lestari[1], M Ismayati[1], D Zulfiana[1], N. P. R. A Krishanti[1], A Zulfitri[1], A. H Prianto[1] and S Yusuf[1]

[1]Research Center for Biomaterials, Indonesian Institute of Sciences (LIPI), Cibinong Science Center, Cibinong-Bogor 16911, Indonesia

E-mail: titikkartika@biomaterial.lipi.go.id

Abstract. Termites and fungi play an important role through promotion of essential ecological processes in ecosystems such as soil modification and rehabilitation, also decomposition of lignocellulosic materials. However, termite is also known as pest toward vegetation, while fungi can also cause bio-deterioration problems. Biodiversity evaluation of termites and fungi are essential for sustainable management and area planning of botanical garden. Hence, this study reviews the importance of recording biodiversity of fungi and termites in botanical garden in Batam, Riau Island Province and Kuningan, West Java province. The research was conducted by active searching method or exploration on termites and fungi on selected plot areas in the aforementioned botanical gardens. The result of this study indicated that biodiversity of termites and fungi were high on selected areas. The presence of termite and fungi in the ecosystem might beneficially support the ecosystem. However, further studies on ecological condition, soil type and climate need to be addressed to determine their influences on biodiversity of termites and fungi.

1. Introduction

A botanical garden is a conservation area dedicated to collection, cultivation, and display of a wide range of plants labelled with their botanical names. The botanical garden may contain specialist plant collection from particular parts of the world. As an ex situ conservation area, botanical garden might serve as botanical research, environmental education, ecotourism and ecosystem services. In addition, a botanical garden may initiate society awareness and appreciation on environment, animals and plants; then protection on the organisms could be approved.

Botanical garden is artificial ecosystem created to support the protection of macro- and microorganisms. As an artificial ecosystem, some organisms may support ecosystem growth and development or even lead to detrimental effect in particular condition. Abandoned population growth of particular organisms could generate the ecosystem destruction.

In this study, we assessed the biodiversity of organisms that frequently observed assigning negative impact on ecosystem. Termite and fungi, mostly Basidiomycetes, are essential components for ecological process in ecosystem. The two organisms play significant roles in decomposing organic materials, being food sources, and recycling process in forest. Furthermore, decomposition of organic materials by termite creates carbon dioxide circulation in the soil; hence contribute to carbon dynamic in ecosystem, particularly in agriculture, of tropical or sub-tropical countries [1-3]. More than 4000 termite species have been recognized and distributed throughout the world. As much as 2600 species classified in some families and genus, and only 10% of termite species is recognized as pest [4].

Basidiomycetes fungi are the most common and important wood-inhabiting fungi in forests. These species can account for the majority of fruit bodies found on woody debris [5]. The fungi degrade cellulose and lignin as the main component of plant cell wall, hence mostly recognized as decomposer or detrivores the organic materials. Three wood-decaying fungi are grouped based on infestation type [6]; white-rot, brown-rot and soft-rot fungi. All fungi decease the wood quality, such as strength, visual appearance, physical and chemical properties.

However, conversion of forest into agriculture, urban areas, plantations, and industrial areas has generated the change of ecosystem, hence it influence the biodiversity of those organisms and their functions as pest and deteriorating agent.

This study explored the biodiversity of termite and Basidiomycetes fungi in two developing Botanical Gardens; Batam and Kuningan. The information is essential for sustainable management and area planning of those botanical gardens.

2. Material and Methods

2.1. Sampling site

Sampling site was in Sundaland area, in which Batam Botanical Garden and Kuningan Botanical Garden are located. The Batam Botanical Garden is located in the eastern part of Sumatera island; Riau Province, while the other one is located in the western part of Java island; West Java Province. In each botanical garden, sampling site was divided into some areas, which contribute to different vegetation and soil type. In Batam Botanical Garden, the sampling site was in three areas, termed as arboretrum, mangrove and forest. Then, sampling in Kuningan Botanical Garden was conducted in four areas, termed as A (dark humic soil), B (red soil/garden), C (red soil near reservoir), and D (boundary forest) area.

2.2. Biodiversity of termite

2.2.1. Termite sampling

Termites were sampled using active searching method. A standardized protocol, belt-transect, was applied for termite sampling [7, 8]. The transect was 100 m long and 2 m wide, and divided into 40 contiguous sections (each sub-transect 5 x 1 m) and numbered sequentially. Each sub-transect was sampled one trained person for each 30 min (**Figure 3**). In order to standardize sampling effort, the collectors worked steadily and continuously for searching the following microhabitats: surface soil, litter and humus at the base of trees and buttress root, inside of dead logs, tree stumps, branches and twigs, the soil within and beneath rotten logs, all termite nests and mounds, and so on. Termite collection should consisted of soldier and worker for further identification. All termite samples were kept in EtOH.

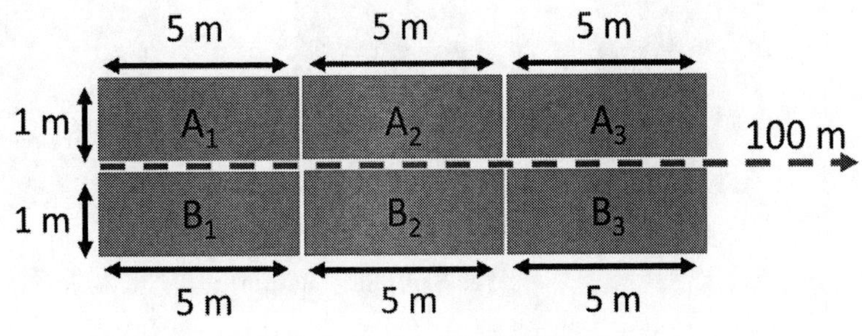

Figure 1. Belt-transect method for termite sampling, which consist of sub-transect area (A1 – A3 and B1 – B3)

2.2.2. Identification of termite
The termite sample was identified based on morphological characters of head, abdomen, mandible, body and antennal shape, tarsal segment, tibia and femur [9-11].

2.3. Biodiversity of fungi
2.3.1. Fungal sampling
Fungi were sampled by active searching method. The fungi were explored in each sampling site for one to one and half hours to search the main target of fungal fruit body. The collected fungi were kept in paper bag for further identification.
2.3.2. Identification of fungi
As fungal identification, the collected fungi were isolated and grown in laboratory before identification. First, the fruit body was sterilized in sodium hypochlorite 5% for five minutes, then rinsed with distilled water before air-drying in ambient temperature for two to 12 hours. The dried sample was transferred into potato dextrose (PDA) agar media in Petri dish. The Petri dish was then incubated in ambient temperature for three to seven days [12]. The grown fungi on PDA media were identified based on characters of macroscopic morphology, of which mushroom body, pileus, stipe color change, texture, odor, and veil presence were characterized [13].

3. Results and Discussion
3.1. Biodiversity of termite
In Batam Botanical Garden, 207 samples of termite were collected from three sampling areas, and 200 samples were collected from four areas of Kuningan Botanical Garden. In identification, trmite samples were classified into functional group, which is based on known feeding habits [14-15] or worker gut content analysis. Protozoa associated with termites indicated the functional group of termites; lower and higher termites. In all species of Lower termite genera harbor throughout most of their lives faunules of flagellate protozoa in the hindgut [16], which affect their feeding habit. The species of Lower termites feed on sound or nearly sound wood. In contrast, the higher termites, Termitidae, that either lack intestinal flagellates have a rather varied diet consisting of wood (much more decayed wood that taken by the Lower termites), dried grass, leaves, or in fact most other vegetable material. In addition, this family may ingest soil and extract from it useful nutrient. The diversity of termite species in each sampling site of botanical garden is described in **Figure 2** and **Figure 3**.

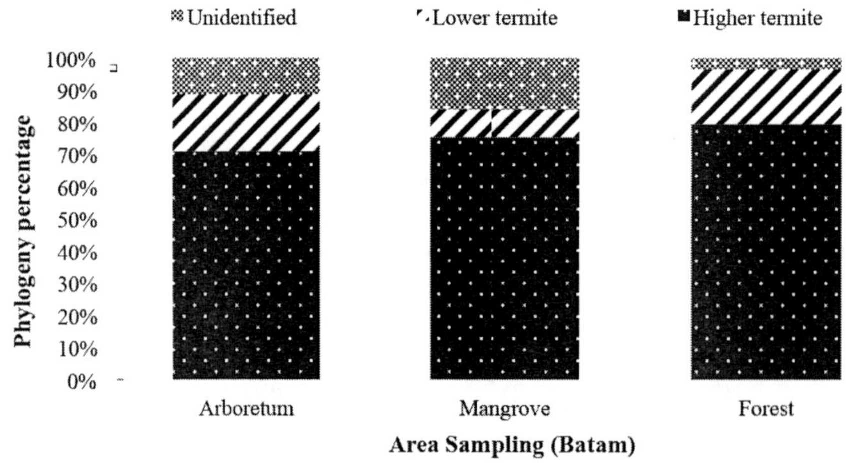

Figure 2. Termite collection in Batam Botanical Garden

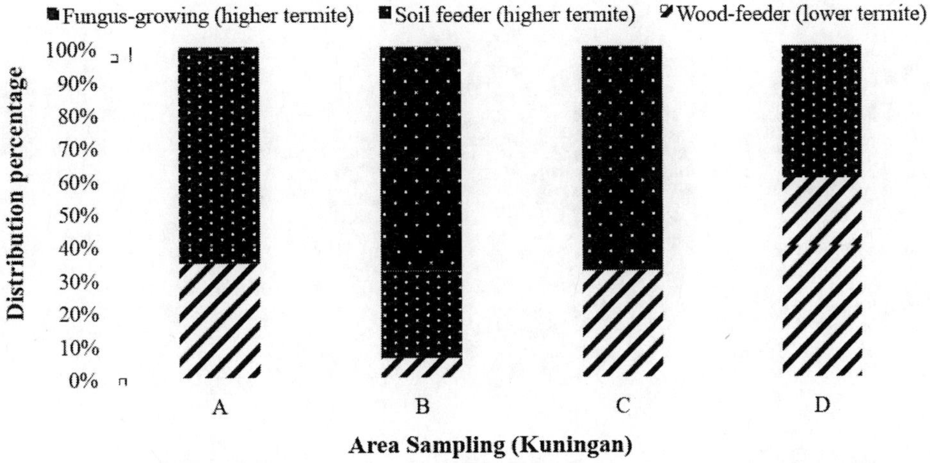

Figure 3. Termite collection in Kuningan Botanical Garden

The higher termite was the dominant termite found in both botanical garden. Based on the feeding habit of higher termite, which feed on mostly decayed materials, termites found in two botanical gardens mainly played their function as detrivores. Detrivores contribute to decomposition and nutrient cycle, hence this roles are an important aspect of many ecosystem. In addition, the existence of higher termite in ecosystem may prove as ecological indicator. In ecology perspective, the higher termite is more vulnerable toward ecosystem changes compared to Lower termites [17]. They are much more susceptible to the fluctuations in microclimate that accompany habitat disturbance, particularly involving forest canopy disruption. Hence, the dominancy of higher termite in an ecosystem might indicate that the two botanical gardens are considered as undisturbed areas with natural composition

3.2. Biodiversity of fungi

Fungal samples from Batam Botanical Garden were collected as much as 36 samples from three sampling areas, and 65 samples were collected from four areas of Kuningan Botanical Garden. Identification of fungi based on morphological characters of the fruit body or the spores, which illustrated according to family group (**Figure 4** and **Figure 5**).

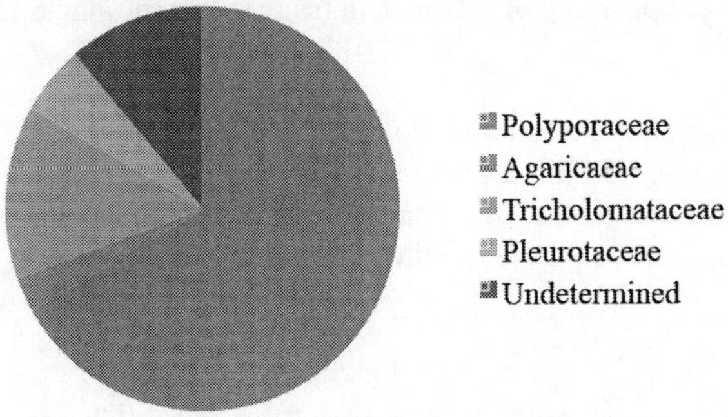

Figure 4. Fungal collection from in Batam Botanical Garden

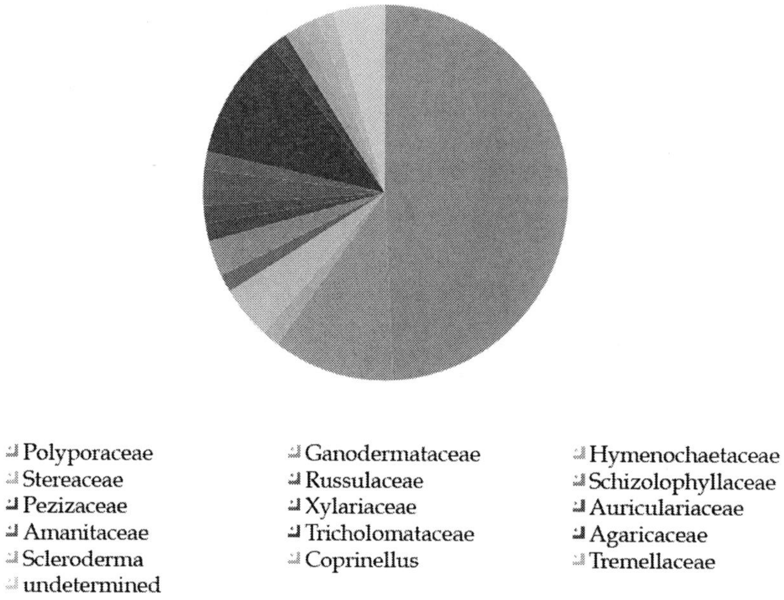

Figure 5. Fungal collection from in Kuningan Botanical Garden

Polyporaceae was the dominant fungi found in both Botanical Garden. The polyporaceae are a family of poroid fungi belonging to Basidiomycetes. The abundance and richness of polyporous fungi in an ecosystem connect with the amount of coarse woody debris, not with fertility gradient of the ecosystem [18]. The amount of large dead wood in one of the most important factors for fungi and fungi-inhabiting species [19]. The polyporous fungi are sensitive to environmental changes and required specified substrate. Thus, the coarse wood debris plays a key role for providing the structure for natural fungus community. The existence of the coarse woody debris are commonly associated with the main characteristic of natural forest. Hence, the abundance of polyporous fungi in an ecosystem might indicate its condition as natural ecosystem. Now, the polyporous fungi are used increasingly in management planning of forest as indicators of naturalness in conservation.

4. Conclusion

The biodiversity of termite and fungi in both botanical garden; Batam and Kuningan Botanical Garden were high in selected areas. The higher termite was found as dominant genera in both botanical garden, while polyporous fungi was the main fungal community in either Batam or Kuningan Botanical Garden. The presence of those termite and fungi in the ecosystem might beneficially support the ecosystem.

Acknowledgment

Indonesian Institute of Sciences through PKT Kebun Raya Bogor Thematic Program 2017 funds this project. The authors acknowledged all collaborators of this program.

5. References

[1] Mill A.E. 1992 Termites as agricultural pests in Amazonia, Brazil Outlook Agric. 21(1): 41-46.

[2] Logan J W M 1992 Termites (Isoptera) - a pest or resource for small farmers in Africa Trop. Sci. 32: 71-79
[3] Kashyap R K and Verma A N 1994 An appraisal of losses caused by termites in agricultural crops. In: O P Agrawal (Editor), Perspectives in Entomological Research. Scientific Publishers, Jodhpur, pp. 307-321
[4] Inward D, Beccaloni G. and Eggleton P 2007 Death of an order a comprehensive molecular phylogenetic study confirms that termites are eusocial cockroaches Biology Letter 3: 331 – 335
[5] de Vries BWL 1990 On the quantitative analysis of wooddecomposing macrofungi in forests. I. Wageningen Agric Univ Paper 90:93–101
[6] Zabel R A and J J Morrell 1992 Wood Microbiology Decay and Its Prevention Academic Press. Inc. San Diego
[7] Davies R G 1997 Termite species richness in fire-prone and fire-protected dry deciduous dipterocarp forest in Doi Suthep-Pui National Park, Northern Thailand Journal of Tropical Ecology 13:153-160
[8] Jones D T, P eggleton 2000 Sampling termite assemblages in tropical forests: testing a rapid biodiversity assessment protocol Journal of Applied Ecology 37: 191-203.
[9] Hill G F 1942 Termites (Isoptera) from the Australian region Council for scientific and industrial research Commonwealth of Australia
[10] Syaukani 2010 A guide to the Nasus termites (Nasutitermes, Termitidae) of Kerinci Seblat national park, Sumatra Nagao Natural Environment Foundation
[11] Krishna K, D A Grimaldi, V Krishna, M S Engel 2013 Treatise on the Isoptera of the world. AMNH Bulletin 377 (1)
[12] Ilyas, M, M Rahmansyah dan A Kanti 2006 Seri Panduan Teknik Isolasi Fungi LIPI Press.
[13] Aurora D 1986 Mushrooms demystified (A comprehensive guide to the fleshy fungi) 2nd edition Ten Speed Press Berkeley
[14] Eggleton P, D E Bignell, W A Sands, N A Mawdsley, J H Lawton, TG Wood, & NC Bignell. 1996 The diversity, abundance, and biomass of termites under differing levels of disturbance in the Mbalmayo Forest Reserve, southern Cameroon. Philosophical Transactions of the Royal Society of London, Series B, 351:51-68
[15] Eggleton P, R Homathevi, D Jeeva, D T Jones, R G Davies, & M Maryati 1997 The species richness and composition of termites (Isoptera) in primary and regenerating lowland dipterocarp forest in Sabah, east Malaysia. Ecotropica, 3:119-128
[16] Krishna K 1969 Biology of termites Volume I New York Academic Press
[17] Davis R G, P Eggleton, D T Jones, F G Gathorne-Hardy, L M Hernandez 2003 Evolution of termite functional diversity: Analysis and Synthesis of local ecological and regional influences on local species richness Journal of Biogeography 30(6):847-877
[18] Sipola A, M Simila, M Monkkonen, & J Jokimaki 2004 Diversity of polyporous fungi (Polyporaceae) in northern boreal forests: Effects of forest site type and logging intensity. Scandinavian Journal of Forest Research 19: 152-163.
[19] Muller J, H. Engel, M Blaschke 2007 Assemblages of wood-inhabiting fungi related to silvicultural management intwensity in Beech forests in southern Germany European of Journal Forest Research 126:513-527

IOP Conf. Series: Earth and Environmental Science **166** (2018) 012009 doi:10.1088/1755-1315/166/1/012009

Laboratory Evaluation of The Selected Entomopathogenic Fungi and Bacteria Against Larval and Pupal Stages of *Spodoptera litura* L

A Zulfitri[1], A. S Lestari[1], N. P. R. A Krishanti[1] and D Zulfiana[1]

[1]Research Center for Biomaterials, Indonesian Institute of Sciences (LIPI)
Jl. Raya Bogor Km 46 Cibinong 16911, Indonesia

E-mail: apriwi.zulfitri@lipi.go.id

Abstract. In the present study, efficacy of selected entomopathogenic fungi and bacteria against *Spodoptera litura* were evaluated. The study used two life phase of the insect, third to fourth larval and pupal stages. Evaluation on larval stage used larvae-dipping method while soil premixing method was used in efficacy test against *S. litura* pupae. Spore solution of five entomopathogenic fungals (*Paecilomyces* sp, *Beauveria bassiana*, T2B, B2 and T4B), cell suspension of 1 entomopathogenic bacteria (*Serratia* sp.) and suspension solution of consortium between T2B and *Serratia* sp. were used. Evaluation on larval stage showed the highest mortality was found in B2 treatment (70%), followed by *Serratia* sp. (60%) and consortium (50%) treatment. While on *S. litura* pupal stage evaluation, consortium treatment resulted in 62,5% of tested pupae failed to become adult moth. Other treatments on pupal stage evaluation showed under 40% failure transformation to adult stage.

1. Introduction

Managing *Spodoptera litura* is complicated because they are actively feeding at nighttime while hiding under the plant-host leaves at daytime. Since regular usage of pesticide causes rapid resistance in *S. litura*, more effective and safer biological ways to control the pest are needed.Controlling insect pest using entomopathogenic microbes including fungi and bacteria has played key role in sustainable pest management program. Entomopathogenicmicrobes usage are lower in cost, higher efficiency and safer for beneficial organisms than chemical pesticide [1]. During their life cycle, the fungi occur as infections in living insect only for short period while may remain dormant as conidia in the soil surrounding host cadaver [2]. Therefore, the soil environment is an important source for isolating entomopathogenic fungi and soil-inhabiting entomopathogenic fungi are an important and wide spread component of most terrestrial ecosystems [3].

Mechanism of fungal infection is different than those of bacteria and viruses. Both microbes are need to be ingested first by the host while fungal invades their host using mycelium through cuticle [1]. Zulfiana [4] reported that based on *Scanning Electronic Microscope* (SEM) observation on fungal infected termites larvae, infective process might be started by invasion of formed mycelium to larvae body via cuticle, then mycelium were germinated, proliferating and take over the larvae body.

Therefore, entomopathogenic fungi are promising biocontrol for non feeding insect life stages such as eggs and pupae [5].

During last decades, *S. litura* have been reported to show increase resistance to commonly used and several newer insectide [6][7]. The studies showed that generally newer insecticides exhibited low and very low resistance may due to different mechanism of action used to targeting *S. litura* than those of conventional insecticides. However, not many farmers can afford these insecticides due to their high prices [7], thus resulted in low application. In the future, this might change since other insecticides company will come up with similar composition with lower price that will lead to increasing application. Continuous use of insecticides has resulted in negative impacts such as: increasing the amount of residues that are harmful to the environment, the killing of natural enemies (parasitoids and predators), and health hazards for users [8]. Utilizing biocontrol agents to control agricultural insect pests is an alternative strategy that is more environmentally friendly than the use of chemical pesticide [9].

In the study we applied indigenous entomopathogenic fungi resulted from soil isolation using bait method and entomopathogenic bacteria isolated from undeveloped *S. litura* pupae against larval and pupal stages of *S. litura*. In addition, we also compared efficacy of those entomopathogenic microbes with other several strains of entomopathogenic fungi. In order to increase the efficacy of entomopathogenic microbes, we tried to apply co-inoculation between entomopathogenic bacteria and fungal againts pest target. Since bacterial and fungal biocontrol agents utilize different mechanisms of action to control pest on crops [10], it is important to investigate antagonistic activity of those microbes. Therefore we did a microbial compatibility assessment on potato dextrose agar (PDA) plate before applied the entomopathogens consortiums to test.

2 Materials and Methods

2.1 Qualitative compatibility assessment antagonistic activity between entomopathogenic fungal and bacteria isolates

Entomopathogenic fungal isolates used in the antagonistic study were isolates from Sumba island and West Sulawesi soil samples (T2B, B2, T4). The fungal were obtained using bait system method by incubating *Tenebriomolitor* larvae on soil sample.If any of the larvae showing surface mycosis, then it will be isolated and the colony formed was purified on PDA [2]. Beside those three isolates, we also applied other entomopathogenic fungal isolates which are collection of Research Center for Biomaterials (*Beauveria bassiana*, *Paecilomyces* sp. dan *Metarhizium* M1570)

Entomopathogenic bacterial used was BLSP4 isolated from dead *S. litura* pupae [9] and a known entomopathogenic bacteria, *Bacillus thuringiensis* (BLBt- IPBCC collection), as comparison was used. *B. thuringensis* is one of the entomopathogenic bacteria that has been widely used as bioinsecticide especially to cotton plant pests. Therefore, there were twelve-combinations for antagonistic activity assay.

Microbial compatibility assay was carried out using dual culture on 4 cm PDA plate according to [11, modified]. Fungal inoculum (Ø: 5mm) was placed 1 cm from one side and bacteria inoculums was streaked 1 cm from the other side(Figure 1). The plates were then incubated at room temperature for 72 hours. The isolates growth was evaluated whether they have inhibiting or supporting relationship. Bacteria and fungi that can grow together on a plate will be used for consortium treatment.

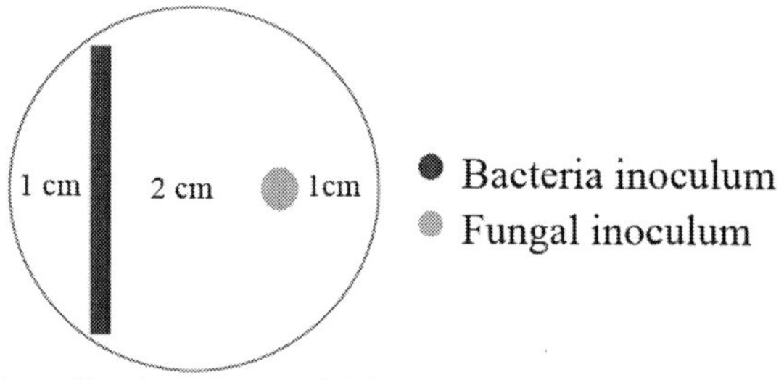

Figure 1. Ilustration of qualitative compatibility assessment betweenentomopathogenic fungal and bacteria isolates

2.2 Entomopathogenic fungus and bacteria treatment before bioassay against S. litura
Isolates of entomopathogenic fungus used in this study were *B. bassiana*, *Paecilomyces* sp. (IPBCC collection), and 3 isolates from exploration of Sumba and Sulbar's soil (T2B, B2, T4B). The fungus are cultured on *potato dextrose agar* (PDA) at 25-27°C for 10 days. Next, into the media, 10-20 ml of sterile 0.01% Tween solution was added as surfactant. The conidia was harvested by gently scraping the surface of media using inoculation needle so the conidia suspended in the Tween solution. The hyphal debris were removed by filtering the mixture through autoclaved fine mesh cloth. Conidial concentration was counted using haemacytometer under 100x magnification.

Entomopathogenic bacteria isolates (*Serratia* sp.) was subcultured on *nutrient agar* (NA). In efficacy test, bacteria cell was harvested from a 24 hours old culture in nutrient broth (NB). Centifugation (10 mins, 8000 rpm) was carried out to harvest bacteria cells then the cells were dissolved in 10 ml of sterile 0.85% NaCl solution and vortexed.

Potato dextrose broth (PDB) was used as fungal growth media in consortium treatment. Entomopathogenic fungal isolate was inoculated into 100 mL of PDB and incubated with shaking at 120 rpm, room temperature. On day-7, a 24 hours old of *Serratia* sp. grown in NB (1% v/v) was added into the fungal culture. The consortium culture was then incubated for 24 hours before applied.

2.3 Rearing of S. litura
Larvae of *S. litura* were reared in plastic containers (34×25×7 cm) provided with freshly picked taro leaf for food supply. The containers were kept clean and the food was changed daily. Fifth and sixth instar were kept in plastic cup individually to avoid crowding.The larvae were reared until pupal stage. Two to three days old of pupae were collected to be used further for bioassay. Before used, the pupae were dipped with sterile 0.5% NaCl solution for 5sec, rinsed with sterile aquadest and air dried.

2.4 Efficacy test of entomopathogenic microbes against S. litura
Efficacy test was conducted against two life stages, larva and pupae. Sterile 0.01% Tween solution was used as control treatment.

2.4.1 Efficacy test of entomopathogenic microbes against S. litura pupae.
Treatments applied were fungal entomopathogen (*B. bassiana*, *Paecilomyces* sp. T2B, T4B), *Serratia* sp and consortium. Method used was premixing using soil as media and zip plastic bag, referred to Anand, R, Prasad, B. & Tiwary, B. N. [5]. Before used, soil was sieved and weighed as much as 30 g for each replication. The conidial suspension (13 mL)was inoculated to the soil and mixed well. Premixed soil was transferred to plastic cup and 2 pupae were transferred in the middle of the soil. The set of experiment were kept in the dark at 22-23°C and examined for emergence of adults up to 14 days post inoculation.Unemerged pupae were considered dead.

2.4.2 *Efficacy test of entomopathogenic microbes against S. litura* larvae**.**
Treatments applied were fungal entomopathogen (*B. bassiana*, *Paecilomyces* sp., B2, T2B, T4B), *Serratia* sp and consortium. Third to fourth instar *S. litura* larvae were dipped in the conidial solution for 2-3 seconds and air dried. The larvae were then incubated for two weeks in plastic cup lined with

filter paper. During incubation, larvae was feed with fresh taro leaf every day. Mortality rate was recorded everyday or until all larva dead [3, modified].

3 Results and Discussion

3.1 Compatibility assessment between entomopathogenic fungal and bacteria

The results of the compatibility test are shown in Table 1. Based on growth patterns on PDA media, from 12 test pairs, only T2B isolates when paired with BLBt or BLSP4 are not inhibited by the bacteria growth (**Figures 2a and b**)

Table 1 Results of compatibility tests between bacteria and entomopathogenic fungi on PDA media.

Fungal isolates	Bacteria	
	BLBt	BLSP4
Paecilomyces sp.	X	X
B. bassiana	X	X
M1570	X	X
T4B	X	X
T2B	O	O
B2	X	X

X: growth inhibition, O: no growth inhibition shown

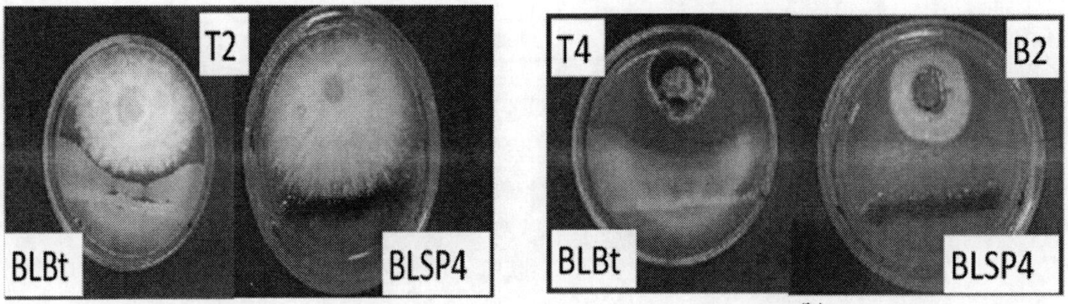

(a) (b)

Figure 2. Ilustration of compatibility test of fungi and entomopathogenic bacteria on PDA media. a. The growth of T2 fungi and entomopathogenic bacteria did not show any inhibitory effect in one PDA plate. b. Growth of fungal isolates (T4 and B2) and entomopathogenic bacteria inhibit each other.

Visual observation of bacterial growth and entomopathogenic fungus on PDA plate showed that there was interaction between fungi and entomopathogenic bacteria. Of the 6 fungal isolates tested, only T2 isolates showed normal mycelium growth and almost covered the plate when it was grown together between entomopathogenic bacteria BLBt and BLSP4. Meanwhile other fungal isolates showed growth inhibition when was grown together with entomopatogenic bacteria. The growth of this fungus appears to be clearly inhibited when compared to the control, ie isolates of fungi grown on individual PDAs in the same incubation period. Therefore, based on qualitative test, it can be concluded that T2 isolate can be applied together as consortium with entomopathogen bacterium BLBt or BLSP4.

Generally, the selection of isolates to be combined is based on the similarity of target pathogens possessed, for the control of pests or plant diseases [12]. But, in this experiment, the consortium is consist of two biocontrol agents with different mechanism of action This compatibility may allow simultaneous mechanism to increase pathogenicity against *S. litura*. Next, consortium between T2 and BLSP4 was used in the bioassay against larval and pupal stage of *S.litura*.

3.2 Efficacy test of entomopathogenic microbes against S. litura pupae

The conidiospores in the suspension were calculated using haemasitometer (**Table 2**) as follows:

Table 2 Spore concentration in suspension of 0.01% tween 80

Isolates	Spore concentration (spora/mL)
B. bassiana	0,9 x 10^7
Paecilomyces sp.	0,2 x 10^7
T2B	1,6 x 10^7
T4B	4,1 x 10^7

While the cell density of *Serratia* sp. in 10 mL 0.85% NaCl was equal to 10^9 cells/mL

Efficacy results (**Figure 3**) showed that the highest mortality (62.5%) in pupae *S litura* was obtained in the suspension treatment of T2B and *Serratia* sp. This may indicate that dual inoculation may be able to increase pathogenicity compared to single inoculation. In some treatments, pupae turn into moths and lay eggs however, the eggs did not hatch into larvae. From the observation result, no mycosis was found, so dead pupae was inoculated on PDA and incubated for further test of cause of death of pupae. After 8 days, symptoms of mycosis in pupae (**Figure 4**) were obtained. The mycelium that grows enveloping pupae exhibited morphological characteristics similar to the inoculated isolates. In the consortium treatment's PDA plates *Serratia* sp growth was also found.

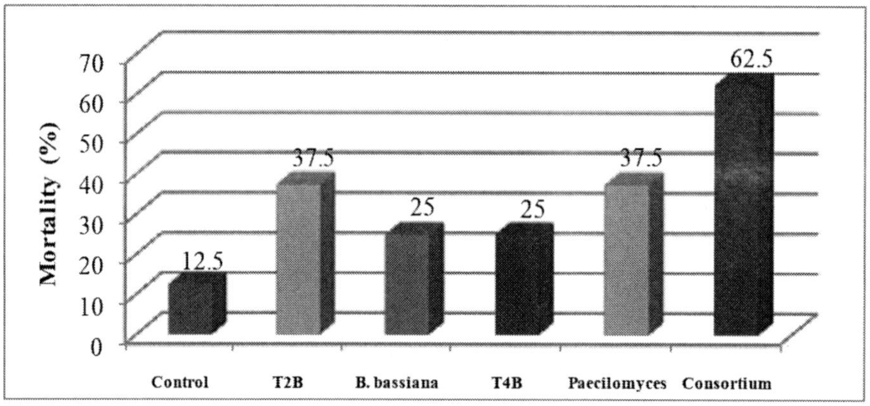

Figure 3. Mortality percentage of *S. litura* pupae 8 days post inoculation with conidial suspension of entomopathogen fungi and bacteria. Values given are mean of 4 replicates

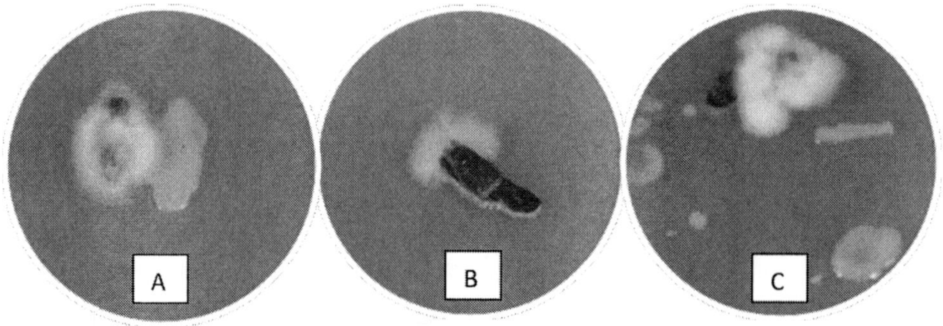

Figure 4. Growth of mycelium in dead pupa isolated with PDA after treatment with A. *B. bassiana*, B. T2B, and C. consortium showed growth characteristics of mycelium and colony that similar to the inoculated isolates

3.3 Efficacy test of entomopathogenic microbes against S. litura pupae

The conidiospores in the suspension (**Figure 5**) were calculated using haemasitometer as follows:

Table 3 Spore concentration in 0.01% tween 80suspension

Isolates	Spore concentration (spora/mL)
B. bassiana	5.6×10^7
Paecilomyces sp.	3.9×10^7
B2	1.4×10^7
T2B	0.6×10^7
T4B	0.5×10^7

While the cell density of *Serratia* sp. in 10 mL 0.85% NaCl was equal to 10^9 cells/mL

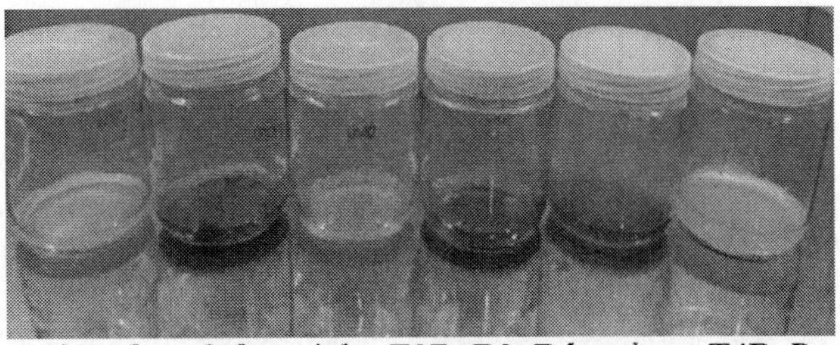

Figure 5. Conidiaspora suspension, from left to right: T2B, B2, B.bassiana, T4B, Paecilomyces, and consortium

Pathogenicity test **(Figure 6)** of microbial entomopathogen isolates showed that the highest mortality (70%) in larvae *S litura* was obtained in the treatment of isolate B2. Treatment of consortium and entomopathogenic bacteria (*Serratia* sp) showed the next highest percentage of mortality, 60 and 50%, respectively. While larval mortality was not found in control treatment, isolate T4B and *Paecilomyces* sp. as found in the pupae test, the consortium treatment was shown to exhibit a sufficiently good pathogenic effect on the third to fourth instar larvae of *S. litura.*

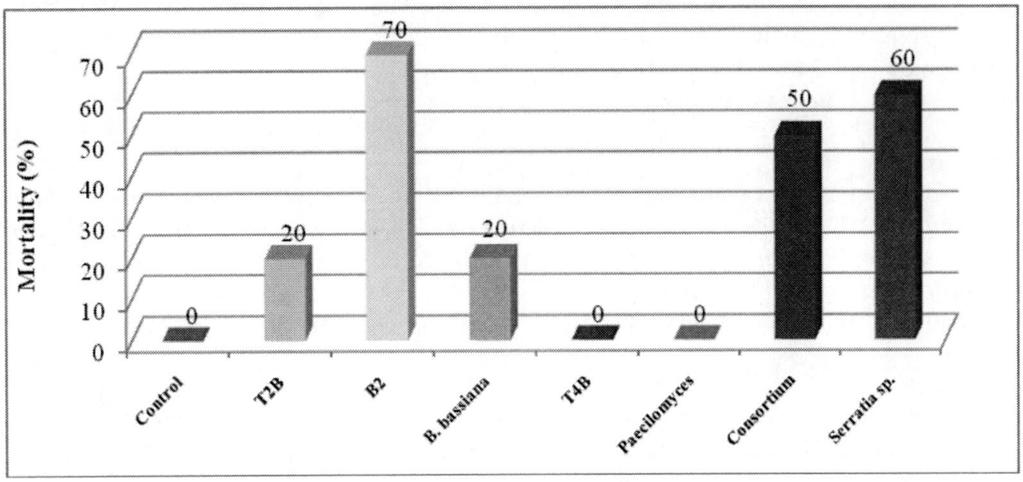

Figure 6 Mortality percentageof *S. litura* larvae 8 days post inoculation with conidial suspension of entomopathogen fungi and bacteria. Values given are mean of 5 replicates

The dead larvae from entomopathogenic fungal treatments showed uniform characteristics of dry and rigid larval bodies whereas in dead larvae due to consortium and *Serratia* sp tend to be wet and reddish (**Figure 7**). From the observation, there were no symptoms of mycosis on all treatments

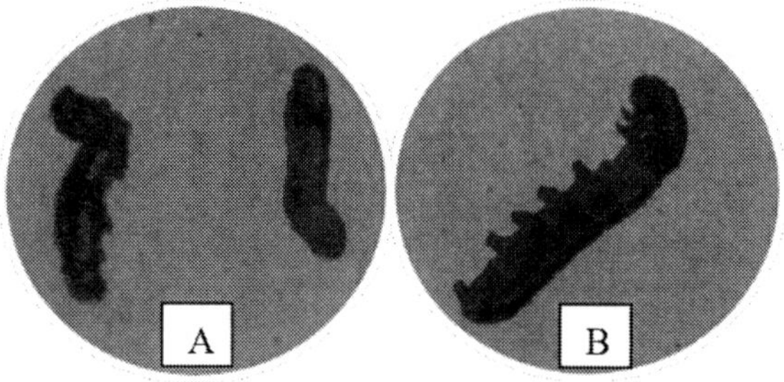

Figure 7. Third to fourth instar larvae of *S. litura*treated with A. suspension of conidia of fungus and B. *Serratia* sp.cell suspension. No symptom of mycosis was found on the surface of the larvae at all treatments

The efficacy results showed that each isolate had different levels of effect on the insect stages tested. At the pupae stage, all treatments demonstrated effect on the mortality of the pupae or the failure of the pupae to grow into moths. While in the larval stage there were some isolates that did not show the effect of pathogenity. This is in contrast to the results of studies suggesting that the lower the instar stage,the more susceptible to pathogenicity of entomopathogenic microbes [1][7]. Differences in results may be due to the condition of the pupae was not too good because there were 12.5% of the control larvae failed to turn into a moth.

The consortium treatment showed a sufficiently good pathogenic effect on the larval and pupae stages. *Serratia* sp. used in this study was isolated from dead *S. litura* pupae and has been identified molecularly showed 94% similarity with *Serratia marcescens* [9]. The authors also reported that the symptoms of larval mortality due to this isolate is the softened and reddish larval bodies and released unpleasant odor when squashed. These symptoms were not found in larvae that were infected with entomopathogenic fungi. Otherwise the larval bodies were shrank and dred. Symptoms of mycosis were not found at the time of observation but when dead pupae that did not develop into moths were grown on PDAs, pupae was overgrown with mycelium that had characteristics and growth patterns similar to those applied. This indicates that the isolates used have pathogenicity effect to *S. litura*.

The use of biopesticides derived from entomopathogenic microbes needs to be accompanied by adequate mass application. The selection of easy to found and inexpensive carrier needs to be further developed for the sustainability of sustainable agriculture.

4. Conclusion

Consortium treatment consisting of T2B and *Serratia* sp. showed the highest effectiveness of pupae mortality. The results of dead pupae isolation on PDA media showed similar mycelium growth characteristics to those of isolates used in the test. While on pathogenicity test to larvae stadia, B2 fungi showed the highest effectivity, followed by *Serrati*a sp and consortium treatment. In general, the consortium has a pretty good antagonistic effect on the two stages of *S. litura* life stages.

5. Reference

[1] Asi MR, Bashir MH, Afzal M, Zia Kand Akram 2013 Potential of entomopathogenic fungi for biocontrol of *Spodoptera litura* fabricius (Lepidoptera: Noctuidae) *Journal of Animal and Plant Sciences* **23** 913-918.

[2] Meyling, NV 2007 Methods for Isolation of Entomopathogenic Fungi from The Soil Environment - laboratory manual Deliverable 5.1 of VegQure under the research programme FØJO III.

[3] Sevim A, Demir I, Hofte M, Humber RA, and Demirbag Z 2010 Isolation and Characterization of Entomopathogenic Fungi from Hazelnut-Growing Region of Turkey *Biocontrol* **55** 279-297.

[4] Zulfiana D, Tarmadi D, Ismayati M and Yusuf S 2010. Pathogenicity of Metarhiziumanisopliae to Subterranian Termites Coptotermes sp. Proceeding of The Seventh Conference of The Pacific-Rim Termite Research Group 7-11.

[5] Anand R, Prasad B, and Tiwary, BN 2009 Relative Susceptibility of *Spodoptera litura* Pupae to Selected Entomopathogenic Fungi *Biocontrol* **54** 84-92.

[6] Ahmad M andMehmood R 2015 Monitoring of Resistance to New Chemistry Insecticides in *Spodoptera litura*(Lepidoptera: Noctuidae) in Pakistan *Journal of Economic Entomology* **108(3)**, 1279–1288.

[7] Tong, H, Su, Q, Zhou, X, Zhou X and Bai L 2013 Field Resistance of *Spodoptera litura* (Lepidoptera: Noctuidae) to Organophosphates, Pyrethroids, Carbamates and Four Newer Chemistry insecticides in Hunan, China J Pest Sci **86 (3)** 599–609

[8] Direktorat Perlindungan Tanaman Hortikultura 2008. *Pengenalan dan Pengendalian Hama Tanaman Sayuran Prioritas* Jakarta : Direktorat Perlindungan Tanaman Hortikultura.

[9] Krishanti NPRA, Wikantyoso B, Zulfitri A, and Zulfiana D 2017 Bakteri Entomopatogen Sebagai Agen biokontrol terhadap larva *Spodoptera litura* (f.) *Berita Biologi* **16 (1)** 13-21.

[10] Mishra D.S., Kumar A., Prajapati C.R., Singh A.K.,Sharma S.D. 2013 Identification of Compatible Bacterialand Fungal Isolate and Their Effectiveness Against Plant Disease. *Journal of Environmental Biology* **34** 183-189

[11] Islam MR, Jeong YT, Ryu YJ, Song CH, and Lee YS 2009 Isolation, Identification and Optimal Culture Conditions of *Streptomyces albidoflavus* C247 Producing Antifungal Agents against Rhizoctonia solani AG2-2 *Mycobiology* **37 (2)** 114-120.

[12] Sicuia O. A., Dinu S., Dinu M., Fătu C., Vălimareanu D., Mincea C., and Constantinescu F 2014 Pests and Diseases Management Using Compatible Biocontrol Bacteria and Entomopathogenic Fungal Strains Scientific Bulletin Series F. Biotechnologies **18** 66-72.

Production of Phytase, Amylase and Cellulase by *Aspergillus*, *Rhizophus* and *Neurospora* on Mixed Rice Straw Powder and Soybean Curd Residue

A Kanti[1] and I. M Sudiana[1]

[1]Research Center for Biology, Indonesian Institute of Sciences (LIPI)
Jl. Raya Bogor Km. 46, Cibinong 16911, Indonesia

E-mail: atityeast@gmail.com

Abstract. There is consensus that the presence of phytate in poultry diet negatively influence protein and energy utilization in poultry, and these influences could be enhanced by hydrolytic enzymes augmentation. The objective of study was to produce phytase, amylase and cellulase by *Aspergillus niger, Rhizopus oryzae* and *Neurospora sitophila* on solid state fermentation. Phytase production ability was first determined on submerge fermentation (SmF) using glucose as the main C-sources and on solid-state fermentation (SSF). The media composition for phytase production were a combination of rice straw powder and soybean curd residue at ratio of 0-100 % w/w. Maximum enzyme activities were observed at 96 h of incubation. SSF produced higher phytase than SmF. Maximum phytase production was achieved by *N. sitophila* (195.66 U/g) followed by *A. niger* and *R. oryzae*. Best media formula was 30:70 w/w of rice straw powder and soybean curd residue,respectively. At the same media formulation, the highest amylase was obtained by *A. niger* (141.85/g) followed by *R. oryzae* and *N. sitophila*. While for cellulase *A. niger* (92.0 U/g) followed by *N. sitophila* and *R. oryzae*. The ability of *A. niger, R. oryzae* and *N. sitophila* to produce phytase, amylase and cellulase on mixed substrate containing rice straw powder and soybean curd residue implies that those isolates are potential for hydrolytic production for poultry feed.

1. Introduction

Reduction of cost of feed in poultry is the main concern of many scientists. The inclusion of phytase in poultry diet has remarkably increased during the past decade. This is due to high concentration of phytate in cereal (barley, maize, sorghum and wheat) ranging from 1.86-2.89 ($g.kg^{-1}$). Higher phytate content is found in oilseed meals (4.0-9.11 $g.kg^{-1}$), and the highest 8.79-24.20 Phytate-P ($g.kg^{-1}$) is found in rice and wheat brands [1-3]. Phytase hydrolyses phytate which further eliminate the intense of phytate bound on mineral, carbohydrate and protein [4]. Thus, augmentation of phytase will reduce feed costs and increase the efficiency of phosphate utilization and other nutrients in cereal-based feed ingredients [5]. It is expected that the inclusion of phytase result in economic and environmental benefits. In addition, not only phytase but other hydrolytic enzymes such as amylase and cellulase are important component of feed ingredients [6].

Up to now, hydrolytic enzymes inclusion in feed ingredients is mostly focused on phytase production, and *Aspergillus niger* is the most popular phytase producer. Several other fungi especially *Neurospora crassa, N. sitophila, Rhizopus oryzae,* and *R. oligosporus* could be important microbes for hydrolytic enzymes production [7]. Those fungi are well known to play major role on traditional fermented foods in Indonesia such as *oncom* and *tempeh*. Solid-state fermentation offers higher enzyme production, less expensive and easier process control. Solid-state fermentation has been effectively carried out to produce phytase. Temperature and carbon nitrogen sources could be important factor influencing phytase production. A rice straw has been used as microbial sources for obtaining cellulolytic microbes such as *Trichoderma reesei* and for ethanol production. *T. reesei* cellulase was produced by the solid-state culture on rice straw medium. The optimal pH and temperature for *T. reesei* cellulase production were 6 and 25°C, respectively.

Rice straw exhibited different susceptibilities towards cellulase to their conversion to reducing sugars. The rice straw bioconversion with cellulase was more current than the general trend by *T. reesei*. This enzyme effectively led enzymatic conversion of acid, alkali and ultrasonic pre-treated cellulose from rice straw into glucose, followed by fermentation into ethanol. The combined method of acid pre-treatment with ultrasound and subsequent enzyme treatment resulted the highest conversion of lignocellulose in rice straw to sugar and consequently, highest ethanol concentration after 7 days fermentation with *Saccharomyces cerevisae* yeast. The ethanol yield in this study was about 10 and 11 g.L(-1) [8].

Rice husk and rice straw samples pre-treated with xylanase prior to treatment with cellulase, produced higher reducing sugar per 100 g of substrate by 29% to 41%. These findings can be adopted for the extraction of reducing sugars from cellulose and xylan-containing waste materials. The purely enzymatic extraction procedure can be substituted by the harsh and bio-adverse chemical methods [9]. Due to high protein and carbohydrate content, soybean curd residues have been used as the material for the production of lipopeptide antibiotic iturin A. *Bacillus subtilis* RB14-CS, cultured using soybean curd residues, after 4 days incubation, produced iturin A for about 3,300 mg/kg wet solid material (14 g/kg dry solid material), which is approximately ten fold higher than that in submerged fermentation [10]. Due to the nutrient composition of soybean curd residue and rice straw, we suspect that those materials can be used as a source of phytase, amylase and cellulase production. The objective of this study was to produce phytase, amylase and cellulase on mixed substrate of soybean curd residues and rice straw.

2. Material and Methods

2.1. Fungi Isolation and identification

The isolation of fungi from traditional fermented food was carried out following methods described by [11]. The *oncom* was obtained from local market in Bogor, West Java, Indonesia. To isolate the fungi from the sample, 1.0 g of sample was diluted in 9 ml sterilized water and vortex-mixed. One-tenth of a ml of successive decimal dilutions was spread on acidified Dichloran Rose Bengal chloramphenicol agar (OXOID, Cat.1076012). This selective medium was used due to the prevention growth of bacteria and the suppression of molds spreading. Plates were incubated for 5 days at room temperature. Strain purification was carried out at least twice by selecting one of each type of fungi colonies and streaking twice for single colony. The plate was incubated at 27°C for 3 days. Representative colonies were selected, purified and maintained on Potato Dextrose Agar (PDA) (OXOID, Cat.CM 0139).

rDNA identification Freshly-grown cells on the Potato Dextrose Broth were used for DNA extraction [12]. PCR amplification of the partial Internal Transcribed Spacer (ITS) ribosomal subunit was performed with primers ITS 4: 5'– TCC TCC GCT TAT TGA TAT GC – 3'and Primer ITS 5: 5'– GGA AGT AAA AGT CGT AAC AAG G –3' [13] using GoTaq master mix (Promega, M7122). PCR products were visualized on 2% agarose and then sequenced with both primers using Big Dye terminator v3.1 Cycle Sequencing Ready Reaction Kit (Applied Biosystems) following the manufacturer's instructions. The partial 26S sequences determined in this study were compared to those in the EMBL/GenBank/DDBJ databases using the nucleotide Basic Local Alignment Search Tool (BLASTn) [14].

Morphology Observations. Aspergillus, Rhizophus and *Neurospora* were grown on PDA plates at 30°C for 7 days to observe its morphological characteristics and to test its single hyphal growth rate. The determination of the single hyphal growth rate was conducted within 24 h of colony development at 30°C under a microscope. Two photographs of growing fungi on Petri dish were taken. The hyphal growth rates were determined. After sporulation has occurred, the dimensions of the developed perithecia, asci and ascospores were separately observed [15].

Culture maintenance. The isolated strains of fungi were grown in PDA (Potato Dextrose Agar) medium for 120 h, at 30°C. These strains were further observed for their ability to produce phytase under submerge culture and solid-state fermentation at various temperatures and media compositions.

2.2. *Fungal strain screening*

The isolated strains of fungi grown in PDA medium. Enrichment culture media containing 0.5 % sodium phytate as the sole phosphorus and glucose were used for the primary screening of phytase producers. The method was based on estimation of phosphate solubilisation from sodium phytate in aqueous media. The strains were grown under agitation at 150 rpm, at 30°C for 96 hours. After 96 hours of incubation time, the fungal biomass was separated by centrifugation at 8000 rpm for 20 minutes and the supernatant was then used for estimation of phytase, amylase, and cellulase production using method described by Liu et al. (1999) [16].

2.3. *Enzyme preparation and assay*

Inoculum preparation for SS and SmF. The culture was grown and maintained on potato dextrose agar (PDA) slants. The slants were stored at 4°C and sub-cultured after 4 days. Five-day-old fully sporulated slant was used for inoculant preparation. For this, 10 ml sterile distilled water containing 0.1% Tween-80 was added to the slant and spores were scraped with a sterile needle. The obtained inoculant contained 4.7 x 10^7 spores per ml.

Substrates preparation for SSF. For the SSF, the media for phytase production was composed of a combination of rice straw powder and soybean curd residue at ration of 0-100% which were obtained from Cibinong Science Centre Rice Filed Trial and local market. Ten grams of the dried mixed substrate taken in a cotton plugged of 250 ml Erlenmeyer flask were supplemented with 6.0 ml of salt solution containing 0.5 % $MgSO_4$.7H2O and 0.1 % NaCl. While for the SmF, 10 grams of the dried mixed substrate taken in a cotton plugged of 250 ml Erlenmeyer flask were supplemented with 100 ml of salt solution containing 0.5 % $MgSO_4$.7H2O and 0.1 %NaCl.

Enzyme extraction. Enzyme extraction was carried out using distilled water with 0.1% of Tween-80. Known quantities of fermented substrates were thoroughly mixed with the required volume of distilled water (so that the final extraction volume was 100 ml) by keeping the flasks on a rotary shaker at 180 rpm for one hour. The suspension was centrifuged at 8000 g for 20 min at 4°C and the clear supernatant was assayed for enzyme activity.

Phytase assay. Phytase activity was assayed by measuring the amount of inorganic phosphorus released from sodium phytate solution using the method of Sing et al. (2013)[17]. One unit of enzyme activity was defined as the amount of phytase required to release one micromole of inorganic phosphorus per minute under the assay conditions.

Amylase assay. Alpha amylase was assayed by adding 0.5 ml of enzyme to 0.5 ml soluble starch (1%, w:v) in 0.1 M phosphate buffer, pH 6.0, for 30 min at 40 °C. The reaction was stopped and reducing sugar was determined with dinitrosalicylic acid according to the method of Bernfeld [18]. An enzyme unit is defined as the amount of enzyme releasing one µMol of equivalent glucose from the substrate per hour at 40°C.

Cellulase assay. Cellulase was assayed by adding 0.5 m; of enzyme to 0.5 ml soluble carboxymethyl cellulose (1%, w:v) in 0.1 M phosphate buffer, pH 6.0, for 30 min at 40°C. The reaction was stopped and reducing sugar was determined with dinitrosalicylic acid according to the method of Bernfeld [18]. An enzyme unit is defined as the amount of enzyme releasing 1 µMol of equivalent glucose from the substrate per hour at 40°C.

Biomass estimation. Fungal biomass estimation was carried out by determining the N-acetyl glucosamine released by the acid hydrolysis of chitin present in the cell wall of the fungi [19]. For this, 0.5 g (dry wt) of fermented matter was mixed with concentrated sulphuric acid (2 ml) and the reaction mixture was kept for 24 h at room temperature (30°C). This mixture was diluted with distilled water to make a 1 N solution, autoclaved for 1 h, neutralized with 1 N NaOH and the final volume was made up to 100 ml with distilled water. One ml of solution was mixed with 1 ml acetyl acetone reagent and was incubated in a boiling water bath for 20 min. After cooling, a volume of 6 ml ethanol was poured followed by the addition of 1 ml Ehrlich reagent and the resulting mixture was incubated at 65°C for 10 min. Once the reaction was cooled, the optical density of the reaction mixture was read at 540 nm against a reagent blank. Glucosamine (Sigma) was used as the standard. The obtained results were expressed as mg glucosamine per gram dry substrate (gds).

3. Results and Discussion

3.1. Isolation and identification of fungi

Based on morphological observation and molecular analyses, isolates originated from traditional fermented food, were identified as *A. niger, R. oryzae* and *N. sitophila.* These isolates produced phytase, amylase and cellulase on submerge fermentation (**Table 1**).

Table 1. Phytase, amylase and cellulase of fungi

No.	Species name	Phytase (U/ml)*	Amylase (U/ml)*	Cellulase (Unit/ml)*
1	*A. niger*	11.1	13.2	8.3
2	*R. oryzae*	6.5	5.6	6.9
3	*N. sitophila*	8.3	9.6	7.5

Amylase, phytase and cellulase were determined at 96 h incubation.

Under submerge fermentation (SmF), *A. niger, R. oryzae* and *N. sitophila* produced phytase enzyme from 6.5 to 11.1 U/ml. The highest phytase, amylase and cellulase production was obtained by *A. niger*. SmF can be used to produce hydrolytic enzymes. But several parameter should be optimized. Contamination problem was less on SmF than that of SSF. But for higher volume of enzymes production, SSF could be more preperable, since it is less exspensive than SmF. Therefore hydrolytic enzymes production by fungi is mostly conducted by SSF. Earlier study on production of hydrolytic enzymes is quite intensive. Of 32 microorganisms, isolates from soil and decayed fruits were tested for phytase production. The most active fungal isolates with phytase activity was *A. niger*. The maximum phytase activity occurred at the end of the growth phase [20]. *R. oryzae* produced high phytase on oil cake supplemented with (w/w) mannitol, 2.05%; ammonium sulfate, 2.84% and phosphate, 0.38%) [21]. Earlier study showed that *N. sitophila* not only produced phytase but also proteinase. At which proteinase production increased on substrate containing sugar beet pulp (SBP), wheat bran (WB) and citrus waste (CW) on solid state fermentation (SSF). Optimum conditions for protein enrichment of these lignocellulosic by products by *N. sitophila* under SSF were optimal moisture content for growth on SBP. CW was 75-80% (w/w) whereas for WB, it was 65% (w/w). The protein content of untreated SBP, WB and CW after 5 days SSF on surface was increased from 15, 13 and 7% (w/w) to 30, 30 and 18.2%, respectively [22]. Our finding shows that *N. sitophila* produce 3 hydrolytic enzymes: phytase, amylase and cellulase (**Table 1**).

3.2. Production of phytase on solid-state fermentation (SSF)

A. niger, R. oryzae and *N. sitophila* produced higher phytase on solid state fermentation (**Table 1** and **Figure 1**). The highest phytase production was achieved by *N. sitophila* (**Figure 1**). This is the first report on the ability of *N. sitophila* to produce very high phytase on rice straw powder and soybean curd residues. The activity of phytase produced by *N. crassa and N. sitophila* was much higher than that of produced by *Pichia anomala* [23]. Best formula for phytase production was 30:70 w/w of rice straw powder and Soybean curd residues, respectively.

3.3. Production of amylase on solid-state fermentation (SSF)

A. niger, R. oryzae, and N. sitophila produced amylase higher on SSF and SmF (**Table 1** and **Figure 2**). The best formula for amylase production was 30: 70 w/w of rice straw powder and soybean curd residues, respectively. *A. niger* produced the highest amylase compare to other isolates (**Figure 2**). Amylase production on submerged fermentation holds tremendous fungal potentiality in high biomass yield of alpha-amylase. This study also shows that amylase could be produced at large amount by *A. niger* using mixed rice straw and soybean curd residue which also implies that this formula could be further developed for hydrolytic enzyme production.

Enzymes production by *A. niger* is affected by pH, temperature and nitrogen sources of the medium for the productivity of α-amylase utilizing *Ipomoea batatas*. The maximum activity of α-amylase was recorded as 450 U/mg after 7 days of submerged fermentation at pH 7.0 and room temperature 28°C. Among the organic and inorganic nitrogen sources, inorganic sources showed maximum yield in which ammonium nitrate showed highest amylase activity of 475 U/mg at the same pH and temperature.

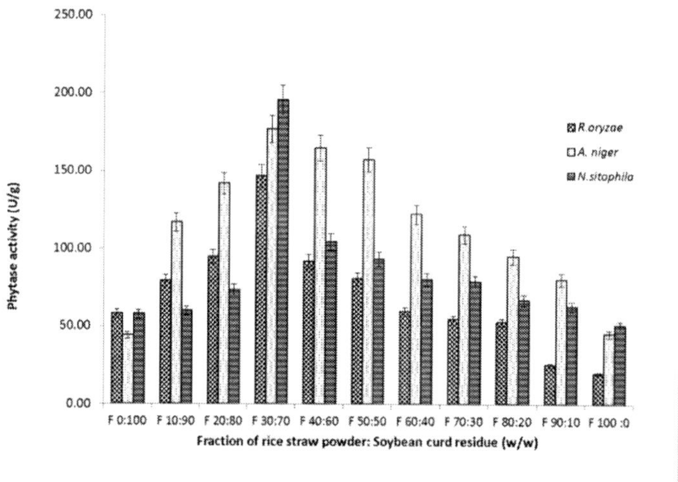

Figure 1. Phytase activity of *A. niger, R. oryzae, and N. sitophila* on solid state fermentation at 30°C

Many works have been conducted on *A. niger*. For instance, *A. niger* can be used in industrial process after characterization [24]. Amylase and protease was synthesized by *A. niger* strain UO-1 in media prepared with brewery (BW) and meat (MPW) wastewaters supplemented with different starch concentrations. The highest amylase (70.29 and 60.12 EU/mL) and protease (6.11 and 6.03 EU/mL) production were obtained in the BW and MPW media supplemented with 40 g of starch/L of medium after 88 h of fermentation, respectively. Chemical oxygen demand (COD) appeared affect enzymes production. High COD input reduced enzymes production by more than 92%. High amylase and protease activities were found in the BW medium supplemented with casaminoacids, peptone or yeast extract, but ammonium nitrate and sodium nitrate were also good nitrogen sources for amylase production. The stabilities of amylase and protease were higher at 50°C and pH 4.95 and at 53.4°C and pH 3.87, respectively, but they were highly sensitive at temperatures of 70 °C or higher [25]. These all means that many factors should be optimized for high hydrolytic enzymes production.

The production of amylase by *A. niger* on three cassava whey media in liquid shake culture was compared. The supplemented cassava whey (SCW) medium has exhibited amylase activity of 495 U/ml. Biomass cropped was 1.63 g/l in the SCW medium. Yeast extract employed as a nitrogen supplement increased biomass yield of *A. niger* to 2.75 g/l with maximum amylase activity of 643 U/ml. Sodium nitrate ($NaNO_3$) as nitrogen supplement had the lowest biomass yield of 0.77 g/l and amylase activity of 206 U/ml. Thus, yeast extract as nitrogen supplement of cassava whey medium supported maximum production of amylase and biomass of *A. niger* [26].

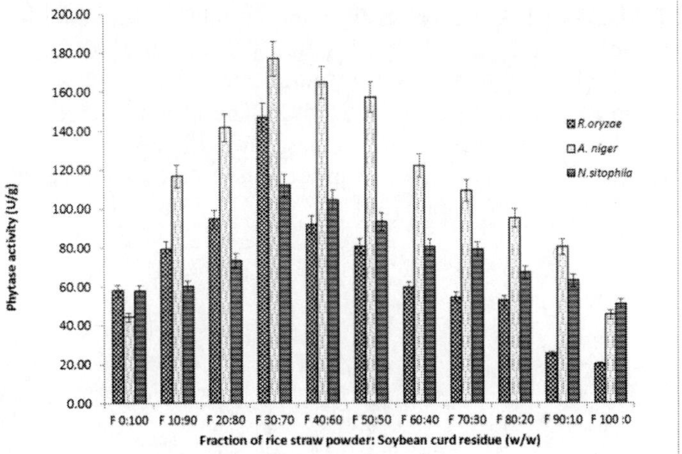

Figure 2. Amylase activity of *A. niger*, *R. oryzae*, and *N. sitophila* on solid state fermentation at 30°C

3.4. Production of cellulase on solid state fermentation (SSF)

A. niger, R. oryzae, and N. sitophila produced cellulase higher on SSF than SmF (**Table 1** and **Figure** 3). The highest cellulase production was achieved by *A. niger* (**Figure 3**). Best formula for phytase production was 30:70 w/w of rice straw powder and Soybean curd residue, respectively. Fermentation technique affects cellulase production. SSF cultures produced higher cellulase activities than those in submerged and solid-state cultures (1,768, 1,165, and 1,174 U l^{-1}, respectively). Although SSF-cultures grew less than the other cultures, they produced significantly higher cellulase yields (370, 212, and 217 U g^{-1} glucose, respectively) and volumetric productivities (24, 16, and 16 U l^{-1} h-^{1}, respectively).

Likewise, endoglucanase and xylanase activities were higher in SSF-cultures. Under the tested conditions, it seems that fungal attached growth on perlite may favour better enzyme production. Biofilms are efficient systems for cellulase production and may replace solid-state fermentation. SSF holds promise for further optimization and development. The results of this work reveal that fungal biofilms may be implemented for the commercial production of cellulase employing the technology developed for submerged fermentation at high cell densities [27]. Fungal cellulases are well-studied enzymes and are used in various industrial processes. *Trichoderma* cellulase system has been the knowledge of enzymatic depolymerisation of cellulosic material. *Trichoderma* strains can produce substantial amounts of endoglucanase and exoglucanase but extremely low levels of β-glucosidase. This deficiency necessitates screening of fungi for cellulytic potential. A number of indigenously isolated fungi were screened for cellulytic potential.

A. niger MS82 was capable of producing moderate to high levels of both endoglucanase and β-glucosidase when grown on different carbon containing natural substrates, for example, grass, corncob, bagasse, and purified celluloses. Furthermore, it was observed that the production of endoglucanase reaches its maximum during exponential phase of growth as opposed to β-glucosidase during the stationary phase. Enzyme production by solid-state fermentation was equally investigated and found to be promising. Highest production of cellulase was achieved at pH 4.0 at 35°C under submerged conditions [28]. *A. niger* KKS, isolated from a farmland near Suwon, was immobilized on Celite and polyurethane foams. Enzyme activities produced by the immobilized cell system in a bubble column were higher than that of shake-flask culture. The enzyme productivities were twice as high as beta-glucosidase, beta-xylosidase, and xylanase activities obtained in a bubble column were significant when the ground rice straw was used as a substrate [29].

A. niger NCIM 548 under same fermentation conditions in submerged fermentation (SmF) and solid state fementation (SSF) used a central composite face centered design of response surface methodology (RSM). As per statistical design, the optimum conditions for maximum production of pectinase (1. 64 U/mL in SmF and 179. 83 U/g in SSF) and cellulase (0. 36 U/mL in SmF and 10. 81 U/g in SSF) were, 126 h, pH 4.6, and carbon source concentration 65 g/L in SmF. By contrast, it was 156 h, pH 4.80, and moisture content 65% in SSF. The response surface modelling was effectively applied to optimize the production of both pectinase and cellulase by *A. niger* under same fermentation conditions. It will result in a cost-effective process in both submerged and solid state fermentation using agro industrial wastes as substrate [30].

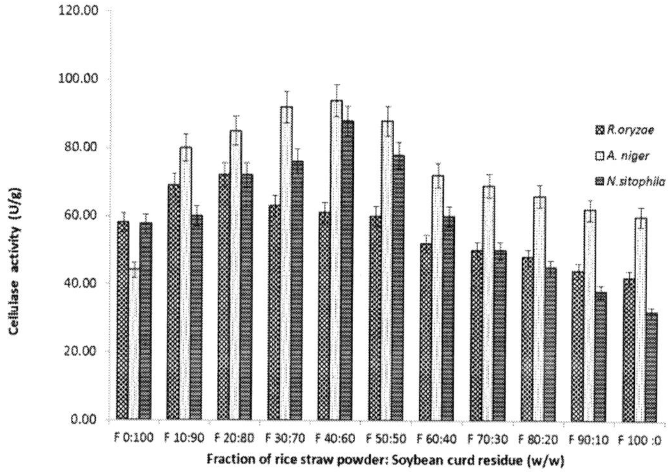

Figure 3. Cellulase activity of *A. niger*, *R. oryzae*, and *N. sitophila* on solid state fermentation at 30°C

Animal feed contains bound phosphorus from 18- 88% of total phosphorus content such as phytate. This phytate phosphorus cannot be directly digested by monogastric animals like poultry and pigs due to lack of intrinsic phytase in their gastrointestinal tracts. Phytate behave as an anti-nutrient through chelating various cations such as Ca^{2+}, Fe^{2+}, Zn^{2+}, and Mg^{2+} and thereby reducing their bioavailability. Supplementations of phytase into animal feed will breakdown phytate into inositol and phosphate. Our finding revealed that *N. crassa* and *N. sitophila* produce phytase under submerge fermentation as well as under solid-state fermentation. Production of phytase under solid-state fermentation was higher than SmF, which implies that SSF is more favourable for phytase production. Development of SSF technology to optimize phytase is furthermore necessary. In addition to temperature, substrate composition influences phytase production. Those parameters will be optimized for future study.

4. Conclusion

SSF produced higher phytase than SmF. Maximum phytase production was achieved by *N. sitophila* followed by *A. niger* and *R. oryzae*. Best media formula was 30:70 w/w of rice straw powder and soybean curd residue, respectively. At the same media formulation, the highest amylase was obtained by *A. niger* followed by *R. oryzae* and *N. sitophila*. Meanwhile, cellulose was highly produced by *A. niger* (92.0 U/g) followed by *N. sitophila* and *R. oryzae*. The ability of *A. niger, R. oryzae* and *N. sitophila* to produce phytase, amylase and cellulase on mixed substrate containing rice straw powder and soybean curd residues implies that those isolates are potential for hydrolytic production for poultry feed.

Acknowledgments
This project is funded by The Indonesian Institute of Sciences through Commercial Product Development Program 2015, and DIPA 2017 of Research Center for Biology-LIPI. The authors acknowledged Dr. Lisman Suryanegara as the program coordinator, and Rosaliana, and Rizka Syahputri for laboratory work.

5. References

[1] Maga JA 1982 Phytate: Its chemistry, occurence, food interactions, nutritional significance and method of analysis *Journal of Agricultural and Food Chemistry* **30** 1–9 http://doi.org/10.1021/jf00109a001
[2] Heaney RP, Weaver CM and Fitzsimmons ML 1991 Soybean phytate content: Effect on calcium absorption *American Journal of Clinical Nutrition* **53** 745–747
[3] Haraldsson AK, Veide J, Andlid T, Alminger ML and Sandberg AS 2005 Degradation of phytate by high-phytase Saccharomyces cerevisiae strains during simulated gastrointestinal digestion *Journal of Agricultural and Food Chemistry* **53** 5438–5444 http://doi.org/10.1021/jf0478399
[4] Liebert F and Portz L 2005 Nutrient utilization of Nile tilapia Oreochromis niloticus fed plant based low phosphorus diets supplemented with graded levels of different sources of microbial phytase In *Aquaculture* **248** 111–119 http://doi.org/10.1016/j.aquaculture.2005.04.009
[5] Leytem AB, Willing BP & Thacker PA 2008 Phytate utilization and phosphorus excretion by broiler chickens fed diets containing cereal grains varying in phytate and phytase content *Animal Feed Science and Technology* **146** 160–168 http://doi.org/10.1016/j.anifeedsci.2007.11.006
[6] Kim EY, Kim YH, Rhee M, Song JC, Lee KW, Kim KS, Park S 2007 Selection of *Lactobacillus* sp. PSC101 that produces active dietary enzymes such as amylase, lipase, phytase and protease in pig *The Journal of General and Applied Microbiology* **53 (2)**, 111–117 http://doi.org/10.2323/jgam.53.111
[7] Zhou XL, Shen W, Zhuge J & Wang ZX 2006 Biochemical properties of a thermostable phytase from *Neurospora crassa FEMS Microbiology Letters* **258 (1)** 61–66 http://doi.org/10.1111/j.1574-6968.2006.00205.x
[8] Belal EB Bioethanol production from rice straw residues 2013 *Brazilian Journal of Microbiology* : **44(1)** 225–34 http://doi.org/10.1590/S1517-83822013000100033
[9] Dutta N, Mukhopadhyay A, Dasgupta AK, Chakrabarti K 2014 Improved production of reducing sugars from rice husk and rice straw using bacterial cellulase and xylanase activated with hydroxyapatite nanoparticles *Bioresource Technology* **153** 269–277 http://doi.org/10.1016/j.biortech.2013.12.016
[10] Mizumoto S, Hirai M and Shoda M 2006 Production of lipopeptide antibiotic iturin A using soybean curd residue cultivated with *Bacillus subtilis* in solid-state fermentation *Applied Microbiology and Biotechnology* **72(5)** 869–875 http://doi.org/10.1007/s00253-006-0389-3
[11] Choi Y, Hyde KD & Ho WWH 1999 Single spore isolation of fungi Fungal Diversity **3** 29–38
[12] Butinar L, Santos S, Spencer-Martins I, Gunde-Cimerman N 2005 Yeast diversity in hypersaline habitats *FEMS Microbiology Letters* **244(2)** 229–34 http://doi.org/10.1016/j.femsle.2005.01.043
[13] White TJ, Bruns S and Taylor JW 1990 Amplification and direct sequencing of fungal ribosomal RNA genes for phylogenetics In: PCR Protocols: A Guide to Methods and Applications eds Pp. 315-322 White Academic Press Inc New York
[14] Stephen F, Altschul Thomas L, Madden Alejandro A, Schäffer Jinghui Zhang Zheng Zhang, Webb Miller, David JL 1997 BLAST and PSI-BLAST: a new generation of protein

database search programs *Nucleic Acids Research* **25** 3389–3402 https://doi.org/10.1093/nar/25.17.3389

[15] Yang X, Huang S, Zhang Y and Zhang X 2015 Identification and characterization of a rare species of *Neurospora International Journal of Agriculture and Biology* **17**(1) 163–168

[16] Liu B, Jong C and Tzeng Y 1999 Effect of immobilization on pH and thermal stability of *Aspergillus ficuum phytase* **25** 517–521

[17] Singh NK, Joshi DK and Gupta RK 2013 Isolation of phytase producing bacteria and optimization of phytase production parameters *Jundishapur Journal of Microbiology* **6(5)** http://doi.org/10.5812/jjm.6419

[18] Saqib AAN and Whitney PJ 2011 Differential behaviour of the dinitrosalicylic acid (DNS) reagent towards mono- and di-saccharide sugars *Biomass and Bioenergy* **35(11)** 4748–4750 http://doi.org/10.1016/j.biombioe.2011.09.013

[19] Jost DI, Indorf C, Joergensen RG & Sundrum A Determination of microbial biomass and fungal and bacterial distribution in cattle faeces *Soil Biology and Biochemistry* **43(6)** 1237–1244 http://doi.org/10.1016/j.soilbio.2011.02.013

[20] Xiong AS, Yao QH, Peng RH, Li X, Fan HQ, Guo MJ and Zhang SL 2004 Isolation, characterization, and molecular cloning of the cDNA encoding a novel phytase from *Aspergillus niger* 113 and high expression in Pichia pastoris *Journal of Biochemistry and Molecular Biology* **37(3)** 282–291

[21] Rani R and Ghosh S 2011 Production of phytase under solid-state fermentation using *Rhizopus oryzae*: novel strain improvement approach and studies on purification and characterization *Bioresource Technology* **102(22)** 10641–9. http://doi.org/10.1016/j.biortech.2011.08.075

[22] Shojaosadat SA, Faraidouni R, Madadi-Nouei A and Mohamadpour I 1999 Protein enrichment of lignocellulosic substrates by solid state fermentation using *Neurospora sitophila* 2014 In *Resources, Conservation and Recycling* **27**, pp. 73–87. http://doi.org/10.1016/S0921-3449(98)00086-X

[23] Olstorpe M, Schnürer J and Passoth V 2009 Screening of yeast strains for phytase activity *FEMS Yeast Research* **9(3)** 478–488. http://doi.org/10.1111/j.1567-1364.2009.00493.x

[24] Sundar R, Liji T, Rajila C, and Suganyadevi P 2012 Amylase Production by *Aspergillus niger* under Submerged Fermentation using Ipomoea Batatas *International Journal of Applied Biology and Pharmaceutical Technology* **3(2)** 175–182

[25] Hernández MS, Rodríguez MR, Guerra NP & Rosés RP 2006 Amylase production by *Aspergillus niger* in submerged cultivation on two wastes from food industries *Journal of Food Engineering* **73(1)** 93–100 http://doi.org/10.1016/j.jfoodeng.2005.01.009

[26] Oshoma CE, Imarhiagbe EE, Ikenebomeh MJ and Eigbaredon HE 2010 Nitrogen supplements effect on amylase production by *Aspergillus niger* using cassava whey medium *African Journal of Biotechnology* **9(5)** 682–686 Retrieved from <Go to ISI>://00027538170001

[27] Gamarra NN, Villena GK and Gutiérrez-Correa M 2010 Cellulase production by Aspergillus niger in biofilm, solid-state, and submerged fermentations *Applied Microbiology and Biotechnology* **87(2)** 545–551. http://doi.org/10.1007/s00253-010-2540-4

[28] Sohail M, Siddiqi R, Ahmad A and Khan SA 2009 Cellulase production from *Aspergillus niger* MS82: effect of temperature and pH *New Biotechnology*, **25(6)** 437–441 http://doi.org/10.1016/j.nbt.2009.02.002

[29] Kang SW Kim SW and Lee JS 1995 Production of cellulase and xylanase in a bubble column using immobilized *Aspergillus Niger* KKS *Applied Biochemistry and Biotechnology* **53(2)** 101–106 http://doi.org/10.1007/BF02788601

[30] Kumar S Sharma HK and Sarkar BC 2011 Effect of substrate and fermentation conditions on pectinase and cellulase production by *Aspergillus niger* NCIM 548 in submerged (SmF) and solid state fermentation (SSF) *Food Science and Biotechnology* **20(5)** 1289–1298 http://doi.org/10.1007/s10068-011-0178-3

IOP Conf. Series: Earth and Environmental Science **166** (2018) 012011 doi:10.1088/1755-1315/166/1/012011

Increased Blood Glucose Level on Pelagic Fish as Response to Environmental Disturbances at East Coast Pangandaran, West Java

D. M Malini[1], Madihah[1], A. F Apriliandri[1] and S Arista[1]

[1]Departement of Biology, Faculty of Mathematics and Natural Sciences, Padjadjaran University, Jatinangor-Sumedang, 45363, Indonesia

E-mail: desak.made@unpad.ac.id

Abstract. Pananjung Pangandaran Nature Reserve Area is one of the tourism destinations in Indonesia, which has large biodiversity. The environmental stressors such as pollutants can reduce the biodiversity of fish. Many physiological changes occur in response to environmental disturbances. One of these responses is blood glucose levels. The objective of this study was to evaluate the quality of the waters environment through the glucose levels of pelagic fish found in the study area. This research was conducted at Pangandaran Nature Reserve. The research was carried out using observational method. The research measured environmental parameters, such as water temperature, dissolved oxygen levels, and carbon dioxide concentrations and blood glucose level of pelagic fish. Five species had high glucose level such as *Lactarius lactarius* (173 mg/dl), *Carangoides Praeustus* (105.5 mg/dl), *Scomberomorus plumieri* (140 mg/dl), *Rastrelliger kanagurta* (202 mg/dl) and *Trichiurus lepturus* (136.3 mg/dl). In conclusion, the study indicated that the quality of East Pangandaran Beach Region, West Java has decreased quality based on the increased blood glucose levels on pelagic fish.

1. Introduction

The coastal area of Pangandaran is one of the tourism destinations in Indonesia which has large fishery resources and provides a major contribution towards the economic activities in West Java Province [1]. This area has been developed for various purposes such as sport, residential, maritime and aquaculture usages. Unfortunately, all activities of tourism and recreation in the marine environment and beach affect the quality of the sea-water, for example plastic, food waste, dish soap and "gray water" from camping and hydrocarbon residues from boat motors or OHV engines. Chemical and physical factors cause body reactions that may contribute to the health of individual fish. Fish and other aquatic animals are subject to a broad variety of stressors because their homeostatic mechanisms are highly dependent on prevailing conditions in their immediate surroundings. Stress causes increases in glucose level in fish. Therefore, the impacts of contamination on the aquatic biota can be evaluated through analysis of glucose level in fish obtained from the contaminated site [2]. One of type fishes that can be used as an indicator of environmental stress is pelagic fish who lives in the water column of coastal, ocean, and lake waters, but not on or near the bottom of the sea or the lake.

The response to environmental stress in fish is characterized by the stimulation of the hypothalamus, which results in the activation of the neuroendocrine system and a subsequent cascade of metabolic and physiological changes [3]. These changes enhance the tolerance of an organism to face an environmental variation or an adverse situation while maintaining a homeostasis status [4].

Under condition of stress, the body of the fish emits immediate responses recognized as primary and secondary responses. The primary response is the perception of an altered state by the central nervous system (CNS) and the release of the stress hormones, cortisol and catecholamines (adrenaline and epinephrine) into the circulation by the endocrine system [5].

Secondary responses occur as a consequence of the released stress hormones [6], causing changes in the blood and tissue chemistry, e.g. an increase of plasma glucose [7]. This entire metabolic pathway produces a burst of energy to prepare the fish for an emergency situation [8]. The blood glucose is the main source of energy and essential substrate for cell metabolism especially the brain cells in the fishes. Normal blood glucose levels of fish are 40-90 mg/dl [9]. Glucose homeostasis is regulated by multiple hormones. Of these, insulin is the only hormone that can decrease glucose concentration; thus it plays a key role in glucose metabolism [10]. The environmental disturbance can changes metabolism glucose of poikilotermic animals such as fish. At the time of stress, insulin inactivation causes the use of glucose by stopping cells, resulting in increased glucose in the bloodstream. The study about the Effect of Chromium, Zinc (II) and Sulphate Heptahydrate ($ZnSO_4.7H_2O$) on Glucose Content of Freshwater Fish, *Heteropneustes fossilis* showed that the values of blood glucose have a positive correlation with concentrations of this metal. The glucose levels were found higher in the exposed fish *Heteropneustes fossilis* when compared to the control [11]. Since stress has been reported to elevate glucose levels in fishes, so the quality of the waters environment can be evaluated through the glucose levels [12]. The objective of this study was to evaluate the quality of seawater through the glucose levels in the blood of pelagic fish at East Coast Pangandaran, West Java.

2. Materials and Methods

The materials used in this study were sea water of east coast of Pangandaran, the blood of pelagic fish, amylum, O_2 Reagent, Concentrated H_2SO_4, $MnSO_4$, and Na-thiosulfate. The tools used in this research were surgical tool, stationery, tray, burette, bucket, measuring cylinder, glucometer, glucose strip, universal pH indicator, fishing net (fishing rod), syringe, camera, erlenmeyer glass, volume pipette, plastic sample, refractometer, gloves, mercury thermometer, stative, tissue, and vinoject.

This study used a quantitative descriptive approach observational method. Sampling technique was used to collect pelagic fish samples on 7 to 13 May 2017. They were taken from different sites along the east coast of Pangandaran. Sampling was primarily conducted with a 30 ft. seine net. Immediately after collecting, fishes were put into a pan of sea water for identification and counting. Blood glucose was quantified following the method of Stefani et al. (2010) where in a drop of fresh blood from caudal severed fish was touched with glucose strip inserted in a standard glucometer (Apex Biotechnology Corp.) [10] The result was obtained *in situ*. Information about fishes and habitat for all sites were collected. Habitat quality for each site was assessed according to a number of physical and water quality criteria established by Plafkin et al. [25]. Data on blood glucose was presented as mean ± standard deviation and analysis of the data was carried descriptively out.

3. Results and Discussion

3.1. The quality of physical and chemical waters at East Coast Pangandaran

The physical and chemical of seawaters at the east coast of Pangandaran was analyzed. Temperature, salinity and pH were measured *in situ* while dissolved oxygen and BOD were analysed in the laboratory. The analysis results of physical and chemical seawater can be observed in **Table 1**.

Based on seawater quality standard criteria listed in Regulation No. 82/2001 [12], the water temperature (28^0C), salinity (33‰), and pH (6) in east coast of Pangandaran was definitely supportive for life of aquatic biota. Meanwhile, the data of BOD (3.8 ppm) and DO (4.5 mg/dl), were recorded in the observation location were less than the standard criteria. According to Tono et al (2010) and Tini et al (2012), the DO levels that are suitable for marine biota is greater than 5 mg/L while for BOD levels should be ≥ 20 ppm [13-14]. Thus, the status of the seawater quality in the east coast of Pangandaran belongs to a moderately contaminated category.

Table 1. The physical and chemical conditions of the sea-water in the East Pangandaran Beach

No.	Parameters	Range
1	Salinity	33‰
2	Air temperature	28°C
3	Water temperature	29°C
4	pH	6
5	Dissolved Oxygen	4.5 mg/dl
6	BOD	3.8 ppm

The temperature in this area includes a good range for fish life (25-31°C) [15]. The water temperature affects the rate of metabolism in the body of the fish, higher temperature can increase the metabolic rate. Water temperatures can also have a negative effect directly to the aquatic biota, the temperature affects the solubility of oxygen in the waters. The higher the temperature will lead to a lower oxygen solubility rate [16]. According to Sastrawijaya et al (2010), Solids concentration suspended in sea-waters comprising phytoplankton, zooplankton, mud, waste human, animal waste, plant waste and industrial waste, so that more concentration component in the water can increase the value of suspended solids (Total Suspended Solid). Suspended solids in east coast Pangandaran come from human waste, fishery waste and residual plants or waste from the community. Suspended solids also come from runoff land that comes from land due to rain.

Based on the quality standard according to the Regulation No. 82/2001 on the Management of Water Quality and Water Pollution Control, the pH value in the research area, belongs to the category good enough for fish life ranging from 6-9 [16]. However, the high BOD indicates that the amount of oxygen needed by microorganisms to oxidize organic matter in the water is high; it means that the water is already in a deficit of oxygen. Meanwhile, the high number of microorganisms multiplies in the water due to the abundance of food available (organic matter). Therefore, BOD always indirectly associated with levels of organic matter in the water. This is supported by the discovery of a lot of rubbish in the surrounding area. Two types of waste (organic and non-organic) can be found on the East Coast Pangandaran. The most common of organic waste are macro algae, sea grass, coconut, and paper, while non-organic waste that is found in the study area is baby diapers, plastic, food waste, and waste. Macro algae and sea grasses are two organisms stranded in the sea because the roots cannot with stand the waves. Meanwhile, paper and coconut found on the east coast of Pangandaran are delivered by the activities of coconut sellers and stalls on the beach. While, inorganic wastes (baby diapers, plastics, food scraps, and cigarettes) could be delivered from the activities of household, visitors, fishermen, and traders. The entry of domestic wastes into the waters will affect physical and chemical conditions of the water.

The garbage is heavily stagnant on the beach and some are drowned in the bottom such as baby's diapers. This can affect the physical condition of the seawater such as decreasing the value of DO and increasing the need for O_2. Domestic waste usually contains high organic matter, which will cause the decrease of dissolved oxygen so that it can cause respiration of disturbed fishes [18]. If dissolved oxygen is unbalanced, then the fish will experience stress and may cause death. This is because the body's tissue cannot bind the dissolved oxygen in the blood and the brain does not get enough oxygen supply, a common condition called anoxia. Low dissolved oxygen can lead to the presence of toxic gases such as ammonia. High levels of ammonia can cause tissue damage to the gills, where the gill plates to swell so that the respiratory system will be disrupted [15].

3.2. The morphologic of fish obtained in east coast Pangandaran

The fish are one of organisms that can be used to measure environmental degradation for some reason. First, fishes are sensitive to the wide array of direct stresses. Second, fishes integrate the adverse effect of complex and varied stresses on other components of the aquatic ecosystem, such as habitat and macro invertebrate, by virtue of their dependence on those components for reproduction, survival and growth [19]. Only fish over 10 cm standard length were used for this study as a sample. A total of 10 species belonging to 7 families were captured, namely *Megalaspsis cordyla, Sardinella lemuru, Scomberomorus plumieri, Rastrelliger kanagurta, Lactarius lactarius, Carangoides praeustus, Pomadasys maculatus, Leiognathus equulus, Trichiurus lepturus.* and *Decapterus macarellus.* The morphology of fishes is presented in **Figure 1**.

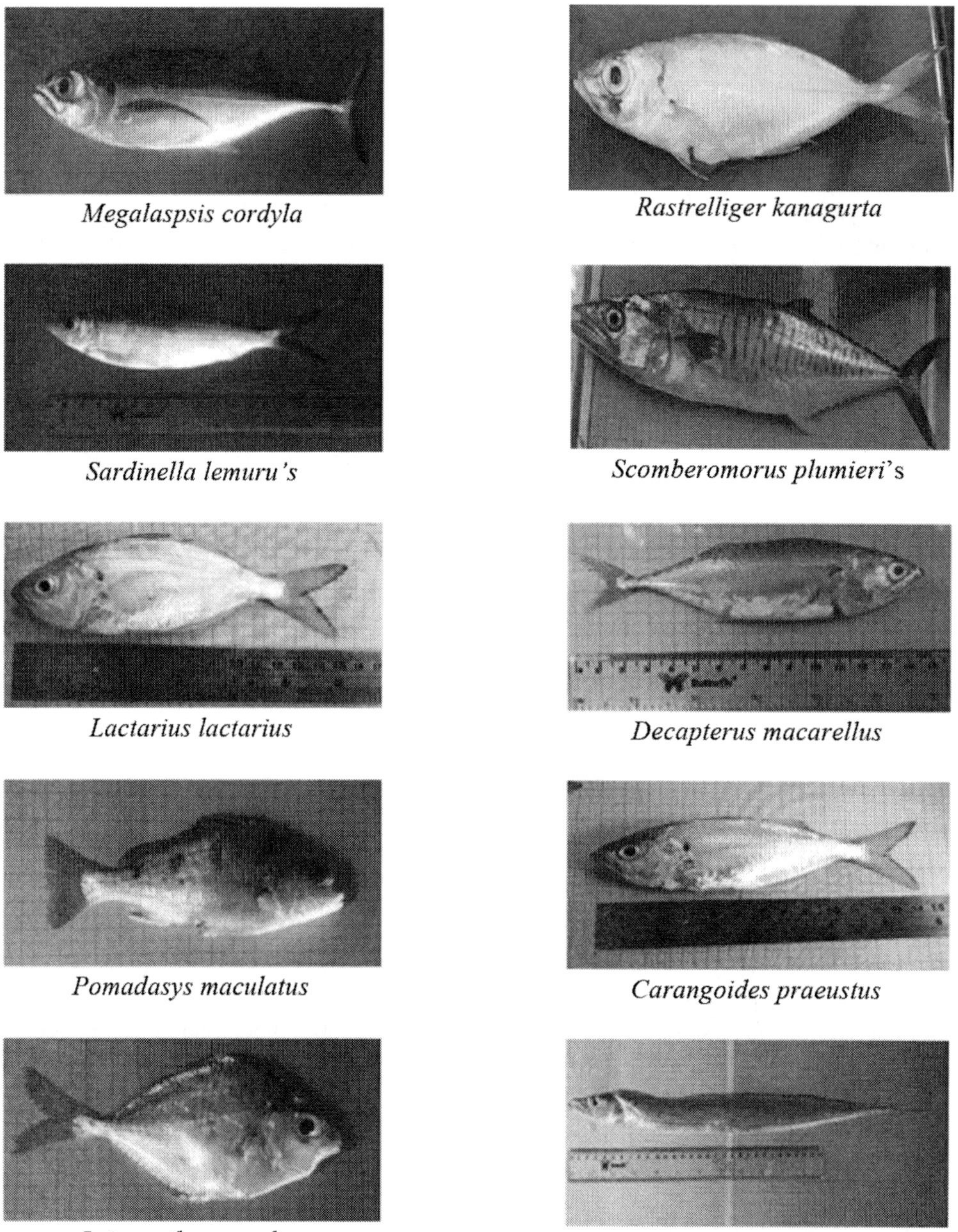

Megalaspsis cordyla — *Rastrelliger kanagurta*

Sardinella lemuru's — *Scomberomorus plumieri*'s

Lactarius lactarius — *Decapterus macarellus*

Pomadasys maculatus — *Carangoides praeustus*

Leiognathus equulus — *Trichiurus lepturus*

Figure 1. The morphology of fishes found in the study area.

3.3. The blood glucose level of fish

Environmental stressors such as contaminants can cause a variety of biological responses in fish ranging from the biomolecular and biochemical to population and community-level effects. To assess fishes's health, the bioindicators technique utilizes a suite of biological responses both as integrators of stress effects and as sensitive response (early-warning) indicators of existing and past environmental conditions. The growth in fish as influenced by principal ecological factors in the environment, such as seawater environment they inhabit, competition and food availability [20]. Based on measurements that have been done using a glucometer, the level of blood glucose is presented in **Table 2**.

Table 2. Blood glucose levels, weight, and length of pelagic fish

No	Species	Blood glucose level (mg/dl)	Total weight (g)	Total length (cm)
1	*Megalapsis cordyla*	51 ± 1.15	88,5	22,7
2	*Decaplerus macarellus*	72.5 ± 1.73	50	15
3	*Scomberomorus plumierri*	160 ± 2.74	226	30
4	*Sardinella lemuru*	77.5 ± 0.58	36,5	16
5	*Restralliger kanagurta*	202 ± 3.26	85	17
6	*Lactarius lactarius*	201 ± 2.56	51	15
7	*Carangoides praeustus*	146 ± 1.76	54	18
8	*Lutjanus rivulatus*	94 ± 1.26	42	14,5
9	*Leiognathus equulus*	46 ± 0.96	31	13,5
10	*Trichiurus lepturus*	136.3 ± 1.69	76	46

Our study found six species with hyperglycaemic condition namely, *R.kanagurta* (202 mg/dl), *L. lactarius* (201 mg/dl), *S. Plumieri* (160 mg/dl), *C. Praeustus* (146 mg/dl), *T. lepturus* (136,3 mg/dl), and *L. rivulatus* (94 mg/dl). According to Patrice [21], normal blood glucose level of fish is 40-90 mg/dl. According to Barton (2002), the physical size of the fish determines blood glucose levels, the larger of fish body size requires more energy to support their life and as a result, equally higher secretion of glucose. Hyperglycaemia condition is not only influenced by the total length or weight but also depends on the environmental conditions and habitat quality.

In east coast Pangandaran was found domestic waste that comes from household waste such as urine, faeces, and garbage. In addition, wastes from aquaculture and agriculture were found. The entry of domestic waste into the waters will affect the physical and chemical conditions of the seawater. According to Effendi [18], domestic waste usually contains organic matter that would cause a decline in dissolved oxygen that can finally generate disturbed fish respiration. Low dissolved oxygen can cause any toxic gases such as ammonia gas. High levels of ammonia can cause damage to gill tissue, where gill plates swell so that the respiratory system will be disrupted [16]. The effect of suspended solids on fish is directly on the gills of the fish [22] and therefore oxygen which enters the body is reduced due to gills that are covered in solids. This can interfere the process of respiration so that fish need more high energy to survive. Suspended Solid is one of stress factor in fish on the east coast of Pangandaran beside ammonia; organic chemicals (demand oxygen) and low dissolved oxygen.

Many physiological changes that occur in response to environmental disturbances are now used routinely for assessing stressed states in fish. Stress responses are mediated through neuronal and endocrine pathways, known as the primary response, following initial perception of the stressor. The response to stress in fish is characterized by the stimulation of the hypothalamus, which results in the activation of the neuroendocrine system and a subsequent cascade of metabolic and physiological changes [3]. These changes enhance the tolerance of an organism to face an environmental variation or an adverse situation while maintaining a homeostasis status [24].

Under conditions of stress, the body of the fish emits immediate responses recognized as primary and secondary responses. The primary responsibility is the perception of an altered state by the central nervous system (CNS) and the release of the stress hormones, cortisol and catecholamines (adrenaline and epinephrine) into the bloodstream by the endocrine system. Secondary responses occur as a consequence of the released stress hormones [4], causing changes in the blood and tissue chemistry, e.g. an increase of plasma glucose [7]. This entire metabolic pathway produces a burst of energy to prepare the fish for an emergency [8]. Normally, the formed glucose will enter the cell will stimulate glycogenesis and lipogenesis. In the stress condition occurs response of receptors that have received stimulation, then the response will be forwarded to the hypothalamus and secrete catecholamine hormone that allows glucose into the cell. In hyperglycaemic conditions insulin inactivation occurs so that the process of glycogen formation is inhibited, glucose cannot enter the cells and cause high blood glucose levels. The rise in blood glucose is primarily generated by cortisol- mediated gluconeogenesis that also inhibits cellular uptake of circulating glucose thus increasing the levels in blood circulation. The increased level of glucose is a manifestation for the higher needs of tissues to fuel the metabolic needs of osmoregulation and an important source of energy for maintaining homeostasis in fish during chronic stress [24]. Thereby, blood glucose levels can be used as an indicator of environmental stresses [19]. In conclusion, the increase in level blood glucose in fish indicates that seawater, where were fish inhabit, were contaminated by pollutants.

4. Conclusion
Based on the results of this study can be concluded that the quality sea water of East Coast Pangandaran, West Java moderately contaminated category and the elevated level of blood glucose level on pelagic fish as a response to environmental disturbances.

5. References
[1] Baransano K and Jubhar CM 2011 Eksploitasi dan Konservasi Sumberdaya Hayati Laut dan Pesisir di Indonesia *Jurnal Biologi Papua* **3(1)** 39-45.

[2] Barton BA 2002 Stress in Fishes: A Diversity of Responses with Particular Reference to Changes in Circulating Corticosteroids *Integ. and Comp. Biol.* **42** 517–525

[3] Lowe and Davison 2005 Plasma osmolarity, glucose concentration and erythrocyte responses of two Antarctic nototheniid fishes to acute and chronic thermal change *Journal of Fish Biology* **67 (3)** 752–766

[4] Barton BA, Schreck CB and Barton LD Effects of chronic cortisol administration and daily acute stress on growth, physiological conditions, and stress responses in juvenile rainbow trout *Diseases of Aquatic Organisms* **2** 173-185

[5] Perry SF, Bernier NJ 1999 The acute humoral adrenergic stress response in fish: facts and fiction *Aquaculture* **177** 285–295

[6] Barton BA, Haukenes AH, Parsons BG and Reed JR 2003 Plasma cortisol and chloride stress responses in juvenile walleyes during capture, transport, and stocking procedures. *North American Journal of Aquaculture* **65** 210–219

[7] Begg K and Pankhurst NW 2004 Endocrine and metabolic responses to stress in a laboratory population of the tropical damselfish *Acanthochromis polyacanthus Journal of Fish Biology* **64** 133–145

[8] Rottmann RW, Francis-Floyd R and Durborow R 1992 The role of stress in fish disease *SRAC Publication* **474**

[9] Rahardjo MF, Sjafei DS, Affandi R and Sulistiono 2011 *Ikhtiologi* Jakarta : Lubuk Agung

[10] Eames SC, Philipson LH, Prince VE and Kinkel MD Blood Sugar Measurement in Zebrafish Reveals Dynamics of Glucose Homeostasis *ZEBRAFISH* **7 (2)** 205-213

[11] Hoq T, Hasan R, Haque N, Md Ashrafuzzaman 2015 The Effect of Chromium on Glucose Content of Freshwater Fish, *Heteropneustes fossilis American Journal of Zoological Research* **3 (1)** 1-3

[12] Peraturan Pemerintah No. 82. 2001 Pengelolaan Kualitas air dan Pengendalian Pencemaran Air Jakarta

[13] David M, Shivakumar R, Mushigeri SB and Kuri RC 2005 Blood glucose and glycogen levels as indicators of stress in the freshwater fish, *Labeo rohita* under fenvalerate intoxication *Journal of Ecotoxicology & Environmental Monitoring* **15** 1-5

[14] Sufianto B 2008 Uji Transportasi Ikan Mas Koki (*Carrasius auriatus* L.) Hidup dengan Sistem Kering Perlakuan Suhu dan Penurunan Konsentrasi O2 *Tesis* Pascasarjana IPB Bogor

[15] Kordi KMGH and Tancung AB 2007 *Pengelolaan Kualitas Air dan Tanah dalam Budidaya Perairan* Jakarta: PT. Rineka Cipta

[16] Kordi KMGH 2012 Budidaya Ikan Konsumsi di Air Tawar Jakarta: Lily Publisher

[17] Sastrawijaya T 2010 Pencemaran Lingkungan Jakarta : Rineka Cipta

[18] Effendi H 2003 Telaah Kualitas Air: Bagi Pengelolaan Sumber Daya Alam Dan Lingkungan Perairan Yogjakarta: Kanisius

[19] Fritsche L, Weigert C, Häring HU, Lehmann R 2008 How insulin receptor substrate proteins regulate the metabolic capacity of the liver–implications for health and disease *Curr. Med. Chem* **15 (13)** 1316–1329.

[20] Adams SM and Jaworska JS 1990 Influence of ecological factors on the relationship between mfo induction and fish growth: Bridging the gap using neural networks. *Marine Environmental Research* **42 (1-4)** 197-201

[21] Patriche T 2009 The Importance of glucose determination in the blood of the Cyprinids Importanţa Determinării Glucozei Din Sângele Ciprinidelor *Biotehnologii 42* **(2)**

[22] Barton BA 2002 Stress in fish: A diversity of responses with particular reference to changes in circulating Corticosteroid1 *Integ And Comp. Biol* **42** 517-525

[23] Stryer L 2000 *Biokimia* Edisi IV **2** Jakarta: EGC

[24] AD Pickering, Pottinger TG and Christie P 1982 Recovery of the brown trout, *Salmo trutta* L., from acute handling stress: a time-course study *Journal of Fish Biology* **20 (2)** 224-244

[25] Plafkin JL, Barbour MT, Porter KD, Gross SK and Hughes RM 1989 Rapid Bioassessment Protocols for Use in Streams and Rivers: Benthic Macroinvertebrates and Fish. *U.S. EPA, Office of Water*. EPA/444/4-89-001, Washington, D.C

IOP Conf. Series: Earth and Environmental Science **166** (2018) 012012 doi:10.1088/1755-1315/166/1/012012

Colonization and Performance of Diazotroph Endophytic Bacteria on Palm Oil (*Elaeis guineensis* Jacq L.) Leaves

I. N. P Aryantha[1,2] and A.R.M Hidiyah[1]

[1]Biotechnology Study Program, School of Life Sciences and Technology
[2]Biosciences and Biotechnology Research Center
Bandung Institute of Technology,
Jl. Ganesa No. 10, Bandung 40132, Indonesia

E-mail: nyoman@sith.itb.ac.id

Abstract. Excessive and continous use of inorganic fertilizer can lead to negative impact on the environtment qualities, therefore microbial biochemical approaches which are more friendly need to be considered. The absorption of nitrogen elements has been widely known through the roots and then transported to the leaves before chlorophyll performing the works on the photosynthesis process. However, the discovery of many diazotroph endophytic bacteria in the leaves suggests that nitrogen is not only supplied through the soil but it can be fixed by the diazotroph endophytic bacteria through the leaves. This study aims to evaluate the colonization and performance of the diazotroph endophytic bacteria on palm oil leaves. A total of 42 isolates of endophytic bacteria isolated by Mardiah (2014) were selected based on their ability to fix nitrogen qualitatively on nitrogen free medium and quantitatively based on Nessler reagent. One of the selected isolates with the highest nitrogen fixation capability is KSD2. The sequencing results showed that KSD2 isolate is closely related to *Bacillus cereus*. Colonization and performance evaluation of *Bacillus cereus* on palm oil seedlings were performed at 0, 2, and 4 weeks after inoculation. Treatments included Co (control/ untreated), Ta (inoculation via soil), and Da (inoculation via leaves) with 7 replications. Data were evaluated using SPSS program with Univariate Analysis of Variance (significance level >95%). The colonization of *Bacillus cereus* was evaluated with PCR method approach using a specific primer (BCFomp1/ BCRomp1). PCR results showed that *Bacillus cereus* successfully colonized palm oil leaves as evidenced by the target sequences of 575 bp long.

1. Introduction

The nitrogen element is absorbed by the plant in the form of nitrate compounds (NO_3^-) and ammonium (NH_4^+). Biologically, this element is obtained by certain microorganisms known as nitrogen-fixing bacteria or better known as diazotroph bacteria. It has been reported that nitrogen fixation by diazotroph microorganisms is a major source of nitrogen addition in forest ecosystems [1]. Diazotroph bacteria have been widely isolated from rhizosphere and rhizoplane of non-leguminous crops [2]. However, the holding efficiency of N_2 is lower when compared with the diazotroph endophytic bacteria. This is because the availability of photosynthates for rhizosphere and rhizoplane bacteria are limited [3].

Based on its potential, this study is more focused on diazotroph endophytic bacteria. Diazotroph endophytic bacteria lives on plant tissues and have the ability to fix nitrogen. The advantages of diazotroph endophytic bacteria is the plants directly provide photosynthates as nutrients for bacterial growth, and also provide an environment with low oxygen levels, thus spurring the expression of the nitrogenase enzyme responsible for nitrogen fixation. In addition, endophytic bacteria also do not have to compete with other soil microbes to obtain root exudates for their survival [3-4]. The advantages possessed by the diazotroph endophytic bacteria make the N potential donated by the bacteria is larger than the diazotroph nonendophytic bacteria. Based on research in sugar cane, diazotroph endophytic bacteria were able to increase root dry weight and plant height compared to rhizosphere bacteria [5]. In addition, the potential of N contributed by diazotroph endophytic bacteria is greater than diazotroph non-endophytic bacteria, due to their presence in the intercellular tissues of plants that are not easily lost, while the free N nutrients in nature are highly labile, rain-washed and eroded, and easily evaporate into the air [5].

Leaves are the plant organ known to be colonized by diazotroph endophytic bacteria. During this time, the absorption of nitrogen elements has been widely described through the roots and will furtherly be transported to the leaves as chlorophyll forming that works on the photosynthesis process. The discovery of many diazotroph endophytic bacteria in the leaves suggest that nitrogen is not only supplied through the soil but can be fixed by the diazotroph endophytic bacteria through the leaves. Further, the nitrogen pathway to the leaves can be shortened and more efficient so it is useful for faster photosynthesis. Inoculation of the diazotroph endophytic bacteria in leaves is expected to increase the N element as chlorophyll forming. In the end, the photosynthesis process will immediately increase the plant growth. Therefore, the colonization of the diazotroph endophytic bacteria in leaves needs to be further evaluated.

Successful inoculation of diazotroph endophytic bacteria in leaves is identified by its colonization in plant tissues. The detection, visualization, and quantification of inoculated bacteria on the surface and into plant tissues during the plant growth cycle are supportive in the direct study of *in vivo* bacterial activity. In addition, the ability to compete with native bacteria can also be revealed. Species and cultivars of host plants, endophytic species and strains, inoculum concentrations, age and growth conditions of host plants are known as limiting factors of endophytic introduction [6-8].

So far, the endophytic bacteria capability resulting from the palm oil network in fixing nitrogen and its role in palm oil nurseries is less understood. Therefore, the ability of colonization and subsequent fixation of nitrogen from endophytic bacteria inoculated on leaves of palm seedlings should be extracted to minimize use of chemical fertilizers. In this study, inoculation of leaves and soil to introduce diazotroph endophytic bacteria into palm oil seedlings was performed. The colonization was detected using PCR. Increased bacterial colonization and the capacity of certain bacteria to modulate plant metabolism are key issues for further study, as this may provide insight into the intercropping-endophytic relationship of plants.

2. Materials and Methods

2.1. Qualitative Selection of Nitrogen Fixation Capabilities with NfB (Nitrogen free Bromtimol) Medium

Forty two endophytic bacteria isolates used for this research isolated from PT Astra Agro Lestari plantation (PT Agro Menara Rahmat, Central Kalimantan). This isolation has been done by Mardiah [9]. Isolation of endophytic bacteria from palm oil is done by taking healthy tissues in the roots, stems, leaves, and fruits aseptically. Each tissues were surface sterilized with 70% ethanol and 2% sodium hypochloride, then planted in Nutrient Agar (NA) medium aseptically. After 48 hours of incubation, bacterial isolates that grew on the media purified and the pure culture stored in a reaction tube containing NA media as stock culture [9]. Qualitative Selection of Nitrogen Fixation Capabilities with NfB started by culturing the isolates on 10 ml of NB medium for 24 h at room temperature. A total of 1 ml bacterial cultures were introduced into Mannitol NfB semi solid medium and incubated for 24 hours. The positive bacteria binding to nitrogen was observed from blue to yellow change of colour in the media and also the formation of the pellicle on the surface of the media [10].

2.2. *Quantitative Selection of Nitrogen Fixation Capabilities with Nessler reagent*

Bacteria that have been grown on NB media for 24 hours were centrifuged for 10 minutes, 4000 rpm. A total of 1 ml of culture supernatant was added with 20 μl of seignette salt and 20 μl of nessler reagent before incubation for 5 minutes. The mixture of a solution exhibiting a yellow color indicates the presence of NH_4^+ is formed. The formed NH_4^+ was calculated by spectrophotometer at a wavelength of 425 nm [11-12]. The supernatant absorbance values obtained were then converted to ppm by the formula (supernatant absorbance = 1.7594 * ppm NH_4Cl + 0.0175) with reference to standardized NH_4^+ curves.

2.3. *In Planta Test with Nessler Reagent*

A total of 1 ml bacterial culture was inoculated on the leaf that has been first injured with a sterile needle. After one week, the part of the leaf that has been inoculated by the bacteria was cut off then surface sterilized and crushed. The scours were centrifuged for 10 minutes at a speed of 4000 rpm. A total of 1 ml centrifugation supernatant was diluted 10x and tested its ammoniac uptake by adding 20 μl of seignette salt and 20 μl of Nessler reagent. Next is calculated the absorbance at λ 425 nm [11-12]. The supernatant absorbance value obtained was then converted to ppm by the formula (supernatant absorbance = 1.7594*ppm NH_4Cl + 0.0175) with reference to NH_4^+ standard curve.

2.4. *Identification of Morphological Character of Colonies and Microscopic Observations*

The morphology of the colony was observed visually in petri dishes against the shape, color, edges, surface, and colony growth patterns. Microscopic observations of bacterial cells were carried out for cell shape and Gram reaction. The technique of gram staining was done in the following protocoles: the first stage created a smear by dripping aquades as much as 3 drops then putting as much as one bacterial colony, leveling it, and passing it on fire quickly to dry. The isolate smears were dabbed with the basic dye of crystal violet and left for 1 minute, then washed with water flowing carefully. The swab is then streaked with Lugol or Gram Iodine and left for 1 minute. The excess reagents were removed and the smear soaked in 96% alcohol for 30 seconds. The swabs were washed with water flowing carefully. The smear was coloured with safranin for 1 minute, washed again with water flowing carefully and then dried. The smear was observed with a microscope.

2.5. *Molecular Identification of 16S rRNA*

The amplification of 16S rRNA gene was carried out using PCR colony method. The pure bacterial isolates were grown on solid NA medium and extracted using ose into sterile deionized water and homogenized and then boiled for template in the PCR process. Five μL of bacterial suspension was used as template and mixed with 2.5 μL primer 785F 16S rRNA (5 '-AGAGTTTGATCCTGGCTCAG- 3'), 2.5 μL reverse primer 907R 16S rRNA (5 '-GGTGACCTTGTTACGACTT - 3'), and 15 μL sterile deion, and 25 μL PCR mix in PCR tube. The PCR tube containing 50 μL of mixed material was then homogenized using a vortex prior to PCR. The amplification process was carried out at an initial denaturation temperature of 95°C for 3 minutes with 1 cycle, 95°C denaturation for 30 seconds, annealing 60°C for 30 seconds, and extension 72°C for 30 seconds with 30 cycles, end extension 72°C for 7 minutes with 1 cycle.

2.6. *Sequencing*

Samples are sequenced using Macrogen services in Korea. Sequencing results were chromatogram data (file with extension ".abl"). The sequencing data was further analyzed by the BLAST (Basic Local Alignment Search Tool) software from the NCBI website (www.ncbi.nlm.nih.gov/BLAST) and BioEdit.

2.7. *Phylogenetic Tree Analysis*

The result of sequencing was edited using Bioedit program. Reference species were searched using BLASTn nucleotide search (BLASTn) of NCBI. The sequence of isolates was aligned with some reference species using the CLUSTAL_W program version 2.0.11. Sequence alignment was re-edited to remove unreadable gaps or nucleotides during sequencing. The phylogenetic tree was constructed based on distance analysis using neighbour-joining method. The phylogenetic tree was constructed using the MEGA 6 program. The trust values of each established branch were determined using bootstrap analysis based on re-sampling of 1,000 times.

2.8. Preparation of Liquid Inoculant

The pure culture of the diazotroph endophytic bacteria on the agar was tilted and rejuvenated on the NA medium. A total of 1 colony was grown into 50 ml of NB under agitation for 24 hours. One ml of the liquid colonies was taken and was poured into 100 ml of NB medium. This culture was grown in reciprocal shaking incubators at room temperature until it reaches the maximum growth rate, for incubation of the used velocity culture of 150 rpm.

2.9. Inoculation of Diazotroph Endophytic Bacteria In Palm Oil

The used Saplings Seeds were 6 months year old of unfertilized palm oil seeds. Palm oil seedlings were first selected based on the parameters ofrelatively uniform with an average height of 20-22 cm and has 5-6 leaf blade. The planting medium was 5 kg of sterilized soil. The inoculation was done by injuring the leaves and by spraying the liquid inoculant on the wounded leaf. Plants inoculated with 10^{11} bacterial cell crops. To remove excess nutrients, the cells were washed twice with sterile water followed by centrifugation of 8000 g for 20 minutes prior to inoculation. The control plants were uniquely inoculated with 5 ml of sterile water. Sampling was carried out on week 0, 2nd, and 4th after inoculation of one plant per replication.

2.10. DNA extraction

The DNA was extracted from 100 mg of leaves using the Dneasy Plant Mini Kit (QIAGEN). The amount of DNA was photometrically calculated at 260 nm wave lengths. The quality of the extracted DNA in photometric checks with A260 / A280 calculation ratio should be above 1.9 and A320 close to 0.

2.11. Evaluation of Colonization by PCR method

Primers used in this PCR targetted gene fragments of *motB* on *Bacillus cereus*, *BCFomp1* forward primers (5'-ATCGCCTCGTTGGATGACGA-3'), and reverse *BCRomp1* primer (5'-CTGCATATCCTACCGCAGCTA-3'). The length of the PCR product was 575 bp. The amplification process was performed at 94°C initial denaturation temperature for 3 minutes with 1 cycle, denaturation 94°C for 30 seconds, annealing 54.5 °C for 1 minute, and extension 72°C for 1 minute with 30 cycles, end extension 72°C for 7 minutes with 1 cycle [25].

2.12. Measurement of Chlorophyll Content

A total of 0.1 gram of leaves were cut into small pieces followed by an addition of 10 ml of 80% acetone, crushed for 5 minutes until smooth for extraction. The extraction results were filtered and 80% acetone were added to the volume limit of 10 ml. Calculation of chlorophyll content was done at λ 645 nm and λ 663 nm, with the following calculation:

Total chlorophyll (mg / L) = 20.2 D645 + 0.02 D663
Chlorophyll a = 12.7 D663 – 2.69 D645 Chlorophyll b = 22.9 D645 – 4.68 D663

2.13. Statistical Analysis

Statistical analysis to determine the absorption of ammonium and total chlorophyll was performed by univariate analysis (ANOVA) in randomized block design (RAK) at 95% confidence level. If there is any real difference, Post Hoc Test DMRT (Duncan's Multiple Range Test) at 95% confidence level is performed. The statistical test was carried out using SPSS statistical (Statistical Product and Servie Solution) program for Window release 16.

3. Results and Discussion

3.1. Qualitative Test Result of Diazotroph Endophytic Bacteria Isolate

Forty-two Mardiah's endophytic bacterial isolates (2014) were selected based on their qualitative nitrogen fixing ability by growing it on a semi-solid Nitrogen Bromtimole (NFB) medium (0.5% agar content). In a qualitative test, of the 42 endophytic bacteria positively capable to fix nitrogen, twenty three isolates or 54.8% showed the change of color medium from blue to yellow (**Figure 1**). The formation of pellicle as well as discoloration of the medium from a bluish green color to yellow was growth characteristics and nitrogenase activity of non-symbiotic nitrogen-blocking (N) bacteria [13]. Bacteria capable of living on nitrogen-free media will secrete organic acids causing the discoloration of the media [9]. In addition, the semi-solid media allow non-symbiotic nitrogen-fixer bacteria to thrive at low partial of O_2 pressures that favorable this bacterium at the time of nitrogen fixation [14].

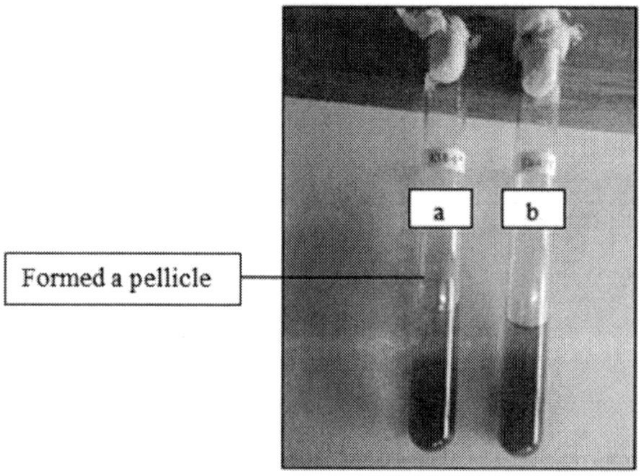

Figure 1. The result of NFB medium test by diazotroph endophytic bacteria, a) the medium becomes yellow due to nitrogen fixation activity, b) the medium unchanged

3.2. Quantitative Test Result of Diazotroph Endophytic Bacteria Isolate

In this research, the result of linear regression for NH_4^+ standard curve was 0.989. From the standard curve, the value of ammonia concentration in each sample can be calculated. The magnitude of the nitrogen fixation rate in the sample within 24 hours based on the calculation obtained the following results (**Table 1**). Based on these data, it is known that the highest level of nitrogen fixation was measured in KSD2 isolate. This is also in accordance with the results of qualitative tests showing that KSD2 isolates have the ability to fix nitrogen.

In this test, there were several isolates capable to fixing nitrogen well on NfB medium but showed low result on test with Nessler reagent or vice versa. It is assumed that the C source used in these two tests is different. In NfB medium used C source of mannitol while in Nessler test used C source of glucose. Both qualitative and quantitative results showed that KSD2 isolate has the highest N fixation level compared to other isolates. This is presumably because the isolates of KSD2 are able to use carbon sources of mannitol and glucose well for optimization of nitrogenase activity. The substrate type affects the maximum activity of nitrogenase enzyme from Nostoc linckia bacteria [15].

Table 1. Nitrogen fixation level of pure culture endophytic bacteria isolated from palm oil

Colony Code	Average of Ammoniac Uptake (ppm)	Colony Code	Average of Ammoniac Uptake (ppm)
KSAK1	49.7420*	KST11	33.5768
KSAK2	20.5807	KST12	49.0378*
KSAK3	33.4168	KST13	28.5832
KSAK4	14.4347	KST14	24.1018
KSAK5	39.4027	KST15	23.3976
KSAK6	34.9853	KST17	45.8688*
KSAK7	37.9942	KST18	30.7279
KSAK8	23.8777	KSBga	36.2337
KSAK9	36.0416	KSBgB	43.9481*
KSAK10	50.3502*	KSB1	13.4744
KSAK11	16.0032	KSB4	24.0058
KST1	35.4974	KSB14	32.4885
KST2	23.8137	KSD2	55.8560*
KST3	45.4526*	KSD3	38.4104
KST4	39.4027	KSD6	0.9264
KST5	18.8201	KSBK1	39.3707
KST6	23.7497	KSBK3	32.3604
KST7	40.1389*	KSBK4	18.5320
KST8	29.5115	KSAn1	21.4449
KST9	33.8329	KSAn3	39.9469
KST10	28.9673	BKA17	33.1287

Note : the above value is ppm of ammoniac uptake compared with negative control (n = 3)
(*) Isolates with ammoniac uptake more than 40 ppm

3.3. In Planta Diazotroph Endophytic Bacteria Test Result

The results of *In Planta* test on 42 isolates of endophytic bacteria in N fixation showed that KSD2 isolate bacteria had the highest nitrogen fixing ability compared to other endophytic bacteria isolates of 46.6562 ppm (**Table 2**). This result is likewise consistent with qualitative and quantitative results of pure culture which also shows that KSD2 isolate has the highest nitrogen fixation potential. Based on the results of nitrogen fixation test qualitatively, quantitatively, and *In Planta*, the KSD2 isolates were further selected as inoculants to be tested for their ability in fixing nitrogen on palm oil leaf saplings.

There are endophytic bacteria in rice and corn plants that are capable for N_2 fixation in the air. A total of 142 isolates of endophytic bacteria were isolated consisting of 95 isolates of endophytic bacteria from rice and 47 isolates from maize from West Java, East Java, and South Kalimantan [16]. A total of 20 isolates of endophytic bacteria were isolated from palm root.Three of these bacteria have the ability to fix nitrogen namely NFB 2, NFB 3 and NFB 4 with ARA 2.10, 2.78, and 3.13 ppm/hour activities respectively. The isolates are also known to increase plant height, number of leaves, and number of root of palm oil seedling [17]. The combination of diazotroph endophytic bacteria isolate inoculation with no nitrogen fertilization gave the best vegetative growth yields for palm oil seedlings, ie, the diameter of congestion, seedling height, and dry weight of palm oil seedlings [18].

Table 2. Levels of Nitrogen Fixation of Endophytic Bacteria in Palm Oil Leaf

Colony Code	Average of Ammoniac Uptake (ppm)	Coloni code	Average of Ammoniac Uptake (ppm)
KSAK1	41.5890 de	KST11	40.7599 cd
KSAK2	41.4033 de	KST12	46.6626 de
KSAK3	46.5122 de	KST13	39.4251 a
KSAK4	44.0218 de	KST14	45.8688 de

KSAK5	43.2791 de	KST15	41.1536 de
KSAK6	42.5429 de	KST17	44.6332 de
KSAK7	46.4257 de	KST18	42.1044 de
KSAK8	42.6453 de	KSBga	46.4962 de
KSAK9	45.0877 de	KSBgB	40.4654 bc
KSAK10	42.6805 de	KSB1	41.5346 de
KSAK11	41.8195 de	KSB4	40.3854 bc
KST1	44.0794 de	KSB14	39.5083 a
KST2	41.3361 de	KSD2	46.6562 e
KST3	42.6229 de	KSD3	46.1280 de
KST4	42.6261 de	KSD6	45.8720 de
KST5	41.0832 de	KSBK1	40.1101 ab
KST6	41.9763 de	KSBK3	44.0570 de
KST7	41.0416 de	KSBK4	43.9738 de
KST8	42.9174 de	KSAn1	42.8438 de
KST9	40.8335 de	KSAn3	40.7119 cd
KST10	40.8784 de	BKA17	40.7439 cd

The above value is ppm of ammoniac uptake compared with negative control (n = 2). The numbers followed by the same letter show no significant difference based on the Duncan test at a 95% confidence interval.

3.4. *Morphological Identification of Diazotroph Endophytic Bacteria Isolate*

Morphological identification was done by visualizing colony character and cell shape observation using light microscope with 10 x 20 magnifications, after gram staining. There is also a KOH test for confirmation. The results showed that KSD2 colony visualized morphology of white milk, irregular colonies, edges of filamentous colony, smooth surface, with a colony diameter of about 2.5 mm. Gram staining results showed that the isolates were a type of gram (+) bacteria, rod-shaped, and not slimy when tested for KOH (**Figure 2**).

3.5. *Molecular Identification and Sequencing of 16S gene rRNA*

The pairwise alignment and multiple sequences were performed on the KSD2 sequence and some of the subject sequences obtained from BLASTn using the Mega 6 program with *E. coli* as an outgroup. The result of phylogenetic tree construction with neighbour-joining test (**Figure 3**) shows that KSD2 isolate belongs to *Bacillus cereus* group. So, it can be assumed that KSD2 isolate closely related with *B. cereus* species. *B. cereus* is a rod-shaped bacteria found in the ground. These bacteria can adapt to extreme environments and resistant to hot temperatures, cold temperatures, and radiation as they can form endospore [19]. These bacteria are both aerobic and anaerobic and can metabolize carbohydrates, proteins, peptides, and amino acids during the production of L-lactate, acetate, formate, succinate, ethanol, and carbondioxide [20]. This bacterium is also a nitrogen fixer on the soil. Nitrogenase enzyme activity was detected in *B. megaterium, B. cereus, B. pumilus, B. circulans, B. licheniformis, B. subtilis, B. brevis*, and *B. firmus* [21]. Biological nitrogen activity was detected in *B. mariflavi* and *Paenibacillus massiliensis* while also identifying nifH gene fragments in *B. megaterium* and *B. cereus* [22]. Some strains of *B. subtilis* and *B. cereus* have been isolated and selected from China based on their ability to stimulate plant growth and disease control caused by fungi affecting wheat roots [23]. The widely use of *B. cereus* A47 strain in the country is able to stimulate wheat production about 11%. Some species of *Bacillus* are also able to invade tissue in various plant species where these bacteria have an important role in plant protection and growth spurts. This ability belongs to a species commonly recognized as a living organism free from soil, including *B. cereus* [24].

3.6. *Colonization of Endophytic Bacteria in Leaves of Palm Oil*

The association ability of KSD2 isolate with palm seedlings is characterized by the presence of this isolate within the plant tissue. The PCR results (**Figure 4**) showed that the inoculation through the leaf at 4th week after inoculation (Lane 2) contained a band at 575 bp corresponding to the reference (Lane 9). The reference is a gDNA isolated from a pure culture of KSD2 isolate.

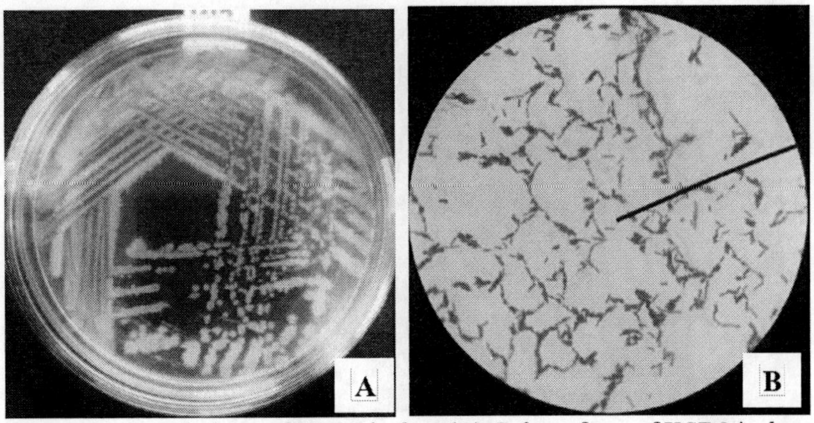

Figure 2. Morphology of KSD2 isolate (A) Colony form of KSD2 isolate, (B) Gram Staining Cell Form

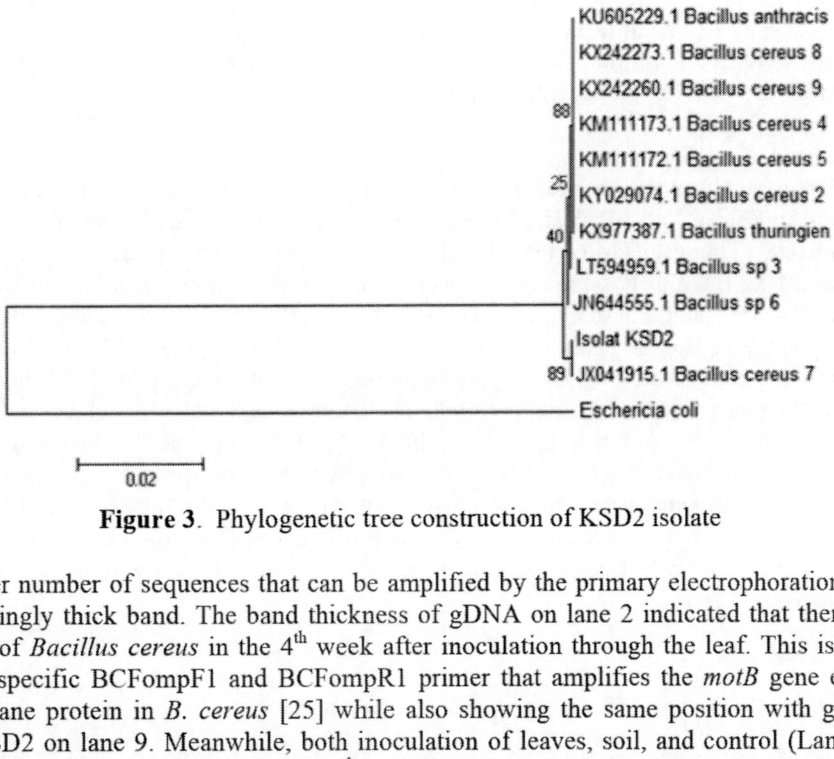

Figure 3. Phylogenetic tree construction of KSD2 isolate

The higher number of sequences that can be amplified by the primary electrophoration results will show increasingly thick band. The band thickness of gDNA on lane 2 indicated that there was much colonization of *Bacillus cereus* in the 4th week after inoculation through the leaf. This is because the PCR uses a specific BCFompF1 and BCFompR1 primer that amplifies the *motB* gene encoding the outer membrane protein in *B. cereus* [25] while also showing the same position with gDNA's pure isolate of KSD2 on lane 9. Meanwhile, both inoculation of leaves, soil, and control (Lane 5, Lane 6, and Lane 7) did not show any bands on the 2th week after treatment. This is due to the gDNA that is not well insulated. Manzano et.al (2003) usedspecific primers in the PCR process to detect *B. cereus* in coffee concentrate samples. The obtained data indicated the specificity and primary sensitivity that can be used to check for the presence of *B. cereus* in different food products, and to avoid the need for labor and time consumption [26]. PCR method was similar to the one used to detect the presence of *B. cereus* in raw milk samples [27]. First, endophytic bacteria usually penetrate through secondary roots

IOP Conf. Series: Earth and Environmental Science **166** (2018) 012012 doi:10.1088/1755-1315/166/1/012012

by secreting cellulase or pectinase enzymes [28], or upper parts of plants such as stems, flowers, sprouts radicals, stomata or cotyledon and leaves torns [29].

575 bp

Information
Lane 1 : Ladder 100 bp
Lane 2 : Leaves week 4th
Lane 3 : Soil week 4th
Lane 4 : Control week 4th
Lane 5 : Leaves week 2nd
Lane 6 : Soil week 2nd
Lane 7 : Control week 2nd
Lane 8 : Before inoculation
Lane 9 : Positive control
Lane 10 : Negative control

Figure 4. Results of PCR of palm leaves gDNA to gDNA of KSD2 isolate using *Bacillus cereus* specific primer

The bacteria colonize at the point where it enters or spreads throughout the plant living in cells, intercellular space or in vascular systems [30]. Compliance of existing compounds on palm leaf sapling is believed to be one of the keys to successful entry of isolate KSD2 into the palm oil network. KSD2 isolates were isolated from palm leaf tissue so that the compounds present in the palm oil leaves were in accordance with their nutritional sources.

3.7. Leaf Nitrogen Fixation Ability

The univariate of analysis variance showed that inoculation of isolate KSD2 (*B. cereus*) on palm oil seedlings was significant ($p> 95\%$) at 4th week after inoculation compared to control and inoculation via soil (**Figure 5**). The results of Hino and Wilson showed that bacteria from the *Bacillus* genus were able to fix nitrogen. Bacillus has a nitrate reductase enzyme; this enzyme acts as an electron acceptor [31]. Treatments at 0 and 2nd week were not significant either in control, inoculation via soil, or inoculation via leaves. The insignificant inactivation of *B. cereus* in leaves at 2th weeks was suspected because bacteria were still in the process of adaptation to the environment and still in the process of penetration into plant cells or in other words, the bacteria had not entered as endophytes. This condition causes the working of nitrogenase enzyme in *B. cereus* to be ineffective because of exposure to free oxygen. The presence of contacts between endophytic bacteria isolates and free oxygen enables the destruction of nitrogenase-enzyme complexes that will inhibit the nitrogen-blocking process by endophytic bacteria [32]. The nitrogenase enzyme will become inactive and experience an irreversible reaction when in contact with free oxygen, because free oxygen will react with the metal group in the enzyme nitrogenase [33].

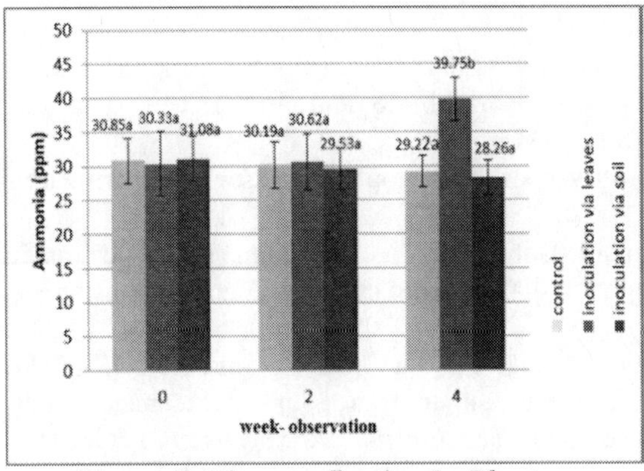

Figure 5. Nitrogen fixation I*n Planta*

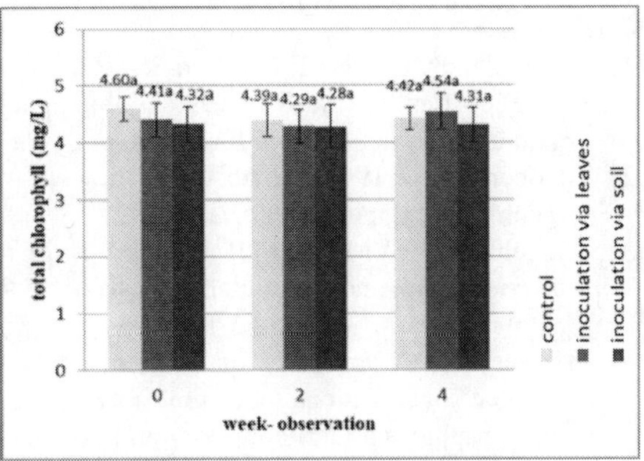

Figure 6. Total chlorophyll *In Planta*

The treatment of *B. cereus* inoculation via soil at week 0 (before inoculation), 2th week, and 4th week was not significantly suspected due to competition with other soil microbes. In addition, the presence of certain organic acids in the soil, either from the metabolite results of other microorganisms in the soil and from the root palm exudate itself, is also suspected to inhibit the work of nitrogenase enzymes that exist in *B. cereus* if absorbed into the cell. The presence of glutamine, glumat acid, asparagine, alanine, and tironin can decrease the activity of nitrogenase 50-98% [34]. Temperature and soil moisture greatly affect the ability of bacteria in mineralizing organic N to ammonia. The maximum mineralization process of organic N by bacteria occurs at a temperature of 30-35°C and its activity is not maximal if dry soil conditions and oxygen quantities are very limited [35].

3.8. Chlorophyll Content

The result of univariate of analysis variance showed that inoculation of KSD2 isolate (*B. cereus*) both via soil and through leaves on palm oil seedlings at weeks 0, 2, and 4 was not significant compared to control ($p < 95\%$) (**Figure 6**). This is assumed because the key of chlorophyll formation is not only determined by nitrogen but also by other elements, whereas in this research there is no fertilization at all so it is assumed that other chlorophyll-forming elements are not fulfilled. Chlorophyll is commonly synthesized in leaves to capture different amounts of sunlight in each species depending on environmental and genetic factors. Factors affecting chlorophyll synthesis include: light, sugar or carbohydrates, water, temperature, genetic factors and elements of nitrogen, magnesium, iron, manganese, Cu, Zn, sulfur, and oxygen.

Based on Susilowati (2003) study, JCBd 2.1 and JLkCN 2.3 isolates significantly influence plant growth height compared to control. However, the inoculation treatment did not differ significantly in stem diameter, leaf number, wet canopy weight, crown dry weight, root dry weight, chlorophyll content, and N content [15]. In soybeans, nitrogen fixation is highly dependent on the number of photosynthesis products and the maximum activity of nitrogenase which is often associated with high photosynthetic activity. Previous reports have shown that the large amount of chlorophyll is one indicator of photosynthesis. Chlorophyll is the key biochemical component responsible for photosynthesis. Chlorophyll biosynthesis is performed by certain genes within the chromosome. These genes play a role in coding enzymes involved in the biosynthetic pathway tetrapirol (porpirin nucleus) as the structural center of chlorophyll [36-37]. The main factor of chlorophyll forming is nitrogen (N).

4. Conclusions

Endophytic bacterial isolates with the best nitrogen fixation ability in palm oil leaf were *B. cereus* of 46,6562 ppm. PCR results showed that *B. cereus* application on palm oil leaves was able to colonize by visualizing 575 bp DNA bands on PCR results. *B. cereus* application in leaves can increase 39.03% of leaf ammonia levels at 4th week after inoculation.

5. References

[1] Cleveland CC, Alan RT, David SS, Hank F, Robert WH, Lars OH, Steven SP, Erika FL, Joseph CVF, Adrien E, Matt F W 1999 Global patterns of terrestrial biological nitrogen (N_2) fixation in natural ecosystems *Glob. Biogeochem Cycle* **13** 623–645

[2] Dobereiner J 1992b History and new perspectives of diazotrophs in association with non-leguminous plants *Symbiosis* **13** 1–13

[3] Kirchhof G, VM Reis, Jl Baldani, B Eckert, J Dobereiner A Hartmann 1997 Occurence, physiological, and molecular analysis of endophytic diazotrophic bacteria in gramineous energy plants *Plant Soil* **194** 45-55

[4] James EK, Olivares FL, de Oliveira ALM, dos Reis Jr FB, da Silva LG and Reis VM 2001 Further observations on the interaction between sugar cane and Gluconacetobacter diazotrophicus under laboratory and greenhouse conditions *Journal of Experimental Botany* **52** 747-760

[5] Sembiring L, Wiwik EW, Sri W, Nuryani HU 2007 Pemanfaatan bakteri diazotrof Endofit Sebagai Agen Pupuk Hayati Pada Tanaman Tebu *Kerjasama Kemitraan Penelitian Pertanian dengan Perguruan Tinggi (KKP3T)* Peneliti Badan Litbang Pertanian

[6] Kessler P, H Matzke and S Keller 2003 The effect of application time and soil factors on the occurrence of *Beauveria brongniartii* applied as a biological control agent in soil *Journal of Invertebrate Pathology* **84** 15-23

[7] Tefera T and S Vidal 2009 Effect of Inoculation Method And Plant Growth Medium On Endophytic Colonization Of Sorghum By The Entomopathogenic Fungus *Beauveria bassiana Biocontrol* **54** 663-669

[8] Parsa S, V Ortiz and FE Vega 2013 Establishing fungal entomopathogens as endophytes: towards endophytic biological control *Journal of visualized experiments*

[9] Mardiah Irma 2014 The Control Of Pathogen Fungi *Ganoderma Boninense* By Chitinase Enzimatic Activity of Oil Palm Endophytic Bacteria Master's Program Thesis, Institut Teknologi Bandung.

[10] Yu SS Latt, ZK Kyew EP and Lynn TM 2011 Accumulation of ammonia in culture broth by wild-type nitrogen-fixing bacterium, *Stenotrophomonas maltophilia IJABPT* **2** 72-77

[11] Houda N, Hanene C, Ines M, Myriam BS, Imen D and Abdennaceur H 2014 Isolation and characterization of microbial communities from a constructed wetlands system: A case study in Tunisia *African Journal of Microbiology Research* **8** 529-538

[12] Nurlita H and Utomo S 2011 Potensi nitrifikasi oleh bakteri yang terdapat di laut aliran kali plumbon, laut aliran kali banjir kanal barat dan laut aliran kali banjir kanal timur *Jurnal Presipitasi* **8** 1-7

[13] Widayati WE 1998 Aktivitas Nitrogenase dan Produksi Fitohormon dari Bakteri Penambat N2 Udara Hasil Isolasi 6dari Rizosfer dan Nira Tebu *Bulletin P3GI* **148** 32-42

[14] Harran S and N Ansori 1992 Bioteknologi Pertanian Pusat Antar Universitas Bioteknologi Institut Pertanian Bogor Bogor

[15] Mirsha DK, Kumar HD and Jha M 1985 Effects of Some Carbon Sources on Growth and Nitrogen Fixation in the *Cyanobacterium Nostoc* linckia *Biologia Plantarum* (Praha) **27** 1-5

[16] Susilowati DN, Saraswati R, Elsanti and Yuniarti E 2003 Isolasi dan Seleksi Mikroba Diazotrof Endofitik dan Penghasil Zat Pemacu Tumbuh pada Tanaman Padi dan Jagung *Balai penelitian Bioteknologi dan Sumberdaya Genetik Pertanian* 128-143

[17] Firrani, Misfalla 2011 Isolasi dan Uji Kemampuan Bakteri Endofit Diazotrof yang Memfiksasi Nitrogen Bebas pada Akar Kelapa Sawit (*Elaeis Guineensis* Jaqc.) Universitas Sumatera Utara Skripsi

[18] Panjaitan A, Iswandi, A Widyastuti, R Widayati WE 2015 Kemampuan Bakteri Diazotrof Endofit Untuk Meningkatkan Pertumbuhan Vegetatif Bibit Kelapa Sawit (*Elaeis guineensis jacq*) *J. Tanah Lingk* **17** 1-7

[19] Mignot T, B Denis, E Couture-Tosi, AB Kolsto, MMock and A Fouet 2001 Distribution of S-layers on the surface of *Bacillus cereus* strains: phylogenetic origin and ecological pressure *Environ Microbiol* **3** 493-501
[20] Mols M, De Been M, Zwietering MH, Moezelaar R and Abee T 2007 Metabolic capacity of *Bacillus cereus* strains ATCC 14579 and ATCC 10987 interlinked with comparative genomics *Environ Microbiol* **9** 2933–2944
[21] Xie GH, Su, BL and Cui ZJ 1998 Isolation and identification of N_2-fixing strains of Bacillus in rice rhizosphere of the Yangtze River valley *Acta Microbiol Sin* **38** 480–483
[22] Ding Y, Wang J, Liu Y and Chen S 2005 Isolation and Identification of nitrogen fixing bacilli from plant rhizospheres in Beijing region *J Appl Microbiol* 99 1271-1281
[23] Ryder MH, Yan Z, Terrace TE, Rovira AD, Tang WH and Correll RL 1999 Use of strains of Bacillus isolated in China to suppress take-all and *Rhizoctonia* root rot, and promote seedling growth of glasshouse-grown wheat in Australian soils *Soil Biology and Biochemistry* **31**19–29
[24] Pleban S, Chernin L, Chet I 1997 Chitinolytic activity of an endophytic strain of *Bacillus cereus* Lett *Appl Microbiol* **25** 284-288
[25] Adley C, Arshak K, Molnar C, Oliwa K, Velusamy V 2009 Design of Spesific DNA Primers to detect the *Bacillus cereus* Group Species IEEE Sensors Applications Symposium New Orleans LA USA
[26] Manzano M, Giusto C, Lacumin L, Cantoni C, Comi G 2002 A molecular method to detect *Bacillus cereus* from a coffee concentrate sample used in industrial preparations *Journal of Applied Microbiology* **95** 1361–1366
[27] Nemeckova Irena, Solichova K, Roubal P, Uhrova B, Svirakova E 2011 Methods for Detection of *Bacillus sp., B. cereus, and B. licheniformis* in Raw Milk Czech *J. Food Sci* Vol **29** Special Issue S55–S60
[28] Agarwal S, Shende ST 1987 Tetrazolium reducing microorganisms inside the root of *Brassica* species Curr *Sci* **56** 187-188
[29] Kobayashi DY and JD Palumbo 2000 Bacterial endophytes and their effects on plants and uses in agriculture 199-233 *In* CW Bacon and JF White (ed.) Microbial endophytes Marcel Dekker, Inc. New York: NY
[30] Hallman J, Quadt-Hallman A, Mahaffee WF and Kloepper JW 1997 Bacterial endophytes in agricultural crops *Canadian Journal of Microbiology* **43** 895-914
[31] Hino S and Wilson PW 1957 Nitrogen fixation by a facultative *Bacillus* Department of Bacteriology University of Wisconsin Madison: Wisconsin.
[32] Dixon R and Kahn D 2004 Genetic Regulation of Biological Nitrogen Fixation *Nature Reviews* **2** 621-631
[33] Ludden PW 2001 Nitrogenase Complex *Encyclopedia of Life Science* 1-8
[34] Reis VM and Dobereiner J *1998* Effect of high sugar concentration on nitrogenase activity of *Acetobacter diazotrophicus* Arch *Microbiol* **171** 13–18
[35] Deenik J 2006 Nitrogen Mineralization Potential in Important Agricultural Soils of Hawaii *Soil and Crop Management* **16** 1-5
[36] Wang WY, Wang WL, Boynton JE and Gillham NW 1974 Genetic Control Of Chlorophyll Biosynthesis in *Chlamydomonas The Journal Of Cell Biology* **63** 806-823
[37] Suzuki JY, Bollivar DW, Bauer CE 1997 Genetic analysis of chlorophyll biosynthesis *Annu Rev Genet* 31 61–89

Antibacterial Activity of Bacterial Endophytes from Kupa Plant *(Syzygium Polycephalum* Miq. (Merr & Perry) Against Pathogenic Bacteria

I Indrawati[1], N Rossiana[1] and T. R Hidayat[1]

[1] Biology Department, Faculty of Natural Sciences and Mathematics
Padjadjaran University
Jl. Raya Bandung –Sumedang Jatinangor Km 21, Sumedang 45363, Indonesia

E-mail: ida.indrawati81@gmail.com

Abstract. The World Health Organization confirms that infectious diseases are normally treated using synthetic antibiotics. The bacterial endophytes from the tropical plant *Syzygium polycephalum*, locally known as Kupa, can be used as an alternative solution to reduce the utilization of synthetic antibiotics. The purpose of this study was to obtain one of the most potent bacterial endophytes to be used as antibacterial agents. In the present study, descriptive analysis was conducted consisting of bacterial isolation by pour plate method, bacterial identification with Vitek 2 compact BIOMERIEUX and antibacterial test by Kirby-bauer method. A total of 9 isolates of bacterial endophytes have been successfully obtained. From these isolates, a total of 4 species of endophytic bacteria were identified: *Bacillus* sp. (1), *Bacillus* sp. (2), *Bacillus pumilus* and *Bacillus amyloliquefaciens*. Our antibacterial tests revealed that *Bacillus* sp. (2) derived from the leaves appeared to be the most potent antibacterial isolates against pathogenic bacteria with 22 and 9 mm of inhibitory zone to Methicillin-resistant *Staphylococcus aureus* (MRSA) and to *Bacilluscereus*, respectively. On the other hand, endophytic isolate *Bacillus* sp (1) derived from stem was able to inhibit *Klebsiellapneumoniae* and *B. cereus* with inhibitory zones as much as 10 and 7 mm, respectively. This result strongly indicated that the antibacterial effect of bacterial endophytes from this study was species-specific and indeed the bacterial endophytes in this study could serve as a potential source of novel natural antibiotics.

1. Introduction

To date, the World Health Organization estimates that the infectious diseases are the world's leading cause of death. Indeed, infectious diseases have caused death of approximately 13 million people worldwide each year, especially in developing countries as Indonesia [1].

Recently,*Syzygium polycephalum Miq. (Merr & Perry),* or locally recognized *as* Kupa, is an indigenous Indonesian plant. This plant has been considered as one of the medicinal plants due to its biological and chemical properties beneficial to cure several diseases. One of the bioactive compounds that was identified in this plant is flavonoid. This bioactive compound can be applied as antibacterial, wound infection inhibitor, anti-fungal, anti-viral, anti-cancer and anti-proliferative agents. In addition, flavonoid could also be utilized as anti-allergic, cytotoxic and anti-hypertensive [2] Apart from plant extracts, the bioactive compounds can be obtained from the bacterial endophytes.

Over the last decades, the bacterial endophytes are known as one of bioactive compounds producers, such as the secondary metabolite compounds with various biological activities [3]. These microorganisms can live in either the vascular or in the intercellular space, roots, stems, leaves and fruit [4]. The bacterial endophytes have several advantages, such as their shorter life cycle than their host plants, which can save production time. Thus, the antibacterial compounds produced can be obtained on a large scale with less space. Another advantage of using bacterial endophytes is to preserve the sustainability of medicinal plants, especially rare or endangered species, hence not to be over-exploited.

Among the pathogenic bacteria, in the present study we mainly focus on four different bacterial species due to their importance pathogenicity and resistance to the synthetic antibiotics. *Pseudomonas aeruginosa* is bacteria that cause urinary tract infections, meningitis, diarrhea, entrokolitis necrosis and pneumonia [5] whereas *Klebsiella pneumonia* causes pneumonia, urinary tract infections and sepsis in patients with vulnerable immune system [6]. *Bacillus cereus* produces diarrhea-causing enterotoxins. On the other hand, *Methicillin-resistant Staphylococcus aureus* (*MRSA*) has a mutated gene that is resistant to almost all beta-lactam antibiotics. Meanwhile, *S. aureus* can cause septicaemia, pneumonia, endocarditis, osteomyelitis, gastroenteritis and abscesses [7]. The present study aims to obtain the bacterial endophytes isolates from *S. polycephalum*, where the most potent isolate could be used as antibacterial agent against bacterial pathogens such as *P. aeruginosa, K. pneumoniae, MRSA* and *B. cereus*. This study could be as an alternative to over-used of synthetic antibiotics thus it can be medically accountable.

2. Materials and Methods

2.1. Isolation of bacterial endophytes

The isolation of bacterial endophytes was conducted by adapting the protocol from Krishnan *et al.*, (2012). Briefly, the stems, fruits and leaves of Kupa plants were washed with clean running water. Afterwards, samples were cut into pieces with the size approximately 2-3 cm. These samples were then washed with sterile distilled water and soaked on 95% ethanol for 30 s, and then rinsed with sterile distilled water and dried. After, the samples were smoothed and isolated (1 g) for serial dilution until 10^{-7}. Furthermore, the last three dilutions were introduced into the petri dish containing the nutrient agar medium (NA) and incubated at $37^{\circ}C$ for 24 h [8].

2.2. Identification of bacterial endophytes

2.2.1. Macroscopic observation

On macroscopic observation, bacterial endophytes colonies were identified according to their shape, colour, elevation, edge and size.

2.2.2. Microscopic observation

The microscopic observation was performed by gram staining method. The positive gram bacteria were indicated by their purple colour. In contrast, the negative gram bacteria were indicated by their red colour.

2.2.3. Biochemical observation

The biochemical observation was conducted using Vitek 2-compact BIOMERIEUX. Briefly, bacterial endophytes isolates were inoculated on Trypticase Soy Agar (TSA), Blood Agar (BA) and Mac Concey Agar (MCA) medium. Furthermore, the positive gram bacteria were analysed using Vitex BCL cards. On the other hand, the negative gram bacteria were analysed with Vitex GN cards.

2.3. Identification of antibacterial activity

A total of one loop wire of bacterial endophytes isolates with 24 h lifetime were inoculated into 3 ml of NaCl physiology sterile to meet 0.5 Mc Farland turbidity with 2 times of dilution (1.5×10^{6} CFU ml^{-1}). Afterwards, a total of 0.1 ml of suspension was inserted into a sterile vial containing paper discs, then incubated at $37^{\circ}C$ for 24 h. Test on antibacterial activity was done by preparing 15 ml TSA medium in a sterile petri dish. Furthermore, a total of 0.1 ml of suspension of the tested bacterial strain with 24 h of lifetime with 0.5 Mc Farland turbidity (1.5×10^{8} CFU ml^{-1}) were spread on the agar medium. Afterwards, the disc paper was placed on the agar surface and incubated at $37^{\circ}C$ for 24 h prior to the measurement of bacterial inhibition zone. The measured parameter was the inhibition zone (mm) formed around the disc paper. The results of the inhibition zone measurement were then categorized according to Pan *et al.* (2009) [8] as follows:

1. Inhibition zone diameter > 20 mm: very strongly sensitive,
2. Inhibition zone diameter 10-20 mm: strongly sensitive,
3. Inhibition zone diameter 5-10 mm: semi-sensitive,
4. Inhibition zone diameter < 5mm: resistant.

The percentage of inhibition (I%) was calculated according to the modified formula from Amadioha (2004) [9] as follows:

$$I\% = \frac{dc-dt}{dc} \times \frac{100}{1}$$

Where dc represents the diameter of inhibition zone formed on control. Whereas dt represents the diameter of inhibition zone produced by the bacterial endophytes isolates from Kupa plant.

3. Results and Discussion

3.1. Isolation of bacterial endophytes

In the present study, the macroscopic observation of the bacterial endophytes isolates according to their colour, size, shape, edge and elevation is presented in **Table 4.1**.

Table 4.1 Macroscopic characteristics of bacterial endophytes from Kupa plants

Kupa body parts	Isolates	Morphology				
		Colour	Size	Shape	Edge	Elevation
Stems	B 5-2	Translucent white	Moderate	Amorphe	Diverged	Flat
	B 6-3	Milky white	Small	Spheric	Flat	Convex
	B 7-4	Translucent white	Large	Amorphe	Smoothly jagged	Flat
Leaves	D 5-1	Milky white	Moderate	Amorphe	Diverged	Flat
	D 5-2	Translucent white	Moderate	Rounded	Smoothly jagged	Convex
	D 6-3	White	Moderate	Spheric	Flat	Convex
Fruits	Bh-5	Milky white	Moderate	Rounded	Smoothly jagged	Convex
	Bh-6	Milky white	Moderate	Amorphe	Wavy	Convex
	Bh-8	Yellowish white	Moderate	Amorphe	Wavy	Flat

The microscopic observation of endophytic bacterial isolates by Gram staining technique at 1000 times magnification is presented in **Table 4.2**.

Table 4.2 Microscopic characteristics of bacterial endophytes isolates

Kupa body parts	Isolates	Cell shape	Gram staining
Stems	B 5-2	Short bacillus	Positive
	B 6-3	Short bacillus	Positive
	B 7-4	Short bacillus	Positive
Leaves	D 5-1	Bacillus	Positive
	D 5-2	Long bacillus	Positive
	D 6-3	Bacillus	Positive

	Bh-5	Short bacillus	Positive
Fruits	Bh-6	Short bacillus	Positive
	Bh-8	Short bacillus	Positive

3.2. Identification of bacterial endophytes

A total of 3 of bacterial endophytes isolates from Kupa stems were successfully identified. The isolate B 5-2 was allegedly identified as *Bacillus sp.* (1). On the other hand, bacterial endophyte isolate B 6-3 was fathomed to be *B. amyloliquefaciens* with 88% probability. Other isolates from the Kupa stems, the B 7-4 expected to be *B. amyloliquefaciens* with 91% of probability.

Endophytic bacterial isolates derived from Kupa leaves consist of 3 different isolates. The first isolate, D 5-1, was suspected to be *Bacillus sp.* (2). The second isolate, D 5-2, was predicted to be *B. amyloliquefaciens* with 91% probability. On the other hand, the isolate D 6-3 was predicted to be *B. amyloliquefaciens* with 91% of probability.

A total of 3 bacterial endophytes isolates from the fruits were identified. The isolate Bh-5 was predicted as *B. pumilus* with 90% probability, whereas isolate Bh-6 was suspected as *B. amyloliquefaciens* with 91% probability. Additionally, isolate Bh-8 was suspected as *B. amyloliquefaciens* with 85% probability.

3.3. Antibacterial activity of bacterial endophyte isolates

The results of antibacterial activity test can be seen in **Figure 4.1** as follows

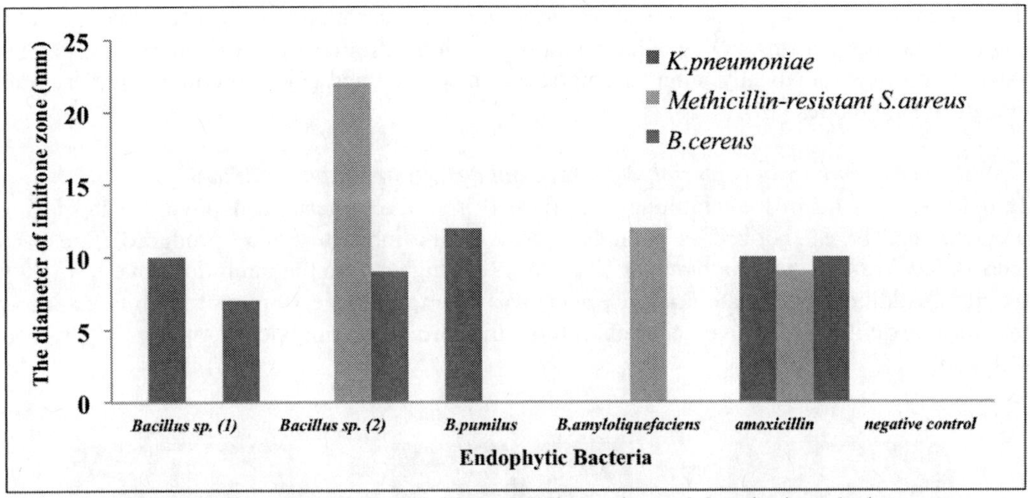

Figure 4.1 Antibacterial activities of different bacterial endophyte isolates

In this study, the antibacterial activities of bacterial endophyte isolates seemed to be species-dependent (**Figure 4.1**). The isolate D 5-1 had the highest activity against MRSA. However, this isolate appeared to be less sensitive against other pathogens. This is probably due to the different bioactive compounds as results from the bacterial endophytes secondary metabolites. In addition, the pathogenicity level of the bacterial strain tested could also influence the sensitivity of bioactive compounds from each isolates

3.4. Antibacterial activity of bacterial endophyte isolates against P.aeruginosa

According to inhibition zone formed from the bacterial endophyte isolates against *P. aeruginosa*, it appeared that all isolates did not show any inhibition zone. Thus, it can be categorized as resistant. The isolate *Bacillus sp.* (1), *Bacillus sp.* (2), *B. amyloliquefaciens* and *B. pumilus* did not show inhibition zone at all around the disc paper. In addition to its resistance to endophyte isolates, *P. aeruginosa* was also resistant to antibiotic such as amoxicillin that was used as positive control.

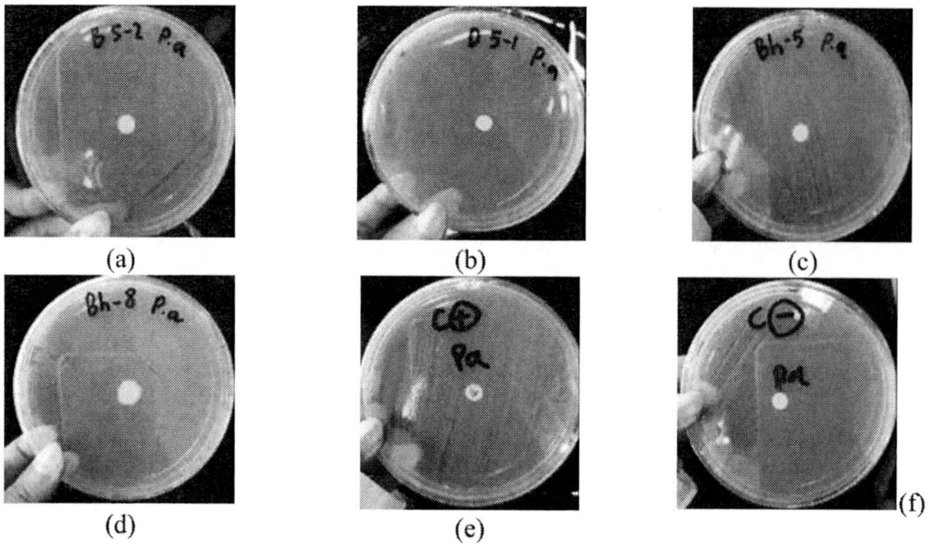

(a) (b) (c)

(d) (e) (f)

Figure 4.2 Antibacterial activities of different bacterial endophyte isolates ((a) *Bacillus sp.* (1); (b) *Bacillus sp.* (2); (c) *B. pumilus*; (d) *B. amyloliquefaciens*; (e) Amoxicillin (positive control) and (f) control negative) against *P. aeruginosa*

To our knowledge, the resistance of *P. aeruginosa* to antimicrobial agents could be achieved through several mechanisms: (1) by reducing the permeability from the outer membrane, (2) by having a system that can automatically pump antibiotic out of the cell and (3) by producing the enzyme that can inactivate antibiotics [10].

3.5. Antibacterial activity of bacterial endophyte isolates against K.pneumoniae

In our study, results of the inhibition zone formed from the bacterial endophyte isolates against *K. pneumoniae* can be categorized as strongly sensitive. The inhibitory zone produced from bacterial isolates *B. pumilus* to *K. pneumoniae*was 12 mm or 20% higher than the inhibition zone from 30 μg of amoxicillin. Additionally, isolate *Bacillus sp. (1)* was able to produce 10 mm of inhibition zone or 0% to that of amoxicillin. Moreover, the inhibition zone formed on amoxicillin against *K. pneumoniae* was 10 mm.

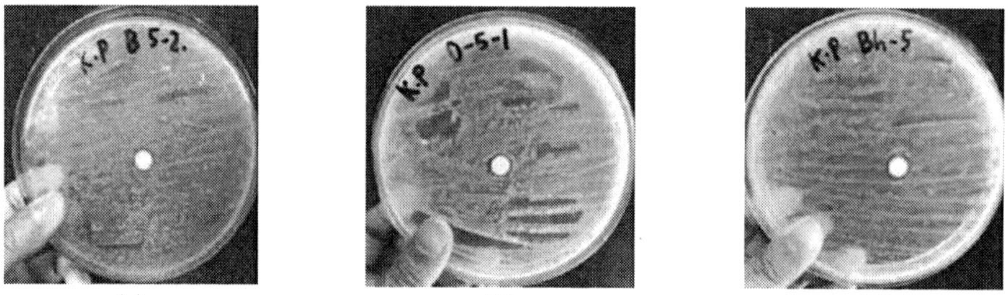

(a) (b) (c)

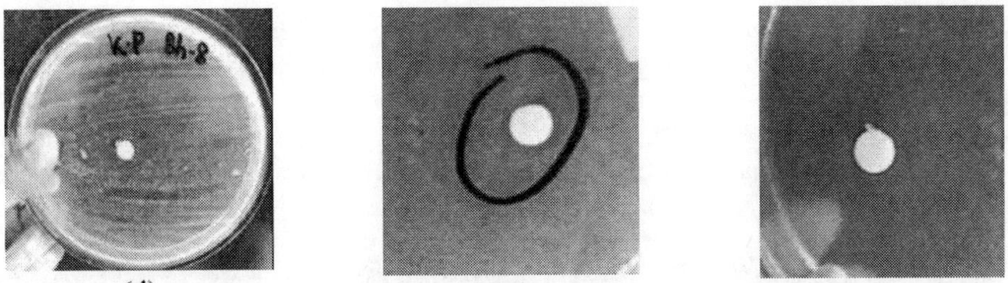

(d) (e) (f)

Figure 4.3 Antibacterial activities of different bacterial endophyte isolates ((a) *Bacillus sp.* (1); (b) *Bacillus sp.* (2); (c) *B. pumilus*; (d) *B. amyloliquefaciens*; (e)Amoxicillin (positive control) and (f) control negative) against *K. pneumoniae*

3.6. Antibacterial activity of bacterial endophyte isolates against Methicillin-resistant S.aureus

According to classification by [8], our findings showed that the sensitivity levels of bacterial endophyte isolates to MRSA could be categorized as strongly sensitive. This result suggests that there was a difference in antibacterial activity. Indeed, the largest inhibition zone diameter was obtained in *Bacillus sp.* (2) with 22 mm or 144.44% to that of amoxicillin 30 µg. Furthermore, *B. amyloliquefaciens* isolate appeared to be the second with 12 mm or 33.33% compared to inhibition zone by amoxicillin. The inhibition zone formed by amoxicillin was 9 mm.

We observed that the inhibition zone generated by *Bacillus sp.* (2) and *B. amyloliquefaciens* was greater than that of amoxicillin as positive control. Therefore, we may assume that the antibiotic produced from endophytic bacteria is more effective than the synthetic antibiotics.

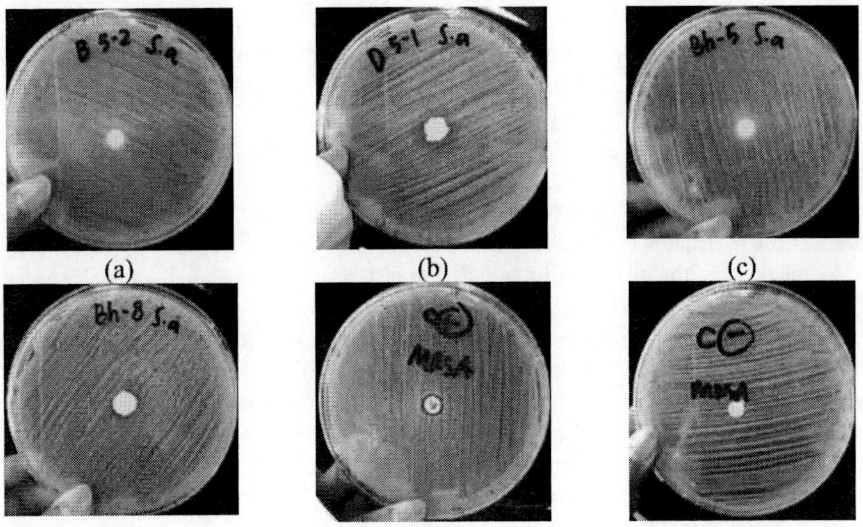

(a) (b) (c)

(d) (e) (f)

Figure 4.4 Antibacterial activities of different bacterial endophyte isolates ((a) *Bacillus sp.* (1); (b) *Bacillus sp.* (2); (c) *B. pumilus*; (d) *B. amyloliquefaciens*; (e)Amoxicillin (positive control) and (f) control negative) against MRSA

These results are in accordance with the study conducted by Happy (2007), which demonstrates no significant difference in ampicillin, erythromycin, amoxicillin and generic tetracycline against *S. aureus*. Additionally, other study by Putri (2014) revealed that either amoxicillin or erythromycin did not produce inhibition zone that formed on the periphery of disc paper against *S. aureus*. This is probably related to the genes possessed by MRSA that can create resistance to beta-lactam antibiotics, such as amoxicillin.

3.7. Antibacterial activity of bacterial Endophyte Isolates Against B. cereus

From the inhibition zone formed against *B. cereus*, the antibacterial activities from the isolated bacterial endophytes could be categorized as sensitive. The inhibition zone produced from *Bacillus sp (2)* against *B. cereus* was 9 mm or 10% to that of amoxicillin 30 µg. Moreover, the isolate *Bacillus sp (1)* produced inhibition zone as much as 7 mm or 30% to that of amoxicillin. Indeed, the formed inhibition zone by amoxicillin against *B. cereus* was 10 mm.

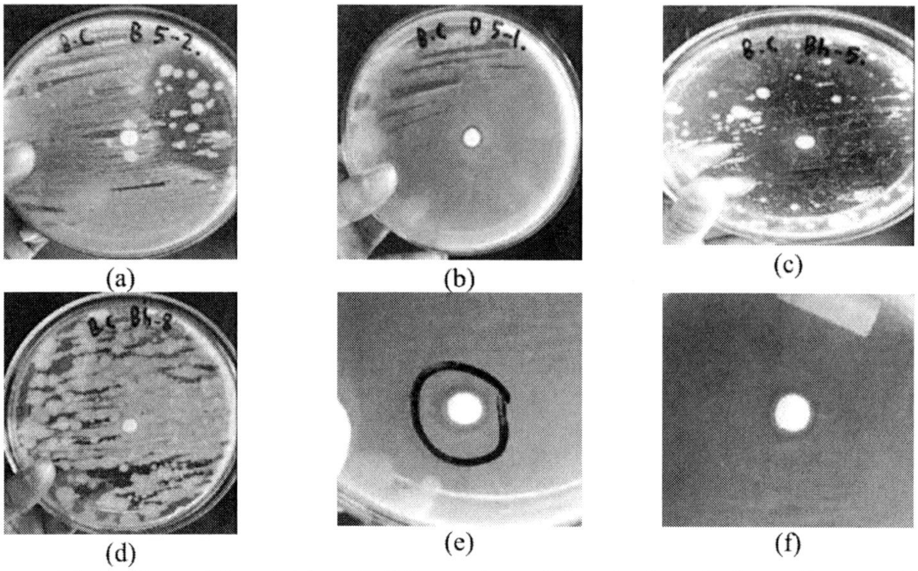

Figure 4.5 Antibacterial activities of different bacterial endophyte isolates ((a) *Bacillus sp.* (1); (b) *Bacillus sp.* (2); (c) *B. pumilus*; (d) *B. amyloliquefaciens*; (e) Amoxicillin (positive control) and (f) control negative) against *B. cereus*

4. Conclusion

In the present study, a total of 9 bacterial endophyte isolates were obtained from the Kupa plant *S. polycephalum*. These 9 isolates consist of 3 bacterial isolates from the stems, 3 isolates from the leaves and 3 other isolates from the fruits. Additionally, we demonstrated that bacterial endophyte isolates displayed inhibitory activity against the bacterial pathogens. The isolate *Bacillus sp.* (1) from the leaves produced inhibition zone as much as 10 and 7 mm against *K. pneumoniae* and *B.cereus*, respectively. Moreover, bacterial isolate *Bacillus sp.* (2) from Kupa' stems produced 22 mm of inhibition zone against MRSA but only 9 mm to that of *B. cereus*. The isolate from Kupa fruits, *B. pumilus* generated 12 mm of inhibition zone against *K. pneumoniae*. On the other hand, other Kupa' endophytic bacterial isolate *B. amyloliquefaciens* produced 12 mm of inhibition zone against MRSA. Our study suggests that the bacterial endophytes from Kupa plants may be used as potential candidate of antibacterial agents against the pathogenic bacteria.

Acknowledgement

The authors wish to thank to the Rector of Universitas Padjadjaran, Prof. Dr. med. Tri Hanggono Achmad, dr. for his financial assistance through the *Academic Leadership Grant* (Program 1-1-6).

5. References

[1] Salni, H Marisa dan RW Mukti 2011 Isolasi Senyawa Antibakteri Dari Daun Jengkol (*Pithecolobium lobatum* Benth) dan Penentuan Nilai KHM-nya *Jurnal Penelitian Sains* 14 1 14109

[2] Purnama WB 2013 Aktivitas Antibakteri Glukosa Terhadap Bakteri *Staphylococcus aureus*, *Pseudomonas aeruginosa*, *Bacillus subtilis* dan *Escherichia coli* Naskah Publikasi Fakultas Farmasi Universitas Muhammadiyah Surakarta

[3] Brooks, Geo F, Janet, S, Butel SA, Morse 2007 Mikrobiologi Kedokteran Jawetz, Melnick & Adelberg ed 23 Jakarta: EGC 251- 254
[4] Agustina, S 2015 Peningkatan Resistensi Kultur Bakteri *Staphylococcus aureus* terhadap Amoxicillin Menggunakan Metode Adaptif Gradual *Jurnal Farmasi Indonesia* **7 (3)** 191
[5] Sunarti, S 2015 Persebaran *Syzygium* endemik Jawa *Prosiding Seminar Nasional Biodiversity Indonesia* **1(5)** 1093-1098
[6] Desriani, Ukhradia, MSP, Maria B, Akhmad R dan Puspita L 2014 Isolat dan karakteristik bakteri endofit dari tumbuhan binahong dan ketepeng china *Jurnal kesehatan Andalas* **3(2)** 19-27
[7] Krishnan PBR, Kush A and Ravikumar P 2012 Isolation and functional characterization of bacterial endophytes from *Carica papaya* fruits *Journal of Applied Microbiology* ISSN 1364-5072
[8] Pan X, Chen F, Wu T, Tang H and Zhao Z 2009 The Acid, Bile Tolerance and Antimicrobial Property of *Lactobacillus acidophilus* NIT *Journal food control* **20** 598-602
[9] Amadioha AC 2004 Control of Black rot of Potato Caused by *Rhyzoctonia bataticola* using some plant extract Arch *Phytopahol. Plant Pret* **37** 111-117
[10] Moore NM, Flaws ML Antimicrobial resistance mechanism *Pseudomonas aeruginosa* Clin *Lab Sci* **24** 47-51
[11] Happy S 2007 Pola kepekaan kuman *Staphylococcus aureus* terhadap antibiotika generik dan non generik (minithesis) Universitas Andalas: Padang
[12] Rasyid PAA, Rahmatini R 2014 Perbedaan Sensitivitas Kuman *Pseudomonas aeruginosa* Penyebab Infeksi Nosokomial Terhadap Beberapa Antibiotika Generik dan Paten *Jurnal Kesehatan Andalas* **3 (3)** 329-331

IOP Conf. Series: Earth and Environmental Science **166** (2018) 012014 doi:10.1088/1755-1315/166/1/012014

Diversity of Local Food in Sasak Ethnic, Lombok Island, West Nusa Tenggara

M Rahayu[1] and L. K Ibo[1]

[1]Research Center for Biology, Indonesian Institute of Sciences, Cibinong Science Center, Cibinong-Bogor 16911, Indonesia

E-mail: mulyati_r@yahoo.com

Abstract. Several species of local plants in Lombok are formerly often used by the community as an alternative food. Community awareness of the food usually comes from life experience, knowledge from generation to generation, and local wisdom. These things need attention for the sake of sustainability. The ethnobotany study is necessary. The aim of this study was to make an inventory of non-rice local food diversity of East Lombok districtWest Nusa Tenggara. This research was executed through open interview and direct observation in the field. This study is one of the local food conservation efforts for strengthening national food security. The results of ethnobotany studies recorded 15 species of non-rice local foods contain carbohydrate. The most widely used plant part is the tubers/rhizomes. Until now, the species of non-rice food plants, remain cultivated but without handling such as rice (*Oryza sativa*) and maize (*Zea mays*).

1. Introduction

Different of plants cultivated by the community as alternative food and a source of carbohydrates in Indonesia is very diverse. To meet the caloric needs of the Indonesian population, the intensive cultivation is still limited to rice and corn. This will pose a risk to food security, because in terms of food balance, especially the balance of carbohydrates, the Indonesian diet is considered less than ideal. This is due to the high consumption of rice as the only source of calories [1]. Carbohydrates (KH) are the main food source because they are the source of energy, followed by proteins and vitamins which are complementary needs [2].

In response to these problems, the government implemented a food diversification program started in 1972, but its implementation still not produce the expected changes in diet [1] Food diversification that includes the diversity of consumption of carbohydrate, protein, fat, vitamin and mineral resources is still less than optimal due to the limited production of diverse agricultural food commodities. Based on these facts, for the non-rice foods can be considered as a new source of carbohydrates.

Lombok island is one of the areas in the province of West Nusa Tenggara, predominantly inhabited by the Sasak ethnic which is a native of this area. In this island, *Dioscorea spp* was commonly used by the community as an alternative food. The community's knowledge of plants generally comes from life experiences and knowledge from generation to generation. The local knowledge and wisdom of the community needs to get the attention of all parties. Therefore, this research needs to be done to make an inventory of the diversity for non-rice local foods producing carbohydrate (KH).

This research- is one of method to increase the role of local food plant as substitute for the main food that its existence has begun to diminish. Besides that, this research is also expected to strengthen and generate local food to strengthen national food security.

2. Methodology

Ethnobotany data of food-producing carbohydrates (KH) was carried out according to methods reported in [3, 4, 5] with some modifications. The research was done by non-structural interviews and "open ended", and direct observation in the field. The interview is aimed at local people who know and use plants as carbohydrate-producing foods. The plant species recorded its local name, parts used and their status (cultivation or wild). Unknown species of scientific name, made herbarium according to Nesbitt reference [6] then identified in Herbarium Bogorienses, Botany Division-Research Center for Biology, LIPI.

This research was conducted in Jeruk Manis and Kembang Kuning villages, East Lombok-West Nusa Tenggara regency, the area bordering Rinjani Mountain National Park. The majority of the population are Muslims and their main livelihood is as farmers.

3. Results and Discussion

Currently, the rapid advances in cultivation technology cannot be obstructed. Slowly but surely, the type of carbohydrate-producing tubers began to be abandoned and replaced with the type of carbohydrate-producing plants that have relatively easy harvest handling and post-harvest process.

Indonesia has 77 types of non-rice food sources, such as cassava (*Manihot esculenta*), maize (*Zea mays*), ganyong (*Canna indica*), gembili (*Dioscorea aculeata*), taro (*Alocasia* spp.), sago (*Metroxylon sagu*), beans, sweet potatoes (*Ipomoea batatas*), etc. [7]. Of various tubers possessed by Indonesia, cassava is the most important source of carbohydrate followed by sweet potatoes in second place [8].

The results of the study showed 15 species of local food plants used by the Sasak community in Jeruk Manis and Kembang Kuning villages, East Lombok regency (Table 1).

Table 1. Food Plants species of non-rice in East Lombok.

Scientific name	Local name	Family	Part of use	Status
Alocasia mocrarrhiza (L.)	Bira	Araceae	tuber	cultivation without maintenance
Canna indica L.	Sebet	Cannaceae	tuber	cultivation without maintenance
Dioscorea alata L.	Uwi piit, egal	Dioscoreaceae	tuber	cultivation without maintenance
*Dioscorea bulbifera*L.	Kalem	Dioscoreaceae	tuber	cultivation without maintenance
Dioscorea esculenta (Lour.) Burkill	Kembili	Dioscoreaceae	tuber	cultivation without maintenance
Dioscorea hispiida Dennst.	boyod, gadung	Dioscoreaceae	tuber	Wild
Dioscore sp.	Gadung kendit	Dioscoreaceae	tuber	Wild
Ipomoea batatas (L.) Lam.	Ambon jama	Convolvulaceae	tuber	cultivation without maintenance
Lablab purpureus (L.) Sweet	Komak	Fabaceae	Fruit	cultivation without maintenance
Manihot esculenta Crantz	Ambon jawa	Euphorbiaceae	tuber	cultivation without maintenance
Maranta arundinacea L.	Marus	Marantaceae	Tuber	cultivation without maintenance

Phaseolus lunatus L.	Komak kuning	Fabaceae	Fruit	cultivation without maintenance
Pueraria Montana var. *Lobata* (Willd.) Sanappe &Pradeep	Babaye	Fabaceae	tuber	Wild
Zea mays L.	Jagung	Poaceae	Fruit	cultivation with maintenance
Sp 1	Kenebele	Sp 1	tuber	wild

Based on interviews and observations in the field, only four species "boyod" or "gadung reket" *Dioscorea hispida*, "gadung kendit" *Dioscorea* sp., and "babaye" *Pueraria montana* var. *lobata* and "kenebele") are wild species that grow in forest areas or shrubs. Meanwhile,11 others species are planted but generally without maintenance except for *Zea mays*. Corn is planted and maintained intensively (cultivation with maintenance) (Table 1).

(a) (b) (c)

Figure 1. Photos of colllected samples, (a) *Dioscorea hispida*, (b) *Dioscorea sp* and (c) *Pueraria montana* var. *lobata*

In the world, *Dioscorea* genus as alternative food consists of 28 species [9], while in Indonesia it is known 6 species [10]. Exploration results and interviews with local communities in the study area identified five species of *Dioscorea* that are used as food source by the Sasak ethnic. Among the five species of *Dioscorea*, *Dioscorea alata* is most commonly cultivated and utilized by local communities at study sites. In Java, this species is known by the local name "uwi / uwimanis / kelapa" [10]. Period of growth ranging from planting to the adult experiences four phases [11]. Research conducted Praptiwi, et al [12] showed that the nutritional composition of sweet Uwi harvested in phase three and four had no difference. Thus, it can be harvested more quickly to shorten the time of planting and more quickly used as a source of carbohydrate.

In contrast to *Dioscorea alata*, *Dioscorea hispida* is not widely used as a foodstuff because it is known tuber only in the famine season. According to Sumunar&Estiasih [13], gadung tuber is one source of high carbohydrate foods. Carbohydrates in gadung are dominated by starch. Besides having carbohydrate content also, it contains as well cyanide which can cause poisoning and death. Some process to remove the remaining HCN content or minimize need to be done so that the gadung tuber becomes safe and feasible to consume. Mean while,the research by Nugroho and Estyaniyana [14]used dried tuber starch *Dioscorea hispida* from Timor Leste, Central Kalimantan and Yogyakarta having the highest toxicity originating from Timor Leste. Thus, this species of dry bulb in Timor Leste is used as a repellent of rice lice.

Garut (*Maranta arundinacea*) is one of the plants in which the tuber is widely used by local communities in Indonesia, especially those living in rural areas. This species comes from Tropical America, and then spreads widely to other tropical countries among others Indonesia, India, Sri Lanka and Philippines [8].This species grows in various regions of Indonesia. In the study sites, this species is often cultivatedin garden or home yards without intensive maintenance, even growing in the shrub area and the edge of the national park. Amylose content of tubers is relatively lower than cassava (*Manihot esculenta*) or sago (*Metroxylon sagu*), but high potassium percentage [15], easily digested for infant and sick foods [8]. The use of medicinal ingredients for tubes has been demonstrated. Among others, tuber porridge is used as a tapel for flatulence and tuber flour mixed with milk or water for antidote poisoning or problems with the stomach [10, 15]. Unfortunately, the Sasak community at the study site does not recognize its utilization as a traditional medicine.

Even though, the morphological variability of hyacinth bean (*Lablab purpureus*) is wide, many cultivars extinct.This species is used most often by the Sasak community as food and can also be used as forage feed, green manure, silage and ground cover. This species is believed to be native from India, Southeast Asia or Africa, and naturalized and cultivated in tropical and subtropical regions, especially in India, Southeast Asia, Egypt and Sudan [16].In the study site, the 'komak' is generally planted flowers are white or violet. The planting is carried outin the bunds of garden [10,11,18].Per 100 gram edibleportion immature pods contains82.4 g water, 4.5 g protein, 0.1 g fat, 10.0 g carbohydrates, 2.0 g fibre, and1.0 g ash. The presence of cyanogenic glycoside has been reported in certain cultivars [16].

The most rarely used species as food by Sasak community is"babaye" (*Pueraria montana* var.*lobata*); because this species lives wildly dan rare to be found.According to local communities, the harvesting process is quite difficult.

Canna indica or commonly known as ganyong is usually grown as an ornamental plant, because the colour of leaves and flowers varies. In the study sites, itisgenerally grown as green leaf and planted in garden or home yard. In Vietnam, the sticks are dried and then processed into noodles, while the young buds are made into green [17]. The Sasak community at the study site only recognizes the use of tubers as non-rice food or snacks.

The results show that the younger generation is also around 20 years old who live in this area are no longer familiar with the species of non-rice food plants mentioned above. It seemsthat modern technology, the flow of globalization and tourist visits has eroded the local wisdom, culture and traditional knowledge of Indonesia's biodiversity. This needs to be immediately anticipated by conducting an inventory and further research, so that the species that can potentially be to develop, so that diversification of food diversity can walk well.

4. Conclusion

The Sasak people in Jeruk Manis and Kembang Kuning villages are still familiar with the species of non-rice food. Fifteen species of non-rice food plants that are used in the famine season or as snacks were recorded. Eleven species have been cultivated even without intensive maintenance. This shows the local wisdom of Sasak people in handling food security for their survival.

5. References

[1] Sumunar SR and Estiasih T 2015 Uumbi gadung (Dioscorea hispida Dennst) sebagai bahan pangan mengandung senyawa bioaktif: Kajian pustaka Wildyam (Dioscorea hispida Dennst) as Bioactive Compounds Containing Food : A Review *3*(1) 108–112.

[2] Anonimous 2014 KekinianKeanekaragamanHayati Indonesia 2014 LIPI Press344pp

[3] Vogl CR, B Volg-Lukasser and RK Puri 2004 Tools and methods for data collection inethnobotanicalstudies of home garden. *Field Methods 16 (3):* 285–306.

[4] Suminguit VJ 2005 *Ethnobotanical documentation: A user's guide* Asia – PacificDatabase on Intangible Cultural Heritage (ICH) by Asia - Pacific Cultural Centre for Unesco (ACCU) Paris.
[5] Nolan JM & NJ Turner 2011 Ethnobotany: The Study of People – Plant relationships *In* Anderson EN, D Pearsall, E Hunn& NJ Turner (eds) *Ethnobotany* Wiley – Blackwell, New Jersey: 133–148.
[6] Nesbitt, M 2014 Use of Herbarium Specimens In Ethnobotany *In* Salick, J, K
[7] Konchar & M Nesbitt (eds) *Curating biocultural collection: A hamdbook* Royal Botanic Garden, Kew: 313 – 328
[8] Hadi P 2014 Kedaulatan Pangan: Kemandirian, Pemberdayaan dan Pembrantasan Kemiskinan *Makalah dalam Pokok Pikiran untuk Diskusi Panel Ikatan Alumni Akademi Pertanian Ciawi* Bogor, 26 Maret 2014.
[9] Anonimous 1977 *Ubi-ubian* Proyek Sumber Daya Ekonomi Lembaga Biologi Nasional–LIPI Bogor 113pp.
[10] Arora RK 1985 *Genetic Resources of Less Known Cultivated Food Plants*National Bureau of Plant Genetic Resources New Delhi 126pp.
[11] Heyne K 1997 *Tumbuhan Berguna Indonesia* Badan Litbang Departemen Kehutanan.
[12] Onwueme, IC & ZN Ganga 1996 *Dioscorea alata* L. *In*: Flach M & F Rumawas (eds) *Plant Yielding Non-Seed Carbohydrat Plant Resources of South-EastAsia No. 9* Bogor Indonesia.
[13] Praptiwi M Harapini & Sumarnie 2001 Komposisi Nutrisi, Produksi Bagian Vegetatif dan Umbi Uwi Manis (*Dioscorea alata*) dengan Pemupukan Organik *Prosiding Seminar Nasional III Pengembangan Wilayah Lahan Kering: 236 – 239* Bandar Lampung, 3 – 4 Oktober 2000 .
[14] Nugroho LH & A Estyaniyana 2014 Toksisitas Kandungan Fitokimia dan Karakter Pati Aksesi Gadung (*Dioscorea hispida* Dennst.) Timor Leste, Kalimantan Tengah dan Daerah Istimewa Yogyakarta *Jurnal Tumbuhan Obat Indonesia 7 (1): 55 – 66.*
[15] Villlamayor, FG & Jukema. 1994 *Maranta arundinacea* L. *In;* Flash M & F Rumawas (eds) *Plants yielding non-seed carbohydrates Plant Resources of South-East Asia No. 9* Bogor, Indonesia.
[16] Shivashankar G & RS Kulkarni 1993 *Lablab purpureus* (L.) Sweet. *In;* Flach M & F Rumawas (eds) *Kacang-kacangan Sumber Daya Nabati Asia Tenggara 1* Jakarta
[17] Ong, HC & JS Siemonsma 1996 *Maranta arundinacea* L. *In*: Flach M & F Rumawas (eds) *Plant Yielding Non-Seed Carbohydrat Plant Resources of South-East Asia No. 9* Bogor Indonesia.
[18] Burkill IH 1935*A Dictionary of Economic Product of The Malay Peninsula*Government of The Straits Settlemens and Federated Malay States. Crown Agents for The Colonies 4 Hill Bank London SW 1.

IOP Conf. Series: Earth and Environmental Science **166** (2018) 012015 doi:10.1088/1755-1315/166/1/012015

Antihyperlipidemic Activity of *Archidendron pauciflorum* Fruit Peel Extract in Streptozotocin-induced Diabetes Female Wistar Rats

Madihah[1], A. R Rahim[1], D. M Malini[1] and W Hermawan[1]

[1]Departement of Biology, Faculty of Mathematics and Natural Sciences, Universitas Padjadjaran, Jatinangor-Sumedang, 45363, Indonesia

E-mail: madihah@unpad.ac.id

Abstract. The relationship between diabetes and hyperlipidaemia is a well-recognized phenomenon. Djengkol (*Archidendron pauciflorum*) fruit peels have been used traditionally as medicinal plant for treats diabetes. The present study investigated the antihyperlipidemic activity of ethanol extract from *A. pauciflorum* fruit peel on the diabetic female Wistar rats. The diabetic condition was induced by intravenous injection of streptozotocin at a dose of 65 mg/kg BW. The ethanol extract of Djengkol at a dose of 385, 770, and 1540 mg/kg BW as well as glibenclamidee at a dose of 10 mg/kg BW as reference administered daily for 14 days to the diabetic rats. The result did not show dose-response manner, which the extract treatment at a dose of 770 mg/kg BW slightly decreased the cholesterol total, triacylglycerol, LDL cholesterol and atherogenic indices, as well as significantly increased HDL cholesterol compared with the reference group ($p<0.05$). Overall, our data suggest that *A. pauciflorum* fruit peel extract has an antihyperlipidemic potency in streptozotocin-induced diabetic female Wistar rats" rather than "In conclusion, the *A. pauciflorum* fruit peel extract has an antihyperlipidemic potency in streptozotocin-induced diabetic female Wistar rats.

1. Introduction

Diabetes mellitus (DM) is a disease with combinations of various disturbances in the body's metabolic system due to chronic hyperglycemia conditions associated with abnormal metabolism of carbohydrates, fats, and proteins resulting from deficiency or decrease the effectiveness of insulin [1]. Diabetes causes microvasvular complications including diabetic retinopathy, nephropathy, neuropathy, and diabetic foot disorders, as well as macrovascular complications include cardiovascular disease [2]. Accumulation of lipids in diabetes are mediated through various impaired in the metabolic and regulatory processes, especially insulin deficiency, leading to diabetic patient more susceptible to hyper-cholesterolemia and hypertriglyceridemia. The major pathogenesis of disrupted lipid metabolism in diabetes is the increased mobilization of fatty acids from adipose tissue and secondary elevation of free fatty acid level in the blood [3].

Hyperlipidemia is characterized by increased total serum cholesterol, as well as low-density lipoprotein (LDL) cholesterol, and very low-density lipoprotein (VLDL) cholesterol levels. Hyperlipidemia-associated lipid disorders are considered as the cause of atherosclerotic cardiovascular disease [4]. Hyperlipidemia is a determinant of the development of atherosclerosis and an important risk factor for cardiovascular disease. This constellation, often refer to as the 'lipid triad', specifically generated atherogenic changes in the arterial wall and promoted the instability of plaque, thus influenced acute cardiovascular events [5]. The risk for the development of atherosclerosis is generally associated inversely with levels of high-density lipoprotein (HDL) cholesterol. The major anti-atherosclerotic effect of HDL is known to be reverse cholesterol transport by scavenged cholesterol from the peripheral vasculature with transport to the liver, where it is excreted in the biliary system. However, HDL exhibits multiple other physiologic effects that may play a role in the reduced risk for atherosclerosis [6].

Studies of the antihyperlipidemic potency of the Fabaceae plant have been carried out using diabetic test animals, e.g. *Cassia fistula* [7], *Clitoria ternatea* and *Vigna mungo* [8], *Tamarindus indica* [9], *Senna auriculata* [10]. Extracts from the Fabaceae plant were known to contain alkaloid compounds, antraquinone, catenin, flavonoids, phenols, saponins, steroids, tannins, terpenoids, xantroproteins and glycosides. The results of the study showed decreased blood glucose, triacylglycerol, total cholesterol, LDL cholesterol, and VLDL cholesterol, as well as increased HDL cholesterol levels after treatment the extract in diabetic animals [7-10].

Archidendron pauciflorum (Fabaceae) fruit peel has been used traditionally as medicinal plant, i.e for diabetes treatment [11]. Our previous study showed that LD50 value from acute toxicity test of *A. pauciflorum* fruit peel extract on female Wistar rats estimated as 15382,412 mg/kg body weight (BW) [12]. In this study, we investigated the antihyperlipidemic potency of Djengkol (*Archidendron pauciflorum*) fruit peel in diabetic rats. The dosages used were 2.5, 5, and 10% of the LD50 value or aproximately 385, 770, and 1540 mg/kg BW, respectively. Diabetic condition in test animals could be induced by diabetogenic agent, eg streptozotocin (STZ). Streptozotocin (2-deoxy-2-[3- (3-methyl-nitrosourea)-D-gluco piranose]) is a toxic molecule causes damage to pancreatic β cells. Streptozotocin is widely used as diabetogenic agent since it could applied at low doses and has higher success rate (up to 95%) than other inducer substances, such as alloxan [13].

2. Materials and Methods

2.1 Collection and extraction of plant materials

Archidendron pauciflorum fruit peels were collected from Karangwangi village, Cianjur West Java. Identification of the plant material was performed by the Taxonomy laboratory in Biology Department, Faculty of Mathematics and Natural Sciences, Padjadjaran University. Dried fruit peels were powdered mechanically using a commercial blender and extracted with ethanol. Dried extract was obtained using a rotary evaporator (EYELA, N1100SWD, Tokyo Rikakiai Co., LTD, Japan) at 40^0 C.

2.2 Experimental design

Twenty-four female Wistar rats (160-180 g) were obtained from the animal house of the Bio System Laboratory, Universitas Padjadjaran. The rats were weighted and sorted into six groups (**Table 1**). They were nurtured in a cage based on a standard environmental condition. In this study, piglet standard diets (CP-551, PT. Charoen Pokphand) were chosen as their daily diet. Before the experiment, those animals were acclimatized for a week. We prepared six treatments i.e., diabetic control (DC); diabetic treated with the extract at a dose of 385 mg/kg BW (DE1); diabetic treated with the extract at a dose of 770 mg/kg BW (DE2); diabetic treated with the extract at a dose of 1540 mg/kg BW (DE3); diabetic treated with glibenclamide at a dose of 10 mg/kg BW as reference group (Ref); and normal non-diabetic control (NC).

2.3 Induction of diabetes and treatment of the extract

Before induction of diabetes by STZ, the animals were fasted overnight. Their baseline fasting glucose level was determined using a glucometer, by collecting their blood via tail cut. Diabetes was induced by intravenous injection of a freshly prepared solution of STZ at dose of 65 mg/kg BW in 10 mM citrate buffer solution (pH 4.5) [14] of five groups, while the negative control rats were injected by the vehicle alone. Then, the animals were fasted overnight for 72 hours after administration of STZ. The blood was collected via their tail cut for determining of their fasting glucose levels. The animals with glucose level more than 250 mg/dL were categorized as diabetic rats and used for a further experiment. Normal non-diabetic control and diabetic control rats were given 0.5% CMC solution, whereas *A. pauciflorum* extract or glibenclamide was administered orally to diabetic rats using intragastric tube once a day for 14 days.

2.4 Blood sampling and biochemical analysis
For these analyses, on the fifth day, the fasted animals were sacrificed by cervical dislocation. The blood sample for biochemical analysis was prepared from intra-cardiac. The blood serum was collected as described previously [15]. Total cholesterol and triacylglycerol levels were determined using multiCareinTM cholesterol strips and triacylglycerol strips [16] at before (day 1) and after 14 days of treatment (day 15). HDL cholesterol level was assayed enzymatically with Randox Commercial Test Kits CH200 and C203 (Randox Laboratories Ltd., Crumlin, England, UK). HDL cholesterol level was calculated as describe by Pusparaj *et al.* [17] using spectrophotometry at wavelength of 500 nm with formulae:

$$\text{HDL cholesterol level} = 219.2\ x\ (\text{OD}_{500})\ \text{sample (mg/dL)}$$

Plasma LDL cholesterol level calculated using the equation from Friedewald *et al.* [18] which is:

$$\text{LDL cholesterol} = \text{Total cholesterol} - \text{HDL cholesterol} - \left[\frac{\text{Triacylglycerol}}{5}\right]\left(\frac{\text{mg}}{\text{dL}}\right)$$

The atherogenic indices calculated using the following formulae as describe by Chigozie and Chidinma [16], which are:

$$Cardiac\ Risk\ Ratio\ (\text{CRR}) = \frac{\text{Total cholesterol}}{\text{HDL cholesterol}}$$

$$Atherogenic\ Coefficient\ (\text{AC}) = \frac{(\text{Total cholesterol} - \text{HDL cholesterol})}{\text{HDL cholesterol}}$$

$$Atherogenic\ Index\ of\ Plasma\ (\text{AIP}) = \text{Log}\ \frac{\text{Triacylglycerol}}{\text{HDL cholesterol}}$$

2.5 Data analysis
Results were expressed as mean ± standard deviation (S.D). The differences of parameters (mention of parameters in this study) were tested by SPSS ver. 21.00 (IBM, Armonk, NY, USA). Significant differences between means were calculated by one-way ANOVA followed by Duncan multiple range test. Significance levels were set at $P < 0.05$.

3. Results and Discussion

The effect of *A. pauciflorum* fruit peel extract on the blood total cholesterol and triacylglycerol level of tested animals at before (1st day) and after 14 days (15th day) treatments is presented in **Table 1**. Our statistical analysis failed to detect the significant different between total cholesterol level in normal non-diabetic control group (NC) and diabetic groups (DC, DE1, DE2, DE3, Ref) at before the treatment period (**Table 1**). We observed the total cholesterol among diabetic groups at the end of incubation were overall similar, suggesting that *A. pauciflorum* fruit peel extract has a little effect on the total cholesterol of tested animals. Although, the present study detected the triacylglycerol levels in diabetic treated groups (DE1, DE2, D3, Ref) were much higher than that in normal non-diabetic control (NC) at before treatment, however, our data are presented here show that the triacylglycerol level in diabetic treated groups decreased compared with normal non-diabetic control (NC) at the end of treatment. This finding suggests that *A. pauciflorum* fruit peel extract has effect on the decreasing of the triacylglycerol level.

Table 1. Blood total cholesterol and triacylglycerol level of normal and streptozotocin treated rats at before and after 14 days of treatment

Treatment Group	Total Cholesterol (mg/dL)		Triacyglycerol (mg/dL)	
	Day 1	Day 15	Day 1	Day 15
Normal non-diabetic control (NC)	167.00 ± 44.75	177.00 ± 37.58	285.25 ± 73.70[a]	251.50 ± 77,71
Diabetic control (DC)	186.75 ± 3.30	189.50 ± 11.82	317.00 ± 83.21[ab]	261.00 ± 39,08
Diabetic, treated with A. pauciflorum extract dose of 385 mg/kg BW (DE1)	191.50 ± 10.26	187.00 ± 29.64	408.00 ± 63.17[bc]	223.25 ± 139,28
Diabetic, treated with A. pauciflorum extract dose of 770 mg/kg BW (DE2)	196.00 ± 22.14	191.25 ± 24.11	373.25 ± 199.28[abc]	267.25 ± 190,04
Diabetic, treated with A. pauciflorum extract dose of 1540 mg/kg BW (DE3)	179.75 ± 21.71	176.25 ± 25.81	431.25 ± 58.14[c]	226.25 ± 118,90
Diabetic, treated with glibenclamide dose of 10 mg/kg BW (Reference)	168.75 ± 27.68	161.25 ± 34.05	408.00 ± 134.31[bc]	202.00 ± 122,79

Note: Values are expressing as mean ± S.D (n=4). Data was analyzed using ANOVA and Duncan's post-hoc test. Different alphabet in same column showed P values less than 0.05 and considered as significance.

In this study, intravenous injection of streptozotocin (STZ) by dosage 65 mg/kg BW mimicks type 1 diabetes, because STZ could be enter specifically to β cell pancreas through glucose transporter protein GLUT-2 and triggering apoptosis of β cell pancreas, which results in decreased insulin production [19]. Glucose and lipid metabolism linked to each other in many ways; the most important are diabetic dyslipidemia, and they are characterized by elevated triglycerides, low HDL cholesterol, and the predominance of small-dense LDL particles [20]. In type 1 diabetes occurs less advance dyslipidemia because of insulin deficiency altered cholesterol metabolism, i.e. high absorption and low synthesis of cholesterol, as also shown in streptozotocin-induced diabetic in experimental animals [21]. The higher levels of cholesterol and triacylglycerol in diabetic rats compared with normal non-diabetic rats as mentioned above due to accumulation of lipids and insulin deficiency induced derangements in lipid metabolism.

As mentioned above, the cholesterol level of the tested animals was slighly decrease as effect of *A. pauciflorum* fruit peel extract. An earlier study suggested that metabolism of cholesterol in type 1 diabetes show relatively normal lipid pattern compared with type 2 diabetes [21]. In addition, elevated level of plasma triacylglycerol is often associated with abnormal lipoprotein metabolism, leads to an increased catabolism of HDL (resulting in low HDL cholesterol) and a shift in the LDL form into a small dense LDL which are more "atherogenic", thus showed a synergistic risk factor for cardiovascular disease [16, 20], as shown by streptozotocin-induced diabetic rats in this study. Therefore, we analysed the effect of *A. pauciflorum* fruit peel extract on LDL and HDL cholesterol levels of the rats (**Table 2**). We found that LDL cholesterol level in diabetic groups (DC, DE1, DE2, DE3, Ref) was significantly higher ($P<0.05$) than that in normal non-diabetic control group (NC). In contrary, HDL cholesterol level in diabetic groups (DC, DE1, DE2, DE3, Ref) was significantly lower ($P<0.05$) than that in normal non-diabetic control group (NC) (**Table 2**). We also detected the extract treatment group at a dose of 770 mg/kg BW (DE2) was lowest LDL cholesterol level, but was highest HDL cholesterol level compared with the other extract-treated groups. In addition, the cholesterol level in that group was significantly higher than that in the glibenclamide-treated group (**Table 2**).

The treatments of *A. pauciflorum* fruit peel extract and glibenclamide for 14 days reduced plasma levels of triacylglycerol, although did not show a dose-dependent manner. Reduced level of LDL cholesterol and elevated level of HDL cholesterol compared with diabetic control and reference

IOP Conf. Series: Earth and Environmental Science **166** (2018) 012015 doi:10.1088/1755-1315/166/1/012015

groups only showed in the extract treated-diabetic rats by dosage 770 mg/kg BW. The reduction in plasma total cholesterol triacylglycerol, and LDL cholesterol as well as elevated HDL cholesterol level was assumed could reduce the risk of cardiovascular disease in diabetes, as shown by calculated atherogenic indices in this study.

Table 2. LDL and HDL cholesterol level of normal and streptozotocin treated rats after 14 days of treatment

Treatment group	LDL (mg/dL)	HDL (mg/dL)
Normal non-diabetic control (NC)	19.40 ± 17.64c	107.29 ± 33.09d
Diabetic control (DC)	72.25 ± 25.35ba	74.05 ± 14.99bc
Diabetic, treated with *A. pauciflorum* extract dose of 385 mg/kg BW (DE1)	91.99 ± 13.62a	50.85 ± 18.21a
Diabetic, treated with *A. pauciflorum* extract dose of 770 mg/kg BW (DE2)	59.87 ± 39.56b	77.93 ± 20.14c
Diabetic, treated with *A. pauciflorum* extract dose of 1540 mg/kg BW (DE3)	94.31 ± 14.20a	36.69 ± 14.77a
Diabetic, treated with *glibenclamide* dose of 10 mg/kg BW (Reference)	66.79 ± 23.71b	54.06 ± 6.37ab

Note: Values are expressing as mean ± S.D (n=4). Data was analyzed using ANOVA and Duncan's post-hoc test. Different alphabet in same column showed P values less than 0.05 and considered as significance.

The effect of *A. pauciflorum* fruit peel extract on atherogenic indices in tested animals after 14 days of treatment is presented in **Table 3**. Cardiac risk ratio (CCR) and atherogenic coefficient (AC) for diabetic groups (DC, DE1, DE2, DE3, Ref) were significantly higher ($P<0.05$) than those for non-diabetic control group (NC), whereas atherogenic indexes of plasma (AIP) were overall similar among all the treatment groups, except the extract treated group at a dose of 1540 mg/kg BW (DE3). The extract treatment group at a dose of 770 mg/kg BW (DE2) showed lowest of CCR, AC, and AIP values compared with the other extract-treated groups, including the glibenclamide-treated group. An earlier study suggested that Atherogenic indices are powerful indicators of the risk of cardiocarcular disease, the higher the value, the higher the risk of developing cardiovascular disease and *vice versa* [24]. In addition, low atherogenic indices are protective against coronary heart disease [16]. It has been well known that therogenic Index of Plasma (AIP) is part of index atherogenic that usually used for patient with type 2 diabetes to predict the risk of cardiovascular disease. A specific data was needed to determine the indication and the complication compared with using formulae, due to the correlation of nitric oxide (NO) as a marker for alkylation that damaging β cell pancreas-related to atherogenic index, which is not considered [25].

Table 3. Atherogenic indices of normal and streptozotocin treated rats after 14 days of treatment

Treatment group	Cardiac Risk Ratio (CRR)	Atherogenic Coefficient (AC)	Atherogenic Index of Plasma (AIP)
Normal non-diabetic control (NC)	1.67 ± 0,30a	0.67 ± 0,30a	0.37 ± 0.25a
Diabetic control (DC)	2.59 ± 0,68b	1.59 ± 0,68b	0.51 ± 0.29a
Diabetic, treated with *A. pauciflorum* extract dose of 385 mg/kg BW (DE1)	3.82 ± 0,75c	2.80 ± 0,57c	0.63 ± 0.26ab
Diabetic, treated with *A. pauciflorum* extract dose of 770 mg/kg BW (DE2)	2.49 ± 0,42b	1.49 ± 0,42b	0.47 ± 0.14a
Diabetic, treated with *A. pauciflorum* extract dose of 1540 mg/kg BW (DE3)	4.98 ± 1,24d	3.98 ± 1,24d	0.78 ± 0.27b

Diabetic, treated with glibenclamide dose of 10 mg/kg BW (Reference)	2.97 ± 0,34b	1.97 ± 0,34b	0.55 ± 0.29ab

Note: Values are expressing as mean ± S.D (n=4). Data was analyzed using ANOVA and Duncan's post-hoc test. Different alphabet in same column showed P values less than 0.05 and considered as significance.

On the other hand, the present study detected the decreasing of the Cardiac Risk Ratio (CCR), Atherogenic Coefficient (AC), and AIP values on the extract treated-diabetic rats by dosage 770 mg/kg BW. Such finding indicates that *A. pauciflorum* fruit peel extract has a potency as anti-hyperlipidemia, although the effect did not show dose-dependent positive manner. This finding is consistent with previous studies by Singh *et al.* [22], and Ezeigbo and Asuzu [23]. We assume that some chemical components in the extract such as alkaloids, flavonoids, saponins, phenolic, and tannins [22] are responsible for such phenomenon. Supporting this, earlier studies suggested that those chemical components have effect on decreasing of cholesterol [3, 10, 16, 17, 25]. Besides that, those components also showed athero-protective activities or the ability to protect against cardiovascular complications in the diabetic [16, 28, 29, 30].

4. Conclusion

Overall, our data suggest that *A. pauciflorum* fruit peel extract has a potency as anti-hyperlipidemia in streptozotocin-induced diabetes in female Wistar rats following oral administration. It was clearly observed that the extract evidently reduce the level of triacylglycerol, LDL cholesterol, and atherogenic indices (cardiac risk ratio, atherogenic coefficient, and atherogic index of plasma), as well as elevate the level of HDL cholesterol. The results of this study well support its use in traditional health care for the management of diabetes mellitus as well as protecting agent against coronary heart disease. Further biochemical and pharmacological investigations related to anti-diabetes effect of the extract are now in progress.

5. Acknowledgement

The authors thank Mr. Deden Deni for isolating the blood samples. This study was supported by Ministry of Research, Technology and Higher Education of Indonesia with the research grant (No. 718/UN6.3.1/PL/2017).

6. References

[1] American Diabetes Association 2010 Diagnosis and classification of diabetes mellitus *Diabetes Care* **33(1)** S62-S69

[2] Fowler MJ 2008 Microvascular and macrovascular complications of diabetes *Clin. Diabetes* **26(2)** 77-82

[3] Vetrivadivelan K, Venkateswaran KV, Selvasubramanian S and Sesh PSL 2012 Antihyperlipidemic effect of *Tinospora cordifolia* extract in streptozotocin induced diabetic rats *Int. J. Pharm. Sci. Res.* **3(5)** 1423-29

[4] Saravanan R, Rajendra PN, and Pugalandi KV 2003 Effect of *Piper beetle* leaf extract on alcoholic toxicity in the rat brain *J. Med. Food* **6(3)** 261–5

[5] Bertolotti M, Maurantonio M, Gabbi C, Anzivinio C, and Carulli N 2005 Review article: Hyperlipidemia and cardiovascular risk *Ailment Pharmacol. Ther.* **22(2)** 28-30

[6] Bandeali S and Farmer J 2012 High-density lipoprotein and atherosclerosis: the role of antioxidant activity *Curr. Atheroscler. Rep.* **14(2)** 101-7

[7] Khan ZI, Nahar B, Jakaria MA, and Rahmatullah M 2010 An evaluation of antihyperglycemic and antinociceptive effects of methanol extract of *Cassia fistula* L. (Fabaceae) leaves in Swiss albino mice. *Adv. in Nat. Appl. Sci.* **4(3)** 305-10

[8] Solanki YB and Jain SM 2010 Antihyperlipidemic activity of *Clitoria ternatea* and *Vigna mungo* in rats. *Pharm. Biol.* **48(8)** 915-23

[9] Shahraki MR, Harati M, and Shahraki AR 2011 Prevention of high fructose-induced metabolic syndrome in male Wistar rats by aqueous extract of *Tamarindus indica* seed *Acta Med. Iran.* **49(5)** 277-83

[10] Shanmugasundaram R, Kalpana DV, Tresina SP, Maruthupandian A and Mohan VR 2011 Antidiabetic, antihyperlipidaemic and antioxidant activity *of Senna auriculata (L.) Roxb.* leaves in alloxan induced diabetic rats *Int. J. Pharm. Tech. Res.* **3(2)** 747-756

[11] Malini DM, Madihah, Kusmoro J, Kamilawati F, and Iskandar J 2017 Ethnobotanical study of medicinal plants in Karangwangi, district of Cianjur, West Java. *Biosaintifika: Journal of Biology and Biology Education* **9(2)** 345-356.

[12] Madihah, Ratningsih N, Malini DM, Faiza AH, and Iskandar J 2017 Uji toksisitas akut ekstrak etanol kulit buah jengkol (*Archidendron pauciflorum*) terhadap tikus Wistar betina. *Pros. Sem. Nas. Masy. Biodiv Indon.* **3(1)** 33-38

[13] Szkudelski T 2001 The mechanism of alloxan and streptozotocin action in β cells of the rat pancreas *Phys. Res.* **50** 536-54.

[14] Meenu J, Sunul S, and Manoj K 2011 Evaluation of antihyperglycemic activity of *Dodonaea viscosa* leaves in normal and STZ diabetic rats. *Int. J. Pharm. Pharm. Sci.* **3(3)** 69–74.

[15] Lee G. and Goosens KA 2015 Sampling blood from the lateral tail vein of the rat *J. Vis. Exp.* **99** 1-5.

[16] Chigozie, I. J. and Chidinma IC. 2013. Positive moderation of the hematology, plasma biochemistry and ocular indices of oxidative stress in alloxan-induced diabetic rats, by an aqueous extract of the leaves of *Sansevieria liberica* Gerome and Labroy. *Asian Pac. J. Trop. Med.* **6(1)** 27-36

[17] Pusparaj P, Tan CH, and Tan BKH 2000 Effects of *Averrhoa bilimbi* leaf extract on blood glucose and lipids in streptozotocin-diabetic rats *J. Ethnopharmacol.* **72(1-2)** 69-76

[18] Friedewald WT, Levy RI, Friedrickson DS 1972 Estimation of the concentration of low-density lipoprotein cholesterol in plasma, without use of the preparative ultracentrifuge. *Clin. Chem.* **18(6)** 499-502

[19] Deeds MC, Anderson JM, Armstrong A, Gastineau DA, Hiddinga HJ, Jahangir A, Eberhardt NL, and Kudva YC 2011 Single dose streptozotocin induced diabetes: considerations for study design in islet transplantation models *Lab Anim.* **45(3)** 131–40

[20] Parhover KG 2015 Interaction between Glucose and Lipid Metabolism: More than Diabetic Dyslipidemia. Diabetes Metab. J. 39(5) 353-362

[21] Gylling H, Tuominen JA, Koivisto VA, and Miettinen TA 2004 Cholesterol metabolism in type 1 diabetes *Diabetes* **53** 2217-22

[22] Singh SK, Kesari AN, Gupta RK, Jaiswal D, and Watal G 2007 Assessment of antidiabetic potential of *Cynodon dactylon* extract in streptozotocin diabetic rats *J. Ethnopharmacol.* **144(2)** 174-9

[23] Ezeigbo II and Asuzu IU 2012 Anti-diabetic activities of the methanol leaf extracts of *Hymenocardia acida* (Tul.) in alloxan-induced diabetic rats. *Afr. J. Tradit. Complement. Altern. Med.* **9(2)** 204-9

[24] Sihombing JR, Dhamrma A, Chaidir Z, Almahdy, Fachrial E, and Mucnaf E 2015. Phytochemical screening and antioxidant activities of 31 fruit peel extract from Sumatera, Indonesia *J. Chem. Pharm. Res.* **7(11)** 190-196

[25] Kumari M and Jain S 2012 Tannins: an antinutrient with positive effect to manage diabetes *Res. J. Recent Sci.* **1(12)** 1-4

[26] Dobiášová M 2004 Atherogenic index of plasma [log(triglyceride/HDL-Cholesterol)]: theoretical and practical implications. *Clin. Chem.* **50** 1113-15.

[27] Nimmanapalli HD, Kasi AD, Devatia PK, and Nuttaki V 2016 Lipid ratios, atherogenic coefficient and atherogenic index of plasma as parameters in assessing cardiovascular risk in type 2 diabetes mellitus *Int. J. Res. Med. Sci.* **4(7)** 2863-69

[28] Florence NT, Benoit MZ, Jonas K, Alexandra T, Desire DD, Pierre K, and Theophile D 2014 Antidiabetic and antioxidant effects of *Annona muricata* (Annonaceae), aqueous extract on streptozotocin-induced diabetic rats. *J. Ethnopharmacol.* **151(2)** 784–90
[29] Soetan KO 2008 Pharmacological and other beneficial effects of antinutritional fact ors in plants-A review. *Afr. J. Biotechnol.* **7(25)** 4713-21
[30] Dillard CJ and German JB 2000 Phytochemicals: nutraceuticals and human health *J. Sci. Food. Agric.* **80(12)** 1744-56.

Utilization of Bark Condensed Tannin as Natural Preservatives Against Subterranean Termite

M Ismayati[1], A Nakagawa-izumi[2] and H Ohi[2]

[1]Research Center for Biomaterials, Indonesian Institute of Sciences, Cibinong Science Center, Cibinong-Bogor 16911, Indonesia
[2]Graduate School of Life and Environmental Science, University of Tsukuba Tsukuba 305-0872, Japan

E-mail : maya_ismayati@biomaterial.lipi.go.id

Abstract. Extractive-based preservative has been isolated from many parts of plant to protect wood from fungi or pest. This is due to the presence of bioactive chemical as a natural preservative. Present study, the condensed tannin was isolated from bark-waste of *Acacia crassicarpa* and *Acacia mearnsii*. The yield of the extracts obtained from *A. crassicarpa* using a 70% acetone aqueous solution (7% based on bark weight) is less than that obtained from *A. mearnsii* (34%). To evaluate the termiticidal activity of condensed tannin from both extracts, no-choice tests were conducted using *Reticulitermes speratus*. After 14 days, mass loss of the dimmer condensed tannin treated was less than 50%, significantly different with control. In addition, a monomer condensed tannin, catechin was presented the similar result. Meanwhile, the termite mortality of catechin is slightly higher than the dimmer. According to both parameter, the medium level of mass loss and termite mortality, the dimmer suggested as antifeedant in mechanism of termiticidal. The feeding activity is decreased after contact with the treated paper disc. In polymer condensed tannin, the result seems similar to dimmer condensed tannin. Condensed tannin and hydrolyzable tannin which were contained in the bark effected termite consumption because it caused inactivation of enzyme in termite digestion system. In fact, many factors effect termite activity from natural extractives such as compound type, quantity and also extract retention activity. The finding of this study should contribute to the developing of natural wood preservative from extractive-based. Especially, the study of structure-active relationship (SAR) of condensed tannin from *A. crassicarpa*, which can improve the utilization of bark-waste from pulp industry.

1. Introduction

Natural wood preservative based on plant extracts, have been known to protect wood from termite. Many studies have been reported about the effectiveness of reduced chemical or even chemical-free treatmenrs which was compared to the use of plant extracts agains termite [1,2]. Due to the effect of synthetic or chemical pesticide into the environment, the use of bio-termiticide has been increase although in still limited. The natural preservative based on plant extrcts can be isolated from any part of plant such as heartwood [3,4], bark [5,6], leaves, fruit and seed [7,8].

In other situation, bark of *A. crassicarpa,* by-product of the pulp industry as a source of extractive has remains unutilized. Indonesia as the third largest manufacturer of pulp and paper in Asia, It is produces a significant amount of bark by-products. Bark contains a high concentration of tannin; as a result, interest in developing methods for utilizing tannin has recently increased. unfortunately, there is no information that reported about studies on tannin from the bark of *A. crassicarpa*. Pietarinen et al. 2005 only reported about the polyphenolics compounds from the knot and stem wood of *A. crassicarpa* [9].

So, this study aims to utilize the bark waste of A.crassicarpa as natural preservative towards subterranean termite

2. Material and Methods

2.1. Isolation of condensed tannin from bark of A. crassicarpa

Bark of *A. crassicarpa* was collected from 4-5 year-old tress grown on Research and Development plantation area of PT. Riau Andalan Pulp and Paper (PT. RAPP) located in Pelalawan area, Riau, Indonesia at 2011. The isolation of condensed tannins were conducted according to procedure that reported in Ismayati, et al., 2017[10]

2.2. No-choice feeding tests against Reticulitermes speratus using condensed tannin samples from A. crassicarpa and A. mearnsii

R. speratus was collected from the Living Sphere Simulation Field, Research Institute for Sustainable Humanosphere (RISH), Kyoto University (Kagoshima Prefecture, Japan) and maintained at the University of Tsukuba (Ibaraki Prefecture, Japan), under controlled laboratory conditions (28°C and 80% relative humidity).

The condensed tannin sample (C-EA-1, C-WS-AW, and M-WS-AW) was applied into no-choice feeding test according to previous report [10,11]. As a control, untreated paper discs, solvents treatment and catechin were evaluated in the same manner as the discs with the extractives in this no-choice feeding test as the control.

To measure termiticidal activity, the mass loss of paper disc was calculated at the end of test periods and the dead termites were also counted. Three replicates were performed for each sample.

$$\text{Mass loss (\%)} = [\text{ODS1-ODS2/ODS1}] \times 100\% \quad \text{..........(1)}$$

ODS1 is oven-dried paper disc before test, while ODS2 is oven-dried paper disc after test

$$\text{Termite Mortality} = [\text{Number of died termite/50}] \times 100\% \quad \text{..........(2)}$$

2.3. Statistical analysis

The effects condensed tannin on mass loss and termite mortality were examined using an analysis of variance (ANOVA), and samples showing significant differences ($P < 0.05$) were analyzed using Duncan's post hoc test ($\alpha = 0.05$). All analyses were performed using IBM SPSS version 22 (IBM Japan, Tokyo, Japan).

(a) (b) (c)

Figure 1. *A. crassicarpa* tree planted at PT. RAPP, Indonesia (a), bark sample of *A. crassicarpa* (b) and No-choice feeding test using *R. speratus* (c)

3. Result and Discussion

The results of the no-choice feeding test to evaluate the effects of condensed tannin treatments are shown in **Figure 3**. After 14 days, mass loss of condensed tannin treatments (C-EA1, C-WS-AW, and M-WS-AW) were less than 50%, significantly different with treatment control. As dimmer control, catechin was used as control that presented the similar result. As second parameter, the termite mortality of catechin is slightly higher than C-EA1. Based on those result, C-EA1 shows a medium level as natural wood preservative, which suggested as antifeedant in mechanism of termiticidal. After contact with extract, the feeding activity is decreased [12].

A similar result shown by polymer condensed tannin as well as in dimmer treatments. Both polymer condensed tannin from *A. crassicarpa* and *A. mearnsii* (C-WS-AW and M-WS-AW) have a same trend of mass loss and termite mortality in the medium level. Based on these result, it was suggested that polymer condensed tannin can caused by the decreasing of feeding activity of termite due to the inactivation of enzyme in the termite digestion system [13,14]

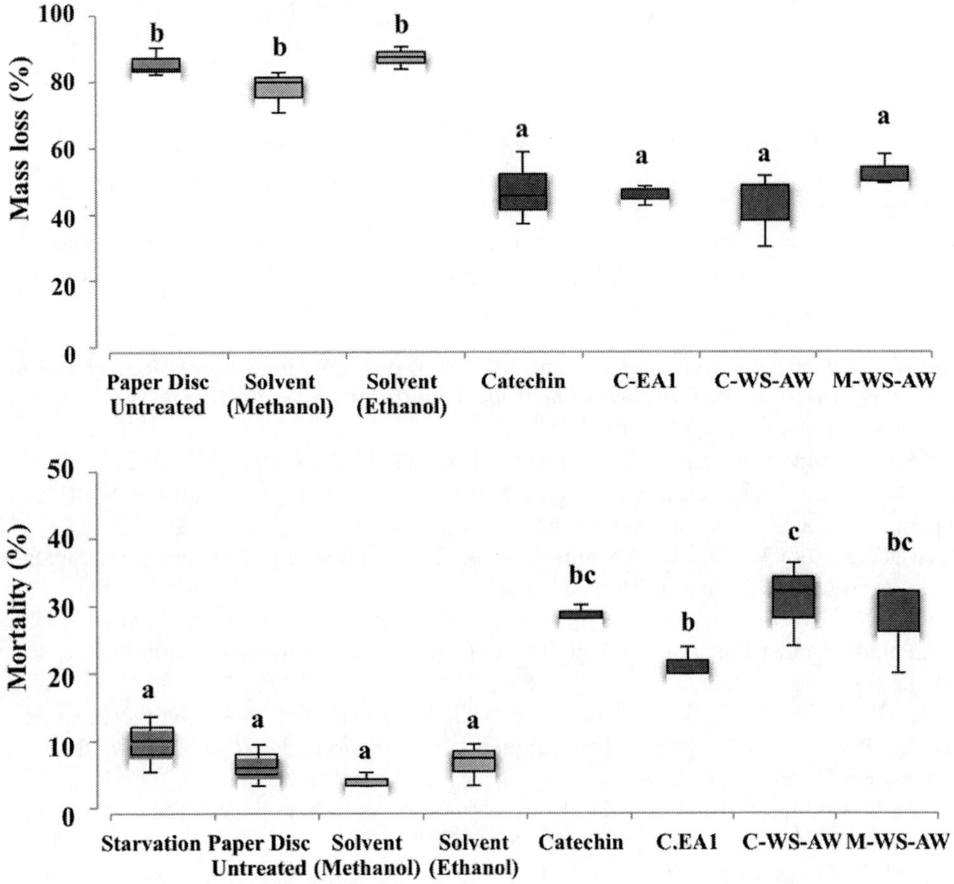

Figure 2. Box plot of mass loss and mortality of *R. speratus* in no-choice feeding tests in *Tectona grandis* crude extractives and MAQ.
Legends: C-EA1= Dimmer condensed tannin from *A. crassicarpa*; C-WS-AW = Polymer condensed tannin from *A. crassicarpa* from; M-WS-AW = Polymer of condensed tannin from *A. mearnsii*

Bio-assay activity of compounds are related to their chemical structures. The structure-active relationship of C-EA1 (flavan-3-ols) that lacks an oxygen at C3 of pyran ring is very interesting for further study. To identify which functional group related into the bioassay activity, the C-EA1 is compared with Quercetin and Fisetin. Previou study, Quercetin that has a hydroxyl group in the C-5 A-ring showed higher antifeedant activity than fisetin [15]. C-5 and C-7 hydroxyl groups in A ring of Quercetin showed higher antifeedant than Fisetin which do not have C-5 This indicates that the C-5 and C-7 hydroxyl groups in the A- ring are important for anti-termite activity [10]. The dimmer of C-EA1 that lacks an oxygen at C3 of pyran ring is very interesting for further study.

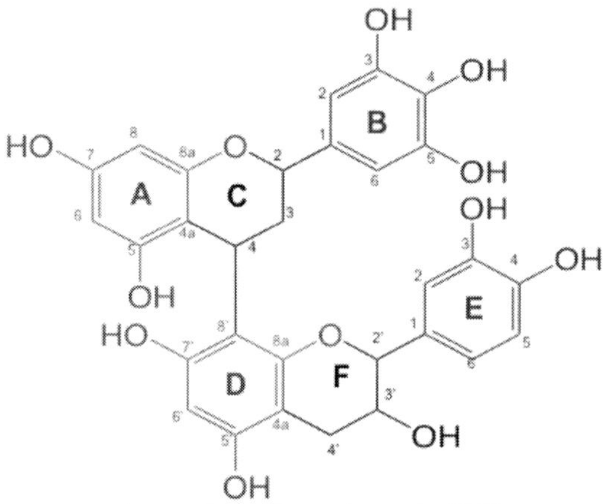

A novel condensed tannin [10]

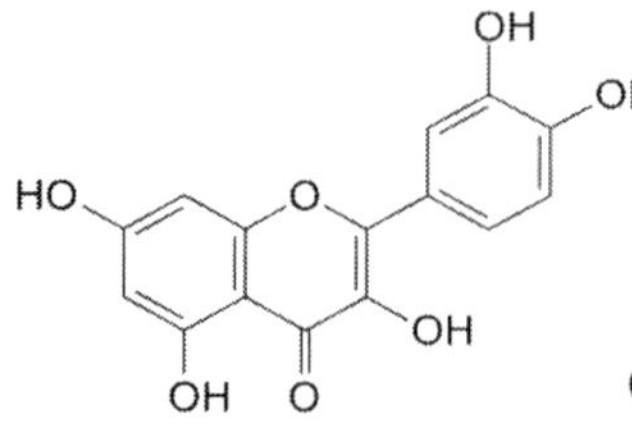

Quercetin

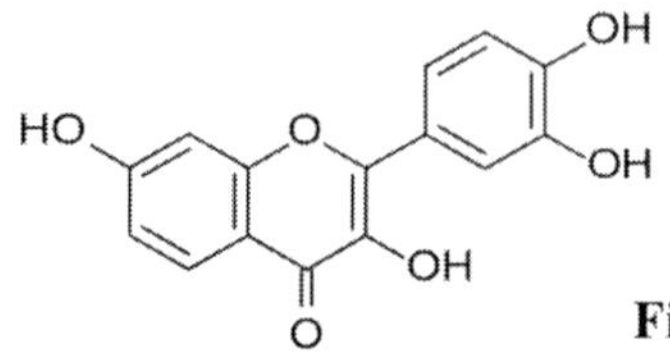

Fisetin

A monomer condensed tannin (ohmura et al., 2000)

Figure 3. A structure activity relationship in condensed tannin

4. Conclusion

The dimmer condensed tannin C-EA1 and polymer condensed tannin from *A. crassicarpa* and *A. mearnsii* are potentially useful as a termite control agent. They act as antifeedant, against *R. speratus*.

5. References

[1] Kim YS, Singh AP, Wong AHH, Eom TJ, Lee KH 2006 Micromorphological and chemical characteristics of cengal (*Neobalanocarpus heimii*) heartwood decayed by soft rot fungi. *Mokchae Konghak* 34(2), 68-77

[2] Rodrigeus AMS, Amusant N, Beauchene J, Eparvier V, Lemenager N, Baudass C, Espindola LS, Stien D 2011 The termiticidal activity of Sextonia rubra (Mez) van der Welff Lauraceae) extract and its active constituent rubrynolide *Pest Manag Sci* 67(11): 1420-1423

[3] Chen K, Ohmura W, Doi K, Aoyama M 2004 Termite feeding deterrent from Japanese larch wood. *Bioresource Technology* 95: 129-134

[4] Ohmura W, Doi S, Aoyama M, Ohara S 2000 Antifeedant activity of flavonoids and related compounds against the subteraanean termite Coptotermes formosanus Shiraki. *J Wood Sci* 46: 149-153

[5] Viviers PM, Botha JJ, Ferreira D, Roux DG, Saayman HM 1983 Synthesis of condensed tannins. Part 7. Angular [4,6;4,8]-prorobinetinidin triflavonoids from black wattle ('Mimosa') bark extract. *J Chem Soc Perkins Trans* 1: 17-22

[6] Yazaki Y, Collins PJ 1994 Wood adhesives from Pinus radiata barks of some pine and spruce species. *Holz als Roh-und Werkstoff* 52: 185-190

[7] Tarmadi D, Setiawan KH, Yusuf S 2004 The Efficacy of the Oleic Acid Isolated from *Cerbera Manghas* L. Seed Against a Subterranean Termite, *Coptotermes gestroi* Wasmann and a Drywood Termite, *Cryptotermes cynocephalus* Light. *Procedia Environmental Sciences* 20: 772-777

[8] Setiawan KH, Tarmadi T, Ismayati I, Yusuf S 2013 Bioefficacy Performance of Neem-based Formulation on Wood Protection and Soil Barrier against Subterranean Termite, *Coptotermes gestroi* Wasmann (Isoptera: Rhinotermitidae). Procedia Environmental Sciences 17: 135-141

[9] Pietarinen SP, Willfor SM, Sjoholm RE, Holmbom BR 2005 Bioactive phenolic substances in important tree species. Part 3: knots and stemwood of *Acacia crassicarpa* and *A. mangium. Holzforschung* 59:94–101

[10] Ismayati M, Nakagawa-izumi, A, Ohi H 2017 Structural elucidation of condensed tannin from the bark waste of *Acacia crassicarpa* plantation wood in Indonesia. *J Wood Sci.* 63(4): 350-359. DOI 10.1007/s10086- 017-1633-4

[11] Ismayati M, Kamaluddin N. N, Nakagawa-izumi, A, Ohi H 2016 Toxicity and Feeding Deterrent Effect of 2- Methylanthraquinon e from Wood Extractives of *Tectona grandis* on the Subterranean Termite *Coptotermes formosanus* and *Reticulitermes speratus*. *Insect* 7(4):63. DOI:10.3390/insec ts7040063

[12] Ismayati, M 2017 Studies on Wood and Bark Extractives as Natural Preservatives towards Subterranean Termites Doctoral's thesis. University of Tsukuba, Tsukuba, Japan.

[13] Ohara S, Suzuki K, Ohira T 1994 Condensed tannins from Acacia mearnsii and their biological activities. *Mokuzai Gakkaishi* 40:1363-1374

[14] Tascioglu C, Yalcin M, Troya TD, Sivrikaya H 2002 Termiticidal properties od some wood and bark extract used as wood preservatives. *Bioresources* 7(3), pp.2960-2969.

[15] Ohmura W, Doi, S, Aoyama M, Ohara S 2000 Antifeedant activity of flavonoids and related compounds against the subterranean termite *Coptotermes formosanus* Shiraki. *J Wood Sci* 46: 149–153

IOP Conf. Series: Earth and Environmental Science **166** (2018) 012017 doi:10.1088/1755-1315/166/1/012017

Insecticidal activities of crude extact of *Metarhizium anisopliae* and conida suspension against *Crocidolomia pavonana* fabricius

Melanie[1], M Miranti[2], H Kasmara[2], S Hazar[3], and A Martina[3]

[1] Student of Environtmental Science Doctoral Programme, Faculity of Post-Graduated School, Padjadjaran University, Jl. Dipatiukur No. 35, Bandung, Jawa Barat 45363, Indonesia
[2]Departement of Biology, Faculty Mathematics and Natural Sciences, Padjadjaran University, Jl. Raya Bandung Sumedang Km. 21, Jatinangor, Jawa Barat 45363, Indonesia
[3]Post Graduate students of Biology Departement, Faculty Mathematics and Natural Sciences, Padjadjaran University, Jawa Barat 45363, Indonesia

E-mail: melie_entobio@yahoo.com

Abstract. Pest resistance, resurgent and the residual effect of conventional insecticides are the basis to apply Insect Pest Management (IPM) strategies. *Metarhizium anisopliae* is an entomophatogenic fungus that has considerable promise for usage in IPM. The spores of fungus had great potential as insect pathogen, in addition the mycotoxins products having current interest as bioactive agent. *M. anisopliae* was exhibited to control pest insect populations of crops especially host plant of *Brassica oleracea* (Cruciferae), i.e. *Plutella xylostella*, *Spodoptera litura,* and *Crocidolomia pavonana*. The insecticidal activities of conidia suspensions and crude extract product containing mycotoxins were investigated, withthe parametersmortality and lethal time of *C. pavonana* 3rd instar larvae. The experimental design of crude extract and conidia suspensions of *M. anisopliae* on mortality effect used Completely Randomized Design (RAL). The bioassay of conidia suspensions mortality test used 6 level spore concentrations (0, 10^5, 10^6, 10^7, 10^8, and 10^9 spores/ml). Analysis of variance (ANAVA) at α 95% followed by Duncan multiple test at α 5% was carried out. The result of crude extract mortality test showed that *M. anisopliae* produced mycotoxins from the third day of fungal growth. The long growth of fungus is directly proportional with the produced mycotoxins. The mycotoxins was optimally produced in 7th days. The extract *M. anisopliae* at highest mortality effect on larvae was tested for 24 to 120 h. One hundred percent mortality of larvae were achieved at 120 h with an average lethal time were 3.6 days shorter than mortality caused by conidia infections. Thus, conidia infections showed that *M. anisopliae* at 10^5 - 10^9 spores/ml concentrations led to mortalities of 75%, 90%, 80%, 85%, and 95 %, respectively ($P<0.05$). The larvae lethal time had no different in all concentration levels, but the shorter lethal time (4.66 days) was showed in the infected larvae with 10^9 spores/ml concentration.

1. Introductions

Nowadays biological control is appreciated and regarded as a desirable technique for controlling pest insects, due to the low environmental impact and the reduce of resistance development [1]. Biocontrol application uses natural agents such as pathogens, parasites and predators, with the use of agrotoxins as a last resort [2]. *M. anisopliae,* most widely used entomopathogenic fungus in biological control, have been formulated and registered as mycoinsecticides [3]. *Metarhizium* grows vegetatively as hyphae producing mycelia and conidia that are infective propagules on insect cuticle [4]. Many researches on *M. anisopliae* have been carried out for use against a wide variety of pests such as insect pests of wood (termites), domestic pests, vector insects (mosquitoes) also insect pests of ornamental plants and cultivated plants (plantations, cereals and vegetables) [5]. *M. anisopliae* is known to be a potential biological agent for controlling *Spodoptera litura*, key pests of tobacco, soybeans and crop from Brassicaceae [6].

C. pavonana is the important insect pest of Brassicaceae [7-9]. This diet can lead to the destruction of head cabbage, which cause attacks by *C. pavonana* as a very serious problem [7,10]. All stages of host plants are susceptible to attack by *C. pavonana* and major yield losses can occur if the pest population is not suppressed [7-8]. Management strategies to control *C. pavonana* beside of selective insecticides also used parasitoid and entomapatotoghenic fungus. *M. anisopliae* is known to be effective in controlling *C. pavonana*. The fungal spores are formulated in pellets and powders. The spores are infectious to *C. pavonana* larvae at spore concentration 10^7 spores/mL for 4 days post infection [11].

The secondary metabolites produced by the fungus are rich source of bioactive chemicals that increase the toxicity potential of entomophatogenic fungus [8]. *Metarhizium* are known to produce mycotoxins such as destruxins (dtxs), cytochalasins C and D, helvolic acid, myroridins, swainsonine, tyrosine betaine, serinocyclins A and B, aurovertins, hydroxyovalicin, viridoxins, fungerins, metacytofilin, hydroxyfungerins A and B, and 12-hydroxyovalicin and hydroxyfungerin, the 7- desmethyl analogs of fusarin C and (8Z)-fusarin C [12-18]. In fact, destruxins are most widespread mycotoxin of the secondary metabolites produced by *Metarhizium* [8]. However, strains of *Metarhizium* differ considerably in their metabolite profiles, with variation in destruxins production as a possible virulence or specificity factor [8].

However, entomopathogenic fungus has numerous benefits over other microbes for insecticide preparation. The development and formulation have used new technology that produce new insecticidal compound for larvicides or adulticides, i.e. in the bio-control of mosquitoes [11]. The fungal spores and metabolites have been significantly demonstrated to show efficacies against adults and its developmental stages of mosquitoes [11]. The weaknesses of using fungal suspensions is that it requires a longer time to reach insect pest mortality because the fungal spores need more time to germinate and penetrate into cuticles [4]. Thus, it was considered to be more feasible for large scale experiments and field implementation [11]. Metabolites of entomomopathogenic fungus were found significant pathogenic after filtration [19]. Efficacy of culture filtrates of five strains of *M. anisopliae* isolated from insects were evaluated against *An. stephensi* and *Cx. Quinquefasciatus* [20].

The extract of *M. anisopliae* has been characterized to determine the secondary metabolites. It contained a rich source of bioactive chemicals, responsible for insecticidal activity [12]. One of the important things to know about *M. anisopliae* extract characterization is the optimization of mycotoxins production. The purpose of this study was to determine the time of optimization of mycotoxins released by fungi products in rough extract which is showed by the effect mortality to *C. pavonana* larvae. In addition, the effectiveness of conidia suspension was compared with the rough extract of *M. anisopliae* on *C. pavonana* larvae.

2. Materials and Methods

2.1. Experimental Insect

The research object was crop caterpillar (*C. pavonana*). It is known as an important secondary pest in cabbage plantation, and other Brassicaceae plants. In this research, third instar of the larvae obtained from HPT Departement, Agriculture Faculty of UNPAD and BALITSA in Lembang were used. Feed bait and testing leaves are cabbage leaves (*B.olereace*) that is a typical food for *C.pavonana*. The larvae was cultured in a rearing box ($40 \times 25 \times 25$ cm^3). The ovitrap was used to collect the eggs at the bottom of the leaves. The group of eggs was reared in the separate rearing box ($20 \times 15 \times 5$ cm^3) with controlled humidity and temperature. Furthermore, the eggs were hatched and become caterpillar/larvae. The instar I-IV larvas were placed in different rearing boxes. The bioassay test used third instar larvae, the stage of active feeding.

2.2. Culture of M. anisopliae

The culture of *M. anisopliae* obtained from Department of Plantation Protection Centre, Bandung. Further prophylactic culture of *M. anisopliae* was carried out on Potato Dextrose agar medium (PDA) of fungal preparation and incubated at room temperature of 22-27°C with pH between 3.3 to 8.5 [5]. It was observed and was made the growth curve, by using colony counter methode.

2.3. M. anisopliae Extraction

The culture of *M. anisopliae* sporas were harvested from growth mediums and collected into the macerator. It was soaked for 24 h using 95% ethanol solvent. Ethanol is a polar solvent and can attract polar compounds including mycotoxins compounds that polar protein molecules. The macerate was concentrated by a vacuum rotary evaporator to evaporate the solvent with a low pressure (40^{0}C) to obtain a viscous extract in the form of a paste [21].

2.4. Prepared Conidia suspensions of and extract solutions

One g of conidia (containing > 10^9 spore/ml) was dissolved into 5 ml aquadest and homogenized by Tween^{80}, then mixed with vortex. Further, the suspension was diluted into 10^9, 10^8, 10^7, 10^6, 10^5 spore/ml of level concentrations. *M. anisopliae* extract was made into a concentration of 1000 ppm (1 mg/1000 mL). Then, the cabbage leaf in Ø 5 cm was dipped into 0.05 ml of extract and the surface was dried for 1 minute.

2.5. Bioassay test

The bioassay of conidia suspensions was performed by topical method (suspension was infected into pro-thorax of larvae) and the bioassay test of extract *M. anisopliae* used leaf dipped method [22]. The *C. pavonana* third instar larvae was used in both tests. Each treatment used 5 larvas with 5 replications. Observations were made daily with parameter of larval mortality and lethal time of larvae.

2.6. Data analysis

The results were analyzed by ANOVA and if the results were significantly different (α 5%), continued with Duncan Multiple Range Test.

3. Result and Discussions

3.1. Mortality effect of crude extract M. anisopliae to the C. pavonana larvae

The growth patterns of *M .anisopliae* cultivated in PDA medium was showed in growth curve (**Figure 1**). By counted the colony of fungal spores , there was an exponential phase on the 1st day to 6th day of growth in PDA. The conidia of *M. anisopliae* declined on the 7th day. It was the phase of fungal death marked by the decreasing population until 10th day.

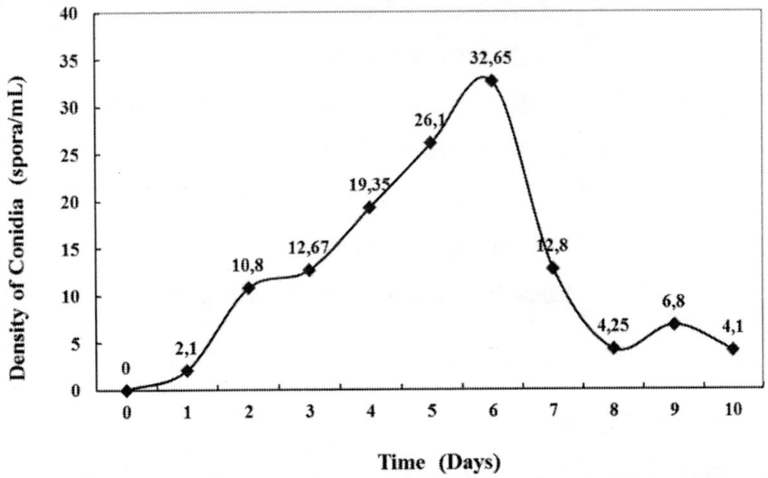

Figure 1. The growth curve of *M. anisopliae* in PDA Culture

M. anisopliae crude extract contained secondary metabolites in the form of mycotoxins known to cause the mortality of *C. pavonana* larvae. Furthermore, the mycotoxins produced from *M. anisopliae* exctract grown on the 1st day until the 7th day, were tested for their mortality effect, in order to determine the most optimal fungal growth time. Finally, mycotoxin with the highest insecticidal effect of *C. pavonana* larvae was then resulted (**Figure 2**).

The results showed that crude extract of *M. anisopliae* in PDA produced mycotoxin starting on the 3rd of fungal growth. The fungal growth was directionally proportional with the production of mycotoxins. The most optimum time of mycotoxin production was up to day 7th, and it was observed from the highest mortality effect on larvae (24 to 120 h): 53.3% mortality in 24 h and 100% mortality of larvae at a 120 h, respectively (**Figure 2**).

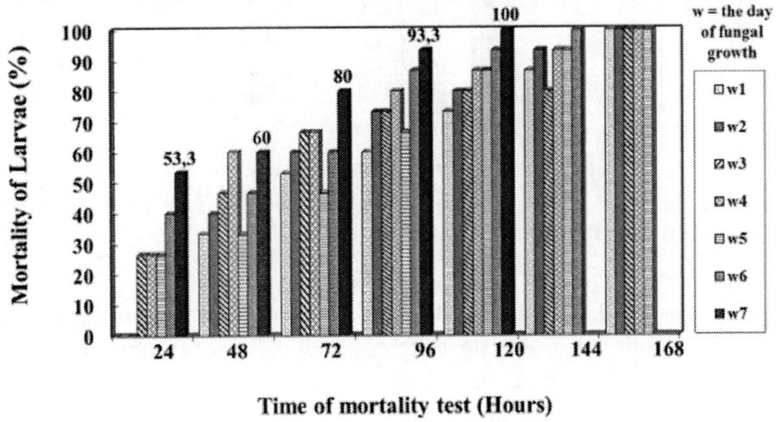

Figure 2. The histogram of mortality effect of *M. anisopliae* extract to *C. pavonana* larvae

The time of fungal growth had an effect on the mycotoxin product observed by lethal time of *C. pavonana* larvae (P <0.05). The mycotoxin product on the 7th day showed significantly different shortened the lethal time of *C. pavonana* larvae for an average of 3.6 days (**Figure 3**). Thus, the optimum time of mycotoxins from *M. anisopliae* extract on PDA media was produced on day 7th.

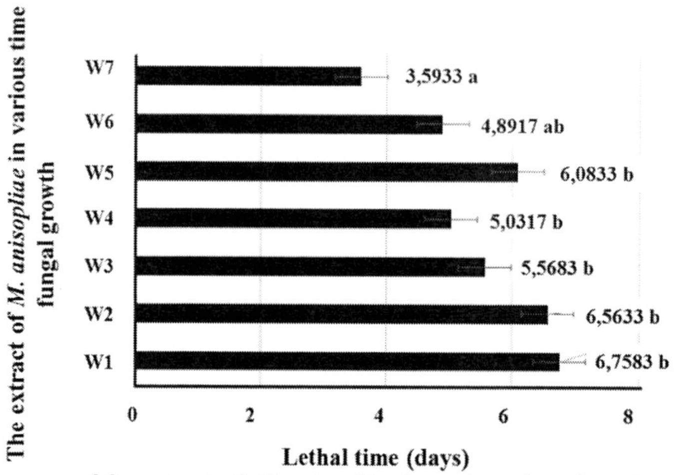

Figure 3. The histogram of the extract of *M. anisopliae* in various time fungal growth effect to lethal time of C. *pavonana* larvae

3.2. Mortality effect of M. anisopliae conidia suspensions to the C. pavonana larvae

The infection of *M. anisopliae* to the larvae showed that conidia suspension at concentration from 10^5 to 10^9 spores/ml had a mortality effect on larvae (P <0.05) with 75%-95% mortality rate, but the difference concentration factor did not significantly effect to the mortality rate of *C. pavonana* larvae (**Figure 4**)

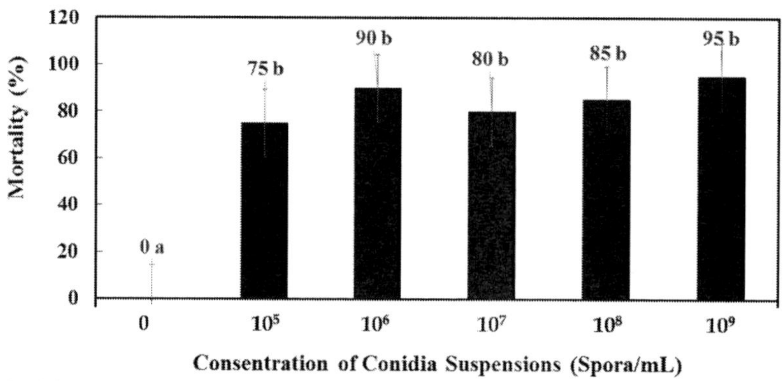

Figure 4. The histogram of *C. pavonana* larvae mortality due to infection of *M. anisopliae* conidia suspensions at various concentration levels

The conidia suspensions of *M. anisopliae* infected *C. pavonana* larvae and significantly affected to lethal time (P <0.05), with there was no different among of level concentrations. The shortest average lethal time of *C. pavonana* larvae was 4.66 days using 10^9 spore/mL conidia suspensions, and the longest was 6.3 days using 10^5 spore/mL of conidia suspensions (**Figure 5**).

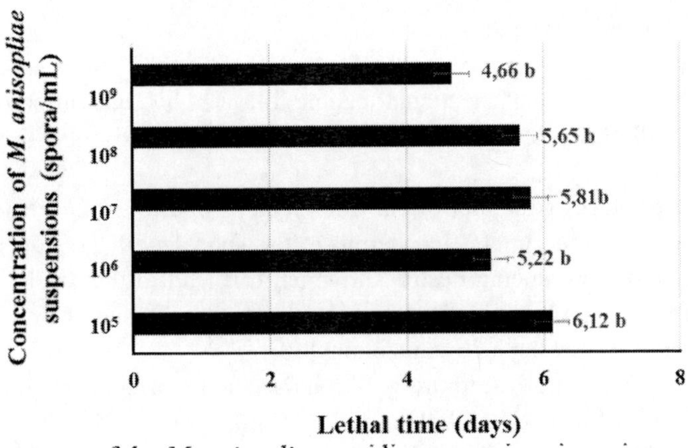

Figure 5. The histogram of the *M. anisopliae* conidia suspensions in various concentrations effect to lethal time of C. *pavonana* larvae

Potato dextrose agar (PDA) media were found most favoured for soil fungal [23], such as *M. anisopliae* that naturally spread in soil [4, 24]. The composition of PDA consists of 20% potato extract and 4% glucose, and low pH (4.5-5.6). It was the ideal conditions for optimal fungal growth in a culture medium containing nutrients with appropriate osmotic pressure and pH [25]. In this study, *M.anisopliae* showed the optimum growth phase in 48-72 hours. Culture media influenced the germination of conidia, appressorial development and mycelial growth of *M. anisopliae* [25]. The 72- hour incubation period is a phase with lest active dividing cells. In this phase, fungal cells obtain nutrient intake and perform metabolic reaction, such as secondary metabolite productions [26]. The product of secondary metabolite was invested by the cells, because it was no longer needed by cell [26]. Determination in incubation period is very important for harvest timing, quantity and quality of the fungal secondary metabolic product, especially for industrial scale process. Penicillin production uses *Penicillium chrysogenum* generally last for 72 hours (about 3-7 days). However, it is necessary to extend the incubation period to 8-10 days to reach the stationary phase for optimize penicillin production. The medium enricedh with a mixture of leachate and sucrose can improve the penicillin product [27]. However, from this study, the first mortality of larva *C. pavonana* (25%) was observed using the extract of 3rd days fungal growth. The highest mortality effect on larvae (100%) was observed using the extract from the 7th days fungal growth.

The study showed that there are no difference at the mortality effect of conidia suspension in all levels concentration. The pathogenic spores were germinated and penetrated to cuticle, as soon as possible, to attack the immune system [28]. The death of the larvae caused by the pathogenicity of the opportunistic fungus and possibly due to the characteristic of spore virulence, the rapid growth of opportunistic fungus, which infects, injures or weakens the insect's immunity [1]. So it depends not only on the number of infected spores.

The fungal spores *M. anisopliae* have virulence levels to kill the infected larvae. The shortest average of lethal time of *C. pavonana* larvae was 4.66 days using 10^9 spore/mL conidia suspensions, and the longest was 6.3 days using 10^5 spore/mL of conidia suspensions. A higher concentration is representative with the possibility of a number of infected spores, possibly increasing contact between host with pathogen and shorter at time of death larvae.. Deaths of larvae are also caused by damage to the intestines due to mycotoxins released by fungi [29]. From this study, the effect of *M. anisopliae* extract succeeded in shortening the time of death of *C. pavonana* lar larvae. The rough extract of *M. anisopliae* from the growth of the 7th day fungus shortened the

deadly time of *C. pavonana* larvae to
3.6 days. The time of death of the larva with infected conidia suspension takes more time for the germination spores to occur on the insecticular cuticle followed by penetration of growing insects in the hemolymph that kill mosquitoes within 7-14 days, depending on dosage, formulation and fungal strain [29].

Penetrate hyphae produces a number of enzymes such as lipase, protease and chitinase enzymes capable of degradation of the cuticle. Furthermore, the spores will develop in the hemocoel by absorbing hemolymph and producing destrucsin which can lead to the death of the larvae. A few days after the larvae dies, the body of the larvae begins to harden and stiffen. This is because the whole body of the larvae is enveloped by mycelium [30].

M. anisopliae secretes large amounts of trehalase acid, a type of enzyme that reduces trehalose in hemolimfa of insects. Trehalose is the main sugar found in hemolimfa insects. The presence of a trehalase enzyme in the body of a fungus infected insect will lead to a decrease in adaptation of insects in the environment [31]. *M. anisopliae* also produces destrucsin cyclic peptide poison. Some of the destrucin, especially destrucin A, B and E show insecticidal activity [4,36].Mycotoxins of *M. anisopliae* inhibits the feeding and mobility of insect larvae [4]. Destrucsin is synthesized via multifunctional peptide synthetic [4], [36] and injection of destrucsin in pest insects causes muscle paralysis [32]. Insect paralysis results in a direct link between insect damage and mortality.Although some studies of *M. anisopliae* are less prospective, one of the fungal application constraints is slowly killing target insects. Furthermore, through this research we found that the crude extract of *M. anisopliae* is more effective than conidia suspension to control *C. pavonana* larvae primarily to slow the death time of the larvae. However, in further research we would like to know the mycotoxin production opportunities in growth media, and the increasing factor of growth medium of mycotoxin production. In the following decade, better knowledge of *M. anisopliae* applications and increased public concern about the effects of reducing pesticide use by using entomophatogenic products for a sustainable environment, especially in controlling *C. pavonana* in vegetable plantations is expected to change.

4. Conclusions

The results of mortality test of crude extract showed that *M. anisopliae* had produced mycotoxin from the 3rd day of fungal growth. The longer the fungal growth, the higher the resulting mycotoxin was obtained. Mycotoxins was optimized on the 7th days. This was demonstrated by the extract of *M. anisopliae* having the highest mortality effect on larvae tested for 24 to 120 hours. One hundred percentage of mortality of larvae was achieved at 120 hours with an average time of 2.5 days shorter rather than mortality caused by conidia infection. The average lethal time of 10^9 spores/mL suspension was 4.66 days. The extraction of *M. anisopliae* proved to be more effective in controlling *S. litura*, observed from the lethal time was shorter than the conidial suspension.

Acknowledgement

We are grateful to Rector & DRPMI Padjadjaran University, that was supported this research with "Hibah Internal Unpad (HIU)" in scheme of Fundamental Research Programme 2017 grant.

5. References

[1] Abd El- Ghany T M, El-Sheikh H H , Abd El-Rahman G A and Abd El-Nasser A M 2012 Biodiversity of entomopathogenic fungi in new cultivated soil with their using to control of

Galleria mellonella Int J Cur.Res. Rev. 04 17-31
[2] Patricia V T, Neiva T O, Elza Á L and Alves L 2014 Biological insect control using Metarhizium anisopliae: morphological, molecular, and ecological aspects Ciência Rural Santa Maria 44 644
[3] Faria M R and Wraight S P 2007 Mycoinsecticides and mycoacaricides: a comprehensive list with worldwide coverage and international classification of formulation types Biol.Control 43 237- 256
[4] Augusto S and Marilene H 2010 Metarhizium anisopliae enzymes and toxins Toxicon 56 1268
[5] Irna M, Suzzana F S and Irda S 2017 Potensi jamur entomopatogen untuk mengendalikan ulat grayak Spodoptera litura F. pada tanaman tembakau invivo J. Agroekoteknologi FP USU 5 484
[6] Nuraida dan A Lubis 2016 Pengaruh formulasi dan lama penyimpanan pada viabilitas, bioaktivitas dan persistensi cendawan Metarhizium anisopliae terhadap Crocidolomia pavonana Fabricius J.HPT Tropika 16 196-2012
[7] Aleni U, Peter M R , Richard S, Yu-rong H, Gurion A, Myron P Z and Michael J F 2014 Trichogramma chilonis Ishii: a potential biological control agent of Crocidolomia pavonana in Samoa Biol. Control 73 31-32. [1]
[8] Morallo R B and Navasero W R 2003 Biology of Crocidolomia pavonana (F.) and population inventions against this recurrent pest of crucifers in the Philippines Philippines Entomologist 17 141-168
[9] Karungi J, Lubanga U K, Kyamanywa S and Ekbom B 2010 Oviposition preference and offspring performance of Crocidolomia pavonana (Lepidoptera: Pyralidae) on different host plants J.Appl.Entomol. 134 704-713
[10] Takeuchi H, Zalucki M P and Furlong M J 2009 Crocidolomia pavonana larval foraging behaviour and feeding site preferences on cabbage Brassica oleracea Entomol. Exp. Appl. 133 154- 164
[11] Gavendra S and Soam P 2014 New prospective on fungal pathogens for mosquitoes and vectors control technology J.Mos. Res. 4 36-52
[12] Lozano T, Garrido J, F. Lafont and Quesada M. 2015. Insecticidal activity of a destruxin-containing extract of Metarhizium brunneum against Ceratitis capitata (Diptera: Tephritidae) J. Econ. Entomol. 108 463
[13] Fuji Y, Tani H, Ichinoe M and Nakajima H 2000 Zygosporin D and two new cytochalasins produced by the fungus Metarrhizium anisopliae J. Nat. Prod. 63 132–135
[14] Ryuji U, Rie I, Yuichi Y, Rokuro M, Kazuro S, Hiroshi T and Satoshi O 2005 New insecticidal antibiotics, hydroxyfungerins A and B, produced by Metarhizium sp. FKI-1079 J. Antibiotics 58 804- 809
[15] Krasnoff S B, Sommers C H, Moon Y S, Donzelli B G, Vandenberg J D, Churchill A C and Gibson D M 2006 Production of mutagenic metabolites by Metarhizium anisopliae J. Agric. Food Chem. 54 7083-7088
[16] Zimmermann G 2007 Review on safety of the entomopathogenic fungus Metarhizium anisopliae Biocontrol Sci. Technol. 17 879-920
[17] Azumi M, Kei-ichi I, Hiroshi K, Takuya N, Fumio I, Tsuyoshi F and Yasuhiro I 2008 Aurovertins F–H from the entomopathogenic fungus Metarhizium anisopliae J. Nat. Prod. 71 278-280
[18] Carlos A C, Ana L C, Letıcia A S, Thais G, Donald W R, Norberto P L and Gilberto U L B 2010 Fungal Biol. 114 473-480
[19] Vyas N, Dua K K and Prakash S 2015 Larvicidal activity of metabolites of Metarhizium anisopliae against Aedes and Culex mosquitoes J Comm Disea. 38 (2) 1-3
[20] Mohanty S S and Prakash S 2008 Laboratory and field evaluation of the fungus Chrysosporium lobatum against the larvae of mosquito Culex quinquefasciatus Parasitol Res.

102 (5) 881-886
[21] Keith B M, Yew-Min T, Johnny M, Liu J, Chun W 2009 Probing inhibitory effects of destruxins from Metarhizium anisopliae using insect cell based impedance spectroscopy: inhibition vs chemical structure Analyst Journal 134 1447-1425
[22] Desak M M, Madihah, Melanie, Hikmat K, Rani M, Fakhrur R and Wawan H 2016 Antifeedant activitiy of ethanol extracts of five selected plants against larvae of taro caterpillar (Spodoptera litura fabricius, 1775) Proc. Int. Conf. 6th ISSH (Bogor) vol 1 (Humanosphere Science School-Biomaterial LIPI) pp 38-48.
[23] Mukesh S, Meenakshi S 2011 Influence of culture media on mycelial growth and sporulation of some soil dermatophytes compared to their clinical isolates J.Microbiol.Anti M.3(8)196-200
[24] Ibrahim L, Butt T M and Jenkinson P 2002 Effect of artificial culture media on germination, growth, virulence, and surface properties of the entomopathogenic fungus hyphomycete Metarhizium anisopliae Mycological Res. 106 705-715
[25] Soundarapandian P and R. Chandra 2007 Mass production of endomopathogenic fungus Metarhizium anisopliae (Deuteromycota; Hyphomycetes) in the laboratory J.Microbiol. 2 (9) 690-695
[26] Barbara C E, Suraia S, Sergio de A and Monica T P 2006 The influence of culture conditions on the biosynthesis of secondary metabolites by Penicillium verrucosum Dierck Microbiol. Res. 161 273-280
[27] Rajan S 2017 Basic of Fermentation Technology (Dehradun & Uttaranchal Institute of Management, India : MedCrave Group LLC) p 54
[28] Prayogo Y, Tengkano W 2004 Pengaruh konsentrasi dan frekuensi aplikasi Metarhizium anisopliae isolat Kendal terhadap tingkat kematian Spodoptera litura J. Ilmiah Sainteks XI 3 233-243
[29] Scholte E J, Bart G J K and Willem T An entomophatogenic fungus (Metarhizium anisopliae) for control of the adult African malaria vector Anopheles gambiae 2008 Entomol. Ber. 68 (1) 21-26
[30] Prayogo Y 2005 Potensi, kendala dan upaya mempertahankan keefektifan cendawan entomopatogen untuk mengendalikan hama tanaman pangan J. Buletin Palawija 10 53-65
[31] Zhao H, Charnley A K, Wang Z, Yin Y, Cao Y Y, Peng G and Xia Y 2006 Identification of an extracellular acid trehalase and its gene involved in fungal pathogenesis of Metarizium anisopliae J. Biochem. 140 319-327
[32] Hu Q B, Ren S X, Chang J H, and Musa J M 2006 Investigation of destruxin A and B from 80 Metarhizium strains in China, and the optimization of cultural conditions for the strain MaQ10 Toxicon 48 491-498

IOP Conf. Series: Earth and Environmental Science **166** (2018) 012018 doi:10.1088/1755-1315/166/1/012018

Prospective Study of *Bacillus sphaericus* and *Pseudomonas aeruginosa* as The Microbial Enhanced Oil Recovery Agents

N Rossiana[1], M Miranti[1], and K Maspudin[1]

[1]Departement of Biology, Padjadjaran University (UNPAD)
Jl. Raya Bandung-Sumedang KM. 21 Jatinangor 45363

E-mail: niarossiana@yahoo.com

Abstract. Microbial Enhanced Oil Recovery (MEOR) is a microbial-based technology to obtain the maximum oil gaining in the oil reservoirs. The adapted bacterial as MEOR agents should be able to produce biosurfactant, high emulsification index values (E_{24}) and capable of lowering the surface-tension in crude oil. The present study undertakes the applicability of two bacterial species *Bacillus sphaericus* and *Pseudomonas aeruginosa* as MEOR agents. Both isolate as indigenous bacteria from West Java oil sludge. The aim of this study is to obtain the best biosurfactant-producing microbe with the highest emulsification index value and the lowest surface-tension in crude oil. Our result shows that *P. aeruginosa* is more suitable than *B. sphaericus* to be used as MEOR agent with the highest emulsification index values (by 65%). This microbe is also able to decrease the surface-tension of crude oil by 7 dyne cm^{-1}. Concurrently, *B. sphaericus* obtains 55% of emulsification index values and is capable of decreasing the surface-tension of crude oil by 2 dyne/cm. *P. aeruginosa* is more suitable than *B. sphaericus* to be used as MEOR agent

1. Introduction

Recently, the annual *crude oil* demand in Indonesia disproportionally increases with its production by 1.7%. In addition, this *crude oil* production has decreased by 1.2% over the last two years. This problem occurs due to the inefficient technique of oil withdrawal that can only obtain 40% of crude oil. Moreover, the remaining crude oil attached to the reservoir can be acquired by tertiary technique using either surfactant or biosurfactant [1]. Due to environmental concern, the demand for biodegradable surfactants is increasing. Moreover, utilization of bio-compounds such as biosurfactant as an alternative to conventional surfactants has been urged due to the rapid development in biotechnology and its advantages, such as lower toxicity and higher biodegradability.

Biosurfactants or surface-active agents are amphiphilic molecules that comprise both hydrophobic and hydrophilic moieties, being the apolar component usually a carbon chain, whereas the polar part, more variable, can be ionic (anionic or cationic) or non-ionic. Those compounds are able to reduce surface and interfacial tensions, as well as to form and stabilize oil in water or water in oil emulsions. Among several potential applications of biosurfactants, its use in MEOR represents one of the most promising methods to recover substantial amounts of the residual oil entrapped in mature oil fields [2]. Indeed, the biosurfactant is able to reduce the surface-tension of oil and water which is a crucial parameter in the Microbial Enhanced Oil Recovery (MEOR) [3].

Oil cannot mix with water due to its hydrophobic characteristic. This occurs due to the difference in interfacial tension between oil and water. Therefore, the interfacial tension of the oil should be lowered to be soluble in water by adding the biosurfactant.

Over the last years, several biosurfactant-producer microorganisms have been used such as *Bacillus subtilis*, *Bacillus sphaericus* and *Pseudomonas aeruginosa* [4]. In this research both bacteria is indigenous microorganism from west java oil sludge. Several characteristics should be met by the MEOR agent, for instance, the ability of producing biosurfactant, high emulsification index values and decrease the interfacial tension of the crude oil. The present study aims to compare the production of biosurfactants produced by *Bacillus sphaericus* and *Pseudomonas aeruginosa* as the MEOR agents.

2. Materials and Methods

2.1. Determination of biosurfactant production

The ability of microorganisms in producing biosurfactant was tested using the blood agar haemolytic method. Briefly, the haemolytic activity of the bacterial strain was observed by inoculation of bacteria on the blood agar medium with the horizontal streak. The inoculum was further incubated at 37 °C in 24 h. The biosurfactant producer bacteria were observed by determining the transparent zone formed around the bacterial colonies [5].

2.2. Determination of microbial growth curve

Bacterial cell growth was monitored by measuring the dry cell weight method. It was determined by centrifugation (10,000 rpm for 30 minutes) of a 1 ml culture broth; the cell pellet was washed with distilled water twice and dried by heating at 50 °C until constant weight was attained.

2.3. Emulsifying activity determination (E^{24})

The emulsifying activity was conducted to determine the emulsification level of total fraction of the biosurfactant against the crude oil. The measurements were made by adding 5 ml of centrifuged supernatant to 5 ml of crude oil. The inoculum was further incubated at 37 °C in 24 h. Afterwards, the mixed solution between the crude oil and the supernatant was vortexed with maximum speed until perfectly mixed, and then it was incubated for 24 h. The Emulsification Index (E^{24}) was finally determined by measuring the height of the emulsified oil solution divided by the total solution after 24 h [7].

2.4. Surface-activity determination

The surface-activity was measured by using the Du-Nuoy method. The measurement was performed in order to determine the surface-tension of the crude oil in the bacterial growth medium. Briefly, the microbe age 5 days Macfarland 3 was inoculated into 100 ml of Salt Medium SS medium in 150 ml glass bottle. After, a total of 10% of crude oil was added into the medium. Moreover, the surface-tension was measured prior to incubation and then re-measured after 7 days of incubation with the Du Noy tensiometer.

2.5. Data analysis

All of the data obtained from the experiment were descriptively analysed. Additionally, general description of the data was obtained by measuring the average, the minimum and the maximum value.

3. Result and Discussion

3.1. Determination of biosurfactant production

Blood haemolysis method is commonly used to detect the presence of biosurfactants produced by the microbes, i.e. indicated by the formation of haemolysis (lysis zone) around the colony. The blood agar haemolysis method was carried out using potential biosurfactant of the lipopeptide antibiotic group causing the lysis of erythrocyte cells and clotting inhibitors as reference [8]. This was because the surfactant serves as a haemolysin substance. Haemolysin could act as an antibody to the erythrocyte membrane antigen, which makes it undergo haemolysis [9]. Our result showed that both *P. aeruginosa* and *B. sphaericus* were able to haemolysed the blood agar.

Table 1. Tests on haemolysis of blood agar

Species name	Transparent zone (cm)	Haemolysis	
		Type	Color
Pseudomonas aeruginosa	0.55	B	transparent
Bacillus sphaericus	0.3	B	transparent

The isolate of *P. aeruginosa* was able to haemolyse the blood agar with the clear zone as much as 0.55 cm (Table 1). On the other hand, *B. sphaericus* isolate produced the clear zone as much as 0.3 cm. Based on clear zone formed, it appears that *P. aeruginosa* has the ability to haemolyse the blood agar more than *B. sphaericus*.

The haemolysis that occurs in both isolates was categorized as beta haemolysis (β) due to its ability to haemolyse the entire red blood cells in the media, thus the colour around the colony became clear (Figure 1).

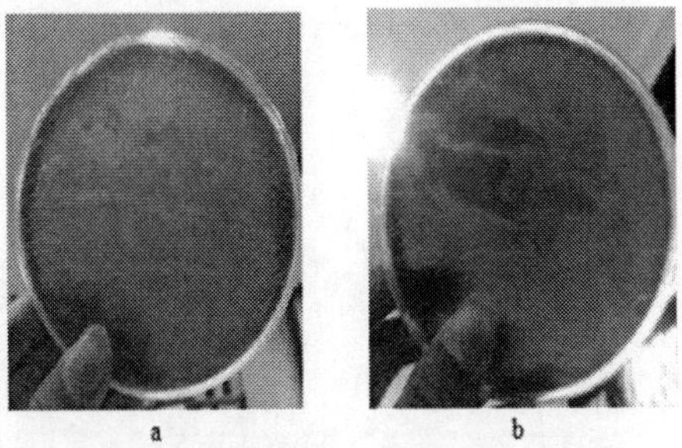

Figure 1. The clear zone formed in haemolysis of *P. aeruginosa* (a) and *B. sphaericus* (b) isolates

Our findings were coherence with the study from Daud et al. [10], who also observed the bacteria that experienced β-haemolysis are the most potential bacteria to produce biosurfactants. In addition, this finding was also confirmed by Ni'matuzahroh [11] who isolated *Pseudomonas* sp. with β-haemolysis type from the coastal area of Surabaya. In contrast, the bacterial isolate without haemolytic activity did not produce biosurfactant [12].

3.2. Bacterial growth curve

In the present study, the initial colony number of *P. aeruginosa* was as much as 1.39 x 10^{12} CFU ml^{-1}. This number was exponentially increased on the third day into 3.90 x 10^{12} CFU ml^{-1}. The optimum number of colonies occurred on day 9 as much as 4.45 x 10^{12} CFU ml^{-1} (Figure 2).

In this study, the exponential phase appeared from day 0 to day 9, indicated by the significant increase of the bacterial colony number. This phase occurs when the bacterial cells utilized the existing nutrient, to the extent of their increase. Moreover, in this phase there is an increase in the number of cleavage and the formation of cell constituents in order to produce an increasing number of cells. In addition, the primary metabolite was also formed in this phase, which is the material of cell constituents and components that increase the amount of metabolism [13].

P. aeruginosa is known as biosurfactant producer in form of rhamnolipid. Indeed, this biosurfactant is an active compound produced by the bacteria that is able to increase the surface tension [14].

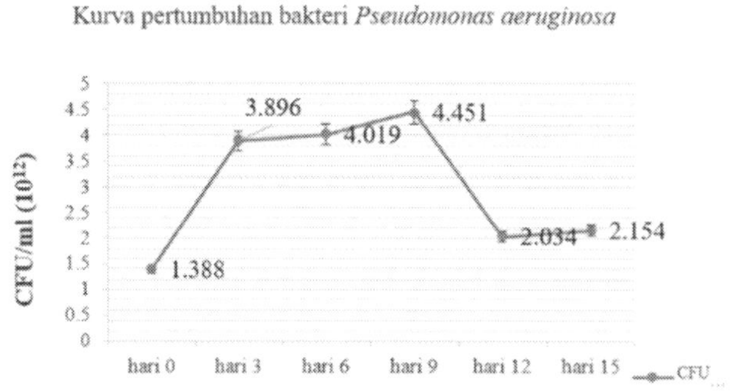

Figure 2. Growth curve of *P. aeruginosa*

In our study, the exponential phase occurred on the day 3, where the number of bacterial cells increased 3 times more than their initial number. However, the division rate on the 6th and 9th days only ranged from 1.03 to 1.11 times from the number of bacterial cells of the previous observation.

Our results revealed that the adaptation phase in *P. aeruginosa* occurred from day 3 to day 6, which subsequently re-increased on day 9. However, observation on day 12 revealed that the bacterial colony increased from 4.45 x 10^{12} to 2.03 x 10^{12} CFU ml^{-1}.

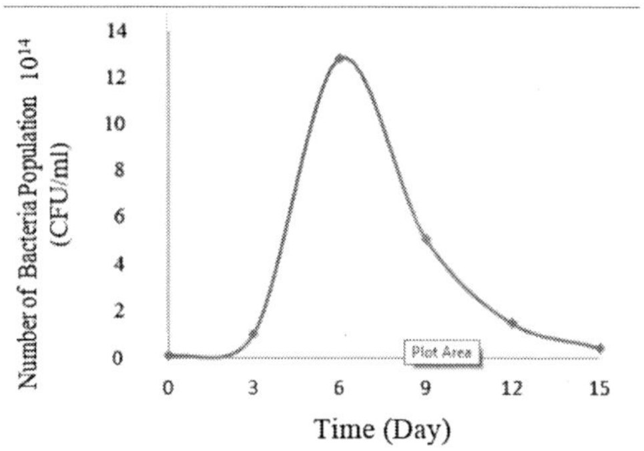

Figure 3. Growth curve of *B. sphaericus*

Growth curve of *B. sphaericus* demonstrated that the exponential phase occurred from day 0 to day 6 (**Figure 3**), where the bacterial population significantly increased. In this phase, the number of bacteria increased from 10.80 x 10^{12} (on day 3) to 106.83 x 10^{12} CFU g^{-1}. Moreover, at the end of exponential phase (day 6), this number increased to 1280 x 10^{12} CFU g^{-1}.

The decline of bacterial population occurred right after day 6 until day 15. The number of bacteria decreased at day 9 from 1,280 x 10^{12} to 511.60 x 10^{12} CFU g^{-1}. Furthermore, the decline on the bacterial population continued from 511.60 x 10^{12} to 155 x 10^{12} CFU g^{-1} (day 12) and further decreased to 49 x 10^{12} CFU g^{-1} (day 15).

To our knowledge, the decline in bacterial population after the exponential phase is relatively common in batch fermentation system due to limiting factors that will be achieved during fermentation. These limiting factors are the decrease of nutrients sources and the accumulated auto-toxic products released by bacteria into the fermentation medium [13].

3.3. Emulsification Index (E_{24})

The emulsification index can be obtained by measuring the height of the emulsified oil solution divided by the height of the total solution [7]. Based on the emulsification index value, we demonstrated that *P. aeruginosa* had better ability to emulsify crude oil than *B. sphaericus* (Table 2).

Table 2. Emulsification index of *P. aeruginosa* and *B. sphaericus*

Species	Emulsification Index value (%)
P. aeruginosa	65
B. sphaericus	55

The emulsification index value of *P. aeruginosa* was 65%, while that of *B. sphaericus* was 55%. Indeed, emulsification capability is one of the requirements of microorganism to be an ideal candidate as MEOR agent. Emulsification could be defined as a process when the two fluids cannot be mixed, thus it needs to be stabilized by the emulsification agent, such as biosurfactant, to form an emulsion. This emulsification capability can be used to increase the oil production by emulsifying the trapped crude oil in the rock thus it can be easily removed with water.

The emulsification index states the precentage of capability of emulsion formation by bacterial isolate. Furthermore, this index can be used to select isolates of biosurfactant-producing bacteria. The large percentage of stable emulsions formed indicate the greater amount of biosurfactant produced by the isolates [15]. The applicability of biosurfactants in several fields depends on their stability at different temperatures. The stability of biosurfactant was tested over a wide range of temperatures. The biosurfactant produced by *Bacillus* and *Pseudomonas* species respectively was shown to be thermo stable as indicated by the stability in the surface tension values and the oil displacement area. Therefore, it can be concluded that these biosurfactants maintains their surface properties unaffected in the range of temperatures between 30 and 100°C.

3.4. Surface tension

The surface tension could be defined as the effort required to expand the liquid surface per unit area. Indeed, the surface tension is a force that arises along the surface line of a liquid. This force arises due to the contact between the two different fluids phase. In our study, the surface tension measurement was performed at the initial and at the end of incubation (Table 3).

Tabel 3. Surface tension on *P. aeruginosa* and *B. sphaericus*

Species	Surface tension	
	Before incubation (dyne cm^{-1})	After incubation (dyne cm^{-1})
P. aeruginosa	27	20
B. sphaericus	27	25

Our results showed that both bacterial species were able to decrease the surface tension of crude oil after incubation. The surface tension for both species prior to incubation was similar 27 dyne cm^{-1}. Afterwards, the surface tension in both bacterial species decreased after 7 days of incubation (20 and 25 dyne cm^{-1} for *P. aeruginosa* and *B. sphaericus*, respectively). In other words, *P. aeruginosa* isolate was able to decrease surface tension as much as 7 dyne cm^{-1}, while *B. sphaericus* isolate was only able to decrease as much as 2 dyne cm^{-1}. Therefore, *P. aeruginosa* had a better ability in lowering the surface tension than *B. sphaericus*.

The effect of sodium chloride addition on biosurfactant produced from *Bacillus* and *Pseudomonas* species was studied. the stability of biosurfactant produced by *Pseudomonas aeruginosa* at NaCl concentration range of 2-20% (w/v) and biosurfactant produced by *Streptomyces sp.* at NaCl concentration range of 1-9% (w/v), respectively.Diminution in surface tension occurs due to the exudates of extracellular compounds by the bacteria with surface active properties (biosurfactant) during the incubation period. Biosurfactant is composed by the hydrophobic and hydrophilic groups on its molecule. It tends to be appeared at the interface between the two different phases of polarity [16]. Biosurfactant is equally able to decrease the surface tension between the two phases. Different polarity properties between the two phases of liquids affect their low solubility. Hence, the presence of surfactant molecules that have a tendency towards both phases, the two liquids can be mixed together.

The fluid molecules that present on the surface undergo the resultant force toward the fluid body. Therefore, these molecules tend to squeeze inward to avoid the surface, where the molecules in the fluid have a balanced resultant force. The tendency of particles to move into the body fluids produce a force, and the amount of power required to break the surface of the liquid thus forming a new extent on the surface, called the surface tension [17].

Surface tension is negatively correlated to emulsification index. Thus, the higher emulsification index results to the lower surface tension. The microorganism as MEOR agent should be able to reduce the surface tension of crude oil. Thus, a low surface tension of crude oil will facilitate its removal from the reservoir. Indeed, low surface tension makes crude oil emulsify with water easily. Therefore, the crude oil will be mixed with water and easier to remove from the reservoir.

The present study demonstrates that *P. aeruginosa* isolate has not only better emulsification ability, but also in reducing the surface tension than that of *B. sphaericus*. To our knowledge, genus *Pseudomonas* has been widely recognized as one of the microbial groups that have favourable ability in degrading crude oil. This genus also has the ability to degrade aliphatic, aromatic and resin fractions. Furthermore, genus *Pseudomonas* can amplify better by using crude oil as its carbon source [11].

4. Conclusion

The present study demonstrates that *P. aeruginosa* has higher emulsification index than *B. sphaericus*. The observed emulsification index of *P. aeruginosa* is as much as 65%, which is higher than that of *B. sphaericus* (55%).

Our findings also observed that both species are able to reduce the surface tension of crude oil in the medium of bacteria. The surface tension decreases as much as 7 and 2 dyne cm^{-1} for *P. aeruginosa* and *B. sphaericus*, respectively. Further research is required to determine an ideal media to produce biosurfactant from both bacterial species.

Acknowledgement

The authors wish to thank to Academic Leadership Grant ALG Prof. Dr. Jetty Nurhajati and to the Ministry of Research and Higher Education of Indonesian Republic (KEMENRISTEKDIKTI) for the National Grant awarded to (Dr. Nia Rossiana). The authors also would like to thank to the technicians in the Biology Department, Universitas Padjadjaran, for their technical assistance during the experiment.

5. References

[1] Agarry S E, Salam K K, Arinkoola A and Aremu M O 2015 Biosurfactant production by indigeneous *Pseudomonas* and *Bacillus* species isolated from auto-mechanic soil environment towards microbial enhanced oil recovery *European journal of engineering Technology* **3** ISSN 2056 5860.

[2] Rossiana N and Wulandari A P 2015 Biosurfactants activity from exogenous fungi to decreasing total petroleum hydrocarbon (tph) of balongan oil sludge *Asian Jr. of Microbiol. Biotech. Env. Sc.* **17** 781-784 ISSN-0972-3005.

[3] Fox S L 1993 Comperative analysis of microbially mediated oil recovery by surfactant produced by *Bacillus licheniformis* and *Bacillus subtilis International Conferences on Microbial Enhanced Oil Recovery.*

[4] Rossiana N, Kusmoro J, and Dhahiyat Y 2018 Utilization of *Talaromyces* sp., *Cladosporium* sp., and *Albizia* (*Paraserianthes falcataria* L. Nielsen) mycorhizae on the phytoremediation of oil sludge: change of lead, nickel total petroleum hydrocarbon (TPH) and polyaromatic hydrocarbon (PAH) *Journal of Petroleum and Environmental Biotechnology* **9** ISSN 2157 7463.

[5] Tabatabaee A, Assadi M M, Noohi A A and Sajadian V A 2005 Biosurfactant producing bacteria from oil reservoirs *Iranian J. Env. Health Sci. Eng.* **2** 6-12.
[6] Capucino J G and Sherman N 1987 *Microbiology: A Laboratory Manual.* The Benjamin Cumming Publishing Company, Inc.
[7] Suryatmana P, Edwan K, Enny R, and Wisnuprapto 2014 Karakteristik biosurfaktan dari *Azotobacter chroococcum*. *Laporan akhir dan Seminar Evaluasi RUT XI* 1-8. Serpong: Kementrian Riset dan Teknologi RI.
[8] Schulz D, Passeri A, and Schmidt M 1995 Screening for biosurfactants among crude-oil degrading marine microorganism from the North-sea *Z naturforch (C)* 197-203.
[9] Yatim W 1999 *Kamus Biologi* Yayasan Obor Indonesia.
[10] Daud M 2005 Performa ayam pedanging yang diberi probiotik dan prebiotik *Jurnal Ilmu Ternak* **5.**
[11] Ni'matuzahroh, Yachya A, and Tanjung M 2006 Studi perbandingan biosurfaktan *Pseudomonas aeruginosa* IA7 d dan surfaktan sintetik Tween-80 dalam *b*iodegradasi solar oleh mikroba perairan pelabuhan Tanjung Perak Surabaya P*enelitian Hayati* 13-18.
[12] Agustiani E D 1998 Penapisan bakteri penghasil biosurfaktan dari sumur minyak bumi *Skripsi* Institut Teknologi Bandung.
[13] Myers D 1946 *Sufactant science & technology*. Third Edition John Wiley & Sons.
[14] Martha M 2015 Low-income cancer survivors' use of health- promoting behaviors *Research for Practice: Medsurg Nursing* 50-62.
[15] Budijono and Hasby M 2010 Efektivitas pemakaian zeolit sebagai media biofilter limbah cair tahu *Jurnal Ilmu Perairan* 64-70.
[16] Vater J, Kablitz B, Wilde C, Franke P, Mehta N and Cameotra S 2002 Matrix assisted laser desorption ionization-time of flight mass spectrometry of lipopeptide biosurfactant in whole cell and culture filtrates of *Bacillus subtilis* C-1 isolated from petroleum slude *J. Appl. Environ. Microbiol.* 6210-6219.
[17] Nitschk M and Pastore G 2006 Production and properties of a surfactant obtained from *Bacillus subtilis* grown on cassava wastewater *Bioresource Technology 336-341.*

IOP Conf. Series: Earth and Environmental Science **166** (2018) 012019 doi:10.1088/1755-1315/166/1/012019

Nitrile Hydrolysing Enzymes-Producing Bacterium Isolated from Gandang Dewata Mountain, West Sulawesi

N Sulistinah[1] and R Riffiani[1]

[1]Microbiology Division, Research Center for Biology,
Indonesian Institute of Sciences (LIPI)
Jl. Raya Jakarta-Bogor Km 46 Cibinong, Indonesia

E-mail: nuniksulistinah@gmail.com

Abstract. Nitrile compounds are widely used and manufactured by the chemical industry for synthesis of important compounds. They are known to be highly toxic because of the CN group onits the molecular structure and most of them are hydrophobic, mutagenic and carcinogenic. Nevertheless, some microbes were reportedly able to hydrolyse nitrile by involving nitrile-hydrolysing enzymes, such as nitrilase, nitrile hydratase and amidase. Nowadays, the application of these enzymes is increasingly recognized because it may be used as a biocatalyst to produce some of important pharmaceutical compounds and/or synthetic chemical that has high economic values. These enzymes also have a significant role in environmental protection due to their capability to eliminate toxic compounds. The probability to acquire potential microbes-degrading nitriles in nature and extreme environments is extremely high. Some bacteria have been isolated from Gandang Dewata mountain areas. SB1D1 bacterial isolate was identified as *Rhodococcus pyridinivorans*. It showed potential activity as bacteria-hydrolysing nitrile. The bacteria was able to grow on CH_3CN (acetonitrile) until the concentration of 1000 mM and also it was demonstrated to degrade 50 mM C6H5-CN (benzonitrile) and produced benzamide, benzoic acid even in low concentrations. SB1D1 isolate is a gram-positive bacterium and the colonies are light orange in colour.Based on the literature review, it is known that *R. pyridinivorans*iswidely studied detail and usually used for bioremediation agent to degrade toxic compounds, such as pyridine, PAH etc.Therefore,*R. pyridinivorans* strain SB1D1 is a good candidate as nitrile hydrolysing-bacteria for degrading aliphatic and aromatic nitrile.

1. Introduction

Nitriles are organic compound that contain CN with the general formula RCN. They are found everywhere in nature, produced and widely used in the manufactures for synthesizing important intermediates in the chemical industry, such asamide, carboxylic acid, amines , ester etc.[1]. Most of synthetic nitriles have been extensively used in the manufacture of extractant, solvent, pharmaceutical, drug intermediate, and pesticides [2]. The excessive use of these compounds triggers environmental problems in the future. Although nitriles are toxic in nature, but many microbes are able to hydrolysenitriles to less-toxic and valuable carboxylic acid using nitrile-hydrolysing enzymes (nitrile hydratase, amidase and/or nitrilase). These enzymes havean important role in the microbial nitriles degradation.Even based on their capability to eliminate toxic compounds, these enzymes also have a significant role in environmental protection [3].Nitrilasehave been reported to be found in various microbes such as *Alcaligenes*, *Bacillus*, *Bulkhorderia*,*Pseudomonas*, and *Rhodococcus*.

The enzymes synthesized by potential microbes are similar to nitrile-hydrolysing enzymes that have been utilized as biocatalysts for the production of important chemical compounds as well as bioremediation agents for the polluted environment [4].

Recently, it has been reported that many industries are interested in the utilization ofnitrile-hydrolysing enzymes isolated from potential microbes as biocatalysts for producing of medicinal or chemical compounds with high economic value. There has been many reports concerning the exploitation of some microbes degrading nitriles and amides to commercially synthesize important compounds, such as for the production of acrylamide, nicotinic acid,benzoic acid, R - (-) - mandelic acid, and S- (+) - Ibuprofen [5]. For example, acrylamide has been commercially produced by Mitsubihi-Rayon Chemical Co., and the application of the utilization of nitrile compounds into drug substances can be found in the process ofbiotransformation of mandelonitrile racemic compounds into R- (-) - mandelic acid and S - (-) - mandelic acidby involving the nitrile enzyme [5]. Its hydrolysis products (amides and carboxylic acid derivatives) have also been widely applied in pharmaceutical, chemical, cosmetic, herbicide, perfume, food [4].

In general the reaction of nitrile hydrolysis can be illustrated in the following:

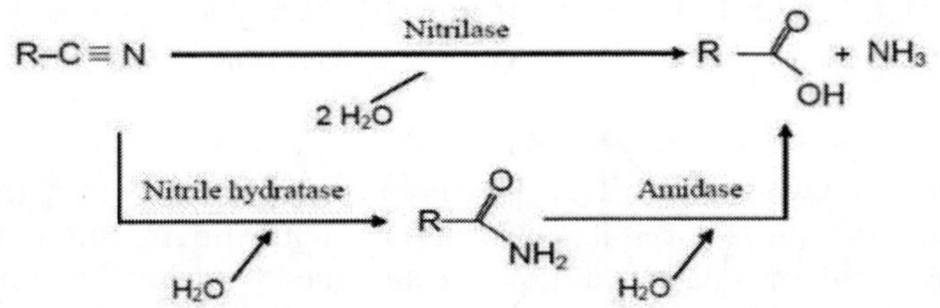

Figure 1. Nitrile-degrading enzymes system (Agarwal, 2012)

Gandang Dewata Mountain, Mamasa Regency is the second highest mountain at Sulawesi. The location is reported to possess mega diversity level and very high uniqueness. With extreme highland conditions (> 1600 m above sea level), it has the potential microbiota for further development (i.e. bioprospecting), associated microorganisms, especially nitrile-hydrolysing bacteria which has the ability as a nitrile producing hydrolysing enzymes as biocatalysts for food production and raw materials of drugs.

The purpose of this study here were to isolate and screen a potential indigenous microbes as a nitrile-hydrolysing microorganisms, especially benzonitrile. It was expected to obtain potential indigenous bacteria from GandangDewata mountain ecosystem which has the high ability and activity to degrade/convert both aliphatic and aromatic nitriles.Thus, itcan be used as a base for further research and provide information and contribution on microbial data of nitrile-hydrolysing bacteria in the region.

2. Materials and Methods

2.1. Collection samples

Soil samples were collected in forest, plantation, surrounding Gandang Dewata area with altitude about 1500-1600 m above sea level and soil pH 6.0-6.5.

2.2.Bacterial growth on nitriles

Nitriles were used as a sole source of energy, carbon, and nitrogen. The culture was inoculatedin minimal medium with certain concentration of nitrile with the following composition : Na_2HPO_4 (0.357 g/L), KH_2PO_4 (0.1g/L), $MgSO_4.7H_2O$(0.1 g/L), $CaCl_2.2H_2O$ (0.01 g/L), $FeSO_4.7H_2O$ (0.001

g/L), yeast extract (0.01 g/L), microelement (1.0 ml/L). The composition of microelement solution was prepared as follows: $ZnSO_4.7H_2O$ (0.1 g/L), $MnCl_2.4H_2O$ (0.03 g/L), H_3BO_3 (0.3 g/L), $CoCl_6H_2O$ (0.2 g/L), $CuCl_2.2H_2O$ (0.01 g/L), $NiCl_2.2H_2O$ (0.02 g/L), $NaMO_4.2H_2O$ (0.9 g/L), Na_2SeO_3 (0.02 g/L).

2.3.Isolation and screening of nitrile-hydrolysing bacteria

About one gram of samples were suspended in 50 ml minimal medium supplemented with various types of nitriles(acrylonitrile, lactonitrile, isobutyronitrile, 3-cyanopiridine andbenzonitrile) in 100 ml Erlenmeyer flask and incubated in rotary shaker (120 rpm) at room temperature (28-30°C) for 7 days under shaking condition. Further enrichment of the medium was carried out by transferring 2% culture to 50 ml fresh minimal medium added with nitrile compound in certain concentration depend on the nitrile toxicity and incubated under similar condition. After three times of transfers, the culture was spread on Nutrient Agar plates and incubated at room temperature. The colony that grows on media was checked by plating on Nutrient Agar plates in several times to obtain a pure isolates.

Screening of isolates were carried out on a sterile macrotitter plates (24 well-macrotitter plate) that contain 1000 μl minimal medium supplemented with 25 mMbenzonitrile or other nitrile. After that, the culture was incubated in rotary shaker at room temperature for 72 hours. Qualitative growth of isolateswasdetermined by transferring 100 μl culture on a sterile microtitter plate (96 well- microtitter plate) and adding with 14 μl INT (Iodonitrotetrazolium). The gradual colour change into pink has indicated growth of isolates.Meanwhile, the activitity was determined using Nessler's reagent [6].

2.4. Confirmation test of nitrile-degradingbactreia

Screened isolates were grown/inoculated in 100 ml Erlenmeyer flask contained 50 ml minimal medium supplemented with nitrile in certain concentration. Furthermore, culture was incubated on the shaker at room temperature for 3 days and growth determined periodically by measuring the optical density of the culture using a spectrophotometer at 436 nm. Nitrile-hydrolysing activity was assayed by measuring the amount of ammonia released. Bacterial isolates that grow well and have the highest activity will be further tested.

2.5. Growth and activity of selected isolates

The inoculum isolateswere prepared by inoculation (4% v/v) cultures grown in the exponential phase into Erlenmeyer 1000 ml containing 500 ml growth medium (pH 7.2) and by adding acetonitrile 200 mM or benzonitrile 15 mM in rotary shaker at room temperature. Samples were periodically analysed for determining the growth and the activity. Nitrile hydrolysing activity was measured by determining the amount of ammonium based on Nessler's method. As much as 1980μl of 0.1N NaOH was added into 20μl of culture, homogenized, then added 20μl of Nessler B reagent and 20μl of Nessler A. Reagent was homogenized, incubated at room temperature ± 28°C for 20 min and measured using spectrophotometer at 420 nm. Ammonium concentrations were determined based on the standard curves. High activity was characterized by the formation of yellow to brownish colour[7].

2.6.Preparation of biomass for nitrile hydrolysis

About2% inoculum was inoculatedintoErlenmeyer flask (vol. 1000 mL) containing 500 ml minimal medium Selected Biomass Isolate Production. Biomass was produced in Erlenmeyer (vol.1000 ml) containing 500 ml of growing medium and benzonitrile as a substrate. Subsequently the culture was incubated on the shaker at room temperature (± 28 °C) for 72 hours. Cells were obtained by culture centrifugation at 10,000 rpm, 4°C for 15 minutes. The pellet or cell obtained was rinsed with 50 mM phosphate buffer (pH 7.2) 2-3 times and then cell/biomass stored at -4 °C before being used for further testing.

2.7.Assay for nitrile-hydrolysing activities

Nitrile-hydrolysing activity was assayed by determining the ammonia released from the nitrile reaction mixtures that contained the cells of bacteria based on Nessler's reagents and measuring the decrease of benzonitrile (substrate) and the increase of hydrolysis product concentration. The substrate used was 50 mMbenzonitrile (dissolved in 10 ml phosphate buffer pH 7.2). About 1.0 grams cell (w/v) was added to the substrate and incubated in rotary shaker at room temperature for 60 min. Sampling of 1.0 ml was done at minutes 0, 30 and 60. Activity of the enzyme was stopped by adding with 250 µl 4N HCL. Then, the samples were centrifuged for 10 min, the supernatant obtained was injected into the chromatography column. The ammonium concentration was measured using spectrophotometer at 420 nm wavelength.

2.8.Analysis of benzonitrilehydrolysis product

Hydrolysis product of benzonitrile were measured by using HPLC Agilent 1100 with column C18 (Supelco, 5 µm: 15 cm x 4.6 mm) and the mobile phase used Ortho phosphoric acid (0.2%): acetonitrile (25:75), flow rate 1mL/minute; λ 254 nm, column temperature was 25°C.

3. Results and Discussion

3.1.Selection, identification and morphological charactericsof a nitrile-hydrolysing bacteria.

By using enrichment nitrile medium culture technique, sixty eight isolates were obtained from soil isolated at GandangDewata. The screening of 68 isolate showed that 17 isolates demonstrated positive reaction. It was characterized by discoloration after the additionof INT (Iodonitrotetrazolium) and the activity was tested based on the formation of ammonium as a release product of nitrile hydrolysis (Fig 3). Identification of 17 isolates was shown in Table 1 as followed:

Table 1. Screened isolat and identification nitrile-hydrolysing bacteria

No	Isolates Code	Identification nitrile-hydrolysing bacteria based on16S RNA	Homology (%)
1	SB3-E6	*Luteipulveratusmongoliensis*strain MN07-A0370 genome	99
2	SB1-B2	*Bacillus mycoides* strain ATCC 6462, complete genome	99
3	SB2-D1	*Roseovariustolerans* strain EL-16	99
4	SB1-C2	*Bacillus cytotoxicus*NVH 391-98, complete genome	100
5	**SB1-D1**	***Rhodococcuspyridinivorans* SB3094**	**100**
6	SB-E2	*Cupriavidusbasilensis* strain 4G11	99
7	SB-MC	*Bacillus pumilus*strain NJ-M2,	99
8	SB2-E4	*Chryseobacterium vrystaatense* strain LMG 22846 c	100
9	SB5-D1	*Paracoccusversutus* strain DSM 582	100
10	SB2-A4	*Micrococcus luteus*NCTC 2665,	99
11	SB2-E3B	*Cupriavidusbasilensis*strain 4G11	99
12	SB2-A8	*Aeromicrobiummarinum* DSM	98
13	SB2-D2	*Arthrobacterarilaitensis* RE117 c	97
14	SB3-A1	*Bacillus horikoshii* strain *FJAT*	100
15	SB8-E9	*Pseudomonas protegens*	99
16	SB3-A6	*Arthrobacterphenanthrenivorans*	99
17	SB2-D2.1	*Cecembialonarensis*LW9	94

Compared to other isolates, SB1D1 was capable of growing well and had the highest activity. The strain grew on several nitriles both aliphatic and aromatic nitrile compound (Table 2). The growth of the isolate on acetonitrile and benzonitrilewas much better than the growth on lactonitrile or isobutyronitrile (data not shown).

The strain SB1D1 was identified as*Rhodococcuspyridinivorans*and was capable of growing on and utilizing acetonitrile and benzonitrile as the sole source of energy, carbon and nitrogen. The morphological characteristics of the strain aregram positive, non-spore forming, non-motile and cocci. The colonies colour appeared light in the early stage on the nutrient agar medium and then became orange on the late development.The strain grew well at room temperature (30 °C).

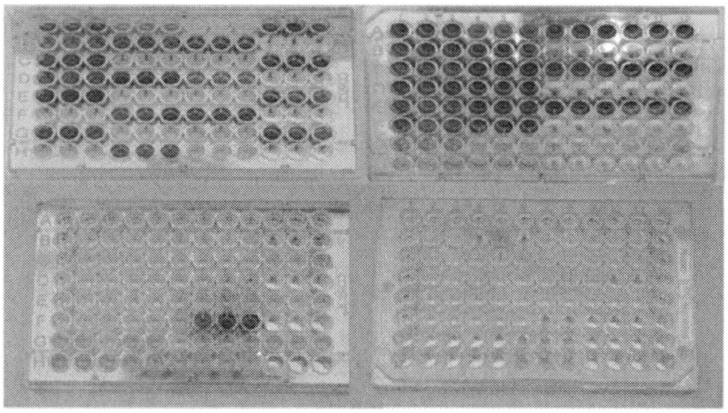

Figure 2. Rapid screening based on the determination of growth using INT and activity based on NH_4 formation

GandangDewata is conservation area reported to have mega diversity and uniqueness, therefore the opportunity to obtain potential microbes as a toxic compound degrader in this area was high. The strain SB1D1 found in this area was expected to have good ability to grow on nitrile. It has been reported that *R. pyridinivorans*was often found in soil, especially in contaminated soil.*R. pyridinivorans*strain TPIK that was isolated from tailing pond of Antam gold mine industry was also reported to have great ability to grow on acetonitrile up to 1000 mM. The isolate was capable of reducing acetonitrile as much as 90% and produced high amount of metabolite product [8]. Members of the genus *Rhodococcus*were often found in many types of environment, especially in extremely contaminated soil. Therefore, a large number of them are known well as toxic degrader strains [9]. It has been reported that *Rhodococci* have great significance in the pharmaceutical industry, environmental and waste management.

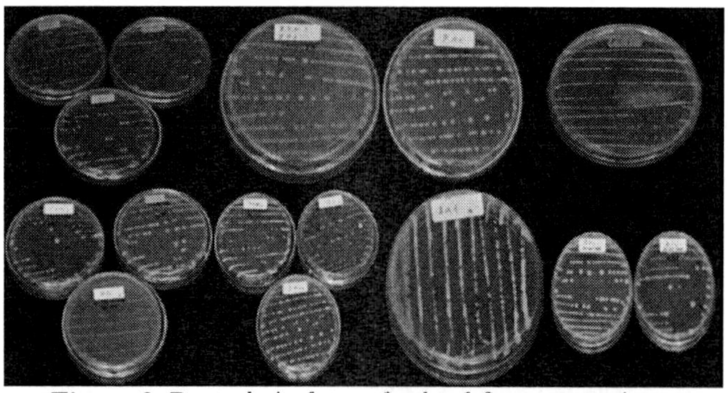

Figure 3. Bacteria isolates obtained from screening

3.2.Growth of R. pyridinivorans strain SB1D1 on nitriles

The strain growth on various nitriles was showed in Table 2. The table showed that *R. pyridinivorans* strain SB1D1 was capable of growing well and had the highest activity on acetonitrile and benzonitrile than others nitriles.

Table 2. Growth and activity test of *R. pyridinivorans*strain SB1D1on various nitrile

Substrates	Growth	Enzyme cctivity
C_2H_3N (acetonitrile)-200 mM	+++	+++
C_3H_3N (akrilonitrile)-10 mM	-	-
C_3H_5NO (lactonitrile)-25 mM	++	+
C_4H_7N (isobutyronitrile)-10 mM	+	+
C_7H_5N (benzonitrile)-15 mM	+++++	++
C_8H_7NO (mandelonitrile)-10 mM	-	-

Note:+++++/++++ = grow well ;+++ = good;++= fair;+ = less;-= not growing

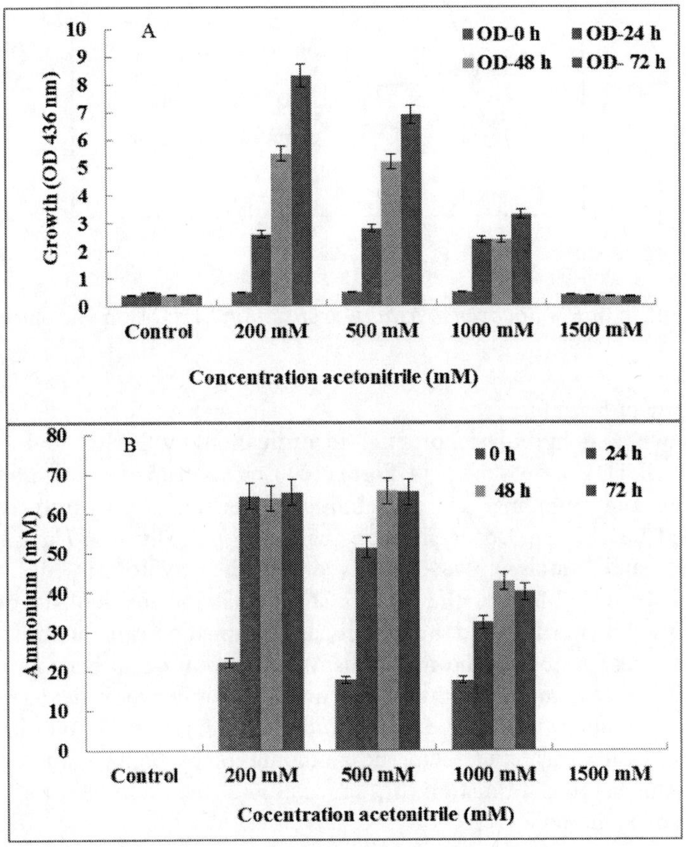

Figure 4. Growth of *Rhodococcus pyridinivorans* strain SB1D1 on acetonitrile in various concentration (A) and ammonium release during growth (B)

The effect of different concentration of acetonitrile and benzonitrile on *R. pyridinivorans* strain SB1D1 was shown in **Figure 4** and **Figure 5**. The highest growth of the strain on CH_3CN was observed at 200-500mM and C_7H_5N at 25 mM. The growth of the strain gradually decreased C_7H_5N on the concentration mentioned above (**Figure 4** and **Figure 5**). *R. pyridinivorans* strain SB1D1 has maximum tolerance ability of up to 40 mM of benzonitrile. Nevertheless, exceeding this concentration, the growth of the strain was completely inhibited due to the toxic effect of benzonitrile.

Benzonitrile is one of the aromatic nitrile compounds usually used as an active or a major component of herbicide like bromoxynil, dichlobenil. Many reports are available on the degradation of aliphatic nitriles and a few on aromatic nitriles. The toxicity level of aromatic nitriles is usually higher than aliphatic nitriles, therefore few microorganisms were able to degrade the compound [10].

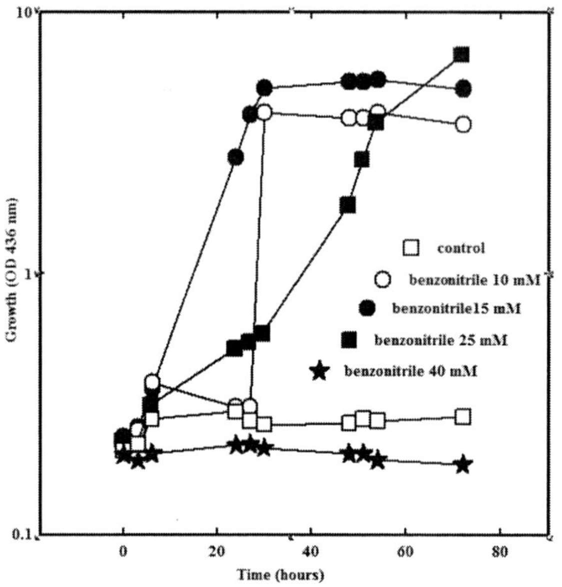

Figure 5. Growth profile of *Rhodococcuspyridinivorans*strain SB1D1 on various concentration benzonitrile

3.3.Catabolic pathway of benzonitrile

The results of growth and hydrolysis product identification by HPLC and enzyme assay in *R. pyridinivorans*strain SB1D1 were shown in **Figure 6**. Benzonitrilewas converted to benzamide and then to benzoic acid and ammonia. The metabolite of benzonitrile hydrolysiswas detected in 30 minutes reaction and benzoic acid concentration increasedgradually in 60 minutes reaction. In this experiment, NHase and amidase was involved in the hydrolysis of benzonitrile by *R. pyridinivorans*strain SB1D1. Most of the reports stated that the metabolism of aromatic nitrile is predominantly catalysed by nitrilase and the hydrolysis of aliphatic nitrile mainly involves NH-ase and amidase. Nitrilase are responsible in the metabolism of benzonitrile to benzoic acid directly [10]. It has reported that NH-ase and amidase pathway was involved for benzonitrile degradation in *Klebsiella* sp. [11].Microbial degradation of nitriles occurs through hydrolysis either by nitrilase, which convert them to their corresponding carboxylic acid and ammonia, or by bienzymatic pathway (NH-ase and amidase). NH-asecatalysis the formation of amides, and then amides was converted to their corresponding acids and ammonia.

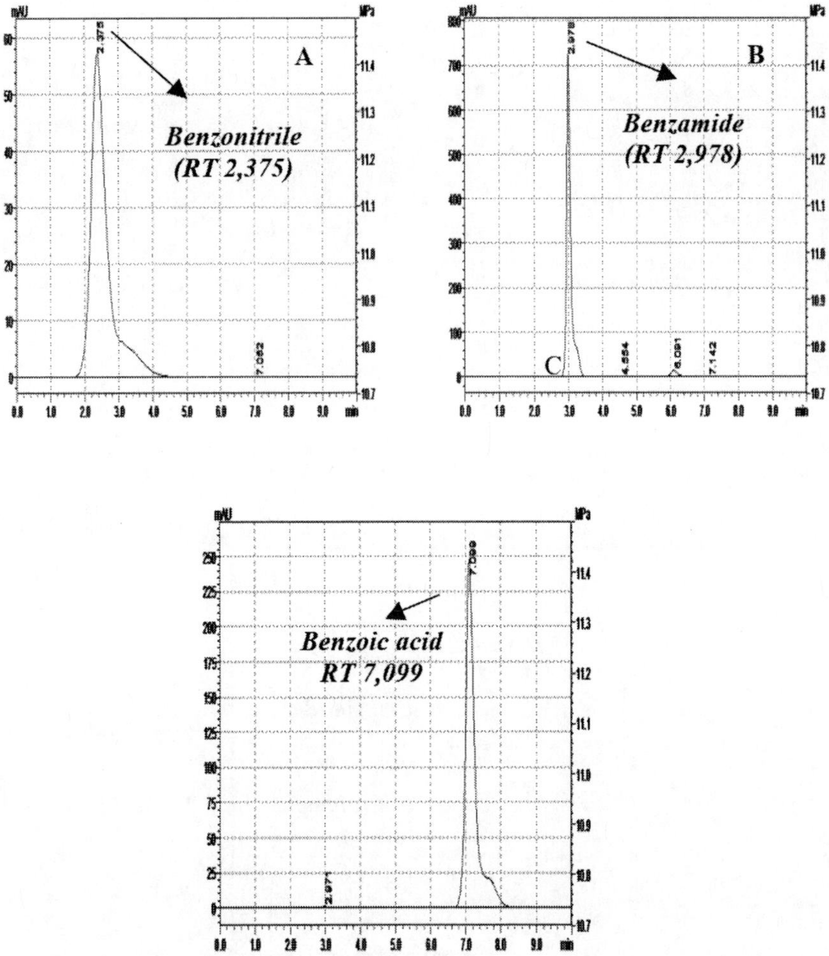

Figure 6. Chromatogram standart of benzonitrile 50 mM (A), benzamide (B),and benzoic acid (C)

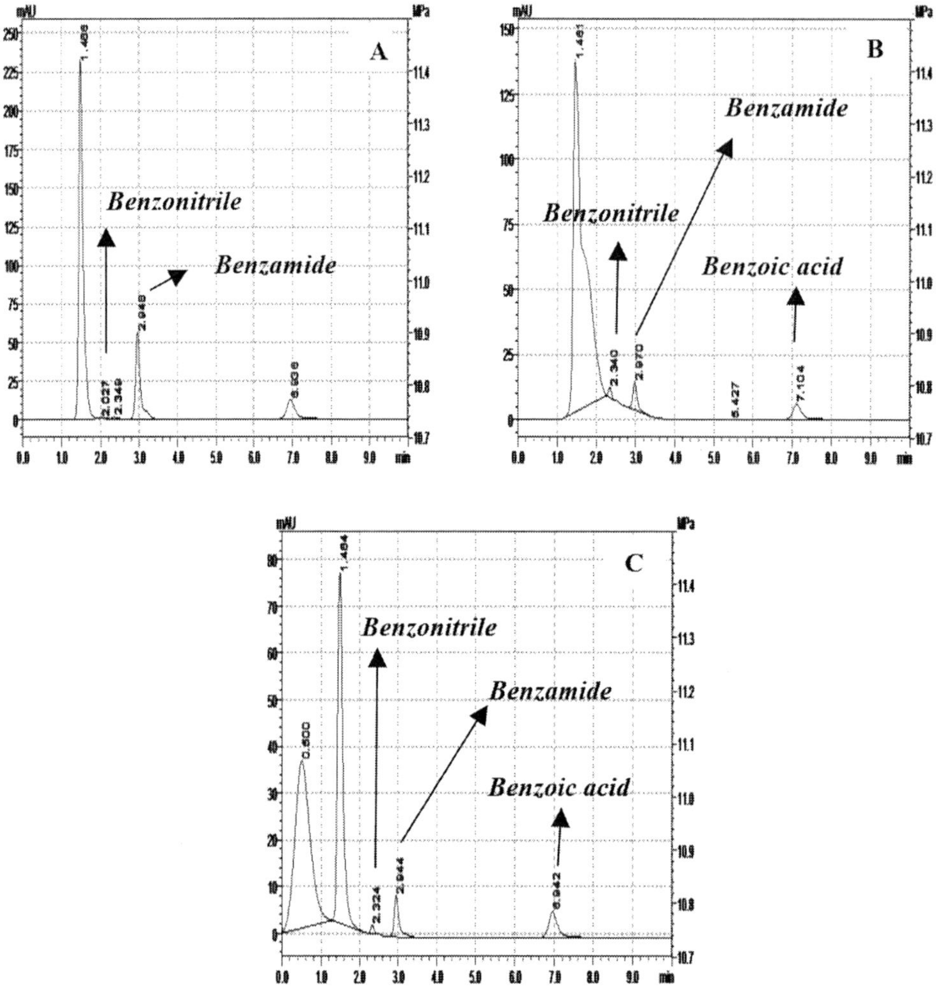

Figure 7. Chromatogram of benzonitrile hydrolysis by *R. pyridinivorans* SB1D1, in the begining reaction (A), 30 minutes (B), and 60 minutes (C) reaction.

4. Conclusions

One potential bacteriumthat was obtained in the research has the ability tohydrolyse of nitrile compound both of aliphatic and aromatic. The bacteria were isolated from soil at GandangDewata West Sulawesi and it wasidentified as *Rhodococcuspyridinivorans* strain SB1D1. The strain was capable of growing and utilizing acetonitrile and benzonitrile as the sole source of energy, carbon and nitrogen.This strain was able to hydrolyse of benzonitrile to benzamide and benzoic acid, although the concentration was relatively low. Therefore, *R. pyridinivoranss*strain SB1D1 is a suitable candidate of indigenous bacteria from GandangDewata West Sulawesi as a potential nitrile degraderwhich can be developed as a biocatalyst in the future for producing important compounds such as food preservative,pharmaceuticaletc. and also may be promising for the remediation of sites contaminated with nitriles. Enzymes involved in the hydrolysis reaction were nitrile hydratase, amidase and not nitrilase enzymes.

Acknowledgments
Authors would like to thank The Expedition Project West Sulawesi - LIPI for financial support in this experiment. We also thank to RestiSitiMuthmainah and IsmuPurnaningsih for their assistance in data analysis.

5. References

[1] Yamada H and Kobayashi M 1996Nitrile hydrataseand its application to industrial productionof acrylamide*Bioscience Biotechnology and Biochemistry***60**1391-1400.
[2] Banerjee A, Kaul P, and Banerjee U C2006 Enhancing the catalytic potential of nitrilase from *Pseudomonas putida* for stereoselective nitrile hydrolysis*App Microbiol. Biotechnol.* **72**77-87.
[3] Santoshkumar M, Nayak A S, AnjaneyaO, and Karegoudar, TB 2010 A plate method for screening of bacteria capable of degrading aliphatic nitriles*Journal Microbiology and Biotechnology***37**111-115.
[4] Agarwal A, Nigam VK, and Vidyarthi AS 2012 Nitrilases-an attractive nitrile degrading biocatalyst*International Journal of Pharma and Bio Science***3**232-246.
[5] Yamamoto K,Oishi K, Fujimatsu I, andKomatsu K I1991 Production of R-(-)-mandelic acid from mandelonitrile by *Alcaligenesfaecalis* ATCC 8750 *Applied and Environmental Microbiology***59**3028-3032.
[6] Sunarko B and Atmosukarto I 2011 Screening of nitrile-degrading endphytic bacteria from biodiversity of Indonesia *Teknologi Indonesia***34** 108-115.
[7] Sulistinah N and Sunarko B 2010 Penapisan mikroba potensial penghasil enzim untuk biotransformasi senyawa nitril *Berkala Penelitian Hayati* **4F**13-18.
[8] Sulistinah N, Riffiani R, and Sunarko B2015*Rhodococcuspyridinivorans* strain TPIK, a nitrile-degrading bacterium isolated from Tailing Pond of Antam Gold Mining Industry, Pongkor*JurnalTeknologi Indonesia***38** 61-67.
[9] Larkin M J, Kulakov L A and Allen C C 2006 Biodegradation by members of the genus *Rhodococcus*: biochemistry, physiology, and genetic adaptation*Adv. Appl. Microbiol.* **59**1-29.
[10] Harper D B 1977 Microbial metabolism of aromatic nitrile *Biochemistry Journal***167**309-319.
[11] Nawaz M S,Heinze T M and Cerniglia C A 1992Metabolism of benzonitrile and butyronitrile by *KlebsiellapneumoniaeApplied Environmental Microbiology***58** 27-31.

Mapping of Tree Damage Classification in The Western Part of Medan City Green Belts Using Geographic Information System

Rahmawaty[1], I. E. P Sembiring[1], R Batubara[1]and P Patana[1]

[1]Forestry Study Program, Faculty of Forestry, University of Sumatera Utara (USU) Jl. Tri Dharma Ujung No. 1 Kampus USU, Medan 20155, Indonesia

E-mail: rahma2107usu@gmail.com

Abstract. The existence of trees in the cities is very important. Trees play an important role as an absorber of CO_2 and a storage facility for carbon (C). This study aimed to classify and map the tree damage in the Western Part of Medan City Green Belts. The experiment was conducted using the Forest Health Monitoring (FHM) and the classification of tree damage classes was determined based on the scoring of index values based on USDA Forest Service. The results showed that there were 46 very healthy trees, 976 healthy trees, 1114 lightly damaged trees, 167 medium trees and 9 dead trees. The level of tree damage was dominated by the lightly damaged trees class. If the health of the tree is well preserved, then the tree can perform its function properly. Therefore, it is necessary to carry outthis research to provide important information about the damage of the trees then it can be used as a reference for the proper trees care activities.

1. Introduction

Medan City is the third largest city in Indonesia. As a metropolitan city, Medan experiences constant transportation activity. Therefore, the existence of trees in the cities is extremely important. Trees have the ability to reduce air pollutants [1].One of the city government efforts to create comfort in Medan City is to build a green belt along the main and secondary roads. The green belts were located either along the edges or median strips of roads or solely along the edges of the road. Green belts along median strips function to assist steering along the road, to assist forming views and restricting glare from other vehicles; those on the edges of roads serve as pollution-absorbent, shade-providing and noise-cancelling barriers as well as wind breakers [1-5]. The primary requirements for landscaping plants in streets as stipulated by the Regulation of the Ministry of Public Works (1996) No. 033are (1) their roots do not damage the road construction, (2) they are easy to maintain, (3) their stems and branches are not easily broken, (4) their leaves do not easily shed and (5) they are safe and offer comfort to motorists and other road users[1].

The trees need to be maintained and should be detected early by knowing the level of damage assessed in terms of damage types, damaged tree location and severity of damage. Early detection allows treatment of unhealthy trees to minimize damage to trees and avoid accidents caused by tree falling [6].The decrease in tree quality is predicted due to air pollution, human activity, biological factors and increasing age of trees. Decrease in quality can be observed from the level of damage suffered by the constituent trees. Damages can be caused by diseases, insect pests, weeds, fire, weather, animals and human activities [6].

Tree damage classification in the Western Part of Medan City Green Belts can implement Mangold modification concept [6]. Trees as the main object of the composition of green belt are expected to be guaranteed its health to provide a sense of security and comfort for the community. Thetree damage in the Western Part of Medan City Green Belts can be presented in the form of maps using Geographic Information System (GIS) technology. The GIS is a system-oriented operation relating to the conventional collection, storage and manipulation of referenced geographical data by computers (hardware and software) capable of handling this data. Among other functions of such mapping is the capacity to show the spatial distribution of geographical phenomena, including the nature and characteristics of their different positions on the surface of the earth [7]. GIS system can be developed a multi techniques and used to compared the field trees mapping survey result accuracy, not just for display a map field data survey result only[8,9]. The model can be develop using another indirect factor such as pixel and whole tree scale, vegetation index, AVIRIS or NDVI statistical analysis[10].

Since information regarding the tree damage in the Western Part of Medan City Green Belts is lacking, conducting this research is of great importance. This study aimed to classify and map the tree damage in the Western Part of Medan City Green Belts. This study is expected to provide information that will assist the government in Medan City in determining a management plan for the environmental development of Medan City.

2. Materials and Methods

This study was conducted from April to October 2016 in the Western Part of Medan City Green Belts. This research was conducted in 11 main roads that were located in five sub-districts, namely: Medan Barat, Medan Baru, Helvetia, Medan Petisah and Medan Sunggal (**Figure 1 and Table 1**).

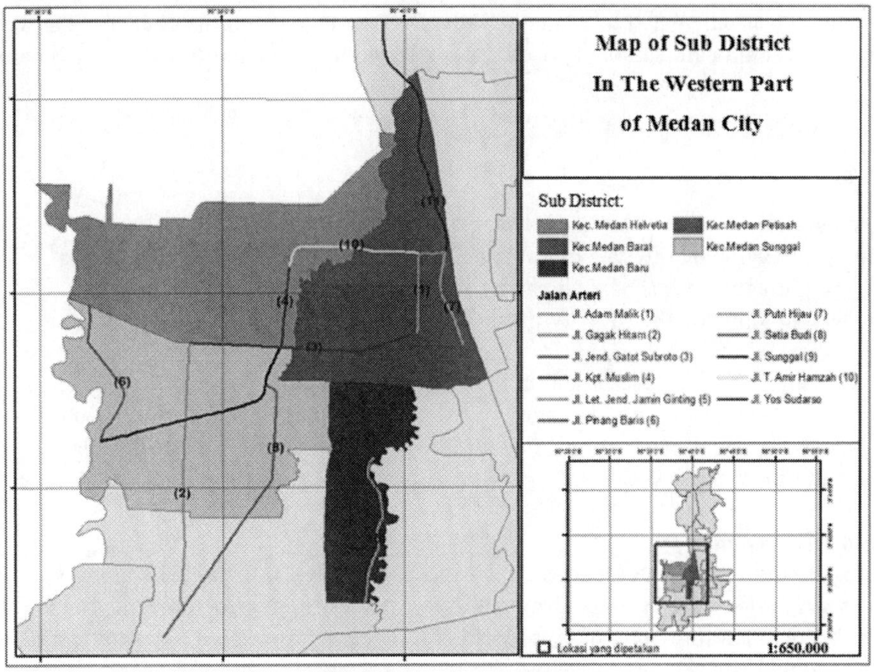

Figure 1. Study site map

Table 1. Name of the road at five sub district in WesternPart of Medan City

No.	Road name	Sub district
1	Jl. Gagak Hitam	Medan Sunggal
2	Jl. Pinang Baris	Medan Sunggal
3	Jl. Sunggal	Medan Sunggal
4	Jl. Setia Budi	Medan Sunggal
5	Jl. Yos Sudarso	Medan Barat
6	Jl. Tengku Amir Hamzah	Medan Barat
7	Jl. Putri Hijau	Medan Barat
8	Jl. Adam Malik	Medan Barat
9	Jl. Kpt. Muslim	Medan Helvetia
10	Jl. Jend Gatot Subroto	Medan Petisah
11	Jl. Let.Jend. Jamin Ginting	Medan Baru

The green belts for each sub-district of Medan City were determined using a purposive sampling technique. A tree inventory was compiled for all tree species (via a census method) located in the selected green belts. The GPS was used to record the coordinate points of the trees and GIS was used to analyse the distribution of the tree damage in green belts [11].

The experiment was conducted using the Forest Health Monitoring (FHM), which is monitoring forest stand conditions to analyse the present and the future, and provide recommendations for improving management. The classification of tree damage classes was determined based on the scoring of index values [12]. The 12 codes and damages type, 10 codes and locations of damage, 10 codes and classes of severity of damage and weights of tree damage index also refers to index values [12].

Damage assessment used criteria based on the FHM method [6,13,14]. Data obtained from the damage assessment was calculated by Damage Index Value (DIV) with the code and the weight of the DIV. Damage assessment calculation (1) used criteria based on previous study [6,13,14],namely:

$$DIV = \sum_{i=1}^{n} (\mathrm{Xi}.\mathrm{Yi}.\mathrm{Zi}) \qquad (1)$$

Where:

DIV : Damage Index Value at the tree level
Xi : The weight value on the type of damage
Yi : The weight value on the part / location of the damaged tree
Zi : The weight value on the severity of the damage

The class of damage for the tree based on the weight of the index value with the following criteria: healthy class: 0 - <5, class of lightly damage: 6 – 10, medium damage class: 11-15, and class of heavy damage: 16 -> 21.The final calculation results (DIV) was classified as: healthy class, lightly class, medium class and heavy class.

3. Results and Discussion

The tree species found on the study site were 13 species (Table 2), with 2,313 numbers of individual trees. *Pterocarpus indiscus* was the dominant tree species found in the study site (54.71%) in accordance to the resultsofprevious study [11,15], followed by *Swietenia macrophylla* (20.66%) and *Swietenia mahagoni* (17.59%).

Table 2. Tree species found at the study site

No.	Name of tree	Scientific name	Number of tree	Percentage
1	Angsana	*Pterocarpus indiscus*	1264	54.71%
2	Mahoni	*Swietenia mahagoni*	407	17.59%
3	Mahoni daun lebar	*Swietenia macrophylla*	478	20.66%
4	Mangga	*Mangivera indica*	9	0.39%
5	Tanjung	*Mimusops elengi*	9	0.39%
6	Glodokan	*Polyalthia longifolia*	60	2.59%
7	Sengon	*Paraserianthes falcataria*	1	0.04%
8	Melur	*Podocarpus elongates*	1	0.04%
9	Petai cina	*Leucaena leucocephala*	7	0.30%
10	Saga	*Adenathera pavonina*	14	0.61%
11	Asam jawa	*Tamarindus indica*	55	2.38%
12	Beringin	*Ficus benjamina*	3	0.13%
13	Waru	*Hibiscus tiliaceus*	4	0.17%
Total			**2313**	100.00%

There are 10 types of damage from 13 types of damage proposed by previous study [6]. Of the 10 types of damage (**Figure 2**), the most common type of damage was vandalism (34%), due to research conducted in urban areas with many advertising activities, such as: sticking banners using nails or wire. The result is in contrast to previous study [16] which found that the most damage type was further decay of the trees. This is because the research was conducted in the Bogor Botanical Gardens where the trees were old.

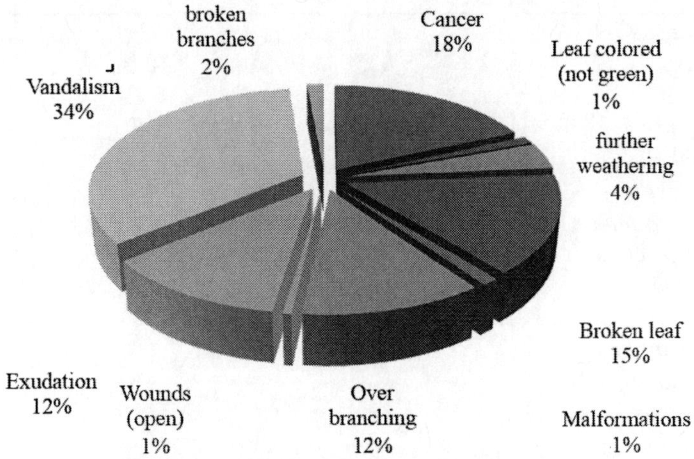

Figure 2. Percentage of damage type

The percentage of tree damage types can be observed in **Figure 3**. A total of 2,312 individual trees on 11 arterial roads found in 5 sub-districts in Medan City, the most affected part of trees was on the bottom of stem (41%), because it was easily accessible to humans. In contrast to previous study [16], the most commonly found tree parts are in the root and stump part, because many trees were old and easily fallen.

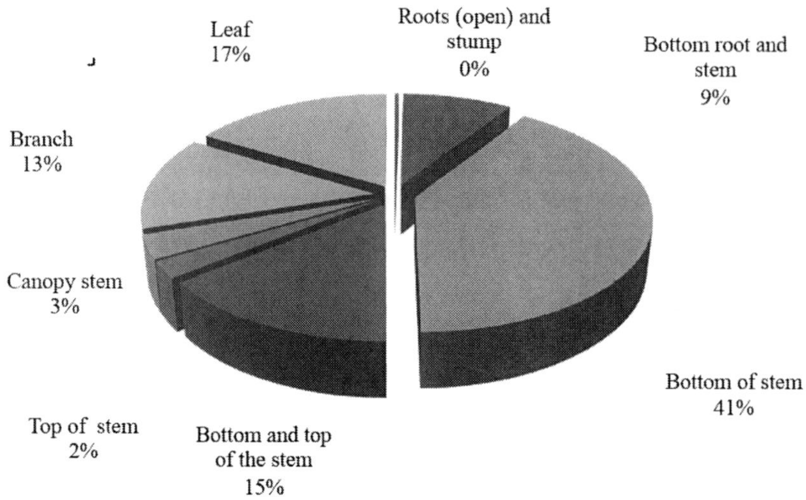

Figure 3. Percentage of damage location

There were three tree damage classifications in the WesternPart of Medan City Green Belts, namely: healthy, lightly damaged, and medium trees. Heavy damage classification was not found in the western part of Medan City Green Belts (**Table 3**). *Pterocarpus indicus*was dominant tree in the study sites (1,232 individual trees). A total of 655 individuals were classified as lightly damage. A total of 455 individuals were healthy and 107 individuals trees were classified as medium.

Table 3. Damage classes based on tree types

No	Name of tree	Very healthy	healthy	lightly	Medium	Heavy	Dead	Number of trees
1	*Pterocarpus indiscus*	8	455	655	107	-	7	1232
2	*Swietenia mahagoni*	13	175	219	32	-	-	439
3	*Swietenia macrophylla*	22	258	180	16	-	2	478
4	*Mangivera indica*	-	7	2	-	-	-	9
5	*Mimusops elengi*	-	4	3	2			9
6	*Polyalthia longifolia*	3	33	21	3	-	-	60
7	*Paraserianthes falcataria*	-	-	-	1	-	-	1
8	*Podocarpus elongates*	-	-	1	-	-	-	1
9	*Leucaena leucocephala*	-	1	5	2	-	-	8
10	*Adenathera pavonina*	-	3	9	2	-	-	14
11	*Tamarindus indica*	-	36	17	2	-	-	55
12	*Ficus benjamina*	-	2	1	-	-	-	3
13	*Hibiscus tiliaceus*	-	2	1	-	-	-	3
Total		46	976	1114	167	0	9	**2312**

Based on FHM criteria, there were 976 healthy trees, 1114 lightly damaged trees, and 167 medium trees. The Percentage of Damage Class can be observed in **Figure 4** while Map of Tree Damage Distribution on 11 arterial roads in Five Sub District in Medan City can be observed in **Figure 5**.

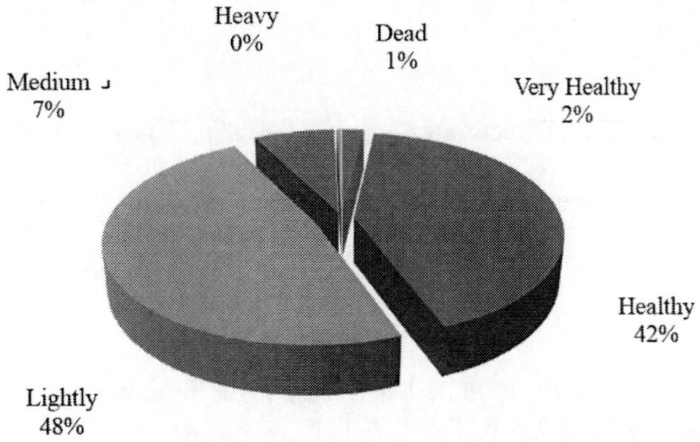

Figure 4. Percentage of damage class

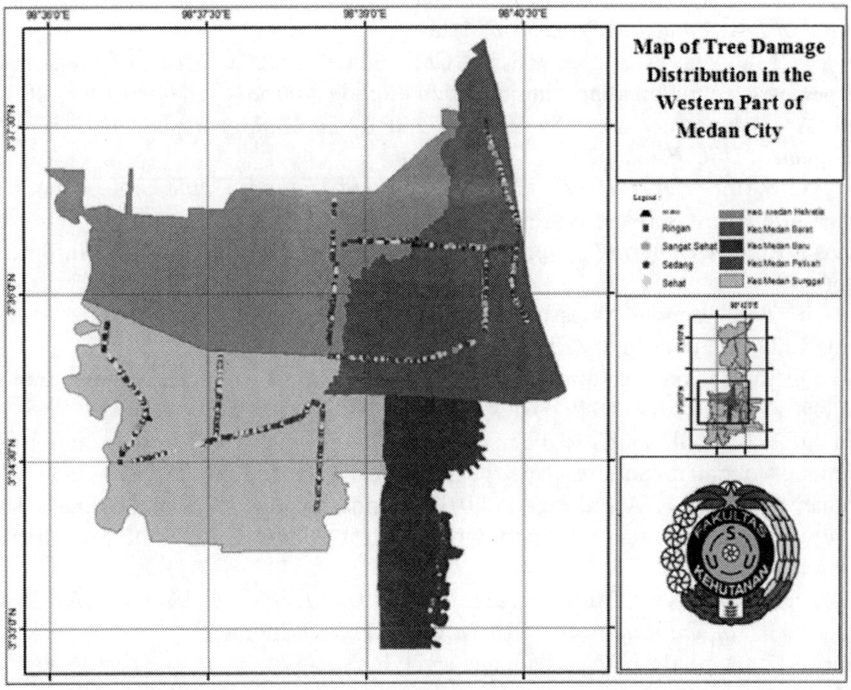

Figure 5. Map of tree damage distribution on 11 arterial roads in five sub district in Medan city

The level of tree damage was dominated by the lightly damaged trees class **(Table 3 and Figure 5)**.Based on these conditions, it will require intensive maintenance, such as watering and trimming. Another thing that that should be a concern in the area was vandalism (destructive trees) that was commonly found in the trees along the western part of Medan City Green Belts.

4. Conclusion

There were three tree damage classifications in the Western Part of Medan City Green Belts, namely: healthy, lightly damaged, and medium trees. Heavy damage classification was not found in this location. The level of tree damage was dominated by the lightly damaged trees class. The most common type of damage was vandalism and most damage locations were found on the bottom of the stem.

Acknowledgement

We are grateful to the Indonesian Directorate General of Higher Education (DIKTI) and the University of Sumatera Utara (USU) for providing financial assistance, as well as to the organising committee of the7th International Symposium for Sustainable Humanosphere 2017, for supporting the presentation of this work on 1-2 November 2017, at the Salak Tower Hotel, Bogor. We also thank the students of the Faculty of Forestry at the USU (especially for Indra) who has assisted with field work.

5.References

[1] Rawat JS and Banerjee SP 1996 Urban forestry for improvement of environment *Energy Environment Monitor* **12** 109-116.

[2] Ministry of Public Works 1996 Planning procedure of technique landscape road No. 033/TBM/1996 *The Directorate General of Highways; Ministry of Public Works Jakarta*pp 58

[3] Sharma SC, Sharga and Roy RK 1994 Abatement of industrial pollution by landscaping *Indian Journal of Environmental Protection***14** 95-97.

[4] Nowak DJ, Stevens JC, Susan MS and Christopher JL 2002 Effects of urban tree management andspecies selection on atmospheric carbon dioxide *Journal of Arboriculture* **28** 113-121.

[5] Rauf A, Rahmawaty and Syofyan A 2016 *Technology-based land use for watershed management USU Press*pp26.

[6] Mangold R 1997 *Forest Health Monitoring: Field Methods Guide* United States Department of Agriculture Forest Service. Washington.

[7] Prahasta E 2004 *Geographic information system: Tutorial ArcView*.CV. Informatika Bandung. pp456.

[8] Diem JE 2002 Remote Assessment of forest health in Southern Arizona, USA: Evidence for Ozone-Induced Foliar Injury *Environ.l Manage.* **29** 373-384.

[9] Xiao QF 2005 Tree health mapping with multispectral remote sensing data at UC Davis, *California Urb. Ecosyst.***8** 349–361.

[10] Xiao QF, Ustin SL and Mc Pherson EG 2004 Using AVIRIS data and multiple-masking techniques to map urban forest tree species. *Int. J. Remote Sens.* **25** 5637–5654.

[11] Rahmawaty, Sitorus NA and Rauf A 2017 Distribution, above-ground biomass and carbon stock of the vegetation in Taman Beringin Urban Forest, Medan City, North Sumatra, Indonesia. *The Malaysian Forester* **80** 73-84.

[12] USDA Forest Service 2001 *Forest Health Monitoring to Monitor the Sustainability of- Indonesian Tropical Rain Forest* SEAMEO BIOTROP Indonesia.

[13] Khoiri S 2004 Study of tree damage level in Srengseng Urban Forest, West Jakarta (*Studi tingkat kerusakan pohon di Hutan Kota Srengseng Jakarta Barat) Thesis* Department of Forest Resource Conservation, Faculty of Forestry Bogor Agricultural University (*Skripsi Jurusan Konservasi Sumberdaya Hutan Fakultas Kehutanan Institut Pertanian Bogor).*

[14] Noviady I and Rivai R R 2015 Identification of shade plants health condition in Ecopark, Cibinong Science Center-Botanic Gardens (*Identifikasi kondisi kesehatan pohon peneduh di kawasan Ecopark. Cibinong Science Center-Botanic Gardens*). *Proceedings of National Seminar of Biodiversity Society of Indonesia (Prosiding Seminar Nasional Masyarakat Biodiversity Indonesia)* **1** 1385-1391.

[15] Rahmawaty, Patana Pand Latifah S 2017 Spatial analysis on distribution of green belt to reduce impacts of climate change in Medan City, North Sumatra *Malays. Appl. Biol.* **46** 67–76.

IOP Conf. Series: Earth and Environmental Science **166** (2018) 012020 doi:10.1088/1755-1315/166/1/012020

[16] Miardini A 2006 Analysis of health tree at Bogor Botanical Garden (*analisis kesehatan pohon di Kebun Raya Bogor*) *Thesis* Department of Forest Resources Conservation and Ecotourism, Faculty of Forestry,Bogor Agricultural University *(Skripsi Departemen Konservasi Sumberdaya Hutan dan Ekowisata Fakultas Kehutanan Institut Pertanian Bogor).*

IOP Conf. Series: Earth and Environmental Science **166** (2018) 012021 doi:10.1088/1755-1315/166/1/012021

Diversity of Indonesian Medicinal Plant in The lowland Forest, Bodogol and Its Surrounding of Mount Gede-Pangrango National Park, West Java

S Susiarti[1], M Rahayu[1] and Rugayah[1]

[1]Botany Division, Research Center for Biology
Indonesian Institute of Science (LIPI).
Jl. Raya Jakarta-Bogor Km 46, Cibinong, Indonesia.

E-mail: susi.etno@yahoo.com

Abstract. Ethnobotanical research of medicinal plant was carried out in the lowland forest at Bodogol, Mount Gede Pangrango National Park, West Java. The study site is inhabited by the Sundanese community in several villages of surrounding National Park. The used methods of research were ethno–directed sampling, open–ended interviews with the local community, and a 'walk in the woods' in four permanent forest plots. A total of 93 medicinal plants were recorded in the Bodogol lowland forest and its surroundings, belonging to 78 genera and 50 families. Those species are used for treatment of 32 different diseases. The species is most frequently used for post-pregnancy. Three of the recorded species namely *Alstonia scholaris, Cinnamomum sintoc,*and *Fibraurea tinctoria* are endangered plants. The community rarely collects medicinal plants from the forest; they prefer to collect from around the village or from cultivated plants.

1. Introduction

Indonesia's key position as a center of the world's biodiversity is reflected in the wealth of natural resources and the variety of ecosystems. The natural diversity of plants and wildlife can be found in conservation areas, which cover a total area of about 28.26 million hectares of nature reserves and national parks [1].

Mount Gede-Pangrango National Park (NP) is one of 50 National Parks in Indonesia. It covers 21.975 ha and located in Bogor, Sukabumi and Cianjur, West Java. The flora of this area was reported containing 1400 species, including 105 species of ornamental plant, 100 species of survival plant and 107 species of medicinal plant [2].

Bodogol is the lowland area of Mt. Gede Pangrango NP that still exists in relatively good condition. This area has been a site for ecological research since 2007. To conserve this area, many activities have been carried out, and ethnobotanical studies have been conducted since 2009. The aims of this study are to create an inventory of species diversity of Indonesian medicinal plants used by the local people and to document the interaction between them and the surrounding forest area.

IOP Conf. Series: Earth and Environmental Science **166** (2018) 012021 doi:10.1088/1755-1315/166/1/012021

2. Materials and Methods

Data were collected using ethno-directed sampling and open-ended interviews with the local Sundanese community. The study was conducted for 2 years, through 4 visits from 2009-2010. Twenty house holds were randomly selected interviewed. The main research questions were as follow: which plant species and which part of the plant are being used for such diseases, for commercial purpose.Meanwhile, the main question was: when does the plant can be harvested and how many times a year.

In addition, the 'walk in the woods' method was used in four permanent forest plots, namely natural forest, Pine (*Pinus merkusii*) forest, Rasamala (*Altingia excelsa*) forest, and Agathis (*Agathis borneensis*) forest. Previous research reported that the numbers of trees per Ha in these forest types are 140 species (natural forest), 20 species (Pinus forest), 133 species (Rasamala forest) and 46 species (Agathis forest) [3].

Several questions concerning medicinal plants were addressed to collect information on the vernacular names of plants, localities where they were collected, parts of plants used and how they are used. Herbarium specimens were collected in the field, further identified at the Herbarium Bogoriense, Botany Division, Research Center for Biology, Indonesian Institute of Sciences (LIPI).

Charateristic of the livelihood respondents in the four villagers around the lowland forest of Bodogol, Sukabumi- West Java consisted of 3status namely farmers (78.26 %), housewives (17.39 %) and shaman (4.35 %). The education grade of these villagers was elementary school. Total percent of respondents of men and women were 78.26 % and 21.74 % wth age 45-54 years (30.43 %), 55-64 years (26.09 %), 65-74 years (30.43 %), and 75-84 years (13.04 %). The study site was presented in Figure 1.

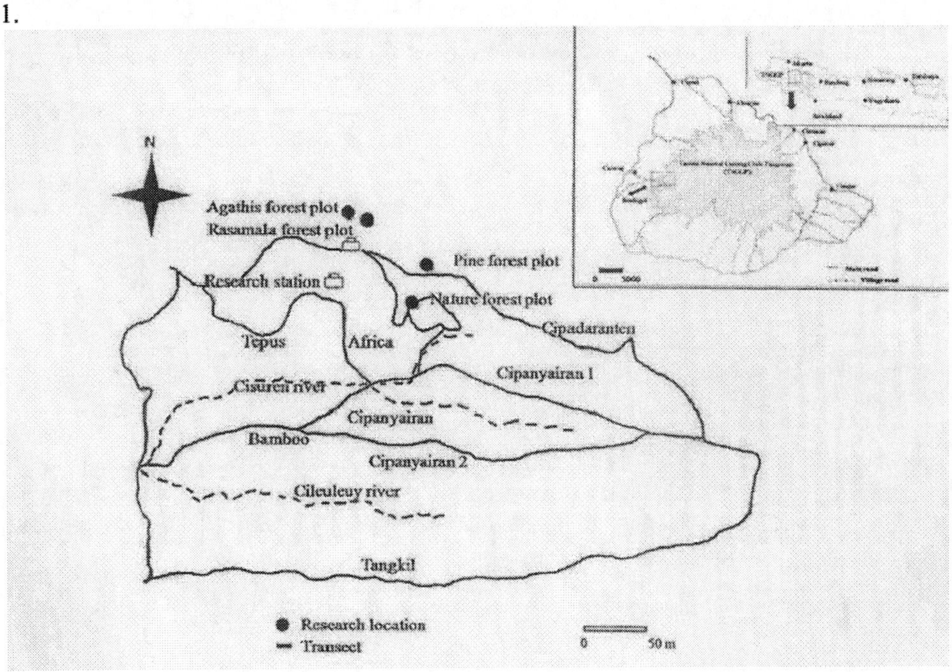

Figure 1. Bodogol area, Mount Gede-Pangrango NP, West Java, Indonesia

3. Result and Discussion

3.1. Diversity of medicinal plant

A total of 93 medicinal plant were recorded in Bodogol, Mt. Gede Pangrango NP, belonging to 78 genera and 50 families (**Table 1**). Those plants are used to treat approximately 32 diseases. Most of them are Angiosperms and only 2 species are ferns. Among 51 families, the highest number of species used by the local people are *Zingiberaceae* (7 species) followed *Asteraceae* (6 species),

IOP Conf. Series: Earth and Environmental Science **166** (2018) 012021 doi:10.1088/1755-1315/166/1/012021

Euphorbiaceae, Moraceae and *Piperaceae* (5 species), *Acanthaceae, Malvaceae* (4 species), *Arecaceae, Poaceae, Rubiaceae* and *Urticaceae* (3 species), *Apocynaceae, Fabaceae, Lamiaceae, Lauraceae, Marantaceae* and *Oxalidaceae* (2 species) with the rest of the families each represented by a single species (**Figure 2**). Palms inventory at Bodogol Forest, Mount Gede Pangrango (NP) found 19 species of palms and *Arenga obtusifolia* is used as medicinal plant [4]. In this area, *Abelmoschus manihot*, that is known by local people as 'kawaro/ mustajab' leaveswas also found. The name recalls the believe that the leaves of this species can heal some diseases such as swelling.

Compare to other locations, the diversity of medicinal plants found in Bodogol (93 species) is lower than in Mount Halimun NP, which has 117 medicinal species [5]. Twenty-eight species found in this research area are similar to those in Mt. Halimun NP [5], five species found in this research area are similar to those reported by Agustini *et al* (2003) [6], 17 species are similar to those reported by Sihotang (2010) [7] and four species the same as representative medicinal plants from Java [8]. Two species, namely *Cinnamomum sintoc* and *Schiima wallichii,* are also found in Cidahu, West Java [9].

Three species for medicinal plant uses were reported in Bodogol *Alstonia scholaris, Cinnamomum sintoc, Fibraurea tinctoria* are include in the endangered plant list of Indonesia [10, 11].

From the 93 species of medicinal plant recorded in this National Park, about 60 % (66 species) are growing wild in the forest (permanent plots), 2 species are found wild in the open area around the village, and the remaining species are cultivated in home gardens.

The more value a plant uses, the greater the importance of the plant [12]. The plants species with high index cultural significance (ICS) values such as *Altingia excelsa, Arenga pinnata*, and *Mikania micrantha* used for medicinal plants are originated from lowland forests of Bodogol [13].

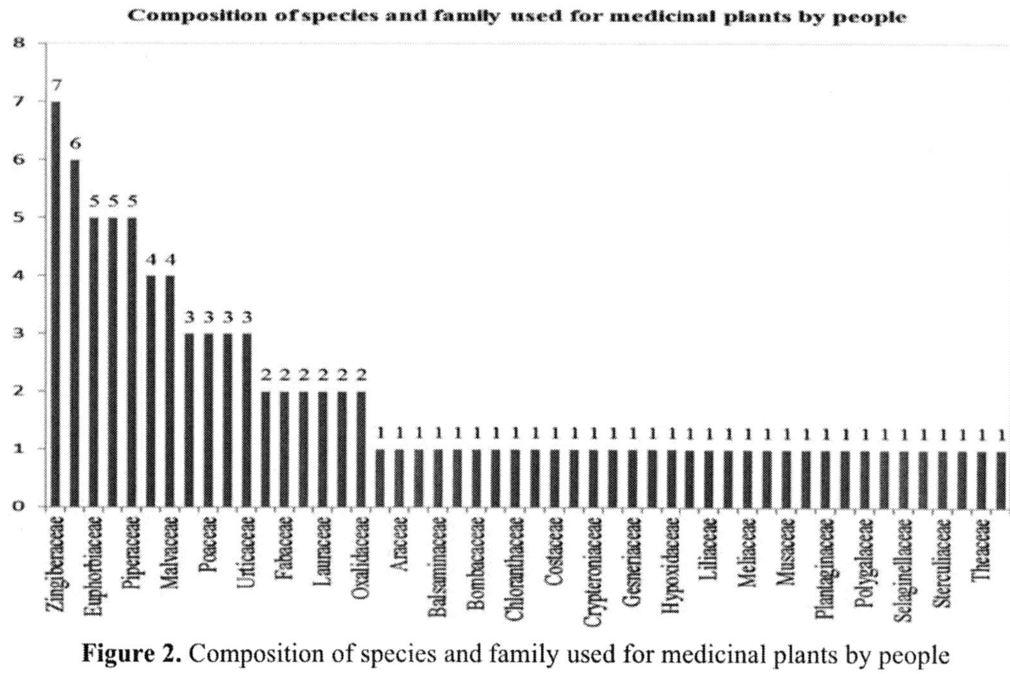

Figure 2. Composition of species and family used for medicinal plants by people

3.2. Diversity of plant parts used

In traditional medicine, different plant parts are used to treat different diseases, and different components of one medicinal plant may have different effects. For some medicinal plants, only one part usually used, while for other species, several parts or the whole plant are used. Of the identified medicinal plants in the research area, the part of the plant used most commonly was leaf only or bark only, while only 2 species were used as the whole plant (**Figure 3**). Leaves are the most commonly used, including young, old or dry conditions. Roots and rhizomes are also commonly used. Parts of the stem used can be either pith or the liquid inside, for example, the liquid from the stem of *Dinochloa scandens* to cure eye problems. Root of *Arenga obtusifolia* (Arecaceae) can be consumed at post pregnancy [4]. In this study, the stem pith of the same species of and the young saray stalk (*Caryota myitis*) can be consumed post-pregnancy.

The local people collect most of the plant used from the village surrounding their house or they cultivate some of them in their garden. Only few species such as *Cinnamomum sintoc, Dinochloa scandens, Fibraurea tinctoria* are collected from forest.

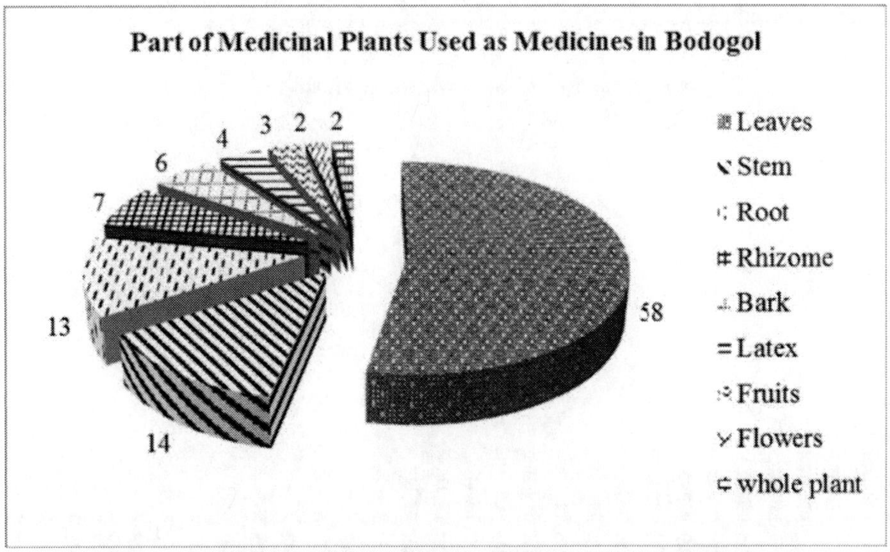

Figure 3. Part of medicinal plants used as medicines in Bodogol

3.3. Diversity of diseases found in Bodogol, Mt. Gede Pangrango NP
The 93 medicinal plant species recorded in Bodogol were used for treatment of 32 different diseases (**Table 1**). Commonly, the local people use one medicinal plant for each disease, but some people also used one medicinal plant for several diseases, for example *Ageratum conyzoides* to treat the stomach, as well as for wound problems, and post pregnancy.

The highest number of species used is for post pregnancy (36), followed by fever (10), backache, problems of the stomach, swelling (7), cough (6), aches, eye, wounds (5), toothache, tonic (4), aphrodisiac, hemorrhoids, liver and warts (3), with other complaints using 1 – 2 species (**Figure 4**).

Pulus (*Laportea stimulans*) is used for asthma treatment. Meanwhile in Mt. Halimun NP area, *Pinanga coronata* is used to treat asthma [5]. A decoction of the plant of *Euphorbia hirta* is also used to treat asthma and its extract has analgesic properties [14]. *Carica papaya, Curcuma longa, C. xanthorriza, Datura metel, Plantago major,* and *Zingiber officinale* are also used for asthma [15].

Fabaceae from Mt. Gede-Pangrango NP have been used extensively as medicinal plants. *Archidendron clypearia* (haruman) and *Euchresta horsfieldii* (local name "ki jiwa"/ or " panjang jiwa") are used as a tonic and for rheumatism, while *Spatholobus littoralis* (kawao) is used as fish poison [16]. *Archidendron clypearia* contains flavonoids which have antiviral activity and polyphenol compounds with anti-inflammatory activity. *Euchresta horsfieldii* contains isoflavone compounds with potential as a lipid-lowering agent. The genus *Spatholobus* contains alkaloids and flavonoids as antioxidants.

The ethylacetate extract of *Eryngium foetidum* showed the strongest antimicrobial activity among the five extracts against the four bacterial strains and *Candida albicans* [17]. In addition, ethanol extract of *Eryngium foetidum* flowers has antibacterial activity against *Staphylococcus aureus* at the consentration of 25 % w/v, with the diameter of growth inhibition is 13.1 mm [18].

Another important species, rasamala (*Altingia excelsa*), is rarely used for traditional medicine in many places in Indonesia. It might be correlated to the narrowly distribution of this species. This species only found in Sumatera and Java [19]. In the research area (Bodogol), the young leaves used by the local people for post pregnancy consumption, and in South East Asia, however, it has been used to cure cough [20]. In China, this plant contains tonic remedy particularly recommended for chest complaints [21].

Plant species in Bodogol such as *Lantana camara* and *Physalis minima*, are used to treat bruises and post pregnancy. Meanwhile, *Lantana camara* and other species of genus *Physalis*, *P. peruviana* have anti cancer potential [22].

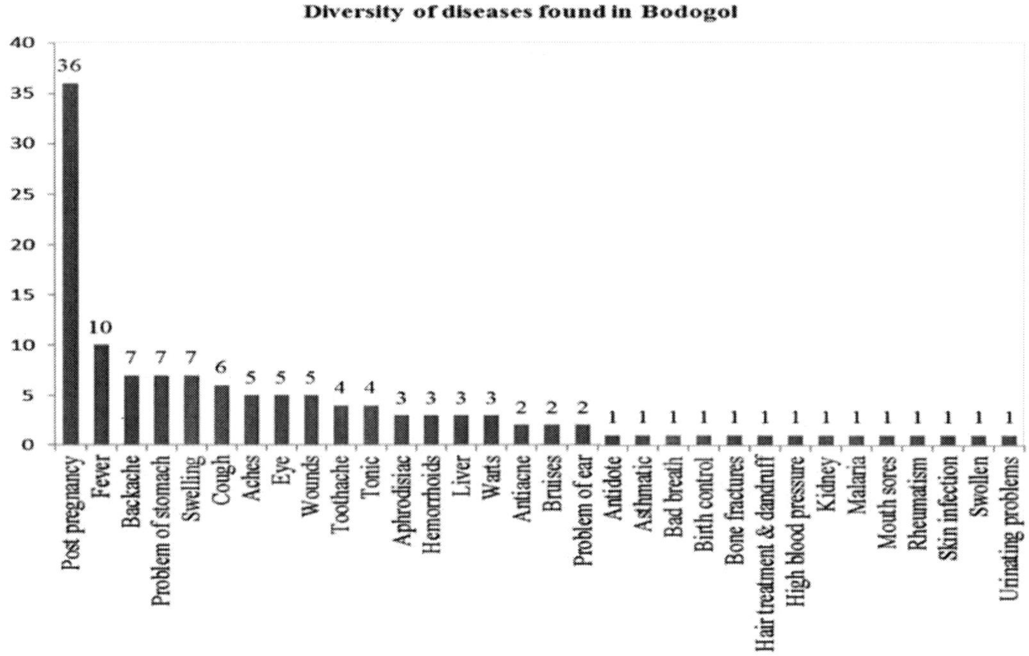

Figure 4. Diversity of diseases found in Bodogol.

3.4. Route of drug administration

Traditional medicinal plants are commonly exploited in simple ways. Medical administrations include oral intake and application to external body parts. Most of the species were orally administrated, mainly as a decoction from plant parts and eaten or chewed (**Table 1**). A number of species were externally applied, namely capituheur (*Mikania micrantha*), jawer kotok (*Plectranthus scutellaroides*) to treat wounds and itches.

Vernacular names usually have meaning, and according to Sundanese in the study area, for example Pulus Rhino means large Pulus (*Laportea* sp.).. This species is also named "daun gatal" (itch leaves). Although the leaves cause itching, but the water from the root (tuak akar) can be used in treatment for asthma and coughs.

3.5 Commercialization of medicinal plant

Amomum compactum and *Orthosiphon aristatus* were commercialized medicinal plants in the research area. Local people in Bodogol, Mt. Gede Pangrango NP have been cultivating these two species in home gardens. The two species are perennial herb, the round cardamom (*Amomum compactum*) harvest after 6 months and they commersialized the fruits for Rp. 40,000 – Rp. 60,000/Kg. In the other hand, Javanese tea (*Orthosiphon aristatus*) usually harvested after 3 months and in interval times 2 weeks. They sold the dried leaves Rp. 5,000/Kg to a collector. They received additional income Rp.

2,880,000 - Rp. 3,600.000/ a year (about US$ 360 from the round cardamom and Rp. 1,680,000 - Rp. 2,400,000,-/ a year (about US$ 200) from java tea.

Indonesian exports Javanese tea to Europe and other countries an average of 170 ton/year of dried leaves in the period 1991-1995 [23]. Meanwhile, there is no statistical data for round cardamom. In Asian market, the round cardomum is still included under true cardamom (*Elettaria cardamomum*) production. The production of round cardamom is unstable and varies for 90-300 ton annual harvest during 1976-1983. The species is only traded in Asia [24].

4. Conclusion

A total of 93 Indonesian medicinal plants were recorded and recognized by the local people in Bodogol lowland forest and the surrounding area in Mt. Gede Pangrango NP, belonging to 78 genera and 50 families. Those species are used for treatment of 32 different diseases. The highest numbers of species are used as post pregnancy consumption. *Alstonia scholaris, Cinnamomum sintoc, Fibraurea tinctoria* are included in the endangered plant list of Indonesia. The community rarely collects medicinal plants from the forest; they prefer to collect from around the village or from cultivated gardens. Further research on ethno pharmacology for post pregnancy of several species (*Altingia excels, Arenga obtusifolia, Beilschmeidia madang, Ficus glabella, Pilea trinervia*) might be very interesting to be carried out.

5. References

[1] Anonim 2007 *Buku Informasi: 50 Taman Nasional Di Indonesia* Dep. Kehutanan R I Lestari Hutan Indonesia dan Jica pp 291.

[2] Anonim 2006 Buku Informasi: Flora Taman Nasional Gunung Gede-Pangrango Balai TN G Gede-Pangrango pp 62.

[3] Alhamd L, Rahajoe J S and Wardi 2008 Biodiversity and Ecosystem Function gunung Gede Pangrango National Park Bodogol West Java Presented in Seminar JSPS- Int. Training Program Andalas University. Padang.

[4] Alandana I M, Rustiami H and Widodo 2015 Inventarisasi Palem di hutan Bodogol Taman Nasional Gunung Gede Pangrango Buletin Kebun Raya 18 81-97.

[5] Harada K, Rahayu M and Muzakkir A 2002 Medicinal plants of Gunung Halimun National Park, West Java, Indonesia BCP-JICA PalMedia Citra Bandung pp 135.

[6] Agustini E, Kindarliah I, Gumilang A R, Djaenudin H, dan Sopian 2003 Buku Informasi: TumbuhanSurvival Dan Tumbuhan Obat Taman Nasional Gunung Gede Pangrango. Mulyana A, Agustini E dan Hendrian (Eds.). Balai TN Gn. Gede Pangrango Dir. Jen. Perlindungan Hutan dan Konservasi Alam pp 66.

[7] Sihotang V B L 2010 Ethnomedicinal study of the Sundanese people at the Bodogol area, Gede Pangrango Mountain National Park West Java In: Wong, Lee, et al. (Eds.), Proc. of The 8th Flora Malesiana Symp. Singapore 23-27 August 2010 pp 519-526.

[8] Paik J, Lee J, Choi S, Park S, Marwoto B, Widjhati R, Irawan D, Juniarti F and Surya A 2012 Medicinal Plants of Indonesia: Representative Medicinal Plants from Java. KRIBB. Korea, Indonesia pp 123.

[9] Royyani M F 2009 Pengetahuan dan pemanfaatan tumbuhan oleh masyarakat Desa Cidahu dan Girijaya, Kecamatan Cidahu, Sukabumi. In: Purwanto dan Walujo (Eds.). Proc. of Seminar Nasional Etnobotani IV LIPI Press. Jakarta pp 145-150.

[10] Hidayat S 2006 Tumbuhan Obat Langka di P Jawa: Populasi dan Sebaran. Pusat Konservasi Tumbuhan Kebun Raya Bogor – LIPI pp 99.

[11] Mogea J P, Gandawidjaja D, Wiriadinata H, Nasution, R E and Irawati 2001 Tumbuhan Langka Indonesia. Seri Panduan Lapangan. Pusat Penelitian Biologi LIPI Bogor Indonesia pp 86.

[12] Turner N J 1988 The importance of a rose: evaluating the cultural significance of plants in Thompson and Lillooet Interior Salish Am Anth. 90 272- 90.

[13] Rahayu M, Purwanto Y and Susiarti S 2012 Nilai kepentingan budaya keanekaragaman jenis tumbuhan berguna di hutan dataran rendah Bodogol Sukabumi Jawa Barat Berita Biologi 11 313-320.
[14] Thin N N and Sosef M S M 1999 Euphorbia L. In: Padua et al. (Eds.) Medicinal & Poisonous Plants 1. Plant Resources of South-East Asia 12 Backhuys Publishers, Leiden pp 263 – 272.
[15] Vaidya G K and Braganza V J 2009 Medicinal Plant solutions to asthmatic problems in: Trivedi P C (Ed). Medicinal Plants, Utilisation and Conservation. 2nd Revised & Enlarged Edition. Aavishkar Publishers Distributors, Jaipur 302 003 9Raj India pp 425-452.
[16] Dwipa A S, Irawan D and Juniarti F 2011 The Uses of Fabaceae from the Mount Gede-Pangrango National Park as medicinal plants. Widjaja et al. (Eds.): Proc. of The 2nd Int. Symp. on Temulawak And The 40th Meeting of National Working Group on Indonesia Medicinal Plant. IPB. Bogor. Indonesia. pp 234 -236.
[17] Lingaraju D P, Sudarshana M S, Mahendra C and Rao K P 2016 Phytochemical screening and antimicrobial activity of leaf extracts of Eryngium foetidum L. (Apiaceae), Indo American J. of Pharmaceutical Research 6 4339-44.
[18] Untari B 2014 Ekstrak etanol dari bunga Ketumbar Laut (Eryngium foetidum L.) sebagai insektisida alami dan antibakteri. Proc. of Seminar Nasional Biodiversitas. 3 128-131.
[19] Vink W 1957 Hamamelidaceae Flora Malesiana 1 363-379.
[20] Soerianegara I, Rifai M A, Martawijaya A and Ilic J 1994 Altingia Norona In: Soerianegara & Lemmens (Eds.), Plant Resources of South-East Asia, No. 5 Timber Trees: Major Commercial Timbers, Indonesia pp 90-94.
[21] Wiart C 2006 Ethnopharmacology of Medicinal Plants, Asia and The Pasific. Humana Press Totowa, New Jersey pp 228.
[22] Arbiastutie Y, Marsono D, Hartati M S, & Purwanto R 2017 The potential of understorey plants from Gunung Gede Pangrango National Park (West Java, Indonesia) as cervixs anticancer agents Biodiversitas 18 109-115.
[23] Padua L S De, Bunyapraphatsara N and Lemmens R H M J 1999 Medicinal & Poisonous plants 1 PROSEA 12 711.
[24] Guzman C C De and Siemonsma J S 1999 Spices, Plant Resources of South-East Asia 13 Backhuys Publishers, Leiden, pp 399.

Table 1. List of medicinal plant in the lowland forest, Bodogol, the mount Gede Pangrango NP, West Java

No	Local name	Scientific name	Family	Utilization		
				Part of plant	Preparation & administration	Medicinal application
1	Kaworo	*Abelmoschus manihot*	*Malvaceae*	leaves	Pounded and applied to body	swelling
2	Babadotan	*Ageratum conyzoides*	*Asteraceae*	Leaves	1. Boiled, drunk. 2. Pounded and applied to the body	1. post pregnancy, problem of the stomach, 2. wounds
3	Ki lame	*Alstonia scholaris*	*Apocynaceae*	Bark	Boiled with other plant then drunk	Aphrodisiac, post pregnancy
4	Rasamala	*Altingia excelsa*	*Hammamelidaceae*	Young leaves	eaten	post pregnancy
5	Kapol	*Amomum compactum*	*Zingiberaceae*	Fruits	Pounded and applied to stomach	post pregnancy
6	Paku kebo	*Angiopteris evecta*	*Marattiaceae*	Stem	Pounded and applied to body	fever
7	Langkap	*Arenga obtusifolia*	*Arecaceae*	Pith of stem	eaten	Post pregnancy
8	Kawung	*Arenga pinnata*	*Arecaceae*	1. pith of young stem and root, 2. Old leaves stalk	1. Soaked then drunk 2. Burned, cooled, applied to face	1. post pregnancy, 2. Anti-acne
9	Putat	*Barringtonia racemosa*	*Lecythidaceae*	Young leaves	eaten	post pregnancy
10	Hariang	*Begonia robusta*	*Begoniaceae*	leaves	Crushed and applied to body	fever
11	Songgom	*Beilschmeidia madang*	*Lauraceae*	Young leaves	eaten	post pregnancy
12	Sembung	*Blumea balsamifera*	*Asteraceae*	leaves	Boiled with other plant, drunk	post pregnancy
13	Temu kunci	*Boesenbergia pandurata*	*Zingiberaceae*	Rhizome	Chewed	post pregnancy
14	Kakatukan	*Breynia microphylla*	*Euphorbiaceae*	leaves	Cooked	post pregnancy

15	Kanyere	*Bridelia glauca*	*Euphorbiaceae*	leaves	Pounded and applied to body	Fever (child)
16	Gedang gandul	*Carica pepaya*	*Caricaceae*	1. Root and Young leaves, 2. Dry leaves	1. Boiled then drunk 2. Pounded and applied in face	1. malaria, post pregnancy 2. Antiacne
17	Saray	*Caryota mitis*	*Arecaceae*	Young stem	Cooked	post pregnancy
18	Temung kanyut	*Cerbera odollam*	*Apocynaceae*	leaves	Crushed, with water and drunk	fever
19	Karas tulang	*Chloranthus officinalis*	*Chloranthaceae*	Stem	Boiled with other plant then drunk	post pregnancy, backache, liver
20	Ki sintok	*Cinnamomum sintoc*	*Lauraceae*	bark	Boiled with other plant then drunk	post pregnancy
21	Hanjuang beureum	*Cordyline fruticosa*	*Liliaceae*	Water of young stem	Drop in eye	Eye (itches, red)
22	Pacing	*Costus speciosus*	*Costaceae*	1. Stem, 2. Water of stem	1. Applied to body, 2. Drop in eye	1. Itches because of plant, 2. Eye (itches, red)
23	Ki banen	*Crypteronia paniculata*	*Crypteroniaceae*	Water of stem	Drunk	Cough
24	Marasi	*Curculigo cardifolia*	*Hypoxidaceae*	Fruit	Eaten	tonic
25	Ki beunyeur	*Cyathea junghuhniana*	*Cyatheaceae*	Latex	Apply to tooth	Tooth ache, warts
26	Seureh	*Cymbopogon citratus*	*Poaceae*	Root	Boiled with other plant then drunk	post pregnancy
27	Reundeu badak	*Cyrtandra picta*	*Gesneraceae*	leaves	Pounded and applied to headache	fever
28	Cangkore	*Dinochloa scandens*	*Poaceae*	Water in stem	Drop into eye	Eye
29	Bangban	*Donax canniformis*	*Marantaceae*	Stem, young leaves	Pounded	Antidote
30	Ki sa'at	*Drypetes longifolia*	*Euphorbiaceae*	Stem and leaves	Boiled then drunk	post pregnancy
31	Kadu	*Durio zibethinus*	*Bombacaceae*	Skin of fruit, bark	Boiled with other plant then drunk	post pregnancy
32	Sintrong	*Erechtites valerianifolia*	*Asteraceae*	leaves and root	Pounded and applied to body. Boiled then drunk	swelling, high blood pressure, post

						pregnancy
33	Walang	*Eryngium foetidum*	*Asteraceae*	leaves	Crushed, mixed with coconut oil and applied to body	aches
34	Ki koneng	*Fibraurea tinctoria*	*Menispermaceae*	Stem	Boiled with other plant then drunk	post pregnancy, stomachache, backache and liver
35	Tabat barito	*Ficus deltoidia*	*Moraceae*	1. latex, 2. Leaves	1. Drop in warts, 2. Boiled then drunk	1. warts, 2. Tonic
36	Bunut	*Ficus glabella*	*Moraceae*	Young leaves	Eaten	post pregnancy
37	Walen	*Ficus ribes*	*Moraceae*	Latex	Apply to tooth	Tooth ache, warts
38	Beunying	*Ficus vasculosa*	*Moraceae*	Latex	Apply to tooth	Tooth ache
39	Ki rapet	*Ficus villosa*	*Moraceae*	Whole part	Boiled with other plant then drunk	post pregnancy, stomachache, backache and liver
40	Hahapaan	*Flemingia strobilifera*	*Fabaceae*	leaves	Boiled with other plants then drunk	tonic
41	Ki besi	*Garcinia rostrata*	*Clusiaceae*	Bark	Boiled then drunk	Problem of stomach, backache
42	Ki cantung	*Goniothalamus macrophyllus*	*Annonaceae*	Whole part	Boiled with other plant then drunk	post pregnancy
43	Handeuleum	*Graptophyllum pictum*	*Acanthaceae*	1. Young leaves, 2. Old leaves	1. Eat, 2. Boiled and drunk	Hemorrhoids
44	Gandasuli	*Hedycium coronarium*	*Zingiberaceae*	Rhizome	Boiled then drunk	Cough
45	Ki sariawan	*Helicia robusta*	*Proteaceae*	Young leaves	Eaten	mouth sores
46	Ki remek tulang	*Hemigraphis alternata*	*Acanthaceae*	leaves	Pounded and applied to body	Bone fractures
47	Remek daging	*Hemigraphis* sp.	*Acanthaceae*	1. bark, 2. Leaves	1. Boiled with other plant then drunk 2. Crushed and applied	1. post pregnancy, 2. Bruises
48	Wera	*Hibiscus rosa-sinensis*	*Malvaceae*	leaves	Crushed with water and drunk	Pregnancy (easier)
49	Pacar tere	*Impatiens platypetala*	*Balsaminaceae*	leaves	Crushed and applied to swollen	swollen (stimulant)

50	Cikur	*Kaemfperia galanga*	*Zingiberaceae*	Rhizome	pounded or chewed and applied to breast	post pregnancy
51	Buntiris	*Kalanchoe pinnata*	*Crassulaceae*	leaves	Pounded and applied to headache	fever
52	Dudurenan	*Knema intermedia*	*Myristicaceae*	leaves	Boiled then drunk	tonic, backache
53	Cente	*Lantana camara*	*Verbenaceae*	leaves	Pounded and applied to body	bruises
54	Pulus badak	*Laportea sp.*	*Urticaceae*	water of root	Drunk	Asthmatic, cough
55	Pulus	*Laportea stimulans*	*Urticaceae*	water of root	Drunk	Asthmatic, cough
56	Kahitutan	*Lasianthus inidorus*	*Rubiaceae*	leaves	Eaten	stomachache
57	Kopi leuweung	*Lasianthus purpureus*	*Rubiaceae*	leaves	Boiled then drunk	Problem of stomach
58	Tangkur	*Lophatherum gracile*	*Poaceae*	Root	Boiled then drunk	Aphrodisiac
59	Kaliki	*Macaranga* sp.	*Euphorbiaceae*	Leaves stalk	Warmed and smoke blown to ear	Problem of ear
60	Kaliki beureum	*Macaranga* sp.	*Euphorbiaceae*	Leaves stalk	Warmed and smoke blown to ear	Problem of ear
61	Patat	*Maranta arundinacea*	*Marantaceae*	Rhizome	Eaten or steamed	hemorrhoids
62	Capituheur	*Mikania micrantha*	*Asteraceae*	leaves	Pounded and applied to body	wounds, fever
63	Cau kole	*Musa acuminata*	*Musaceae*	Pith of stem	Pounded and applied to headache	fever
64	Kumis kucing	*Orthosiphon aristatus*	*Lamiaceae*	leaves	Boiled then drunk	Backache, post pregnancy
65	Calincing	*Oxalis corniculata*	*Oxalidaceae*	Root	Pounded and applied to tooth	Tooth ache
66	Babawangan	*Oxalis* sp.	*Oxalidaceae*	leaves	Pounded and applied to body	swelling
67	Menter	*Peperomia pellucida*	*Piperaceae*	leaves	Pounded and applied to headache	fever
68	Cecenet	*Physalis minima*	*Solanaceae*	leaves	Boiled	post pregnancy
69	Pohpohan	*Pilea trinervia*	*Urticaceae*	Young leaves	Eaten	post pregnancy
70	Seuseureuhan	*Piper aduncum*	*Piperaceae*	Distillation of root	drunk	Cough
71	Seureuh	*Piper betle*	*Piperaceae*	leaves	Boiled then drunk	Birth Control
72	Gedebong	*Piper umbellatum*	*Piperaceae*	leaves	Pounded and applied to stomach	stomachache (child)

IOP Conf. Series: Earth and Environmental Science **166** (2018) 012021 doi:10.1088/1755-1315/166/1/012021

73	Seureuh hideung	*Piper ungaramense.*	*Piperaceae*	leaves	Soak in water, drop into eye	Eye (red)
74	Jengkol	*Pithecelobium jiringa*	*Fabaceae*	Young leaves	Pounded and applied to headache	wounds and skin infection in headache
75	Ki urat	*Plantago major*	*Plantaginaceae*	leaves	1. Pounded and applied to body 2. Boiled with other plant then drunk	1.wounds, swollen (stimulant) 2. post pregnancy, aphrodisiac
76	Jawer kotok	*Plectranthus scutellaroides*	*Lamiaceae*	leaves	1. Pounded and applied to body 2. Boiled with other plant then drunk	1. wounds 2. post pregnancy
77	Baluntas	*Pluchea indica* (L.) Lees.	*Asteraceae*	leaves	Eaten or Boiled with other plants then drunk	post pregnancy
78	Ki encok	*Plumbago zeylanica*	*Plumbaginaceae*	leaves	Pounded, mixed with coconut oil and applied to foot	Rheumatism
79	ki remason	*Polygala venenosa*	*Polygalaceae*	Root	pounded ,mixed with coconut oil and applied to foot	Aches
80	Panggang badak	*Schefflera sp.*	*Araliaceae*	Fruit	Boiled then drunk	Aches
81	Puspa	*Schima wallichii*	*Theaceae*	Flower	Crushed ,mixed with coconut oil and applied to foot	Aches
82	Pakis rane	*Selaginella plana*	*Selaginellaceae*	Young leaves	Eaten	post pregnancy
83	Sandagori	*Sida rhombifolia*	*Malvaceae*	1. leaves , 2. Root	1.Crushed and applied to body 2.Boiled and drunk	1. swelling, 2. rheumatism
84	Reundeu	*Staurogyne elongata*	*Acanthaceae*	1. leaves 2. Flower	1. Eat 2. dried, pounded ,mixed with coconut oil and applied to foot	1. Cough, kidney 2. aches
85	Hantap	*Sterculia oblongata*	*Sterculiaceae*	leaves	Boiled	fever
86	Mahoni	*Swietenia mahagoni*	*Meliaceae*	Bark	Boiled then drunk	hemorrhoids
87	Jirak	*Symplocos fasciculata*	*Symplocaceae*	leaves	Boiled	post pregnancy
88	Jukut sisiletan	*Taccarum* sp.	*Araceae*	leaves and root	Boiled with other plant then drunk	Backache, post pregnancy
89	Pungpurutan	*Urena lobata* L.	*Malvaceae*	1. leaves, 2. root	1. Pounded and applied to body ,2. Boiled and drunk	1. swelling, 2. Urinating problems

IOP Conf. Series: Earth and Environmental Science **166** (2018) 012021 doi:10.1088/1755-1315/166/1/012021

90	Cecengkehan	*Urophyllum arboreum*	*Rubiaceae*	Young leaves	Eaten	Bad breath
91	Tongtak	*Zingiber odoriferum*	*Zingiberaceae*	Water on young leaves	Applied to hair/ head	Hair treatment and dandruff
92	Jahe	*Zingiber officinale*	*Zingiberaceae*	Rhizome	Grated, applied to vagina	post pregnancy
93	Lampuyang	*Zingiber aromaticum*	*Zingiberaceae*	Rhizome	Boiled with other plant then drunk	post pregnancy

IOP Conf. Series: Earth and Environmental Science **166** (2018) 012022 doi:10.1088/1755-1315/166/1/012022

The Effect of Plant Growth Promoting Rhizobacteria (PGPR) on Germination and Seedling Growth of *Sorghum bicolor* L. Moench

S Widawati[1] and Suliasih[1]

[1]Micobiology Division, Research Center for Biology
Indonesian Institute of Sciences (LIPI)
Jl. Raya Bogor Km 46, Cibinong 16911, Indonesia

E-mail: widadomon@yahoo.com

Abstract. Germination is the first important step of the plant growth in the land. The germination and growth of sorghum in marginal land can be optimized by inoculating with PGPR (inoculants), hydrogel and CMC (carriers). The aim of the research was to evaluate the response of germination and seedling growth of sorghum using PGPR inoculation with hydrogel and CMC as carriers *(in vitro)* and pots containing sterile sand (*in vivo*). The experiment used PGPR (*Azospirillum lipoferum* 1103D, *Azotobacter chroococcum* 4103D, and *Bacillus* sp. 3D), and hydrogels as the reserve water, as well as *carboxymethyl cellulose* (CMC) as the adhesive media. The results showed that the combination of PGPR-mix (*Azospirillum, Azotobacter, Bacillus*) with hydrogels as bio-organic fertilizer and CMC as carriers has promoted the growth of the roots, shoots and vigour index of *in vitro* sorghum germination, as well as increased the root length, shoot length and total dry weight of sorghum seedlings in pots containing sterile sand. The best result of *in vitro* experiment (root length = 8.67 cm; shoot length = 12.6 cm; and vigor index = 2127.00) was obtained by sorghum seed inoculated with single PGPR inoculant (*A. lipoferum*) with carrier of CMC. The root length, shoot length and total dry weight of the highest sorghum seedlings were obtained by PGPR-mix inoculants without carriers (46.5 cm, 12 cm, and 0.477 g), PGPR-mix with carrier of CMC (48.67 cm, 15.67 cm, and 0.431 g), and PGPR-mix with carrier of hydrogel (48.67 cm, 15 cm, and 0.430 g).

1. Introduction

The soil plays an extremely important role to plant life. Soil is a culture medium to a plant which provides 3 nutrients (C, O, and H) from atmosphere and 13 nutrients (N, P, K, Ca, Mg, S, Cl, Fe, Mn, Cu, Zn, B, and Mo) from soil [1]. Soil is a habitat to functional microorganisms' life that directly and indirectly provides nutrition to a plant. The fertility and health of soil physically, chemically, and biologically depend on an organic content of soil. If an organic content of soil is lower than 1%, then the soil falls in a category of infertile soil.

The fertility of soil is determined by several components including soil structure, type of soils, microbe communities, type of microbes, and type of plants. If one of the components is damaged due to the continuous use of chemical fertilizer, the environment of soil will become infertile (marginal). This condition disturbs a germination process of seed in soil. The germination process started from imbibitions and terminated on the process of elongated radicle [2]. Velocity and diversity of radicle growth on germination are early indications of seed quality and crop production for optimization [3]. Germination process from dormant seed to radicle release until growth phase of seed depends on seed viability, suitability of environment condition, and breaking effort of seed dormant [4]. Optimization of germination on soil with poor nutrient and microorganisms (marginal) can be promoted with a functional bacteria or plant growth promoting rhizobacteria (PGPR) as a stimulant, CMC (*carboxymethyl cellulose*) as an absorbent agent of water and an adhesive, and hydrogel as a material of water reserves and plant nutrition. The three materials are safe for soil and environment because they do not leave residues on crop products, environment, and human health [5]. PGPR, hydrogel and CMC are raw materials for bio-organic fertilizers. Wu, Cao, Li, Cheung, and Wong (2005) reported that organic biofertilizers have competence in increasing the soil health, promoting plant growth, and increasing crop production [6].

PGPR are a group of functional bacteria that have the ability as a growth promoter (biostimulant). The process occurs by synthesising and regulating the concentrations of various *Growth Regulator* (GR) or *Plant Growth Regulator* (PGR) such as *Indole acetic acid* or IAA [7]. The functional bacteria (PGPR) produce IAA hormone and Acc-deaminase, and can provide nutrient on soil by fixating N_2 from atmosphere with symbiosis and asymbiosis. They are equally able to dissolve P nutrient bound with Al, Fe, and Ca on soil by synthesising PMEase enzyme (*Phosphomonoesterase*). PGPR can control the activity of pathogens of plant-disturber organisms (biocontrol) and decomposers of agrochemical compounds [8], and may control pathogens derived from soil (bioprotectant) by multiple compounds or anti-pathogenic metabolites such as siderophore, chitinase, antibiotics, and cyanide [9]. The group of PGPR will instantly occupy and colonize the plant roots (rhizosphere) aggressively and fast after inoculation in seeds, germs, and soil. This PGPR will immediately provide nutritional inputs needed for plant growth [10]. Some of the bacteria belonging to PGPR are *Flavobacterium, Herbaspirillum, Acetobacter, Acinetobacter, Alcaligenes, Arthrobacter, Bacillus, Beijerinckia, Burkholderia, Serratia, Enterobacter, Erwinia, Pseudomonas, Azospirillium, Azotobacter,* and *Rhizobium* [11]. Bacteria groups may associate with plants with carrier materials in the form of liquid, hydrogel, and CMC media.

Hydrogels are the product of a collection of polymeric materials (macromolecular gel) that have a three-dimensional hydrophilic structure [12], while hydrogel is superabsorbent which is also used as a water absorbing agent [13]. Hydrogels are often used in agricultural environments, especially in dry areas to support water requirements during germination [14]. Wang and Gregg [15] have used hydrogels to support water availability in horticulture. *Carboxymethyl cellulose* (CMC) is a water absorbent agent and an adhesive that can bind water so that water molecules are trapped on gel structure previously formed by CMC [16]. This material is highly soluble in hot and cold water [17] so that inoculated seeds and germs will be attached and inoculated directly to plant roots.

Sorghum (*Sorghum bicolor* L.) is used as a test plant on the effect of organic biofertilizer based on PGPR, CMC, and hydrogel. Sorghum originates from semi-arid tropical regions and is usually sensitive to low temperatures and is mostly grown in tropical and subtropical regions with an average temperature of over 18°C [18]. Some uses of sorghum are as a source of feed, animal feed and industrial raw materials. Sorghum can grow on marginal land; even the growth of the root does not show symptoms of poisoning in the former mining ground [19].

The objectives of this study were to (1) evaluate the germination response and growth of sorghum seeds to inoculation of PGPR with hydrogel and CMC as the carriers in vitro and pots containing sterile sand (in vivo), and (2) to determine the effect of inoculant PGPR with hydrogel and CMC as a carrier on the development of population of PGPR in sterile sand. The results of this study are expected to increase germination, improve soil health, promote plant growth, and increase crop production on marginal soils continuously in the future.

2. Materials and Methods

2.1. Inoculant of PGPR

PGPR inoculants used in the germination of sorghum seeds in vitro are functional bacteria inoculum (phosphate solubilising bacteria and non-symbiotic nitrogen-fixing bacteria). The bacteria have been analysed for their ability to produce IAA, nitrogenase, PMEase, and available P. All the bacteria used in this study were isolated from tin mining land on Bangka Island which has been analyzed for its effectiveness and identified as *Azotobacter chroococcum, Azospirillum lipoferum* (1103D isolate), and *Bacillus* sp. (3D isolates), (4103D isolate) based on the Bergey Systemic Bacteriology method [20-22]. Furthermore, the stored bacteria were cultivated again in Pikovskaya, Mannitol Ashby [24], and Caceres media [25]. After 3 days incubation, all bacteria were cultured in Erlenmeyer flask that contained liquid Pikovkaya (+ $Ca_3(PO_4)_2$ as a source of P), Caseres, and Ashby Mannitol media as the carriers. The bacteria were then shaked in room temperature for 24 hours until bacterial population reaches 10^{10} cfu/mL (± 3 days). The ratio of liquid inoculant combinations of 3 Rhizobacteria is 1:1:1. The ratio of hydrogel and CMC (carriers) combination with each single bacterium inoculum and mixture is 1:1.

2.2. Germination (in vitro)

Super-2 variety sorghum seeds were taken from Sorghum Research Central in Maros, South Sulawesi. The water-absorbent agent and adhesive materials used carboxymethyl cellulose (CMC) and hydrogel for water reserves and carriers. Sorghum seeds were sterilized using sodium hypochlorite 0.02% for 2 minutes, then washed with sterile aquades for 3 times. Clean seeds were immersed in Erlenmeyer flask, each containing 25 mL of liquid inoculant of *Bacillus* sp.3D, *A. lipoferum* 1103D, and *A. chroococcum* 4103D (single and mixed) combined with hydrogel and CMC as the carriers. Seeds soaked with 25 mL of sterile aquadest, 100 mg of CMC in 25 mL of aquadest, and 100 mg of hydrogel in 25 mL of aquadest were used as treatment controls. All treatments were incubated on rotary shaker for 12 hours with medium velocity (60 rpm) at room temperature. The next step, 100 sorghum seeds were inoculated with *Rhizobacteria*, hydrogel and CMC then were arranged on filter paper in sterile Petri dish (15 cm diameter) moistened with 10 mL of sterile aquadest. A volume of 5 mL of extract of each treatment from seed immersion in Erlenmeyer flask was added. All set of treatments, including control were arranged in 4 replications. All works were carried out in a sterile laminar air flow. Petri dishes were taped and incubated at 27 ° C for 7 days in the incubator in a dark room. The moisture of the media during the germination process was maintained by providing a sterile aquadest. Seeds are considered to germinate at the time of the appearance of radicals [26]. Percentages and strength indexes were calculated after 7 days. The Vigour Index was calculated by the method of Abdul and Anderson [27] with the formula: VI = (average root length + average shoot length) (% germination). All data were analysed with analyses of variance (ANOVA) using SPSS software version 12.0. Mean ratio of data were obtained after performing Duncan test ($P < 0.05$).

2.3. Green house experiment (in vivo)

The experiments on the effects of PGPR with hydrogel and CMC as the water-absorbent agent and the adhesive materials on the growth of sorghum seedlings were done in the greenhouses (in vivo) in the Microbiology Division, Research Center for Biology, Indonesian Institute of Sciences (LIPI). In this experiment we used pots containing 300 grams of sterile sand as the planting medium, super sorghum type 2 seeds as the experimental objects, *Bacillus* sp. 3D, *A. lipoferum* 1103D, *A. chroococcum* 4103D as the bacteria which will be tested for its effectiveness in an in vivo experiment, CMC and Hydrogel as the adhesive and absorbent water or carriers.The sorghum seed was washed and inoculated by soaking the seeds in a glass backer containing the extracts from Bacillus sp. 3D, A. lipoferum 1103D, A. chroococcum 4103D (PGPR).The culture of three bacteria were then mixed for 30 minutes. Other treatments were hydrogel and CMC applied to the planting hole before the sorghum seeds were planted. A total of 5 seeds of sorghum that have been inoculated were planted in the pot. After germination and sprouting, 2 plants were removed but 3 plants were kept in the pot. This was done in all pot experiments. The moisture of the planting media was maintained by watering it daily with Müller solution. The dose of the solution was in accordance with the dose of field capacity for sand [28].

This experiment has 15 combinations of treatments and 1 control, each treatment was repeated 3 times. The combination of the treatments in this experiment was arranged as follows: **1). K (Controls)**: The sorghum seeds and planting medium were without bacteria inoculation and without carriers (hydrogels and CMC); **2). A**: The sorghum seeds and planting medium were inoculated with *Bacillus* sp. 3D; **3). B**: The sorghum seeds and planting medium were inoculated with *A. chroococcum* 4103D; **4). C**: The sorghum seeds and planting medium were inoculated with *A. lipoferum* 1103D; **5). D**: The sorghum seeds and planting medium were inoculated with *Bacillus* sp. 3D; *A. chroococcum* 4103D; *A. lipoferum* 1103D (Mix); **6). E**: The sorghum seeds and planting medium were given a CMC; **7). F**: The sorghum seeds and planting medium were inoculated with *Bacillus* sp. 3D plus CMC; **8). G**: The sorghum seeds and planting medium were inoculated with *A. chroococcum* 4103D plus CMC; **9). H**: The sorghum seeds and planting medium were inoculated with *A. lipoferum* 1103D plus CMC; **10). I**: The sorghum seeds and planting medium were inoculated with Mix PGPR plus CMC; **11). J**: The sorghum seeds and planting medium were given a hydrogel; **12). K:** The sorghum seed and planting medium were inoculated with *Bacillus* sp. 3D plus hydrogel; **13). L:** The sorghum seeds and planting medium were inoculated with *A. chroococcum* 4103D plus hydrogel; **14). M**: The sorghum seeds and planting medium were inoculated with *A. lipoferum* 1103D plus hydrogel; **15). N**: The sorghum seeds and planting medium were inoculated with mixed PGPR plus Hydrogel; **16). O:** The sorghum seeds and planting medium were inoculated with mixed PGPR plus Hydrogel and CMC. Sorghum seedling was harvested after 30 days of the planting and their parameters were measured, namely: The shoot length (Cm), the root length (Cm), and the total dry weight (gram) of the seedling. Population of the bacteria in the rhizosphere of the pot plant was calculated after harvesting by plate count method. The data obtained were analyzed statistically using SPSS software version 12.0. The average data comparison was performed by Duncan P test <0.05.

3. Results and Discussion

The results of PGPR (*Bacillus* sp. 3D, *A. chroococcum* 4103D, *A. lipoferum* 1103D), hydrogel and CMC showed that there were significant different effect (P<0.05) on germination (*in vitro*) and seedling growth of sorghum *(in vivo)* (**Table 1**, **Table 2**, **Figure 1**, and **Figure 2**).

3.1. Germination (in vitro)

Treatment of inoculation of PGPR bacteria and carriers (hydrogel and CMC) had different effects on germination of sorghum seed (**Table 1** and **Figure 2**).

Table 1. The effect of PGPR on germination of sorghum seed (*in vitro*)

No.	Treatment	Shoot Length (cm)	Root Length (cm)	Vigor Index
1	K	3.7 ± 0.17^{d}	2.84 ± 0.41^{ef}	683.00 ± 13.61^{e}
2	A	3.60 ± 0.15^{cd}	0.84 ± 0.05^{a}	444.00 ± 19.29^{c}
3	B	3.59 ± 0.04^{cd}	1.31 ± 0.16^{ab}	490.00 ± 14.44^{c}
4	C	5.54 ± 0.16^{f}	2.76 ± 0.17^{def}	916.00 ± 46.60^{g}
5	D	6.40 ± 0.42^{g}	6.00 ± 0.60^{h}	1154.00 ± 74.85^{i}

6	E	1.83± 0.05^{a}	1.21±0.07^{ab}	304.00±4.37^{a}
7	F	3.13±0.11^{bc}	1.60±0.13^{b}	473.00±17.68^{c}
8	G	4.52±0.08^{e}	2.25±0.09^{c}	677.00±1.76^{e}
9	H	8.67±0.26^{h}	12.60±0.68^{i}	2127.00±80.03^{j}
10	I	6.13±0.27^{g}	4.69±0.45^{g}	1082.00±59.80^{h}
11	J	2.25±0.05^{a}	1.26±0.13^{ab}	351.00±7.69^{ab}
12	K	3.05±0.25^{b}	0.93±0.02^{a}	398.00±26.24^{bc}
13	L	3.45±0.40^{bcd}	2.59±0.05^{cde}	604.00±40.41^{d}
14	M	4.55±0.19^{e}	2.27±0.09^{cd}	682.00±25.01^{e}
15	N	5.45±0.38^{f}	3.13±0.08^{f}	829.00±78.79^{f}
16	O	5.56±0.44^{f}	4.63±0.02^{g}	910.00±36.10^{g}

Notes: ± SD is value from 3 replications.The number followed by the same letter are not significantly different at ($p<0.05$) level of Duncan's test.

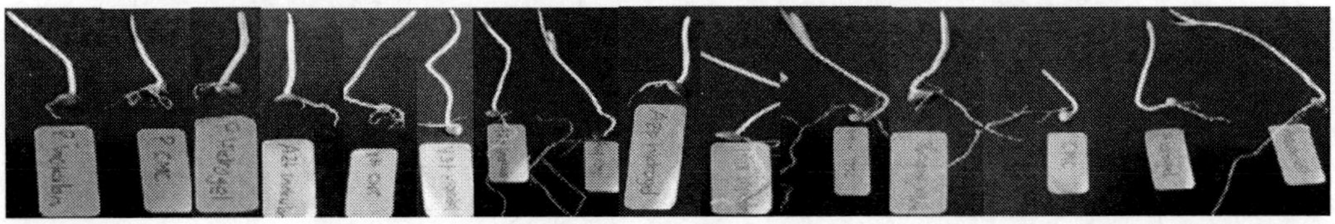

Figure 1. The effect of PGPR with hydrogel and CMC on germination of sorghum seed (*in vitro*)

The mean results of the inoculant effects of PGPR (*Bacillus* sp. 3D, *A. chroococcum* 4103D, *A. lipoferum* 1103D), hydrogel and CMC on sorghum germination (*in vitro*) showed statistically significant differences ($P<0.05$). The shoot length, root length, and vigor index of all treatments had significant effect, except the hydrogel and CMC treatments without PGPR bacterial inoculants that had lower yields than controls, i.e. the sorghum with PGPR, without carriers (hydrogel and CMC), but treated only with aquades (Table 1 and Figure 1). This occurs because of the lower water viscosity properties of hydrogels and CMC, so the imbibition process of hydrogel and CMC is slower than water. Imbibition is the initial process of seed germination (2), so if the imbibition is disturbed, the germination process will be disrupted. The inoculation of sorghum seeds is higher than control. A different result was reported by Alexeychuk et al [14] that the hydrogel stimulate proteolytic activity in seeds during germination. Another reported by Pazderu and Koudela [29] the hydrogel application was influenced germination of lettuce and anion seeds.

The average value of the shoot length, root length, and vigor index on seeds treated with *A. lipoferum* plus CMC (H: 8.67 cm; 12.60 cm; 2127.00), *Bacillus* sp. + *A. lipoferum* + *A. chroococcum* (D: 6.40 cm; 6.00 cm; 1154.00), and mix PGPR plus CMC (I: 6.13 cm; 4.69 cm; 1082.00) was the highest yield and the lowest yield was in the CMC (1.83 cm; 1.21 cm; 304.00) and hydrogel (2.25 cm; 1.26 cm; 351.00). This is reasonable that giving inoculants that contain mixed PGPR will have a positive effect on germination compared to single bacteria-containing inoculants. The influence of the inoculant with single bacteria on germination, the result (shoot length, root length, and vigor index) was lower than control, except in germination which was inoculated with bacteria *A. lipoferum* 1103D. The results of this study support the results of research from Mathivanan et al. [30], Hameeda et al. [31] in *Pennisetum glaucum* L, and Sengupta et al. [32] in corn seeds. The results of this research were supported by the reports of Lenin and Jayanthi [33], that the bacteria *A. lipoferum, A. chroococcum*, and *Bacillus megaterium* as PGPR have the effect in increasing seed germination and vigor index better than control treatment, even showing significant differences.

3.2. Green house experiment (in vivo)

The success in the germination of *Sorghum bicolor* as indicated by the high value of Vigor Index is triggered by the involvement of PGPR that is able to infect seedlings with IAA, one of the Auxin plant growth hormones groups [34]. IAA is produced by the bacteria from the PGPR group. The effect of PGPR inoculation supported by hydrogel and CMC in the growth of *Sorghum bicolor* seedlings and the population of bacteria in a pot of sterilized sand can be seen in **Figure 2** and **Table 2**.

Figure 2. The effect of PGPR with hydrogel and CMC on seedling of sorghum (in vivo)

Table 2. The effect of PGPR on seedling growth of sorghum seed (*in vivo*)

No.	Treatment	Shoot Length (cm)	Root Length (cm)	Total Dry Weight (g)	Bacteria Population (cfu/g of sand/pot)
1	K	26.66±1.45^{a}	4.667±0.88^{a}	0.083±0.01^{a}	0
2	A	41.00±3.12^{cde}	6.83±0.93^{b}	0.150±0.01^{bc}	1.75 x 10^{7}
3	B	40.67±3.48^{cd}	6.83±0.60^{b}	0.220±0.01^{d}	1.50 x 10^{7}
4	C	44.33±0.88^{cdefg}	10.66±0.33^{f}	0.337±0.03^{f}	3.25 x 10^{7}
5	D	46.50±0.87^{g}	12.00±2.08^{g}	0.477±0.01^{g}	5.25 x 10^{8}
6	E	31.67±1.67^{b}	4.83±0.17^{a}	0.223±0.01^{d}	0
7	F	46.33±2.73^{fg}	8.33±2.03^{e}	0.270±0.02^{e}	3.75 x 10^{8}
8	G	41.40±2.96^{cdef}	8.166±0.44^{de}	0.197±0.01^{cd}	2.25 x 10^{8}
9	H	45.83±1.92^{efg}	12.00±1.53^{g}	0.337±0.03^{f}	6.25 x 10^{8}
10	I	48.67±1.86^{g}	15.67± 4.81^{i}	0.431±0.02^{g}	2.75 x 10^{9}
11	J	45.67±2.33^{defg}	7.16±0.93^{bc}	0.237±0.01^{de}	0
12	K	31.00±2.08^{ab}	8.00±1.53^{de}	0.130±0.02^{ab}	1.03 x 10^{8}
13	L	39.33±7.88^{c}	7.66±1.76^{cd}	0.127±0.02^{ab}	3.75 x 10^{8}
14	M	43.83±7.10^{cdefg}	7.16b±0.73^{c}	0.227±0.01^{d}	5.00 x 10^{8}
15	N	48.67±3.34^{g}	15.00±0.58^{h}	0.430±0.01^{g}	7.00 x 10^{9}
16	O	47.55±0.86^{g}	14.89± 1.30^{h}	0.473±0.02^{g}	2.75 x 10^{9}

Notes: ± SD is value from 3 replications. The number followed by the same letter are not significantly different at ($p<0.05$) level of Duncan's test.

The results in the Table and Figures above show that the values of the root length, shoot length, and total dry weight of sorghum seeds are statistically significant. The values of the measurements of the parameters with moderate to highest effect on the treated seeds are shown consecutively: Mix PGPR (**D**: shoot length = 46.50 cm, root length = 12.00 cm; total dry weight = 0.477 g; bacterial population = 5.25 x 10^{8} cfu / g sand / pot), *A. lipoferum* plus CMC (**H**: shoot length = 45.83 cm; root length = 12.00 cm; total dry weight = 0.337 g; bacterial population = 6.25 x 10^{8} cfu / g sand / pot), mix PGPR plus hydrogel and CMC (**O**: 47.55 cm; 14.89 cm; 0.473 g; 2.75 x 10^{9} cfu / g sand / pot), mix PGPR plus hydrogel (**N**: shoot length = 48.67 cm; root length = 15.00 cm; total dry weight = 0.430 g; bacterial population = 7.00 x 10^{9} cfu / g sand / pot), and mix PGPR plus CMC (**I**: shoot length = 48.67 cm; root length = 15.67 cm; total dry weight = 0.431 g; bacterial population = 2.75 x 10^9 cfu / g sand / pot). While the lowest measuring values were obtained on seeds which were not inoculated, i.e. in pots with control treatment (shoot length = 26.66 cm, root length = 4.667 cm; total dry weight = 0.083 g; bacterial population = 0 cfu / g sand / pot), hydrogel (shoot length = 45,67 cm; root length = 7.16 cm; total dry weight = 0,237 g; bacterial population = 0 cfu / g sand / pot); and CMC (shoot length = 31,67 cm; root length = 4.83 cm; total dry weight = 0.223 g; bacterial population = 0 cfu / g sand / pot). Thus the group of PGPR bacteria (*Bacillus* sp. 3D + *A. chroococcum* 4103D + *A. lipoferum* 1103D) which were single bacterium, mixed bacteria, and mixed with hydrogel and CMC as carriers inoculated on seeds of sorghum (*in vivo*) are proven to bring positive impact to the germination of the seeds. The results indicate that bacteria supported by hydrogel and CMC as carriers are effectively infected the sorghum seeds, thus the imbibitions process, the development of radicle, and development of roots and seedlings can normally develop to eventually produce healthy sorghum seedlings. This is in supporting the previous studies by Harris [35] that the contribution of hydrogel was increased the growth of sorghum under semiarid condition.

The success of PGPR bacteria in supporting the healthy growth and development of sorghum is also supported by the volume of bacteria population that is inoculated into growth media and seeds. Soil and seeds are inoculated with bacteria extract with population of 10^{10} cfu/mL. After inoculation usually the number of bacteria population will decrease. Nevertheless, the population will continue to grow and colonize the root zone. This is proven by the measurement of population in sorghum's root areas. The result of average measurement is 10^7 cfu / mL to 10^9 cfu / mL. The amounts are higher from the minimum level of the number of bacteria in fertile soil. According to Obaton [36] the fertile soil at least has to have microbial population of $\geq 10^7$ cfu per soil gram. A similar results was reported by Suman et al. [37] that the bacteria (*A. chroococcum*, *Pseudomonas fluorescence*, and *Trichoderma viride*) based bio-inoculant hydrogel can produce maximum population of 99.6×10^8, 3.9×10^7, and 2.9×10^7 cfu per mL which are highly potential in increasing of shoots and roots growth of wheat crops. In other word, the number of bacteria population 10^7 cfu per ml to 10^9 cfu per ml will be able to support the developments of shoot and root length. The success is also supported by hydrogel which involves in the increase of seed and plant developments, also the number of bacteria populations in media of sands. This is in supporting the previous studies by Seshadri et al. [38] and Akhtar et al. [39]. The sand media itself has large porous, thus need the involvement of hydrogel in aeration improvement, water and nutrition supplies.

Thus the results of all studies on *in vitro* and *in vivo* germination were due to bacteria consortia (PGPR) with hydrogel and CMC as carriers. Tiwari et al. [40] reported that the plants showed the increase of biomass, the growth of shoot length and the growth of root length after inoculation with PGPR bacteria that induced IAA hormone production directly [41]. In particular, the PGPR group of the genus *Bacillus, Azospirillum*, and *Azotobacter* which has been considered as an important component of biofertilizer [42], because after inoculated on the plant, it immediately spreads and dominates the rhizospheres and colonizes the root surface area [43], particularly on some of the roots of tropical grass plants [44]; they may also occur in the roots of the sorghum plant. This phenomenon indicates that the participation of PGPR with hydrogel and CMC as the absorbent and the adhesive were very important and the possibility of such combinations is a novelty that is easy to do with relatively low cost to help eco-friendly agriculture in the future.

4. Conclusion

The PGPR treatment combined with CMC (adhesive) in sorghum seeds can improve germination index and increase growth of root and shoots in *in vitro*. Mixed-PGPR (*Bacillus* sp. 3D + *A. chroococcum* 4103D + *A. lipoferum* 1103D), mixed-PGPR with CMC, and *A. lipoferum* with CMC are the best combinations to increase the germination of sorghum seeds *in vitro*.

Single and mix-PGPR (*Bacillus* sp. 3D, *A. chroococcum* 4103D, *A. lipoferum* 1103D) with hydrogel and CMC (carrier) on sorghum seeds can increase the roots length, shoots length, and total dry weight of sorghum seeds in pots containing sterile sand *(in vivo)*. *Bacillus* sp. plus CMC, *A. lipoferum* plus CMC, mixed PGPR plus hydrogel, mixed PGPR plus hydrogel and CMC were the best combination for sorghum seedling growth and bacteria population (107 cfu / mL to 109 cfu / mL). The combination is an easy thing to do with a relatively low cost to help eco-friendly farming in the future.

Acknowledgments
This research was financially supported by a grant-in-Aid for Scientific Research from the TEMATIK DIPA (*Daftar Isian Pelaksanaan Anggaran*) program, Research Center for Biology, Indonesian Institute of Sciences

5. References

[1] Marsono and Lingga P 2005 *Petunjuk Penggunaan Pupuk* (Jakarta: Penebar Swadaya Press) 8-13
[2] Salisbury FB and Ross CW 1995 *Fisiologi Tumbuhan Jilid I* (Bandung: ITB Dwidjosepoetro Press)
[3] Parera CA and Cantliffe DJ 1994 Pre-sowing seed priming *Hortic. Rev.* **16** 109-141
[4] Harjadi SS 1993 *Pengantar Agronomi* (Jakarta: PT Gramedia Press)
[5] Musnamar EI 2003 *Pupuk Organik Cair and Padat, Pembuatan dan Aplikasi* (Jakarta: Penebar Swadaya Press) 7-13
[6] Wu SC, Cao Zh, Li ZG, Cheung KC, and Wong MH 2005 Effect of biofertilizer containing N-fixer, P and K solubilizer and AM-fungi on maize growth: a greenhouse trial *Geoderma* **125** 155-166
[7] Bottini R, Cassán F, and Piccoli P 2004 Gibberellin production by bacteria and its involvement in plant growth promotion and yield increase *Appl. Microbiol Biotechnol* **65** 497-503
[8] Gunalan 1996 Penggunaan Mikroba Bermanfaat pada Bioteknologi Tanah Berwawasan Lingkungan *Majalah Sriwijaya* **32** (2) Universitas Sriwijaya
[9] Glick BR 1995 The enhancement of plant groth by free-living bacteria *Can. J. Microbiol* **41** 109 17
[10] Vessey JK 2003 PGPR as biofertilizer *Plant and soil* **255** 571-586
[11] Sudhakar P, Chattopadhyay GN, Gangwar SK, and Ghosh JK 2000 Effect of foliar application of *Azotobacter, Azospirillum* and *Beijerinckia* on leaf yield and quality of mulberry (*Morus alba*) *Journal of Agricultural Sciences* **134** 227–234
[12] Narjary AP, Singh A, Chakraborty D, and Singh R 2012 Water availability in different soils in relation to hydrogel application. *Geoderma* **187** 94-101
[13] Buchholz FL and Peppas NA1994 Superabsorbent polymers: *science and technology* **573** 2-9
[14] Alexeychuk G N, Domash VI, Zabrejko SA 1999 Effect of swelling conditions on activity of proteolytic system during seed germination. *Doklady Akademii Nauk Belarusi* **43** (2) 83–86
[15] Wang YT and Gregg LL 1990 Hydrophilic polymers -Their response to soil amendments and effect on properties of a soilless potting mix *J. Am. Soc. Hort. Sci.* **115** 943-948
[16] Food and Agriculture organization (FAO) 1989 (internet) available from http://www.fao.org/fileadmin/user_upload/jecfa_additives/docs/Monograph1/additive-396-m1.pdf
[17] Tranggono 1991 *Bahan Tambahan Pangan (Food Additives)* (Yogyakarta: PAU Pangan Gizi UGM Press)
[18] Singh SP 1985 Sources of cold tolerance in grain sorghum *Can. J. Plant Sci.* **65** 251-257
[19] Nurcholis M, Wijayani A, and Widodo A 2013 Clay and organic matter applications on the coarse quartzy tailing material and the sorghum growth on the post tin mining at Bangka Island *Journal Of Degraded And Mining Lands Management* **1** (1) 27-32

[20] Brenner DJ, Krieg NR, Staley JT 2005 the Proteobacteria (*Bergey's Manual of Systematic Bacteriology* vol 2C 2nd edition) ed GM Garrity (New York: Springer) p 1388. ISBN 978-0-387-24145-6.
[21] Brenner DJ, Krieg NR, Staley JT 2005 The Gammaproteobacteria (*Bergey's Manual of Systematic Bacteriology* vol 2B 2nd edition) ed MG Garrity (New York: Springer) p 1108. ISBN 978-0-387-24144-9.
[22] Krieg NR and Dobereiner J 1984 The genus *Azospirillum* (*Bergey's manual of systematic bacteriology* vol 1) ed NR Krieg and JG Holt (Baltimore: The Williams and Wilkins Co.) 194-104
[23] Vos P, Garrity G, Jones D, Krieg NR, Ludwig W, Rainey FA, Schleifer KH, Whitman WB 2009 The Firmicutes (*Bergey's Manual of Systematic Bacteriology* Vol 3 2nd edition) ed GM Garrity (New York: Springer) p 1450. ISBN 978-0-387-95041-9.
[24] Rao NSS1995 *Soil Microorganisms and Plant Growth* 3th edition (New Hampshire: Science Publishers Inc.)
[25] Caceres, Enrique A, and Rodriguez 1982 Improved medium for isolation of *Azospirillum* spp. *Applied and Environmental Microbiology* **44** 990-991
[26] Vrbničanin S, Božić DR, Sarić M, Pavlović, and Raičević V 2011 Effect of Plant Growth Promoting Rhizobacteria on *Ambrosia artemisiifolia* L. Seed Germination *Pestic. Phytomed. (Belgrade)* **26** (2) 141–146.
[27] Abdul-Baki AA and Anderson JD 1970. Viability and leaching of sugars from germination barley *Crop Sci.* **10** 31-34
[28] Saono S, Karsono H, and Suseno D 1976 Studies on the effect of different rhizobial strains on *Phaseolus lunatus* in sand culture. *Annales Bogoriense* **6** (3) 143-154.
[29] Pazderu K and Koudela M 2013 Influence of hydrogel on germination of lettuce and onion seed at different moisture levels. *Acta Universitatis Agriculturae Et Silviculturae Mendelianae Brunensis* LXI 202 (6). http://dx.doi.org/10.11118/actaun201361061817
[30] Mathivanan S, Chidambaram ALA, Sundaramoorthy P, Baskaran L, and Kalaikandhan R 2011 Effect of Plant Growth Promoting Rhizobacteria on *Ambrosia artemisiifolia* L. Seed Germination *Pestic. fitomed. (Beograd)* **26** (2) 141–146
[31] Hameeda B, Rupela OP, Gopal R, and Satyavani K 2006 Application of plant growth-promoting bacteria associated with composts and macrofauna for growth promotionof Pearl millet (*Pennisetum glaucum* L.). *Biol. Fertil. Soils* **43** 221–227
[32] Sengupta C, Amita, B, Malusare S 2015 Effect of plant growth promoting rhizobacteria on seed germination and plants development of *Zea mays International Journal of Research in Advent Technology Special Issue National Conference on "ACGT 2015"* 13-14
[33] Lenin G and Jayanthi M 2012 Efficiency of Plant growth promoting rhizobacteria (PGPR) on enhancement of growth, yield and nutrient content of *Catharanthus roseus International Journal of Research in Pure and Applied Microbiology* 2 (4) 37-42
[34] Vikram A 2007 Efficacy of phosphate solubilizing bacteria isolated from vertisols on growth and yield parameters of sorghum. *Res. J Micobiol.* **2** 550-559
[35] Harris J 2006 The effects of manure, genotype, seed priming, depth and date of sowing on the emergence and early growth of *Sorghum bicolor* (L) Moench in semi-arid Botswana. *Soil and Tillage Research* **40** (1–2): 73–88. ISSN 0167-1987.
[36] Obaton M 1977 *In Biological Nitrogen Fixation in Farming System of the Tropics* ed A Ayanaba and PJ Dart (London: Willey) 173-214
[37] Suman A, Verma P, Yadav AN, Srinivasamurthy R, Singh A, and Prasanna R 2016 Development of Hydrogel Based Bio-Inoculant Formulations and their Impact on Plant Biometric Parameters of Wheat (*Triticum aestivum* L.) *Int.J.Curr.Microbiol.App.Sci.* **5** (3) 890-901

[38] Akhter J, Mahmood K, Malik K, Mardan, A, Ahmad M, and Iqbal M 2004 Effects of hydrogel amendment on water storage of sandy loam and loam soils and plants growth of barley, wheat and chickpea *Plant Soil Environ* **50** 463-469.

[39] Seshadri S, Ignacimuthu S, and Lakshminarsimhan C 2002 Variations in heterotrophic and phosphate solubilizing bacteria from Chennai, southeast coast of India. *Indian Journal of Marine Sciences* **31** 69-72

[40] Tiwari DK, Pandey P, Giri SP, and Dwivedi JL 2011 Effect of gibberellic acid (GA3) and other plant growth regulators on hybrid rice seed production *Asian J. Plant Sci.* **10** 133–139

[41] Maor RS, Haskin H, Levi-Kedmi, Sharon A 2004 In planta production of indole-3-acetic acid by *Colletotrichum gloeosporioides* f. sp. Aeschynomene. App. *Environt.Microbiol* **70** 1852-1854

[42] Bhattacharjee R and Dey U 2014 Biofertilizer a way towards organic agriculture a review. *African Journal of Microbiology Research* **8** (24) 2332-2342.

[43] Kyuma K 2004 *Paddy soil science*. (Kyoto: Kyoto Univ. Press and Trans Pacific Press)

[44] Jolly SN, Shanta NA, and Khan ZUM 2010 Phytoplankton quantification of heterotrophic bacteria and *Azospirillum* from the rhizosphere of taro (*Colocasia esculenta* L. Schott) and the nitrogen fixing potential of isolated *Azospirillum*. *Intern. J. Botany* **6** 117-21

IOP Conf. Series: Earth and Environmental Science **166** (2018) 012023 doi:10.1088/1755-1315/166/1/012023

Coloring Properties Assessment of Dyes Produced by Mixed *Aspergillus* and *Paecilomyces*

Suciatmih[1], Nurfianti[2] and S. V Magfirani[2]

[1]Research Center for Biology, Indonesian Institute of Sciences (LIPI)
Jl.Raya Bogor Km 46 Cibinong 16911, Indonesia
[2.]JurusanBiologi FMIPA UniversitasTadulako
Jl. Soekarno Hatta Km 9, Palu

E-mail: suciatmih2008@yahoo.ca

Abstract. Fungi can be used as an alternative source of natural dyes for thetextile industry. This research was conducted using mixed *Aspergillus* and *Paecilomyces*. The physical and chemical properties of the fungal dye were tested using 1% w/v of detergent for 15 minutes. Filtrate color and colors range developed on dyed materials were measured by RHS color chart. Of six different mordants tested, alum gave a brighter color than other mordants, so alum used for further research. The optimum mass of the fungal dye was adsorbed about 0.0023 g without mordant and the color was 82A purple-violet, while alum gave the optimum mass of the fungal dye adsorbed of 0.0013 g, which with 82C purple-violet color. Percentage the fungal dye absorption on cotton cloth without mordant was 9.6276%, while alum gave the fungal dye absorption of 3.6068%. Cotton cloth mass reduction after soaking in adetergent solution of the dyed cotton cloth using the fungal dye without mordant was 0.0036 g and the color was 88A violet, while alum gave the fungal dye reduced of 0.0018 g, which with 88C violet color. The obtained result showed dyes produced by mixed fungi could dye cotton cloth and resistant to detergent solution.

1. Introduction

The increasing awareness of health and pollution hazards of chemical dye stuffs has led to a resurgence of interest in natural dyes. Natural dyes are derived from plants, animals,and microbes. Microbes can be used as an alternative source of natural dyes to develop eco-friendly dye production technologies for textile industries. Microbial dyes have meaningful advantages over animals and plants dyes. These proliferatevery fast and are capable of growing in the cheap culture medium, independence from weather conditions and colors of different shades [1].

Fungi are reported to be potential color producer [2]. Various natural dyes from fungi were reported. These include yellow dyes by*Aspergillus sydowii* (CML2967), two isolates of *A. aureolatus* (CML2964 and E.4.1), two isolates of *A. keveii* (CML2968 and ONI75), three isolates of *Penicillium flavigenum* (CML2965; E.2.7 and 3.1.a) and *P. chermesinum* (CML2966) [3]; and red dyes by *Paecilomyces farinosus* (*Isaria farinosa*) [4]. Previous studies confirmed that fungal dye exhibit non-toxic and biodegradability [5].

Natural dyes, however, still face a huge disadvantage against synthetic dyes. Dyeing of textile with natural dyes usually involves issues of limited shade range and lower fastness properties of the dyed cloths. These problems have been overcome applying a pretreatment to the textile with mordants. Mordants help in the binding of dyes to the cloth by forming a chemical bridge from dye to fiber thus improving the staining ability of a dye with increasing its fastness properties [6]. Selection of mordants is important in natural dyeing processes. Shahid et al. [7] informed that mordantscan increase the depth of shade or drastically alter the final color of the dyed cloth.

The present study aimed to evaluate the potential application of the dyes produced by mixed *Aspergillus* and *Paecilomyces* as natural dyes in the coloring of cotton cloth.

2. Materials and Methods

Chemical materials used in this study includes alum (KAI $(SO_4)_2.12H_2O$), $CaCl_2$. $2H_2O$, $CaCO_3$, $CuSO_4.5H_2O$, $FeSO_4.7H_2O$, glucose, H_3BO_3, KCl, $K_2Cr_2O_7$, $MgSO_4.7H_2O$, $MnSO_4.H_2O$ (monohydrate), NaH_2PO_4, $NaNO_3$, $Na_2MoO_4.2H_2O$, *Potato Dextrose Agar* (PDA) and $ZnSO_4.7H_2O$,while other materials used are beaker glass, cotton cloth, detergent, Erlenmeyer, measuring cup, muslin cloth, Petri dish, stove and test tube.*Aspergillus* and *Paecilomyces* were obtained from the Laboratory of Environmental Microbiology, Research Center for Biology, Indonesian Institute of Sciences (LIPI). The research was conducted in the Laboratory of Environmental Microbiology, Research Center for Biology, LIPI, Cibinong.

2.1. Inoculation process

Mixed fungi consisting of *Aspergillus* and *Paecilomyces* was used for this study. The mixed fungi were grown in Petri dishes containing PDA medium and incubated at room temperature (27-28°C) for 5 days. It was then printed with a straw of pop ice (10 mm) for further inoculation. Five mycelial prints of the mixed fungi [8] were inoculated into Erlenmeyer flask containing mineral salts-glucose medium [9] and then incubated at room temperature (27-28°C) in stationary cultures for 4 weeks. After the incubation period, the mycelium was harvested and the dye was filtered with a muslin cloth for further assay.

2.2. Coloring process

Before dyeing, cotton cloth was washed with detergent, rinsed with water and then dried [8]. The cotton cloth mordanting process was carried out using pre-mordanting technique [8]. To assess the most appropriate mordant for the fungal dye, six different mordants such as $CaCO_3$, $CuSO_4.5H_2O$, $FeSO_4.7H_2O$,KAI $(SO_4)_2.12H_2O$, $K_2Cr_2O_7$,and $MnSO_4.H_2O$ were tested. Samples of cotton cloth (4 cm × 4 cm or 0.24 g) were treated with or without pre-mordanting using 1.2% of each mordant with a ratio of 1: 30 (material: liquor). The mordanting process was carried out at 90° C for 30 minutes. The samples were then squeezed and air dried at room temperature. Mordant which produces the best color on the cotton cloth was used for further research. As acontrol treatment, cotton cloth with code A was weighed and the base color recorded. Cotton cloth was then dipped in the fungal dye solution for 30 minutes at 90° C, lifted and dried under the sun. The dried cotton cloth was weighed again and the color recorded. Finally, the resistance of the cotton cloth was tested by soaking it in detergent 1% solution for 15 minutes, removed, dried and weighed again. As a pre-mordanting treatment, cotton cloth with code B was soaked in mordant alum solution for 30 minutes at 90° C, removed, dried and weighed until the mass was constant. The dried cotton cloth was treated like for coloring without mordant (control treatment). The color of cotton cloth after dyeing from each treatment was determined with the RHS color chart [10]. Each treatment was done with three repetitions.

2.3. Dye uptake by cotton cloth

Dye uptake by cotton cloth was measured both with the adsorption and percentage absorption of dyes by cotton cloths. The adsorption of dye on cotton cloth can be determined by determining themass of cotton cloth before and after staining [11], while percentage absorption of dye by cotton cloth was calculated by measuring the optical density of the dye solution samples at a wavelength of 530 nm [12 modified]. Percentage absorption of dye (%) was calculated using the following equation:

$$\frac{\text{OD before dyeing - OD after dyeing X 100}}{\text{OD before dyeing}} \quad (1)$$

3. Results and Discussion

Mixed *Aspergillus* and *Paecilomyces* in the present study produced pigmentation in mineral salts glucose medium. The dye color produced by the fungi was 187 A greyed-purple (**Figure 1**). Several studies of dyes produced by fungi reported that *Aspergillus niger* produced brown dyes[13]; yellow dyes were produced by *A. terreus* [14], *A. calidoustus* 4BV13 [15], *A. versicolor* [16] and *Aspergillus* sp. [17]; *Aspergillus* sp. produced green dyes [18];*Eurotium* spp. (asexual morph: *Aspergillus*) produced yellow and red dyes [19]; some strains of *Aspergillus* (*A. glaucus* and *A. repens*) produced yellow and red dyes[20]; *E. nidulans* (asexual morph: *Aspergillus*) produced reddish brown dyes [4]; and red dyes by *Paecilomyces sinclairii* [21].Pigments or dyes are generally produced in the cell cytoplasm as a response to disadvantageous environmental conditions, such as nutrient limitation and this process is controlled by a complex regulatory network [22].

PH for the medium in this study before inoculation was 6, however, pH before dyeing (after culturing) for the fungi became 5.7. Hence the dye liquor became more acidic in nature. This result is in accordance with the dyes produced by *Trichoderma* [12].It has been reported that many kinds of fungi have more acidic pH optima during submerged culture for dyes production [23].

Mordants application of natural dyes produced by the fungi generated various shades of cotton cloth (**Table 1**). The fungal dye can stain cotton cloth from 155 D white into 88A & C violet, 87 C violet,and 201 A, B & D grey. Satyanarayana and Chandra [24] informed that this is presumably due to the influence of the chemical content available on mordant materials, namely the existence of Al^{3+} from $KAl(SO_4)_2.12H_2O$, Ca^{2+} from $CaCO_3$, Cu^{2+} from $CuSO_4.5H_2O$, Fe^{2+} from $FeSO_4.7H_2O$, K^+ from $K_2Cr_2O_7$ and Mn^{2+} from $MnSO_4.H_2O$.Mordants play avery important role in imparting color to the cotton cloth in the form of themetal complex formation.Previous studies found that *A. niger* produced a reddish brown color on wool [13] and yellow colors on silk, cotton,and silk-cotton [25]; and *Aspergillus* sp. produced green colors on cotton and silk [18]. The fungi can then be used as the source of natural dyes for various purpose, vis. dyeing of yarns and dyeing of cloths [26].

The intensity of color produced on the dyed cotton cloth using the fungal dye without mordant (88 C violet) and with mordant alum (87 C violet) was found brighter than that obtained for other mordants, so mordant alum used for further research (**Table 1**). The dyed cotton cloth using the fungal dye without mordant was brighter than the dyed cotton cloth using the fungal dye added with mordant alum. Satyanarayana and Chandra [24] informed that this is presumably due to the influence of Al^{3+} from $KAl(SO_4)_2.12H_2O$.

Table 1. The color variations of cotton cloth dyed with the fungal dye using different mordants

Mordant	Cotton cloth color
Without mordant	88A violet
$CaCO_3$	87 C violet
$CuSO_4.5H_2O$	201 B grey
$FeSO_4.7H_2O$	201 A grey
$KAl(SO_4)_2.12H_2O$	88C violet
$K_2Cr_2O_7$	201 D grey
$MnSO_4.H_2O$	201 A grey

Figure 1. Fungal dyes. Mineral salts glucose medium (right) and mixed *Aspergillus* and *Paecilomyces* (left)

The magnitude of dyes adsorption and absorption by cotton cloth treated without mordant can be seen in **Table 2**, **Table 3** and **Figure 2**. The mass of dyestuff adsorbed on cotton cloth without mordant was 0.0023 ± 0.0016 g (**Table 2**) and percentage dye absorption by cotton cloth without mordant was 9.6276 ± 0.2524 % (**Table 3**). The fungal dye was able to change the cotton cloth color from 155 D white to 82A purple-violet. The adsorption and absorption processes are possible because the OH group of cellulose present in the cotton cloth fiber is capable of forming hydrogen bonds with OH or other polar groups of the dyestuff. This bond is generally easy to break because it is not so strong [11]. Therefore, to clarify the color of the cotton cloth and to strengthen the bond between the dye and the cotton cloth it is necessary to add mordant in the staining process.

Table 2. Adsorption of the fungal dye on cotton cloth without and with mordant alum (n = 3)

Code	Cotton cloth mass before dyeing (g)	Cotton cloth mass after dyeing (g)	Average dye adsorption(g)	Cotton cloth color
A	0.2234	0.2257	0.0023 ± 0.0016	82A purple-violet
	Cotton cloth mass + mordant before dyeing (g)	Cotton cloth mass + mordant after dyeing (g)	0.0013 ± 0.0006	
B	0.2096	0.2109		82C Purple-violet

Table 3. Absorption of the fungal dye by cotton cloth without and with mordant alum (n=3)

Code	Treatment	OD before dyeing	OD after dyeing	% Average dye absorption
A	Without mordant	4.8565	4.388933	9.6276 ± 0.2524
B	Mordant alum	4.8565	4.68133	3.6068 ± 0.2062

IOP Conf. Series: Earth and Environmental Science **166** (2018) 012023 doi:10.1088/1755-1315/166/1/012023

(a) (b) (c)

Figure 2. Dyed cotton cloth using the fungal dye without mordant alum. Cotton cloth before dyeing (a); cotton cloth dyed with the fungal dye without mordant alum (b) and cotton cloth after soaked withdetergent 1% (c)

Dyed cotton cloth using the fungal dye added with mordant alum can be seen in **Figure 3**. The fungal dye was able to change the cotton cloth color from 155D white to 82C purple-violet. Cotton cloth dyed using the fungal dye added with mordant alum gave more faded color (82C purple-violet) than cotton cloth without mordant alum (82A purple-violet) (**Figure 2** and **3**). This is presumably due to theinfluence of the chemical content available on themordant material (the existence of Al $^{3+}$ from KAI $(SO_4)_2.12H_2O$). This result is also in accordance with the result of calculating the dyestuff mass adsorbed on cotton cloth, and the result of percentage dyestuff absorption by cotton cloth. The mass of dyestuff adsorbed on cotton cloth with mordant alum (0.0013 ± 0.0006 g) **(Table 2)** was smaller than the mass of dyestuff adsorbed on cotton cloth without mordant alum (0.0023 ± 0.0016) (**Table 2**). Percentage absorption of the dye by cotton cloth treated with mordant alum (3.6068 ± 0.2062 %) (**Table 3**) was also smaller than percentage absorption of the dye by cotton cloth without mordant alum (9.6276 ± 0.2524 %) (**Table 3**). The results indicated that mordant has not provided more dye sites and bears high color depth of shades as compared to those without mordant so that the dyed cotton cloth using the fungal dye does not require theaddition of mordant.

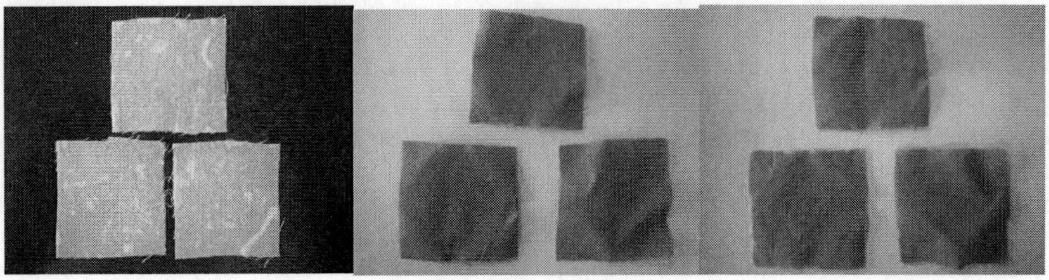

(a) (b) (c)

Figure 3. Cotton cloth dyed using the fungal dye added with mordant alum. Cotton cloth before dyeing (a); cotton cloth dyed using the fungal dye added with mordant alum (b) and cotton cloth after soaked withdetergent 1% (c)

The mechanism of binding on the dyed cotton cloth in the presence of mordant is described in **Figure 4**.The ability of both the dyestuff adsorption and the dyestuff absorption in each cotton cloth in the presence of mordant have to be increased due to functional group as hydroxy on the cotton cloth can occupy the unoccupied sites on metal ion interaction with the dye so that a ternary complex is formed by the metal salt on which one site is with the cotton cloth fiber and the other site is with the dye [27]. In this study, however, both the dyestuff adsorption and the dyestuff absorption by cotton cloths treated with mordant alum was smaller than both the dyestuff adsorption and the dyestuff absorption by cotton cloths without mordant alum (**Table 2&3**). The dyed cotton cloth color using the

IOP Conf. Series: Earth and Environmental Science **166** (2018) 012023 doi:10.1088/1755-1315/166/1/012023

fungal dye with mordant alum was also more faded than the dyed cotton cloth color using the fungal dye without mordant alum (**Figure 2&3**). The results indicated that mordant has not provided more dye sites and bears high color depth of shades as compared to those without mordant so that the dyed cotton cloth using the fungal dye does not require theaddition of mordant.

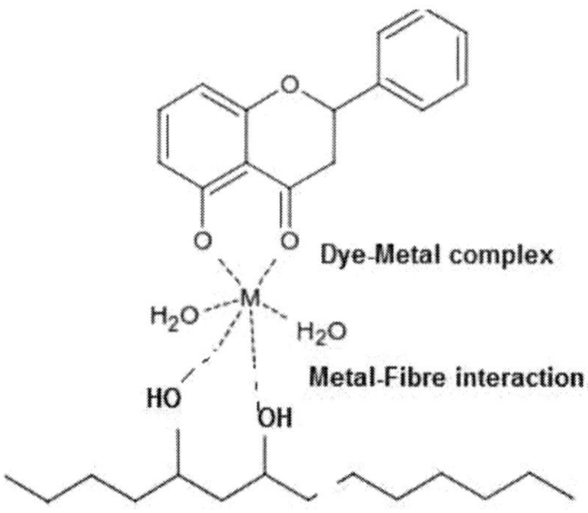

Figure 4. Schematic representation of dye-metal-cellulose fiber interaction (from Kechi et al. [28])

The reduction of average mass of the dyed cotton cloth with the fungal dye without mordant alum after soaking in adetergent solution (0.0036 ± 0.0007) was greater than with mordant alum (0.0018 ± 0.0005) (**Table 4**). Similarly, water OD after the immersion of the dyed cotton cloth using the fungal dye without mordant alum (0.2240 ± 0.0283) was greater than water OD with mordant alum (0.10955 ± 0.0279) (**Table 5**). These results are in accordance with the result of discoloration of detergent 1% solution after immersion test. Detergent solution color after soaking of the dyed cotton cloth using the fungal dye with mordant alum gave more faded than detergent solution color after soaking of the dyed cotton cloth with the fungal dye without mordant alum (**Figure 5**). The results indicated that addition of mordant alum 1.2 % on the dyed cotton cloth using the fungal dyeable to strengthen the bonding of the fungal dye to the cloth.Manurung [11]informed that the OH-group of cellulose present in the dyed cotton cloth fibers using the fungal dye with mordant alum capable of forming covalent bonds with metal ions from mordant as bridges with OH or other polar groups of dyes, so that the adsorbed and absorbed dyes from the cotton cloths were not easily removed or lost. Mordants play avery important role in imparting color to the cotton cloth in the form of metal complex formation or the metal ions of mordants act as electron acceptors from groups of dye donating anelectron to form coordination bonds with the dye molecule, making them insoluble in water [27]. Kamel et al. [29] also reported that color strength is dependent on the metal salt used.

Although the reduction of average mass; and water OD after the immersion on detergent solution of the dyed cotton cloth using the fungal dye without mordant was greater than with mordant alum, but the dyed cotton cloth color using the fungal dye without mordant alum was still brighter than the dyed cotton cloth added with mordam alum (**Figure 2&3**). This result indicated that the dyed cotton cloth using the fungal dye does not require theaddition of mordant. Unless for color fastness to washing, however, the results have not been tested for color fastness to perspiration; and color fastness to light.Furthermore, the fungal dye will be further tested forcolor fastness to perspiration and to light.

Table 4.The result of immersion test with detergent 1% to the color resistance of cotton cloths dyed with thefungal dye (n = 3)

Code	Method	Mass of cotton cloth after staining (g)	Mass of cotton cloth after detergent 1% test (g)	The reduced average cotton cloth mass (g)	Cotton cloth color
A	Without mordant	0.2257	0.2221	0.0036±0.0007	88A violet
B	Mordant alum	0.2109	0,2091	0.0018±0.0005	88C violet

Table 5. The value of water OD after dyeing of the cotton cloth stained with the fungal dye (n=3)

No	Treatment	Water OD
1.	Detergent solution 1%+ the dyed cotton cloth using the fungal dye without mordant (A)	0.224 ± 0.0283
2.	Detergent solution 1%+ the dyed cotton cloth using the fungal dye with mordant alum (B)	0.1096 ± 0.0279

Figure 5. Discoloration of detergent 1% solution after immersion test. Detergent solution color after soaking of the dyed cotton cloth using the fungal dye without mordant (left), detergent solution color after soaking of the dyed cotton cloth with the fungal dye added with mordant alum (middle) and detergent 1% solution colour (right)

4. Conclusion

This study showed that mixed *Aspergillus* and *Paecilomyces* produced natural dyes that can be used to dye textile.Different mordants produced different shades of color on cotton cloth. Addition of mordant alum 1.2% on the dyed cotton cloth using the fungal dyeable to strengthen the bonding of the fungal dye to the cotton cloth.Cotton cloth dyed using the fungal dye does not require theaddition of mordant.

Acknowledgment

We would like to thanks to Research Center for Biology, Indonesian Institute of Sciences (LIPI) that has been providing laboratory facilities; and EtySuryati and NurmaNurjanah (the technicians of Microbiology Division, Research Center for Biology – LIPI) who have helped the study.

5. References

[1] Venil C K and LakshmanaperumalsamyP 2009 An insightful overview on microbial pigment, Prodigiosin *Electronic Journal of Biology* **5**(3)49-61

[2] Ebrahim W, El-Neketi M, LewaldL I, Orfali R S, Lin W, Rehberg N, Kalscheuer R, Daletos Gand Proksch P 2016 Metabolites from the fungal endophyte *Aspergillus austroafricanus* in axenic culture and in fungal–bacterial mixed cultures *Jurnal of Natural Products* **79** (4) 914-922 doi:10.1021/acs.jnatprod.5b00975

[3] Da Costa Souza P N, Grigoletto T L B, de Moraes L A B, Abreu L M, Guimaraes L H S, Santos C, Galvao L R and Cardoso P G 2016 Production and chemical characterization of pigmentsin filamentous fungi *Microbiology* 162 12-22doi: 10.1099/mic.0.000168

[4] Velmurugan P, Lee Y H, Venil C K, Lakshmana perumalsamyP, Chae J – C and Oh B – T 2010. Effect of light on growth, intracellular and extracellular pigment production by five pigment-producing filamentous fungi in synthetic medium *Journal of Bioscience and Bioengineering* **109**(4) 346-350 doi:10.1016/j.jbiosc.2009.10.003

[5] Youssef M S, El- Maghraby O M O and Ibrahim Y M 2008 Production of textile reddish brown dyes by fungi *International Journal of Botany* **4** 349-360

[6] Gunasekaran S and Poorniammal R 2008 Optimization of fermentation conditions for red pigment production from *Penicillium* sp. under submerged cultivation *African Journal of Biotechnology* **7**(12) 1894-1898 doi: 10.5897/AJB2008.000-5037

[7] Shahid M and Mohammad F 2013 Recent advancements in natural dye applications: A review *Journal of Cleaner Production* **53** 310-331

[8] Suciatmih and Soeka Y S 2017 Pengaruh pewarna *Penicillium* yang dipekatkan terhadap hasil optical density dan warna kain katun Prosiding Seminar Nasional Fakultas Pertanian Universitas Nasional 779-792

[9] Baker R A and Tatum J H 1998 Novel anthraquinones from stationary cultures of *Fusarium oxysporum Journal of Fermentation and Bioengineering* **85**(4) 359-361

[10] Anonymous 1966*R.H.S. Colour chart* (London: The Royal Horticultural Society)

[11] Manurung M 2012 Aplikasi kulit buah manggis (*Garcinia mangostana* L.) sebagai pewarna alami pada kain katun secara pre-mordanting *Jurnal Kimia* **6** (2) 183-190

[12] Gupta C, Sharma D, Aggarwal S and Nagpal N 2013 Pigment production from *Trichoderma* spp. for dyeing of silk and wool *International Journal of Science and Nature* **4** (2) 351-355

[13] Atalla MM, El-khrisy E A M, Youssef Y A and Mohamed A A 2011 Production of textile reddishbrown dyes by fungi *Malaysian Journal of Microbiology* **7** (1) 33-40

[14] Akilandeswari Pand Pradeep B V 2017 *Aspergillus terreus* KMBF1501 a potential pigment producer under submerged fermentation *International Journal of Pharmacy and Pharmaceutical Sciences* **9** (4) 38-43 doi:10.22159/ijpps.2017v9i4.16176

[15] Dos Reis Celestino J, de Carvalho L E, da Paz Lima M, Lima A M, Ogusku M M and de Souza J VB 2014 Bioprospecting of Amazon soil fungi with the potential for pigment production *Process Biochemistry* **49** 569-575 https://doi.org/10.1016/j.procbio.2014.01.018

[16] Miao F – P, Li X – D, Liu X – H, Cichewicz R H andJi N - Y 2012 Secondary metabolites from an algicolous *Aspergillus versicolor* strain *Marine Drugs* 10 131-139 doi:10.3390/md10010131

[17] Narendrababu B N and Shishupala S 2017 Spectrophotometric detection of pigments from *Aspergillus* and *Penicillium* isolates *Journal of Applied Biology and Biotechnology* **5** (01) 053-058doi:10.7324/JABB.2017.50109

[18] Sarkar S L, Saha P, Sultana N and Akter S 2017 Exploring textile dye frommicroorganisms, an eco-friendly alternative *Microbiology Research Journal International* **18** (3) 1-9. doi:10.9734/MRJI/2017/29861

[19] Gessler N N, Egrova A S and Belozerskaya T A 2013 Fungal anthraquinones *Applied Biochemistry and Microbialogy* **49** 109-123 doi: 10.1134/S000368381302004X

[20] Caro Y, Anamale L, Fouillaud M, Laurent P, Petit T and Dufosse L 2012 Natural hydroxyl anthraquinoid pigments as potent food grade colorants: an overview *Natural Products and Bioprospecting* 2 174-193doi 10.1007/s13659-012-0086-0
[21] Cho Y J, Park J P, Hwang H J, Kim S W, Choi J W and Yun J W 2002 Production of red pigment by submerged culture of *Paecilomyces sinclairii Letters in Applied Microbiology* 35 195-202 doi: 10.1046/j.1472-765X.2002.01168.x
[22] Hailei W, Ping L, Yufeng L, Zhifang R and Gang W 2012 Overproduction of a potential red pigment by a specific self-immobilization biomembrane-surface liquid culture of *Penicillium novae-zeelandiae Bioprocess and Biosystems Engineering* **35** 1407-416
[23] Bae J T, Sinha J, Park J P, Song C H and Yun J W 2000 Optimization of submerged culture conditions for exo-biopolymer production by *Paecilomyces japonica Journal of Microbiology and Biotechnology* **10** 482-487
[24] Satyanarayana D N V and Chandra K R 2013 Dyeing of cotton cloth with natural dye extracted from Pomegranate peel and its fastness *International Journal of Engineering Sciences and Research Technology* **2** (10) 2664-2669
[25] Aishwarya A d 2014 Extraction of natural dyes from fungus – An alternate for textile dyeing *Journal of Natural Sciences Research* **4** (7) 1-6
[26] Grover N and Patni V 2011 Extraction and application of natural dye preparations from the floral parts of *Wood for diafruticosa* (Linn.) Kurz. *Indian Journal of Natural Products and Resources* **2** (4) 403-408
[27] Mongkholrattanasit R, Krystufek J, Wiener J and Vikova M 2011 Dyeing, fastness and UV protection properties of silk and wool fabrics dyed with *Eucalyptus* leaf extract by the exhaustion process *Fibers and Textiles* **19** (3) 94-99
[28] Kechi A, Chavan R B andMoeckel R 2013 Ethiopian dye plants as a source of natural dyes for cotton dyeing *Universal Journal of Environmental Research and Technology* **3** (4) 501-510
[29] Kamel M M, Helmy H M and Hawary N S 2009 Some studies on dyeing properties of cotton fabrics with *Crocus sativus* (Saffron) flowers using an ultrasonic method *AUTEX Research Journal* **9** (1) 29-35

IOP Conf. Series: Earth and Environmental Science **166** (2018) 012024 doi:10.1088/1755-1315/166/1/012024

The Effect of Biofertilizer Combined with Organic or Inorganic Fertilizer on Growth of *Caesalpinia pulcherrima* and Bacterial Population in Soil

Suliasih[1] and S Widawati[1]

[1]Micobiology Division, Research Center for Biology, Indonesian Institute of Sciences (LIPI),
Jl. Raya Bogor Km 46, Cibinong 16911, Indonesia

E-mail: lishadari@yahoo.co.id

Abstract. The greenhouse experiment was carried out to study the effect of biofertilizer combined with organic or inorganic fertilizer on the growth of *Caesalpinia pulcherrima* and bacterial population in the soil. Biofertilizers including *Azotobacter* sp. (mixture of 5 isolates) *Azospirillum* sp. (mixture of 5 isolates) and *Rhizobium sp.* (mixture of 5 isolates), were applied with organic fertilizer (Palm Oil Mill Effluent/POME) or inorganic fertilizer (Muller Solution/MS) were laid out in Randomized Complete Block Design (RCBD) with three replicates. The result showed that plants which were inoculated with biofertilizer (*Azotobacter, Azospirillum,* and *Rhizobium*) showed better growth compared than control treatment. Plants treated with a combination of biofertilizers with POME or MS had higher growth than plants treated with organic or inorganic fertilizers individually. The application of an organic fertilizer on plant showed more excellent plants' growth than the usage of organic fertilizer. However, there were no significant differences in plant growth between POME and MS combined with biofertilizer. Biofertilizer combined with organic fertilizer combination produced higher bacterial population than inorganic fertilizer combination.

1. Introduction

Fertilizer is among the most utilized substance in agriculture. Based on the production process, fertilizer is categorized into three types; chemical fertilizer (inorganic), organic and biofertilizer. In modern agriculture, inorganic fertilizers especially nitrogen and phosphorus are widely used in soil management plan. This phenomenon resulted in many detrimental implications to the environment. The use of chemical fertilizer contributes to the contamination of soil, water, and air and is also harmful to soil microorganisms and fertility [1-3]. Organic fertilizer and biofertilizer can be alternative solutions to reduce the usage of chemical fertilizer with the purpose of maintaining and improving the quality of the soil.

Biofertilizers are a nutritious, renewable and eco-friendly substance which contain living microorganisms that are able to colonized rhizosphere and internal tissue of plants [4-6]. *Rhizobium, Azotobacter, Azospirilium* are some examples of biofertilizer that has been used for a good period of time to provide nutrients through nitrogen fixation and induce plant's growth through growth hormone synthesis and dissolve phosphate [7-10].

Organic fertilizers have an important role in maintaining soil sustainability and improve both soil physicochemical characteristics and microorganisms activity [11,12]. One sample of widely utilized organic fertilizer is wastewater of palm oil industry. Palm Oil Mill Effluent (POME) is the waste of oil winning process in palm oil industry which contains a high level of organic matter and plant nutrients especially potassium (K), nitrogen (N), magnesium (Mg) and calcium (Ca) but relatively low phosphorus (P) [13]. Wu et al. reported that POME has been extensively applied in irrigation and liquid organic fertilizer in palm plantations [14]. The application of organic and biological nutrients is important for plant growth and shows positive interaction with chemical fertilizer, thus increase fertilization efficiency.

Caesalpinia pulcherrima is flowering species that frequently planted as ornament and shades trees. Plant component such as flower, leaves, bark and root is recognized contain various types of a bioactive compound and have been used for medicinal purposes. The application of fertilizer is recommended for excellent plant growth. The objectives of the research was to investigate the impact of organic, inorganic, biofertilizer and its combinations' application on *C. pulcherrima* growth and soil microbial population

2. Materials and Methods

2.1. Bacteria

The bacteria used consisted of *Rhizobium* (a mixture of 5 isolates), *Azospirillum* sp. (5 isolates) and *Azotobacter* sp. (5 isolates). They were isolated from soil of former tin mine in Bangka, Batu Belubang village, Pangkalan Baru, Bangka (S. 02° 11' 409' and E. 106° 11' 265'), Bangka Belitung province. The bacteria's ability to solubilize phosphate and produce IAA has been tested. Phosphate solubilization by bacteria was tested using [15]. Solubilization of inorganic phosphate was qualitatively performed by inoculating culture of bacteria isolates on pikovskaya agar medium containing 0.5% $Ca_3(PO_4)_2$ as a source of P, 0.5 g $(NH_4)_2SO_4$; 0.2 g NaCl; 0.1 g $MgSO_4$. $7H_2O$; 0.2 g KCl; 10 g Glucose; 0.5 g yeast extract; 20 g agar; $MnSO_4$ and a few $FeSO_4$, distilled water 1000 ml [16]. Media in the petri dish was inoculated with 10 µl (10^6 cfu/ml) of bacterial culture using drop plate method and incubated for 7 days at room temperature. The bacterial colonies capable to solubilize tricalcium phosphate were characterized by halo zone around the colonies. The obtained bacterial isolates were then tested quantitatively on liquid Pikovskaya media. The amount of soluble phosphate was measured on 72 h. Indole Acetic Acid production by bacterial culture was analyzed by Pattern and Glick method. It was performed in a liquid medium of TSB (20 ml) containing L Tryptophan (200 µg/ml) at 72 h based on [17].

2.2. *Effect of fertilizer on growth of C. pulcherrima on pot in greenhouse*

The experiment was carried out using completely randomized design (CRD), consisted of Plant without fertilizer nor inoculant as control (C), *Azotobacter* inoculant (At), *Azospirillum* inoculant (Ap), *Rhizobium* inoculant (R), Organic fertilizer (POME/P), Inorganic fertilizer (Muller solution/M), *Azotobacter* inoculant + POME (At + P), *Azotobacter* inoculant + Muller solution (At + M), *Azospirillum* inoculant + POME (Ap + P), *Azospirillum* inoculant + Muller solution (Ap + M), *Rhizobium* inoculant + POME (R+ P), *Rhizobium* inoculant + Muller solution (R+M). Three replications were prepared for each treatment.

The experiment was carried out in the greenhouse of Microbiology department, Research center for Biology, Cibinong. Growing medium used for the experiment was sterilized sand. The sand was sterilized in an autoclave for an hour. As much as 300 g of the sand was filled into the pot. Seeds of *Caesalpinia pulcherrima* was sterilized by being soaked in a solution of 1% sodium hypochlorite for 30 seconds and rinsed with aquadest three times. The seeds then were germinated in wet (using 10 mL of sterile aquadest) and sterile filter paper on a petri dish. The seed germination was observed every day. Three seedlings were placed on a pot after seven days germination. The application of organic (POME) and inorganic fertilizer (Muller solution) was performed by pouring 10 mL fertilizer every day. The concentration of POME was 5 %. Muller solution was consisted of mixture of A solution

($CuSO_4$ $2H_2O$ 5g; Na; $MgSO_4$ $7H_2O$ 5g; distilled water /DW 20L) + 10 mL B solution (Ferric citrate 30g; DW 1L) + 100 mL C solution (KH_2PO_4 34g; KOH 1.96g; DW 1L) + 10 mL D solution ($MnSO_4$ H_2O 1g; Na; $ZnSO_4$ $7H_2O$ 0.25g $CuSO_4$ $5H_2O$ 0.25g; Na_2MoO_4 $2H_2O$ 0.06g; H_3BO3 0.50g CaCl2 $6H_2O$ 0.05g. *Azotobacter* (a mixture of At1, At2, At3, At4, At5), *Azospirillum* (a mixture of Ap1, Ap2, Ap3, Ap4, Ap5) and *Rhizobium* (a mixture of R1, R2, R3, R4, R5) isolates which used as biofertilizer. The shoot length, the number of branches and fresh weight were measured when the plants were 30 and 60 days old. Rhizosphere soil sample from each treatment was analyzed to obtain bacterial population (CFU/ g soil).

2.3. *Effect of fertilizer plant on bacterial population in soil*

Bacterial population count was performed by solving 1 gram of soil in an Erlenmeyer filled with 90 ml of 0.8% sterile NaCl. The Erlenmeyer was then put in a 120 rpm shaker for 30 minutes and was diluted into 10^1-10^7 dilution. As much as 0.2 ml of each dilution was dropped in a sterile petri dish then respective media was poured over. The media used for *Azotobacter, Azospirillum,* and *Rhizobium* consecutively were Mannitol Ashby [18], Caceres medium [19] and YEMA agar [18]. The number of viable cells was measured by CFU (colony forming unit/ soil (g) after seven days incubation.

The acquired data were analyzed using SPSS 16.0 software, and significant differences of treatments were determined by LSD ($P<0.05$).

3. Results and Discussion

3.1. *The potential of bacteria*

Table 1 represents the ability of *Azotobacter, Azospirillum,* and *Rhizobium* to solubilize phosphate and produce IAA growth hormone. The activity of phosphate solubilization was varied in the range of 9.61–11.56 µg/ ml, while IAA yield ranged from 0.3 – 6.82 µg/ml. Some species of *Azotobacter, Azospirillum,* and *Rhizobium* were reported to show the ability to solubilize phosphate and produce IAA [20-22].

Tabel 1. P solubilization Activity and IAA production of Nitrogen fixing bacteria

Isolates	P Solubilization		IAA (µg/ml)
	Solid medium	Liquid medium (µg/ml)	
At1	+	9.61	4.04
At2	+	10.01	3.21
At3	+	10.45	0.30
At4	+	10.0	0.30
At5	+	10.65	6.82
Ap1	+	11.01	1.56
Ap2	+	10.21	2.81
Ap3	+	11.01	1.82
Ap4	+	11.56	0.30
Ap5	+	10.45	0.30
R1	+	10.09	4.39
R2	+	10.37	1.35
R3	+	11.01	1.99
R4	+	10.51	4.92
R5	+	4.81	4.81

Note: += halozone

3.2. Effect of fertilizer on growth of Caesalpinia pulcherrima on pot in greenhouse

Table 2 presents the data of each fertilizer and combinations' effect on plant growth (shoot length, the number of branches and fresh weight) 1 and 2 months after transplanting. Plants which were inoculated with biofertilizer (*Azotobacter, Azospirillum,* and *Rhizobium*) showed better growth compared than control treatment. It was supported by the rise observed. The shoot length was increased 12-31%. While 16-30% rise in the number of branches and 9-75% increase in fresh weight were measured. Correspondingly to this result, [23] reported that biofertilizer could improve plant height of various species. Yasari and Patwardhan, Hungria et al., Qin et al.) reported that plant growth can be increased by the inoculation of *Azotobacter, Azospirillum,* and *Rhizobium* [24,25,10].

The utilization of nitrogen-fixing bacteria as biofertilizer showed a positive result. It was indicated by the solubilization of phosphate and production of IAA growth hormone. The research by [8,26,10,2] showed that *Rhizobium* (symbiotic nitrogen-fixing bacteria) and *Azotobacter* or *Azospirillum (*nonsymbiotic nitrogen-fixing bacteria) as inoculant were able to improve plant growth due to phosphate solubilization. Nitrogen-fixing bacteria capability to generate growth hormone (IAA) was reported by [7,27-29]. The presence of IAA increased the formation of lateral roots. This phenomenon raised the amount of roots exudate and the nutrients absorbed by the roots which in turn stimulate bacterial growth thus enhance inoculation effect [30].

The application of both POME (organic fertilizer) and Muller solution (inorganic fertilizer) increased plant growth compared to control treatment one month and two months after transplanting. In the first month, the growth of fertilized plant was better than control treatment although the difference was insignificant. Two months after transplanting, the growth of plant fertilized with POME showed a significant difference to that of control treatment for shoot length, the number of branches and fresh weight respectively; 26%, 27,9%, and 21%. A similar result was obtained from Muller solution treatment. There were; 31,7% rise in shoot length, 38,9% rise of the number of branches and 75% rise of the fresh weight. In accordance with [31], reported that POME and inorganic fertilizer (Nitrogen, Phosphorus, and Potassium) significantly increased the growth and yield of *Amaranthus cruentus* compared to control treatment. Notable impact of POME and inorganic fertilizer (Nitrogen, Phosphorus, and Potassium) on soybean was reported by [32].

Plants that were fertilized with inorganic fertilizer presented better plant growth than those fertilized with POME. Falodun et al. stated that inorganic fertilizer increased the growth and yield of soybean higher than POME [32]. Inorganic fertilizer could provide nutrients more rapidly than organic fertilizer. Organic fertilizer released nutrients gradually [33].

The combination of biofertilizer with POME or inorganic fertilizer significantly increased plant growth. There was no significant difference in the plant growth between the plant fertilized with biofertilizer + POME and biofertilizer + inorganic fertilizer. As much as 9-60% rise was observed on the plant fertilized with the combination of biofertilizer + POME or inorganic fertilizer. It was not seen on plant fertilized with single fertilizer. A similar result was acquired by [34] which stated that the combination of biofertilizer and 50% inorganic fertilizer escalated lettuce yield as much as 25%. Whereas the combination of biofertilizer and 50% organic fertilizer generated 34% higher result compared to the application of single organic fertilizer. Shanmugam and Veeraputhran Nwangburuka et al. reported that biofertilizer and organic fertilizer were able to improve growth and yield of the plant [35][36].

Table 2. Effet of fertilizer on growth of *C. pulcherrima*

Treatment	Shoot length (cm)		No. of branches		Fresh weight (g)	
	1 MAT	2 MAT	1 MAT	2 MAT	1 MAT	2 MAT
C	4.8^{a}±0.74	15.6^{a}±0.50	3.2^{a}±0.41	11.7^{a}±0.54	0.25^{a}±0.03	0.42^{a}±0.04
At	7.6^{bc}±0.22	20.6^{bc}±1.10	4.8^{b}±0.29	15.2^{abc}±0.67	041^{ab}±0.07	080^{ab}±0.09
Ap	6.0^{ab}±0.22	19.4^{ab}±0.38	4.7^{b}±0.41	14.9^{abc}±0.35	0.47^{b}±0.01	0.89^{ab}±0.02
R	4.7^{a}±0.07	17.8^{ab}±1.88	4.0^{ab}±0.12	13.55^{ab}±1.42	0.33^{ab}±0.03	0.56^{ab}±0.15
P	5.1^{a}±0.45	19.9^{abc}±1.51	3.0^{a}±0.41	15.0^{abc}±0.75	0.31^{ab}±0.01	0.78^{ab}±0.09
M	5.1^{a}± 0.49	20.8^{bc}±0.92	3.7^{ab}±0.41	16.3^{bc}±0.91	0.32^{ab}±0.05	0.51^{ab}±0.27
At+P	8.3^{c}±0.08	24.6^{c}±3.60	4.4^{b}±0.12	17.6^{c}±2.71	0.28^{ab}±0.06	0.84^{ab}±0.13
At+M	7.4^{bc}±0.41	20.0^{abc}± 0.43	4.3^{ab}±0.41	15.2^{abc}±0.62	0.27^{ab}±0.01	0.50^{ab}±0.07
Ap+P	8.5^{c}±0.05	20.3^{abc}±1.59	4.5^{b}±0.53	14.6^{abc}±0.93	0.44^{ab}±0.01	0.94^{b}±0.04
Ap+M	8.2^{c}±0.42	20.02^{bc}±0.41	5.0^{b}±0.01	16.4^{bc}±0.82	0.42^{ab}±0.03	1.01^{b}±0.04

R+P	5.9^{a}±0.57	19.27^{ab}±1.03	4.3^{ab}±0.01	14.3^{abc}±0.61	0.30^{ab}±0.01	0.66^{ab}±0.06
R+M	6.1^{ab}±0.83	22.1^{bc}±0.84	5.0^{b}±0.01	16.3^{bc}±0.84	0.37^{ab}±0.02	0.86^{ab}±0.26

Notes: MAT (a month after transplanting). ± SD is a value of 3 replications. Means in the same group followed by the same letter in the coloum are not significantly different (p ≤0.05) as determined by the least significant difference (LSD) test.

3.3. Effect of fertilizer on bacterial population in soil harvesting (2MAT)

The application of biofertilizer and organic fertilizer significantly increase the soil bacterial population which was counted on harvesting (**Figure.1**). The highest number of population was seen on the soil inoculated with the combination of *Azospirillum* and POME (576x10^{5}cfu/g soil). There was no significant difference between the mentioned result with the number of population on the soil inoculated with *Azotobacter* + POME (573 x10^5. CFU/g soil). This result was in accordance with the research of [37,38] which indicated that organic fertilizer impacted the microorganisms population at the rhizosphere positively. Muzaffar et al. reported that the application of biofertilizer and organic fertilizer increased the number of *Azospirillum* population [39]. According to [40], organic fertilizer could increase microorganism population since it provided adequate nutrients for soil microorganisms.

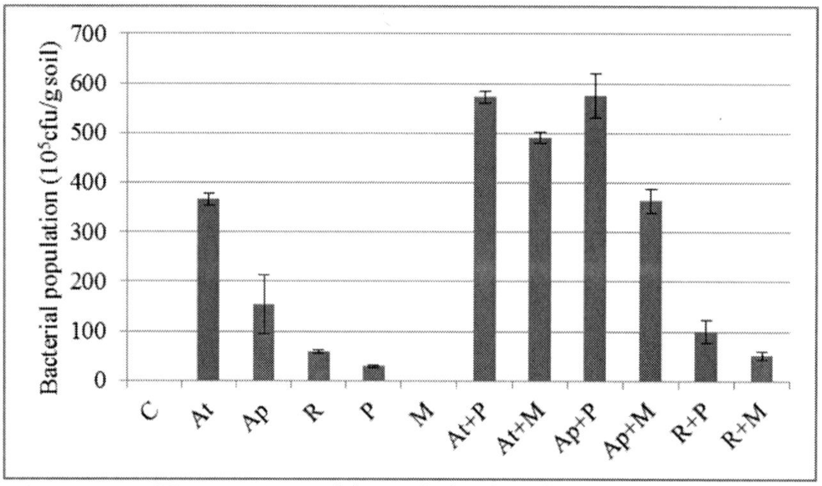

Figure 1. Bacterial population in soil at 2MAT

4. Conclusion

Plants which were inoculated with biofertilizer (*Azotobacter, Azospirillum,* and *Rhizobium*) showed better growth compared than control treatment. The application of inorganic fertilizer on plant showed more excellent plants' growth than the usage of organic fertilizer. The application of biofertilizer and inorganic or organic fertilizer combination resulted in higher plants' growth than single fertilizer usage. Biofertilizer and organic fertilizer combination produced higher bacterial population than inorganic fertilizer combination.

Acknowledgments

This research project was funded by TEMATIK DIPA (*Daftar Isian Pelaksanaan Anggaran*) program, Research Center for Biology, LIPI by Microbial Diversity Standard of recalcitrant compounds hydrolyzes for bioremediation and Producing Materials and Renewable Energy

5. References

[1] Macit I, Koc A, Guler S, Deligoz I 2007 Yield, quality and nutritional status of organically and conventionally-grown strawberry cultivars *Asian J. Plant Sci.* **6** (7) 1131-6

[2] Bhattacharya PN and Jha DK 2012 Plant growth-promoting rhizobacteria (PGPR) emergence in agriculture *World J. Microbiol. Biotechnol.* **28**:1327–1350
[3] Youssef MMA, Eissa MFM 2014 Biofertilizers and their role in management of plant parasitic nematodes *E J Biotechnol. Pharm. Res.* **5** 1-6.
[4] Vessey JK 2003 Plant growth promoting rhizobacteria as bio-fertilizers *Plant Soil* **255** 571-586
[5] Richardson AE, Barea JM, McNeill AM, Prigent-Combaret C 2009 Acquisition of phosphorusus and nitrogen in the rhizosphere and plant growth promotion by microorganisms *Plant and Soil* **321** 305–339
[6] Raghuwanshi R 2012 Opportunities and challenges to sustainable agriculture in India *Nebio* **3** 78–86
[7] Zahir ZA, Abbas SA, Khalid M, Arshad M 2000 Substrate-dependent microbially derived plant hormones for improving growth of maize seedlings Pak. J. Biol. Sci. **3** 289–291
[8] Peix A, Rivas-Boyero AA, Mateos PF, Rodriguez-Barrueco C,Martínez-Molina E, Velazquez E 2001 Growth promotion of chickpea and barley by a phosphate solubilizing strain of *Mesorhizobium mediterraneum* under growth *Soil Biol. Biochem.* **33** 103–110
[9] Remans R, Bebee S, Manrique MB, Tovar E, Rao I, Croonenborghs A, Torres-Gutierrez R, El-Howeity M, Michiels J, Vanderleyden J 2008 Physiological and genetic analysis of root responsiveness to auxin-producing plant growth-promoting bacteria in common bean (*Phaseolus vulgaris* L.) *Plant Soil* **302** 149–161
[10] Qin L, Jiang H, Tian J, Zhao J, Liao H 2011 Rhizobia enhance acquisition of phosphorusus from different sources by soybean plants *Plant Soil* **349** 25–36
[11] Walia S S and Kler DS 2009 Organic versus chemical farming A review *J. Res. Punjab Agric. Univ.* **46** 3&4 114-126
[12] Hiwale S 2015 *Sustainable Horticulture in Semiarid Dry Lands* Springer India
[13] Lim CH, PNG TC, Chan KW and Chooi SY 1993 *Land Application of Digested palm oil Mill effluent (POME) by sprinkle system*. Procedings of the seminar on land, application of palm and rubber factory effluents (Lim KH, Bachik AJ and Poon YC eds) Malaysia Soiety of Soil Science 72-79.
[14] Wu TY, Mohammad AW, Jahim J. and Anuar N 2009 A holistic approach to managing palm oil mill effluent (POME) Biotechnological advances in the sustainable reuse of POME *Biotechnology Advances* **27**(1) 40-52
[15] Olsen SR and Sommers LE 1982 *Phosphorusus* In Page AL et al. (eds.) Methods of soil Analysis part 2 Agron. Monogr. 9 403-430 2nd ed. ASA and SSSA Madison WI
[16] Gaur AC 1990 *Phosphate solubilizing microorganisms as biofertilizers* Omega Scientific Publishers New Delhi India 176 pages
[17] Pattern CL., Glick BR 2002 Role of Pseudomanas putida indo lactic acid in development of the host plant root system App. Eniron. Microbial. **68** 375-380
[18] Rao NSS 1995 *Soil Microorganisms and Plant Growth* 3th edition (New Hampshire: Science Publishers Inc.)
[19] Caceres, Enrique A, and Rodriguez 1982 Improved medium for isolation of *Azospirillum* spp. *Applied and Environmental Microbiology* **44** 990-991
[20] Deshwal VK, Pandey P, Kang SC, Maheshwari DK 2003 Rhizobia as a biological control agent against soil borne plant pathogenic fungi *Indian J. Exp. Biol.* **41** 1160-116
[21] Tank N, Saraf M 2010 Salinity-resistant plant growth promoting rhizobacteria ameliorates sodium chloride stress on tomato plants *J. Plant Interact.* **5** 51-58
[22] Ahemad M, Khan MS 2012 Ecological assessment of biotoxicity of pesticides towards plant growth promoting activities of pea (*Pisum sativum*)-specific *Rhizobium* sp. strain MRP1 *Emirates J. Food Agric.* **24** 334-343
[23] Rokhzadi A and Toashih V 2011 Nutrient uptake and yield of chickpea (*Cicer arietinum* L.) inoculated with plant growth promoting Rhizobacteria *Aus J Crop Sci* **5**(1) 44-48

[24] Yasari E and Patwardhan AM 2007 Effects of Aztobacter and Azospirillum inoculations and chemical fertilizers on growth and productivity of Canola *Asi.J.Plant.Sci.* **6** (1) 77-82
[25] Hungria M, Campo RJ, Souza EM, Pedrosa FO 2010 Inoculation with selected strains of *Azospirillum brasilense* and *A. lipoferum* improves yields of maize and wheat in Brazil *Plant Soil* **331** 413–425
[26] Kumar V, Behl RK, Narula N 2001 Establishment of phosphate solubilizing strains of *Azotobacter chroococcum* in the rhizosphere and their effect on wheat cultivars under greenhouse conditions *Microbiol. Res.* **156** 87–93
[27] Thakuria D, Taleekdar NC, Goswami C, Hazarika S, Boro RC, Khan MR 2004 Characterization and screening of bacteria from rhizosphere of rice grown in acidic soils of Assam *Curr. Sci.* **86** (7) 978–985
[28] Baudoin E, Lerner A, Mirza MS, Zemrany HE, Prigent-Combaret C, Jurkevich E, Spaepen S, Vanderleyden J, Nazaret S, Okon Y, et al 2010 Effects of *Azospirillum brasilense* with Genetically modified auxin biosynthesis gene *ipdC* upon the diversity of the indigenous microbiota of the wheat rhizosphere *Res. Microbiol.* **161** 219–226
[29] Shilev S 2013 *Soil Rhizobacteria Regulating the Uptake of Nutrients and Undesirable Elements by Plants* In: Arora NK (ed.) Plant Microbe Symbiosis Fundamentals and Advances Springer India 147-50
[30] Spaepen S, Vanderleyden J and Remans R 2007 Indole-3-acetic acid in microbial and microorganism-plant Signaling *FEMS Microbiol. Rev.* **31** 425–448.
[31] Jombo EO, Remison S U, Law-Ogbomo KE and Osaigbovo AU 2012 Effect of POME and NPK on the vegetative growth and total yield of *Amaranthus cruentus* on ultisols of Benin City, Nigeria Nigerian Journal of Agriculture, Food and Environment **8** (3) 14-20
[32] Falodun EJ, Osaigbovo AU and Remison SU 2010 Effect of Palm Oil Mill Effluent and Npk 15:15:15 Fertilizers on the Growth and Yield of Soya Bean *Advances in Natural and Applied Sciences* **4** (2) 112-116
[33] Leo E 2001 Effect of fertilizer on grain yield of soya bean. *Eur. J. Agron.* **13** 212-221.
[34] Young CC, Lai WA, Shen FT, Huang WS and Arun AB 2004 *Characterization of Multifunctional biofertilizer from Taiwan and biosafety considerations.* International Symposium on Future Development of Agricultural Biotechnology Park. The symposium series for celebrating the establishment of the Agricultural Biotechnology Park, Council of Agriculture, Executive Yuan, & the 80th Anniversary of National Pingtung University of Science and Technology 373-388
[35] Shanmugam P M and Veeraputhran R 2000 Effect of organic manure, biofertilizers,inorganic nitrogen and zinc on growth and yield of rabi rice *Madras Agric. J.* **2** 87-90.
[36] Nwangburuka CC , Olawuyi OJ, Oyekale K, Ogunwenmo KO, Denton, OA and Nwanko E 2012 Growth and yield response of Corchorus olitorius in the treatment of Arbuscular mychorrhizae (AM),Poultry manure (PM), Combination of AM-PM and inorganic Fertilizer (NPK) *Advances in Applied Science Research* **3** (3) 1466-1471
[37] Kuzyakov Y 2010 Priming effects: interactions between l iving and dead organic matter *SoilBiol Biochem* **42** (9) 1363-71
[38] Kaur H, Gosal SK and Walia SS 2017 Synergistic effect of organic, inorganic and biofertilizers on soil microbial activities in rhizosphere soil of Green Pea *Annual Research & Review I Biology* **12** (4) 1-11
[39] Abdullahi R, Sheriff HH and Lihan S 2013 Combine effect of bio-fertilizer and poultry manure on growth, nutrients uptake and microbial population associated with sesame (*Sesamum indicum* L) North-eastern Nigeria *IOSR Journal of Environmental Science, Toxicology And Food Technology (IOSR-JESTFT)* **5** (5) 60-65
[40] Muzaffar Mir, Hassan GI, Abid Mir, Amir Hassan and Muzamil Sulaimani 2013 Effects of bio-organics and chemical fertilizers on nutrient availability and biological properties of pomegranate orchard soil *African Journal of Agricultural* **8** (37) 4623-4627

IOP Conf. Series: Earth and Environmental Science **166** (2018) 012025 doi:10.1088/1755-1315/166/1/012025

Influence of Light Intensity on Leaf Photosynthetic Traits and Alkaloid Content of Kiasahan (*Tetracera scandens* L.)

T Setiawati[1], A Ayalla[1], M Nurzaman[1]and A. Z Mutaqin[1]

[1]Departement of Biology, Faculty of Mathematics and Natural Sciences, Padjadjaran University, Jatinangor-Sumedang, 45363, Indonesia

E-mail:tia@unpad.ac.id

Abstract. Kiasahan (*Tetracera scandens* L.) is one of the plants that people use as atraditional medicine for various diseases. This study was conducted to determine the characteristics of anatomy, morphology, chlorophyll content, and alkaloids of Kiasahaan leaf (*T. scandens*) based on differences in light intensity in the Pananjung Pangandaran Nature Reserve. The survey method was used to determine the location of sampling.The location was determined based on the existence of *T. scandens* plant. The shaded and unshaded area had the light intensity of 5516 and 922,000 lux, respectively. Leaf samples were taken as many as three strands of three different individuals on the shaded and unshaded areas. Parameters measured were the thickness of leaf, leaf area, stomatal density, chlorophyll,and alkaloid content. Measurement of chlorophyll and alkaloids used chlorophyll meter and Mayer's reagent method, respectively. The results showed that the average leaf thickness, leaf area, stomatal density, chlorophyll content in theshaded area were 0.048 mm, 7.086 cm^2, 50.958 cells / mm^2, 15.38 CCI, respectively whereas in the unshaded area were 0.076mm, 6.496 cm^2, 62.678 cells / mm^2, 8.86 CCI, respectively. Alkaloid content of *T. scandens* leaves in the unshaded area was higher than in the shaded area.

1. Introduction

Plants produce secondary metabolites that are beneficial to humans such as food flavoring, insecticides, perfumes, and sources of medicinal materials [1]. As a medicinal substance, the secondary metabolites of plants have biological properties such as antioxidant activity, antimicrobial, detoxification enzyme modulation, immune system stimulation, decreased platelet aggregation and modulation of hormones metabolic, and anticancer [2]. One type of secondary metabolite that is widely used as a medicine is alkaloids. Role of alkaloids in plants is still unknown, but in the health aspect can be used as a medicine for neurosystem or as an additive compound [3].

There are several factors that affect the growth and development of plants, such as the intensity of light. According to Li et al. [5], light is an important environmental factor affecting growth, development, and secondary metabolism of plants. Spectrum, intensity,and quality of light have an important role in the process of plant development and metabolism [5]. The intensity of light that is too high or too low has an impact on morphology, physiology of photosynthesis, and the production of secondary metabolites [6]. As also stated by Coelho et al.[7] that the secondary metabolite synthesis and accumulation of plant is influenced by the light. Light affects the oxidative metabolism in the chloroplast further affecting light absorption and CO_2 diffusion which induce structural and physiological modifications to the leaf [8]. According to Vendramini et al. [9] leaf characteristics usually reflect adaptive strategies plants to important environmental factors, including light, temperature, air, and nutrition. The research conducted by Kong et al.[10] showed that differences of light intensity affecting the external morphology of leaves, internal anatomy, and physiological characteristics; even induces changes in secondary metabolite production of *Mahonia bodinieri*.

Tetracera scandens commonly known to the local community as Kiasahan is one of the plants that people use for traditional medicine such as hypertension, rheumatism, inflammatory diseases, internal pains, urinary disorders, gout and hepatitis [11]. To the best of our knowledge, there has not been any publication about influence of light intensity on leaf photosynthetic traits and alkaloid content of kiasahan (*T. scandens* L.).

The purpose of this research was to know the traits of photosynthetic leaf and alkaloid content in Kiasahan plants (*T. scandens* L.) in areas with different light intensity.

2. Materials and Methods

The materials used in this study were leaves of *T. scandens* L. plant, distilled water, transparent nail polish, ammonia chloroform, sulfuric acid (H_2SO_4) 2N, and Mayer's reagent.

This study used a quantitative descriptive approach.The sampling technique used a survey method to determine the presence of *T. scandens* plants in two different locations i.e. the shaded area and the area exposed to full sunlight (unshaded area). A leaf that taken as a sample was the third, fourth, and fifth leaves from shoots that already fully open. Leaf samples were taken as many as three pieces of three different individuals in the shaded area and unshaded area. Leaf characteristics observed were the thickness of leaves, leaf area, stomatal density, chlorophyll,and alkaloids content. The data were analyzed descriptively.

Leaf thickness was measured using a micrometer screw. The measurement was done five times the repetition and then averaged.Measurement of leaf area was conducted using the gravimetric method [12] i.e. by comparing the weight between leaf patterns made on paper with other paper patterns of extension known [13].Observations of stomata density used a replica method. Transparent nail polish applied to the leaf surface, allowed to dry, and then spread transparent nail polish peeled away slowly. Replica of stomata placed on a glass slide and examined under a light microscope then density of stomata was calculated. Chlorophyll content was measured with chlorophyllmeter Opti-Science CCM-20. Measurement on each leaf sample was taken as many as fiverepetitions, thereafter were averaged.

Alkaloid content was determined using Mayer's reagent. The leaf sample was weighed 2 grams and mashed, then put in a test tube, added chloroform and an ammonia solution of 5 mL each. The mixture was heated, shaken, and filtered. H_2SO_4 2 N solution of 5 drops was added to the filtrate and shaken. The top of the filtrate was taken and tested with Mayer's reagent. The positive test results contain alkaloids when there was white precipitate [14].

Environmental parameters measured include light intensity, temperature, humidity, and pH. Lux meter was used to measure light intensity, Thermo hygrometer was used to measure temperature and humidity, and soil tester was used to measure soil pH. Environmental parameters measurements were performed at two locations (shaded and unshaded areas).

3. Results and Discussion

3.1. Environmental parameter measurement

Environmental parameters showed that the intensity of light in the shaded area was 5,516 lux, while in the unshaded area was 922.000 lux. Under these conditions, the temperature of the shaded and unshaded area indicated 28° C and 26° C, respectively.The result ofhumidity measurement showed thatin the shaded area of 55% while the unshaded area of 50%. This condition indicates that in the higher the shade level, the air humidity also increases. Soil pH measurements showed that the shaded area was 4, while in the unshaded area was 4.8. The acidity (pH) of the soil was classified as acid (<7).

3.2. The thickness of leaf and leaf area

The average leaf thickness and leaf area of *T. scandens* in the shaded and unshaded areas can be seen in **Figure 1** and **2**.

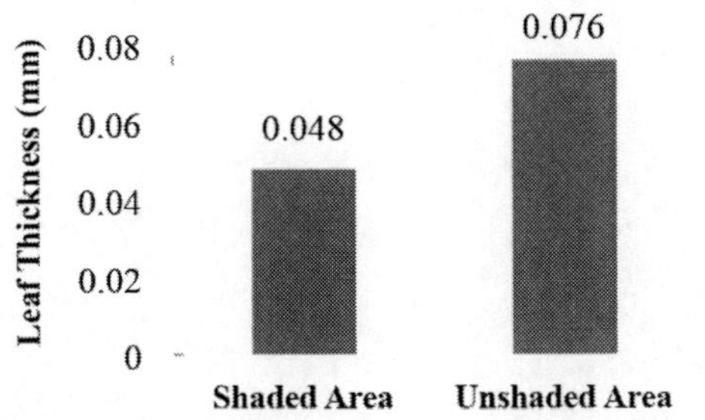

Figure 1. Average leaf thickness of *T.scandens*in shaded and unshaded areas

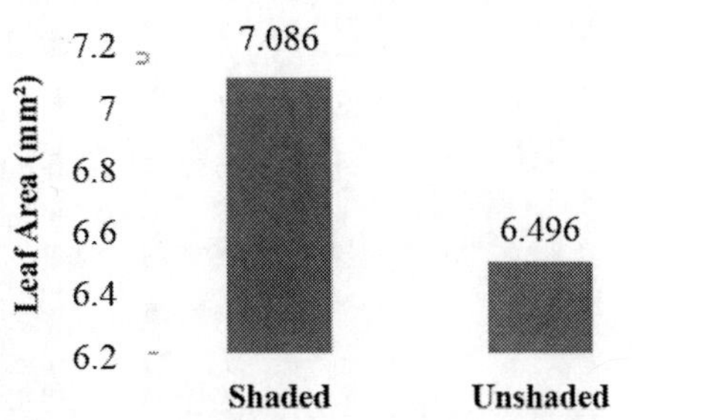

Figure 2. Average leaf area of *T. scandens* in shaded and unshaded areas

Figure 1showed that the average thickness of *T. scandens* leaf was 0.048 mm while in the unshaded area of 0.076 mm. Average leaf thickness in the shaded area was lower than in the unshaded area. In accordance with a recent report [15] that leaves that grow in unshaded areas or commonly called the sun leaves are thicker than the leaves in shade areas or shade leaves. Tucić et al. [16] reported thatmany plants growing in exposed areas of full sunlight produce thicker leaves to minimize excess light interception. Leaf thickness is affected by the thickness of the tissue layer on the leaves, such as mesophyll tissue.Likewise, according to [17] that characteristic changethe leaves on the shaded condition becomes wide and thin. Leaf thinning is caused by a decrease in the palisade layer in leaf mesophyll cells [18].

Figure 2showed that the average leaf area of *T. scandens*in the shaded area washigher than in the unshaded area. Average leaf area in the shaded and unshaded area was7,086 mm² and6,496 mm², respectively.Under low light, plants often get features to capture themore optical energy, such as leaf area enlargement [19].As wellKong et al.[10] reported that the leaves in sun area are thicker and smaller with low stomatal density on both leaf surfaces. Similar to the results of research conducted by[6] which indicates that *Epimedium pseudowushanense*seedlings grow under relatively low light intensity have larger leaves than those grown under high light intensity. According to Fitter and Hay[8], plants under the shading adapt to low light intensity conditions by increasing the leaf area to obtain a larger surface for light absorption. Moreover, adecrease in leaf size in full sunlight area will reduce leaf temperature, potential water loss and damage to leaf photosystems [20].

3.3. Density of stomata

The average stomatal density of *T.scandens* leaf in the shaded and unshaded areas can be seen in **Figure 3**, which was calculated based on the distribution of stomata number on the abaxial longitudinal incision of *T. scandens* leaf (**Figure 4** and **5**).

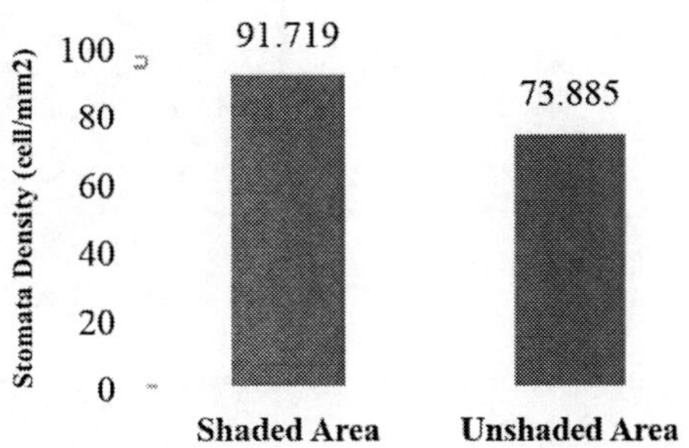

Figure 3. Average stomatal density of *T. scandens*in shaded and unshaded area

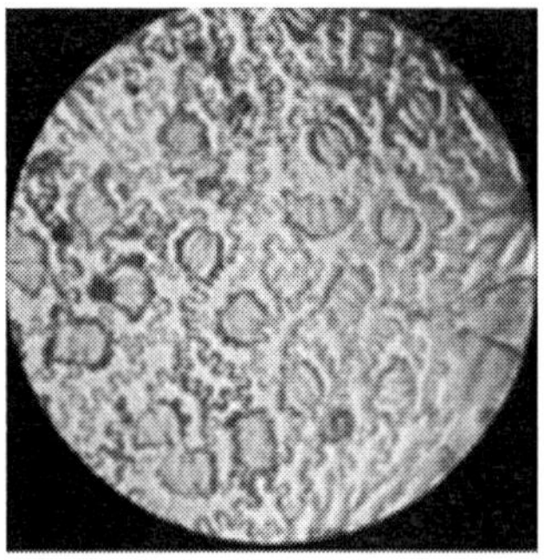

Figure 4. Distribution of stomata number of *T. S scandens*in the unshaded area

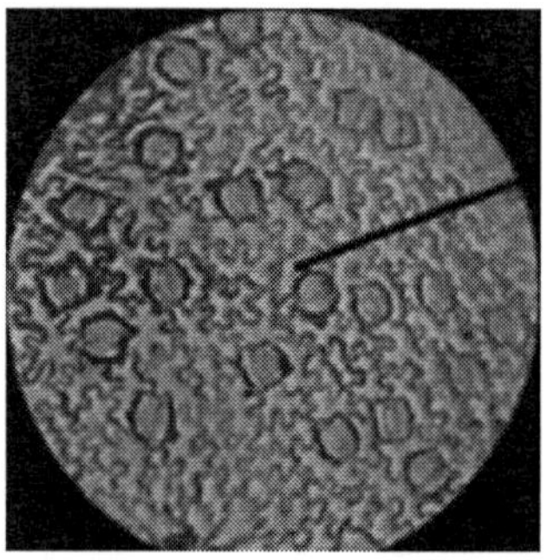

Figure 5. Distribution of stomata number of *T. scandens* in the shaded area

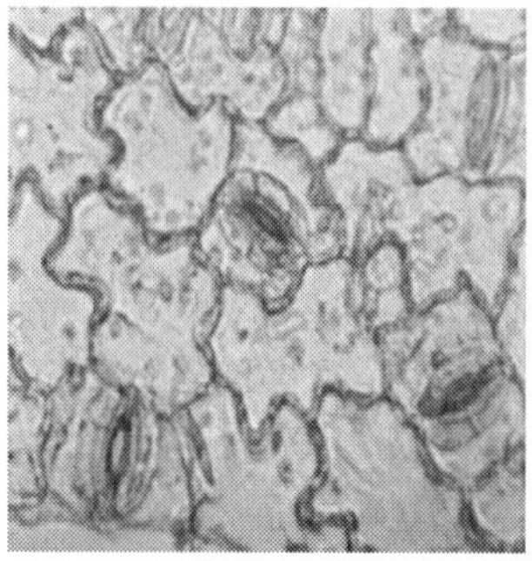

Figure 6. Paracytic stomata of *T. scandens* leaf

Figure 3 indicated that the average stomata density of *T. scandens* shaded leaves of 50.958 cells /mm^2, whereas in the unshaded area of 62.678 cells/mm^2. These results indicated that the average density of stomata in the shaded area was lower than the unshaded area. Stomata control gas exchange, water loss, and temperature of leaves. Stomata will be found more in leaves less exposed to sunlight to reduce evaporation or water loss. Species with higher stomatal density tend to be more responsive to the increase in CO_2, so the rate of photosynthesis is greater [21]. In general, stomatal density decreases with increased CO_2 concentration and compared with leaf developed under low light intensity, the sun leaves have ahigher density of stomata [22].

Observation of stomata on *T. scandens* leaf indicated the type of parasitic (**Figure 6**). According to [23], Kiasahan (*T. scandens*) plants including the Dilleniaceae family, usually have anomocytic stomata type in which each guard cellis surrounded by a number of cells that are not different in size and shape from other epidermal cells; but exceptions to the species of *Tetracera*, the type of parasitic stomata [24]. Type of paracytic / Rubiaceous stomata where each guard cell is accompanied by one or more neighboring cells with the long axis of the neighboring cell is parallel to the guard cell and pore [25].

3.4. Chlorophyll content

The result of chlorophyll content analysis of *T. scandens* leaf on the shaded and unshaded areas can be seen in **Figure 7**.

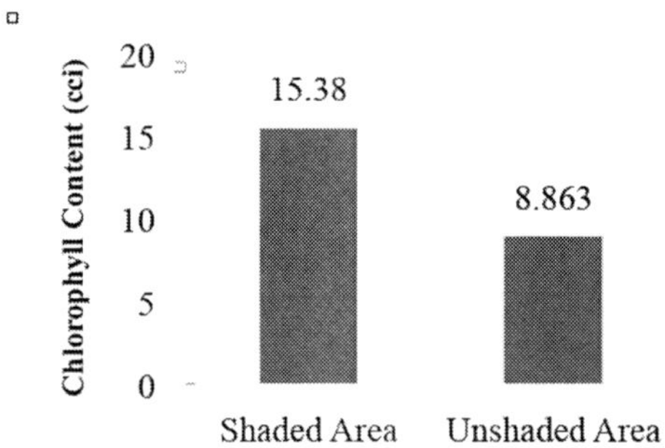

Figure 7.Average chlorophyll content of *T. scandens* in the shaded and unshaded area

Figure 8. The color of *T. scandens* leaves in the shaded(left) and unshaded area (right)

In **Figure 7** appeared that the average chlorophyll content of *T. scandens* leaves in the shaded area of 15.38 CCI higher than in the unshaded area of 8.863 CCI.A higher amount of chlorophyll in plants under shade serves to maximize light absorption under low light conditions. Chlorophyll in shaded plants is arranged in photo-taxis state [17]. According to Watson & Dallwitz [26] one of the characteristics of the adjustment to low irradiation due to shade is an increase in leaf chlorophyll content. This increase is related to the increase of light harvesting complex (Light Harvesting Complex II) and the enlargement of the antenna in photosystem II which resulted in increased light capture efficiency. Light provides an energy source for photosynthesis but excess light can damage photosynthetic apparatus and cause photo-oxidation of chlorophyll. Photo-damage (photo-oxidation of chlorophyll and cell death) occurs only when the plant is exposed to extreme or prolonged pressure. Similar research reported that decreased light intensity followed by an increase in chlorophyll a (Chl. a), chlorophyll b (Chl. b) and total chlorophyll (Chl. a + b) on *Lithocarpus litseifolius*[5] and on *Mahonia bodinieri*[10]. The high light intensity causes chloroplast damage on *E. pseudowushanense* plant [6] resulting in a decrease of chlorophyll content.

Figure 8 indicated differences in leaf color of *T. Scandens* which grow in the shaded and unshaded areas. The leaves of *T. Scandens* which grow on the shaded area have more green color than in the unshaded area. This may be due to the leaves on the shaded have higher levels of chlorophyll than the unshaded area. The study conducted by Stanton [27] revealed that shading of 40% treatment on *Spiraea alba* plant resulted in darker green leaf.

3.5. Alkaloid content

The results of the analysis on the presence of alkaloids on the leaves of *T. scandens* showed that alkaloid found in the leaves of *T. scandens* both in the shaded and unshaded areas. In this study, alkaloid analysis is qualitative using Mayer's precipitation method, however alkaloid content is quantified based on the amount of precipitate formed. In accordance with the statement by Verma et al. [28] that the hydrochloride salt deposits from the leaf extract show the number of alkaloids formed.

In the leaves of *T. scandens* at the unshaded area are found more sediment than in the shaded area, so it can be assumed that the level of alkaloid in the unshaded area is higher than in the shaded area. The intensity of light is a major factor affecting the synthesis and accumulation of secondary metabolites [7]. According to Pompelli et al. [29] in the *Coffee arabica* plant, high-intensity stress triggers increased alkaloid production, such as caffeine, allantoin,and theophylline. Stress in the form of high intensity of light triggers the increase in the presence of free nitrogen. Nitrogen is the source for the passage of metabolism. Plants that live at low concentrations of nitrogen will experience caffeine degradation (xanthine alkaloids), so the resulting caffeine will be lower than at high concentrations of nitrogen. The similar research in *Iris pumila* plants showed that secondary metabolite levels of anthocyanin and phenol were detected higher in areas exposed to full sunlight [30].

4. Conclusion

Based on the results of this study can be concluded that the intensity of the light affects the thickness ofleaves, leaf area, stomatal density, chlorophyll, and alkaloid content of *T.scandens*. There were different characteristics of leaf photosynthetic of *T. scandens* in the shaded and unshaded areas. In the shaded area, average the leaf thickness (0.048 mm), stomatal density (50.958 cells / mm^2) and alkaloid content of *T. scandens* leaves were lower than in the unshaded area. While the average leaf area (7.086 cm^2) and chlorophyll content (15.38 CCI) of *T. scandens*leaves were higher in the shaded area than in unshaded ones.

5. References

[1] Ganapathi B and Kargi F 1990 Recent advances indole alkaloid production by *Catharanthus roseusJ. Exp. Bot.***41** 259-67

[2] Saxena S, Wahl J, Huber-Lang MS, Stadel D, Braubach P and Debatin KM 2013 Generation of murine sympathy adrenergic progenitor-like cells from embryonic stem cells and postnatal

adrenal glands. *PLoS ONE.*8 e64454 https://doi.org/10.1371/journal.pone.0064454

[3] Science Jrank 2016 *Alkaloid* http://science.jrank.org/pages/229/Alkaloid-Alkaloids-pain-pleasure.html

[4] Zoratti L, Karppinen K, Escobar AL, Häggman H and Jaakola L 2014 Light-controlled flavonoid biosynthesis in fruits *Front. Plant Sci.***5** 534

[5] Li A, Li S, Wu X, Zhang J, He A, Zhao G and Yang X 2016 Effect of light intensity on leaf photosynthetic characteristics and accumulation of flavonoids in *Lithocarpus litseifolius* (Hance) Chun. (Fagaceae) *Open J For.***6** 445-59

[6] Pan J and Guo B 2016 Effects of light intensity on the growth, photosynthetic characteristics, and flavonoid content of *Epimedium pseudowushanense* B.L.Guo *Molecules* **21** 1475

[7] Coelho GC, Rachwa MFG, Dedecek RA, Curcio GR, Nietsche K and Schenkel EP 2007 Effect of light intensity on methylxanthine contents of *Ilex paraguariensis* A. St. Hil *Biochem. Syst.Ecol.***2** 75–80

[8] Fitter AH and Hay RKM 1981 *Environmental Physiology of Plants* (London: Academic Press.)

[9] Vendramini F, Díaz S, Gurvich, Wilson PJ, Thompson K and Hodgson JG 2002 Leaf traits as indicators of resource-use strategy in floras with succulent species *New Phytol.* **154** 147–57

[10] Kong DX, Li YQ, Wang ML, Bai M, Zou R, Tang H and Wu H 2016 Effects of light intensity on leaf photosynthetic characteristics, chloroplast structure, and alkaloid content of *Mahonia bodinieri* (Gagnep.) Laferr *Acta Physiol. Plant* **38** 120

[11] Thanh TB, Thanh HN, Minh HPT, Le-Thi-Thu H, Ly HDT and Duc LV 2015 Protective effect of *Tetracera scandens* L. leaf extract against CCl4-induced acute liver injury in rats *Asian Pac. J. Trop. Biomed.* **5** 221-7

[12] Frear DEH 1935 Photoelectric apparatus for measuring leaf areas *Plant Physiol.***10** 569–74

[13] Chaudhary P, Godara S, Cheeran AN and Chaudhari AK2012 Fast and an accurate method for leaf area measurement *International Journal of Computer Applications* **49** 22-5

[14] Seniwaty, Rihanah, Nugraheni IK and Umaningrum D 2009 Phytochemical screening of *Imperata cylindrica* L. Beauv and *Hedyotis corymbosa* L. Lamk *Sains dan Terapan Kimia* **3** 124-33

[15] Murchie EH and Horton P 1997 Acclimation of photosynthesis to irradiance and spectral quality in British plant species: chlorophyll content, photosynthetic capacity,and habitat preference *Plant Cell Environ.* **20** 438–48

[16] Tucić B, Tomić V, Avramov S and Pemac D 1998 Testing the adaptive plasticity of *Iris pumila* leaf traits to natural light conditions using phenotypic selection analysis *ActaOecol.* **19** 473–81

[17] Salisbury FB and Ross CW 1992 *Plant Physiology* 4th Edition (Belmont: Wadsworth Publishing)

[18] Taiz L and Zeiger E 1991 *Plant Physiology* (Tokyo: The Benyamin/Cumming Publishing Company Inc.)

[19] Johnson DM, Smit WK, Vogelmann TC and Brodersen CR 2005 Leaf architecture and direction of incident light influence mesophyll fluorescence profiles *Am. J. Bot.***92** 1425–31

[20] James SA and Bell DT 2000 Influence of light availability on leaf structure and growth of two *Eucalyptus globulus* ssp. globulus provenances Tree *Physiology.***20** 1007–18

[21] CassonS and Grey JL 2008 Influence of environmental factors on stomatal development *NewPhytol.***178** 9–23

[22] Givnish TJ 1988 Adaptation to sun and shade: a whole-plant perspective *Aust. J. Plant Physiol.***15** 63-92

[23] Watson L and Dallwitz MJ 1992 *The families of flowering plants: descriptions, illustrations, identification, and information retrieval.* Version: 13th March 2017 http://delta-intkey.com/angio/www/dillenia.htm

[24] Hoogland RD 1953 *The Genus Tetracera (Dilleniaceae) In The Eastern Old World*

Reinwardtia. Herbarium Bogoriense Kebun Raya Indonesia vol 2 part 2 185-224
[25] Essau K 1965 *Plant Anatomy* (New York : John Wiley & Sons)
[26] Rotundo AM, Forlani and Divaio C 2004 Influence of shading net on vegetative and productive characteristics, gas exchange and chlorophyll content of the leaves in two blackberry (*Rubusul mifolius* Schott) http:/www.actahort.org/books/457/457-42.htm
[27] Stanton KM 2010 Light exposure and shade effects on growth, flowering, and leaf morphology of *Spiraea alba* Du Roi and *Spiraeato mentosa* L*Hortscience* **45** 1912–16.
[28] Verma M, Sharma A, Naidu S, Bhadra AK, Kukreti R and Taneja V 2012 Curcumin prevents formation of polyglutamine aggregates by inhibiting Vps36, a component of the ESCRT-II complex *PLoS One* 7(8) e42923
[29] Pompelli MF, Pompelli GM, de Oliveira AFM and Antunes WC 2013 The effect of light and nitrogen availability on the caffeine, theophylline and allantoin contents in the leaves of *Coffe aarabica* L *Aims Environmental Science* **1** 1-11
[30] Vuleta A, Manitašević-Jovanović S and Tucić S 2011 Light intensity influences variations in the structural and physiological traits in the leaves of *Iris pumila* L.*Arch. Biol. Sci.* **63** 1099-110

IOP Conf. Series: Earth and Environmental Science **166** (2018) 012026 doi:10.1088/1755-1315/166/1/012026

Ability of *Penicillium Griseofulvum* Inacc F 14 in Producing Cellulase Enzyme for Composting Media Plant of White Oyster Mushroom (*Pleurotus Ostreatus* Jacq. Ex Fr.) P. Kumm and Ear Mushrooms (*Auricularia Auricula* J.)

Y. S Soeka[1] and M Ilyas[1]

[1]Research Center for Biology, Indonesian Institute of Sciences (LIPI)
Jl. Raya Bogor Km. 46, Cibinong 16911, Bogor, Indonesia

E-mail: ceuceu_lipi@yahoo.com

Abstract. The aim of this study is to determine the characteristics of cellulase enzymes produced by the fungus LIPIMC selection. Results were then analyzed qualitatively and quantitatively. Enzyme activity includes incubation time, substrate concentration Carboxy Methyl Cellulose (CMC), temperature, optimum pH and the effect of the addition of some metal ions as activators or inhibitors were examined. Further molecular identification of fungus was examined through 16S rDNA sequencing. The optimum conditions for the enzyme activity were the conditions with an incubation period of two days, at pH 6, at the temperature 50°C, at substrate concentration 1.25 %. The cellulase enzyme was activated by cations $CoCl_2$, $CuCl_2$, and NaCl. Meanwhile $MgCl_2$ and $ZnCl_2$ were acted as an inhibitor. Results of molecular identification by 16S rDNA sequencing confirmed that LIPIMC 0575 was *Penicillium griseofulvum*. After identified, the tested supernatant of cellulase enzyme has the ability to inhibit the growth of *Bacillus cereus* and will be used for composting on the media plant of oyster mushrooms (*Pleurotus ostreatus* J.) and ear mushrooms (*Auricularia auricula* J.)

1. Introduction

Enzyme is a protein that has biochemical activity as a catalyst of a reaction. Because of a protein, this enzyme very susceptible to an environmental condition [1]. Enzyme could be isolated from various of source such as plants, animal tissues, and microbes [2]. At this time, microbe enzymes have been proven economically beneficial because microbe culture is far simpler, faster and can be genetically manipulated easier than plants and animals [3, 4]. Cellulase can be produced by fungi, bacteria or actinomycetes, but the most common producer is fungi. Fungi have the potential that good to be used in the cellulase production, considered to be the most active to degrade the polymers that abundantly available in the nature so that can decrease a pollution [5, 6, 7, 8]. Cellulase (EC 3.2.1.4) is one of the extracellular enzyme that can catalyze 1,4-β-D-glucoside bonds in a cellulose molecule glucose monomer [9, 10, 11, 12, 13]. Currently cellulase application is the most useful in industries include pulps and papers, textiles, and laundry [when cellulase enzyme added into detergent, then that detergent reacts with cellulose fibers to produce a bright color and can soften], the tannery, the biofuel production (such as bioethanol), the bread in food industry (cellulase enzyme can increase crumb structure to produce crumb that more regular and smooth to result increasing whiteness in foods and drinks, and animal food), beer production, and agriculture as organic fertilizers [14, 15, 16, 17, 18, 19,

20, 4, 21, 22]. Some fungi that used to cellulase industries were *Trichoderma*, *Humicola*, *Aspergillus*, and *Penicillium* [23, 24, 25].

The aim of this research was to characterize the cellulase enzyme activity of *Penicillium griseofulvum* InaCC F 14 which will be used for composting media plant of white oyster mushroom (*Pleurotus ostreatus* J.) and ear mushroom (*Auricularia auricula* J.).

2. Materials and Methods

Equipment and materials used were include water bath, oven, digital balance, incubator, centrifuge, vortex, shaker incubator, spectrophotometer, measuring pipette, measuring glass, test tube, erlenmeyer, magnetic stirrer, magnetic stirrer bar, micropipette, micropipette tips, plastic heater, test tube rack, petri dish, fungi strains of LIPIMC collections of Research Center for Biology - Indonesian Institute of Sciences Cibinong, Potatoes Dextrose Agar (PDA), paper disc, aquadest, yeast extract, bacteriological peptone, Carboxy Methyl Cellulose (CMC), KH_2PO_4, $(NH_4)_2SO_4$, $MgSO_4.7H_2O$, $CaCl_2$, NaCl, bacto agar, 3,5-dinitrosalcylic acid (DNS) reagent, and Bradford reagent.

Fungi strains used in this experiment were LIPIMC 0544, LIPIMC 0546 a and LIPIMC 0575. LIPIMC 0544 collection was isolated from library air-borne of Research Center or Biology-Indonesian Institute of Sciences Cibinong. LIPIMC 0546 collection was isolated from Research Center for Biology - Indonesian Institute of Sciences Cibinong. LIPIMC 0575 collection was derived from *Magnolia condolii* rhizosphere of Taman Nasional Bodogol, Sukabumi District, West Java. Fungi was cultivated in Potatoes Dextrose Agar (PDA) slant media, disc paper 6 mm, *Bacillus cereus* InaCC B325.

Media used in qualitative selection were specific media contained CMC 1% and composed of 2.0 g of KH_2PO_4 (Merck), 1.4 g of $(NH_4)_2SO_4$ (Merck), 0.3 g of $MgSO_4.7H_2O$ (Merck), 0.1 g of $CaCl_2$ (Merck), 1.0 g of peptone (Difco) and 2.0 g of agar (Bacto). All materials were dissolved in aquadest 100 mL and prepared in accordance to Sohail et al [6].

Media used for microbe culture were the yields of strains selection that the biggest clear zone and have been grown in PDA slant media, then incubated in incubator until aged 3-4 days. Cellulase production media that used were prepared and composed similarly to the qualititive selection media as previously described in Sohail et al [6]. The supernatant were incubated in shaker incubator with 120 rpm velocity, at 37 °C for 5 days. They were centrifuged at velocity 8000 rpm, temperature 4°C for 10 minutes to separate enzyme solution (supernatant) from their substrate particles and cells. The acquired supernatant was saved in a freezer (-10°C) and was further measured for its cellulase activity The cellulase activity of the culture was analyzed with spectrophotometer at λ 540 nm everyday as in accordance to Jahangeer et al [26].

Cellulase enzyme activity was tested according to DNS method [27]. It was tested by measuring the quantity of reductor sugar that was produced by enzyme hidrolysis activity in CMC 1% subtrate.

2.1. *Qualitative examination of cellulase activity*

Each of fungi strain was inoculated in selective agar media of CMC 1%. Each of fungi strains was grown in the surface of media, then incubated for 72 hours. Grown colony was dropped with a pipette congo red 0.1 % indicator and let it for 30-60 minutes, then it washed with NaCl 2 %. The formed clear zone around the colony was measured in accordance Teather and Wood [28]. The formed clear zone around a colony was an indication of the existence of cellulase enzyme activity. Cellulolytic index was measured value of the diameter of the colony that devided by the diameter of the formed clear zone. Isolates that showed existence of cellulase activity with high relative value were then examined quantitatively.

$$\text{Cellulolytic index} = \frac{\text{clear zone diameter}}{\text{colony diameter}}$$

2.2. *Quantitative examination of cellulase activity*

In order to measure cellulase activity quantively, the reaction was prepared in accordance to Kar et al 2006 [29]. An equal amount of enzyme supernatant (i.e. 0.125 mL) was added into 0.125 mL of substrate (i.e. phosphate buffer solution with pH 7.0). The reaction was incubated at 40°C temperature for 30 minutes. To stop the reaction, 0.25 mL of 3,5 dinitro salisilat (DNS) acid was added into reaction [29]. Then it was boiled for 5 minutes and was cooled under some flowing water. Aquadest was then added as much as 2.5 mL. The mixed solution was measured with spectrophotometer at λ 540 nm.

Calculation of cellulase activity was carried out with a formula:

$$\text{Cellulose activity} = \frac{(\text{Glucose content x dilution factor}) \text{ x } 1000}{(\text{Glucose molecular weight x incubation time})}$$

One unit of cellulase activity is defined as enzyme quantity that produced 1 μmol of glucose in 1 minute at examination condition. It is multiplied by 1000 as a convertion from mmol to μmol.

2.3. Characterization of cellulase enzyme

Characterisation of cellulose enzyme in this study was assessed by determining the optimum conditions for enzyme production, including incubation time, pH, temperature, CMC substrate concentration and metal ion addition consecutively. The effect of incubation time against enzyme activity was carried out by reacting enzyme supernatant in period of time, i.e. one to five days. The effect of pH against enzyme activity was carried out by reacting enzyme solution with the concentration of CMC 1%, then incubated in acetic buffer 0.05 N at pH 5, and phosphate buffer 0.05 N at various pH. The tested pH were 5.0, 6.0, 7.0, 8.0, and 9.0. The effect of temperature against enzyme activity was carried out by reacting enzyme solution with the concentration of CMC 1% in buffer solution with the obtained optimum pH from the previous analysis at various temperature, i.e. 35, 40, 45, 50, 60 and 70°C. The effect of CMC substrate concentration against enzyme activity was carried out by reacting enzyme solution with the concentration of 0.5%, 0.75%, 1.0%, 1.25% , 1.5% and 1.75% at the obtained optimum pH dan optimum temperature. The effect of metal ion with final concentration 1 mM was carried out by reacting enzyme solution with addition of various metal at the obtained optimum conditions pH, temperature and CMC concentration from the previous analysis. The tested ions were Na^{+}, Co^{2+}, Cu^{2+}, Mg^{2+} and Zn^{2+} which were available in the form of NaCl, $CoCl_2$, $CuCl_2$, $MgCl_2$ and $ZnCl_2$ respectively. The samples were also compared with reacting enzyme without metal addition.

Examination of antibacterial activity from cellulase enzyme supernatant was assessed by measuring inhibiton zone diameter that was formed around the tested disc [30]. The 6 mm diameter of tested disc was dipped into cellulase enzyme extract supernatant. For a positive control, the tested disc was dipped into Chloramphenicol (400 μg/mL). Meanwhile for a negative control, the tested disc was dipped into steril water. The tested disc papers were placed in Nutrient Agar (NA) media that has previously inoculated with *Bacillus cereus* bacteria. The tested petri dish was incubated at 37°C, for 1-4 days.

3. Results and Discussion

Qualitative selection of cellulolytic microbe was aimed to identify isolates, which have cellulase enzyme activity or have not. Colony that can produce cellulase enzyme will be detected by the formation of a clear zone. The results of research showed that a clear zone was seen and existed at selective CMC 1% agar media around the colony of LIPIMC 0575 after incubated for 48 hours. Meanwhile, a clear zone was not seen and existed either around the colony of LIPIMC 0544 nor the LIPIMC 0546 isolates (**Figure 1**).

A clear zone around the colony of LIPI-MC 0575 isolate (**Figure 2**) is measured at the next research step. Cellulase enzyme can hidrolyze a cellulose become a glucose. Different isolates can produce different clear zone because of their internal factor , such as gen, and external factors, such as nutrition, incubation time, pH, temperature, and supplementation of metal ion as activator.

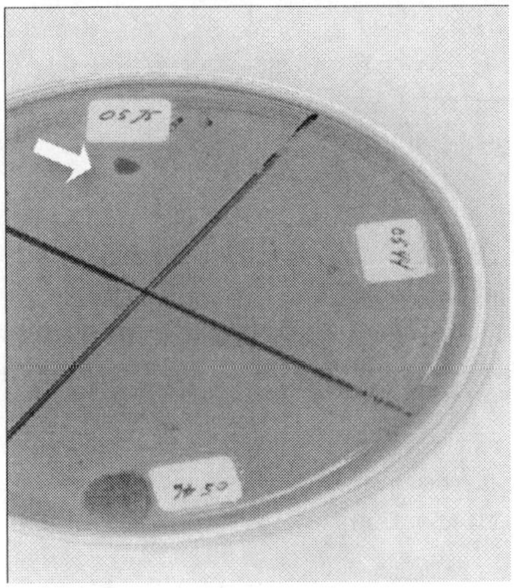

Figure 1. LIPIMC 0554, 0546 and 0575 strains in selective media of CMC 1%

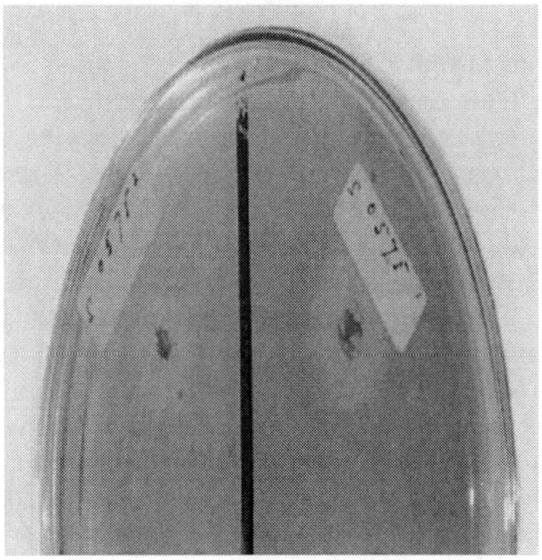

Figure 2. A clear zone of LIPIMC 0575

The ratio of a clear zone diameter of LIPI-MC 0575 isolate with its qualitative colony diameter is expressed as a cellulase index (semi qualitative) (**Table 1**). Enzyme activity depends on the content and structure of different substrates. Carboxy methyl Cellulose (CMC) substrate was much used as a substrate to examine the cellulase activity independently or dependently during cellulose enzymatic hidrolysis [31]. The CMC as a substrate produces enzyme activity that is far different compared with the other substrates, because of CMC is a material that contains a pure cellulose, and does not contain lignin that can inhibit enzyme activity to break down a substrate [32]. One of the qualitative test that commonly used is through staining with congo red solution 0.1%. Congo red solution is a acidic stain that has a negative charge that can not bind with a negative charge in cell wall, cytoplasm and a membran of microorganism cell, so that this stain does not color microorganism, but color the backround of preparations (culture media) [33]. pH of congo red solution influences on media color (**Figure 1** and **Figure 2**). The formation of a clear zone around a colony caused by interaction with (1, 4) - β – D glucan and (1, 3) β D-glucan with a cellulose hidrolysis reaction by a fungus strain that produce cellulase enzyme [34]. Determination quantitatively was carried out by measuring index of a colony diameter and a clear zone diameter that were formed [35, 36].

Table 1. The results of cellulase activity examination qualitatively and semi qualitatively

No. of Collection	Source of Isolate	Incubation (h)	Diameter of colony (mm)	Diameter of a clear zone (mm)	Cellulolytic Index
LIPIMC 0544	Air-borne of Library of Research Center For Biology -LIPI Cibinong	48	-	-	-
LIPIMC 0546	Air-borne of Library of Research Center For Biology -LIPI Cibinong	48	-	-	-
LIPIMC 0575	*Magnolia condolii* Rhizosphere of Taman Nasional Bodogol, Sukabumi Jawa Barat	48	0.8	0.3	2.67

The qualitative testing result was continued with quantitative analysis. Analysis of quantitative is a confirmation and the result of quantitative is not necessarily accurate. Qualitative hidrolysis activity is a picture of cellulolytic microbe ability to form a clear zone around an isolate that was grown in cellulose-contained media and measured with method of 3,5 dinitro salicylic (DNS) acid by calorimetry technique. It is measuring the reductor sugar content of enzyme hidrolysis yield against substrate [37-41].

Characteristic of microbe cellulose enzyme could be carried out by measuring the optimum condition in several different parameters such as pH, temperature, substrate concentration and the existence of metal ion cofactor as activator or inhibitor [42].

The effect of incubation against cellulase enzyme activity was determined by examining the cellulose activity for 5 days incubation period. Enzyme activity against incubation time of *P. griseofulvum* ranges between 1.3- 7.1 U/mL (**Figure 3**). The optimum time of incubation was at the second day. At this period, the enzyme activity was at its optimum activity (i.e. 7.1 U/mL) compared to the activity at other days of incubation period. At the first day, the cellulase activity was the lowest at 1.3 U/mL. At the third day until the fifth day, the cellulase activity was 1.5 U/mL, 3.6 U/mL and 2.9 U/mL respectively.

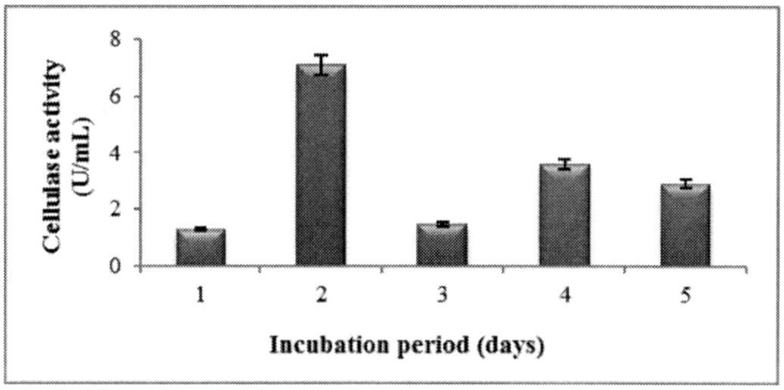

Figure 3. The effect of incubation time against the cellulase enzyme activity

The optimum incubation time of *P. griseofulvum* is 2 days to produce enzyme with the optimum activity. This incubation time to reach optimum enzym activity is more faster than the optimum activity of *Penicillium* sp. that was obtained after 7 days of incubation period [43]. According to the result of research of Sharma 2012 [44], the optimum activity of cellulase enzyme could be different when using different isolate though the medium was the same.

The examination against pH was carried out at enzyme from supernatant that incubated at optimum incubation time (i.e. 2 days, **Figure 4**). The supernatant enzyme has a varied activity at different pH. The optimum activity of enzyme was reached at pH 6.0 (i.e. 5.4 U/mL.) Meanwhile, the enzyme activity at pH 5.0, pH 7.0, pH 8.0 and pH 9.0 were 2.5 U/mL, 3.7 U/mL, 2.7 U/mL and 2.0 U/mL respectively.

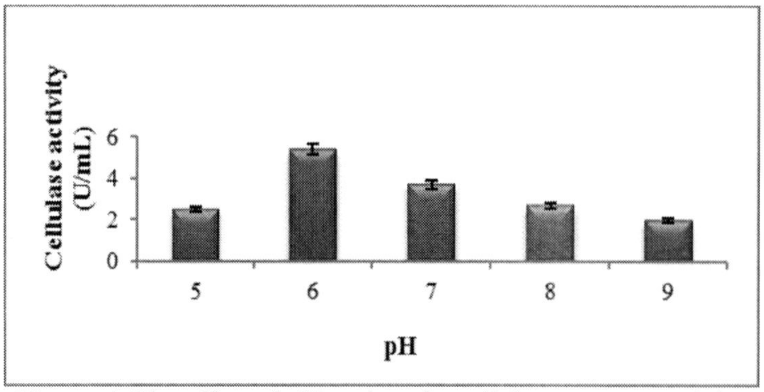

Figure 4. The effect of pH against the cellulase enzyme activity

Availability of the concentration change of substrate or environmental pH will result the enzyme activity. Also undergo changes although it can influence the enzyme activity, for example temperature or media composition. Every enzyme has a particular pH and the particular temperature will cause the enzyme activity reaches the optimum condition. Activity of enzyme will reach optimum condition influenced by certain pH and temperature. At the optimum pH, an enzyme will undergo the ionization changes of ionic group on the active side that cause the active side become more effective in binding with substrat [45]. The result of research of Nagah et al [43] showed that pH 5.0 was an optimum pH optimum of *Penicillium* sp. The result of research of Lu et al [46] showed that *Penicillium citrinum* can produce cellulase that tolerant to the alkali condition and resistent to the high temperature so that have a potential to be applied in detergent industry as an additive material. The changes of pH was not only influence the enzyme activity, but can also change the form or the characteristic of charge of its substrate [1].

The optimum condition of pH and temperature will support the enzyme to catalyze the reaction properly. Whereas the poorly temperature and pH will damage the result or protein is not becoming active in the enzym, so the function and activity of the enzyme will be decreased.

The examination against the effect of temperature was carried out at enzyme from supernatant that incubated with optimum time against temperature that is 2 days and at optimum pH 6.0 (**Figure 5**). The enzyme activity started increasing at 40°C (x U/mL). The activity was optimum at 50°C (5.2 U/mL). It was decreasing at a particular point of the increased temperature. It was x U/mL at 60°C and was not detected at 70°C. By this increased temperature, the enzyme was damaged or denaturated.

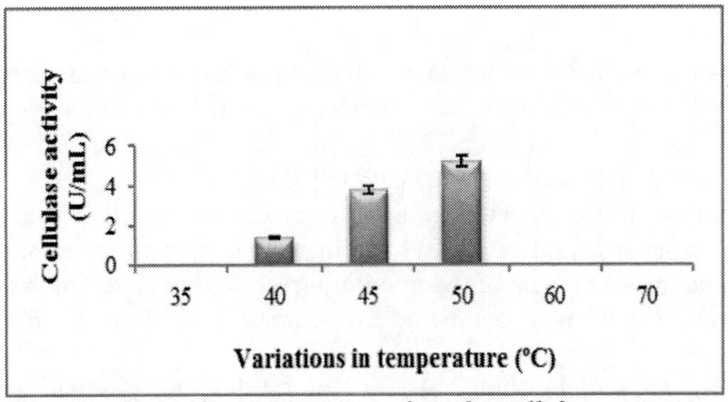

Figure 5. The effect of temperature against the cellulase enzyme activity

Temperature greatly influences on the enzyme activity. Every enzyme has optimum activity and stability at a particular temperature. Temperature really influences the growth and enzyme activity of microorganism. The enzyme activity increases in line with an increasing temperature until it reaches the optimal temperature. In this case, the optimum temperature of *P. griseofulvum* was 50°C (Figure 4). At the optimal temperature, the collision between enzyme and substrate is really effective so that the formation of the enzyme-substrate complex is easier and the formed product increases [43]. According to the result of research of Nagah et al [43] the cellulase of *Penicillium* sp. has the optimum temperature at 30°C. The result of research of Martins et al [47] showed that the optimum temperature to produce cellulose depend on a variation of isolate of its microorganism. According to Sethi and Gupta [48] cellulase production from *Penicillium chrysogenum* can be profitable because the level of enzyme production is higher than *Aspergillus niger*. The result of research of Gusakov and Sinitsyn [49] showed that *Penicillium echinulatum* was identified as a producer of potential cellulase at the process of bioconvertion against CMC. In general, mesophylic microbe will produce the mesophylic enzyme too [50]. The enzyme activity increases in harmony with increasing temperature until reaches the optimum temperature. The optimum temperature of the enzyme depends on the characteristics of the isolates [51]. The activity of carboxymethyl on cellulase enzyme at low temperature and high temperature could be seen as the enzyme mayor physiologic adaptation [52, 53]. Low temperature or temperature under the optimum temperature will result in enzyme experiencing energy deficiencies that are unable to perform maximum activity. The increasing temperature until reaches optimum temperature will increase the enzyme activity. After exceed optimum temperature then enzyme will undergo thermal denaturation and changes enzyme structure that cause denaturation so that enzyme activity decrease [54].

Examination against the effect of CMC substrate concentration was carried out at enzyme from supernatant that incubated within its optimum conditions, that are for 2 days incubation, at pH 6.0 and 50ºC (**Figure 6**). At variation of CMC substrate concentration, the optimum enzyme activity was 7.4 U/mL with substrate concentration at 1.25 %.

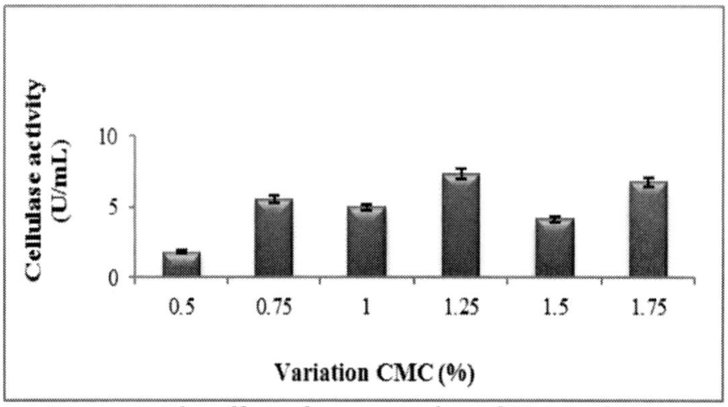

Figure 6. The effect of concentration of CMC substrate

Concentration of substrate is one of the factor that can influence enzyme activity. Supplementation of the excess substrate concentration will result in decreasing velocity of enzymatic reaction [43]. This reaction takes place because of abundance CMC and its high viscosity could decrease the probability of substrate to bind with the active side of the enzyme [55].

Fermentation technique was carried out in the solid media fermentation with *Aspergillus niger* and *P. chrysogenum* according to Jayant et al 2011 [8] that incubation time, carbon source and the early pH of fermentation medium were maintained in optimum condition simultaneously. The condition was at pH 5.0 and temperature 40°C as it was the best to increase the cellulase production after incubated for 8 days.

Several species of *Penicillium* have shown the ability to produce the cellulase enzyme, lignocellulolytic and β-glucosidase, endoglucanase, and xylanase [56]. *Penicillium verruculosum* 28K mutant, *P. echinulatum* 9A02S1, *P. chrysogenum* and *P. funiculosum* [57, 58, 8, 59] can produce carboxy methyl cellulase (CMCase), xylanase and β-glucanase by using the submerged fermentation technique and the solid fermentation technique. Cellulase enzyme activity of *P. nalgiovense* SS240 on oil palm cluster substrate is as much 0.027 mol/mL [60].

Examination against the effect of metal ion addition as activator was carried out at enzyme from supernatant that incubated within its optimum conditions that were for 2 days incubation, at pH 6.0, 50ºC temperature and with 1.25% of substrate concentration (**Figure 6**). There is a particular chemical substance that can increase the enzyme activity (activator) and can inhibit the enzyme activity (inhibitor). LIPIMC 575 strain was activated by divalent cations of $CoCl_2$ and $CuCl_2$ by 25.64% and 38.46% respectively compared to the negative control (**Figure 6**). Whereas monovalent cation of NaCl and divalent cations of $MgCl_2$ and $ZnCl_2$ have decreased the enzyme activity by 12.82% and 14.10% and 21.79% respectively (**Figure 7**). They have acted as inhibitor. In the recent years, several of results of published research reported on commercially producing cellulase and β-glucosidase from *Aspergillus* sp. and from the genus *Penicillium* [61, 62, 63]. Those enzymes also increase their potential application in producing bio-energy and bio-fuel, in textile, pulp and paper industry [64]. Previous research on utilisation of potential indigenous bioresources diversity has been done to give impact on development of domestic-industry [65].

This fungus of LIPIMC 0575 derived from *Magnolia condolii* rhizosphere of Taman Nasional Bodogol, Sukabumi District, West Jawa. It can produce cellulase enzyme. According to Hanum and Kuswytasari [66], strain of member of the genus *Penicillium* produces high cellulase content, to isolate virtually all the fibrous substrates investigated have found a *Penicillium* genus mold, including from a moldy woody substrate (obtained from pine wood), leaving garbage, sawdust and vegetable waste in isolating virtually all of the studied wood substrate found in the *Penicillium* genus, among them from the decayed wood substrates (obtained from Pinus wood), leaf litter, sawdust and vegetable waste. This indicates that the genus *Penicillium* has good cellulolytic and lignolytic capabilities that can grow directly on woody substrates.

According to Prasanna et al [42], there are several metal ions that could have positive or negative effect against cellulase enzyme activity (**Figure 7**). The positive effect happens when enzyme can interact with metal ion as cofactor (activator) so that the enzyme activity can increase. Whereas the negative effect can decrease the enzyme activity as metal ion is acted as an inhibitor. According to Gandjar et al [67] the phosphate requirement in the fungus growth process is not much explained, but the balance between manganese, zinc, and phosphate is one of the determining factors that can provide the advantage as a cofactor in regulating the amount of enzymes involved in the reaction.

In accordance with the research of Pereira et al [68], that numerous metal ions from monovalent, divalent, and trivalent such as Na^+, K^+, Ca^{2+}, Mg^{2+}, Mn^{2+}, Fe^{2+}, Co^{2+}, Cu^{2+}, Ni^{2+}, Zn^{2+}, Hg^{2+} and Fe^{3+} were generally examined as inhibitor and activator in cellulase characterization.

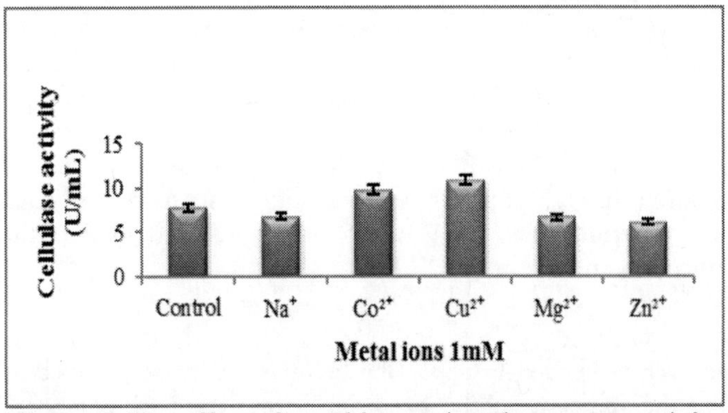

Figure 7. The effect of metal ion against the enzyme activity

After LIPIMC 0575 strain was identified and known as *P. griseofulvum* with the collection number of InaCC F 14, its cellulase supernatant was further examined whether it has the ability as an antibacteria. Supernatant of cellulase enzyme was examined into media contained *Bacillus cereus* bacterium. **Figure 8** shows the formation of a circle around the disc with cellulase enzyme 0575 and around positive control of Chloramphenicol solution with relative values of 2 mm and 2.33 mm respectively, whereas in the negative control of the sterile water disc does not form circle.

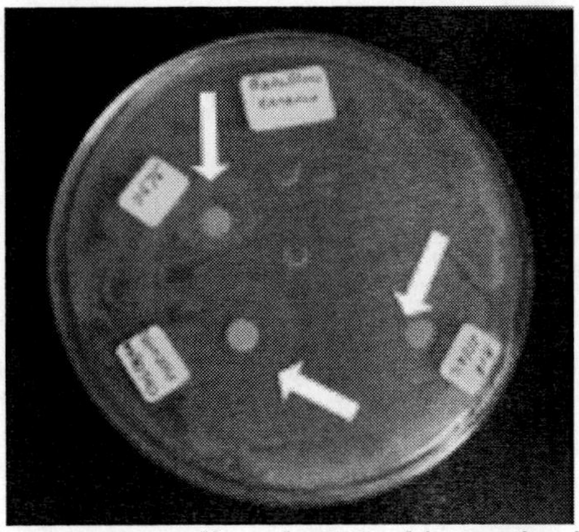

Figure 8. Circles around supernatant disc of 0575, chloramphenicol and the negative control of sterile water discs are not formed circle

Circles formed around the paper disc, supernatant cellulase *P. griseofulvum* and around the paper disc with Chloramphenicol solution, it is possible that the enzyme and antibiotic supernatant concentrations are still low so that they are bacteriostatic and possibly if at high concentrations are bactericidal may form a clear zone around paper disc. According to Pelczar and Chan [69] the higher the concentration of extracts of antimicrobial active compounds contained more and more so that the ability to inhibit the growth of microbes is also higher. The cellulase enzyme supernatant of *P. griseofulvum* InaCC F 14 may inhibit bacterial growth and will be used in composting sawdust as one of the mixed ingredients on the medium of white oyster mushroom (*Pleurotus ostreatus* J.) and ear mushroom (*Auricularia auricula* J.).

According to Saskiawan [70] the addition of cellulolytic microbial inoculants can be used in composting rice straw for white oyster mushroom media (*Pleurotus ostreatus* J.) can affect the full grown time and the weight of the oyster mushroom harvest.

4. Conclusion

The result of semi qualitative analysis showed a clear zone around colony of LIPIMC 0575 fungus with cellulolytic index of 2.67. The optimum enzyme activity consecutively with incubation time for 2 days, at pH 6.0, 50°C temperature and with 1.25 % of substrate concentration were 7.1 U/mL, 5.4 U/mL, 5.2 U/mL and 7.4 U/mL. On concentration of 1 mM of cellulase enzyme extracted from LIPIMC 575 fungus, divalent cations of $CoCl_2$ and $CuCl_2$ have activated and increased the enzyme by 25.64% and 38.46% respectively. Whereas, monovalent cation of NaCl and divalent cations of $MgCl_2$ and $ZnCl_2$ were acted as inhibitors. They have decreased the enzyme activity to 12.82% and 14.10% and 21.79% respectively. The LIPIMC 0575 strain was identified as *Penicillium griseofulvum* by molecular identification through 16S rDNA sequencing. Its supernatant found to be inhibited the growth of *B. cereus* bacterium. The LIPIMC 0575 is registered in the culture of Microbiology collection with the collection number of InaCC F 14.

5. Recommendation

Necessary further research is needed to apply the cellulase enzyme produced from *P. griseofulvum* InaCC F 14 which has the ability to inhibit bacterial growth used for composting sawdust used as a medium material for planting white oyster mushroom (*Pleurotus ostreatus* J.) and mushroom (*Auricularia auricula* J.) so that the production of mushrooms can be accelerated and abundant.

Acknowledgment

This research was funded by DIPA of thematic of Research Center for Biology- Indonesian Institute of Sciences.

6. References

[1] Robinson PK 2015 Enzymes: Principles and Biotechnological Applications *Essays Biochemistry* **59** 1–41
[2] Acharya S and Chaudhary A 2012 Bioprospecting thermophiles for cellulase production: a review *Brazilian J. Microbiol* **43**(3) 844–856
[3] Demain AL and Vaishnav P 2009 Production of recombinant proteins by microbes and higher organisms *Biotechnology Advances* **27** 297–306
[4] Adrio JL and Demain 2014 Microbial Enzymes: Tools for Biotechnological Processes. *Biomolecules* **4**(1) 117–39
[5] Ariffin H, Abdullah N, Kalsom MSU, Shirai Y and Hassan MA 2006 Production and Characterization of Cellulase by *Bacillus pumilus* EB3 *Int. J. Engineer. Tech.* **3**(1) 47-53
[6] Sohail M, Siddiqi R, Ahmad A and Khan SA 2009 Cellulase production from *Aspergillus niger* MS82: effect of temperature and pH *New Biotechnology* **25**(6) 437–41
[7] Sharada R, Venkateswarlu G, Venkateshwar S and Rao MA 2013 Production of cellulase – A Review *Int. J. Pharma. Chem. Biol. Sci.* **3**(4) 1070-90

[8] Jayant M, Rashmi J, Shailendra M and Deepesh Y 2011 Production of cellulase by different co-culture of *Aspergillus niger* and *Penicillium chrysogenum* from waste paper, cotton waste and baggase *J. Yeast and Fungal Res* **2**(2) 24-7
[9] Morana A, Maurelli L, Ionata E, Cara FL and Rossi M 2011 Cellulases from fungi and bacteria and their biotechnological applications In: *Cellulase: Types and Action, Mechanism and Uses* 1-79 Nova Science Publishers, Inc.
[10] Demers A, Doane R, Guzman S and Pagano R 2012 Enzymatic Hydrolysis of Cellulosic Biomass for the Production of Second Generation Biofuels A Major Qualifying Project Report Submitted to the faculty of Worcester Polytechnic Institute In: *Partial fulfillment of the requirements for the Degree of Bachelor of Science*
[11] Mtui GYS 2012 Lignocellulolytic enzymes from tropical fungi: Types, substrates and applications *Sci. Res. Essays* **7**(15) 1544-55
[12] Kim DM, Cho EJ, Kim JW, Lee YW and Chung HJ 2014 Production of cellulases by *Penicillium* sp. in a solid-state fermentation of oil palm empty fruit bunch *African J. Biotechnol* **13**(1) 145-55
[13] Bianco L and Perrotta G 2015 Methodologies and Perspectives of Proteomics Applied to Filamentous Fungi: From Sample Preparation to Secretome Analysis *Int. J. Mol. Sci.* **16**(3) 5803–29
[14] Bhat MK 2000 Cellulases and related enzymes in biotechnology *Biotechnol. Adv.* **18**(5) 355-83
[15] Dienes D 2006 Effect of cellulase enzymes on secondary fiber properties *Ph.D. Thesis* Budapest University of Technology and Economics, Budapest, Hungaria
[16] Vynios DH, Papaioannou DA, Filos G, Karigiannis G, Tziala T and Lagios G 2009 Enzymatic production of glucose from waste paper *BioResources* **4**(2) 509-21
[17] Whitehurst RJ and van Oort M 2010 *Enzymes in Food Technology* Second edition Blackwell Publishing Ltd. A John Wiley & Sons, Ltd., Publication
[18] Hasan F, Shah AA, Javed S and Hameed A 2010 Review Enzymes used in detergents: Lipases *African J. Biotechnol.* **9**(31) 4836-44
[19] Kuhad RC, Gupta R and Singh A 2011 Microbial cellulases and their industrial applications *Enzyme Res* Volume 2011 Article ID 280696 10
[20] Biswas R 2014 Production of Cellulolytic Enzymes In: *Bioprocessing of renewable resources to commodity bioproducts* First Edition Bisaria VS and A Kondo (Eds.) pp 105-32. John Wiley & Sons, Inc.
[21] Irvan, Prawati P and Trisakti B 2015 Pembuatan bioetanol dari tepung ampas tebu melalui proses hydrolysis termal dan fermentasi: pengaruh pH, jenis ragi dan waktu fermentasi *J. Teknik Kimia* **4**(2) 27-31(in Indonesian)
[22] Imran M, Anwar Z, Irshad M, Asad MJ and Ashfaq H 2016 Cellulase production from species of fungi and bacteria from agricultural wastes and its utilization in industry: A Review *Adv. Enzyme Res* **4** 44-55
[23] Sukumaran RK, Singhania RR and Pandey A 2005 Microbial cellulases—production, applications and challenges *J. Sci. Indust. Res.* **64**(11) 832–44
[24] Zhang Z, Liu JL, La JY, Dua CJ, Ma QS and Feng JX **2014** Predominance of *Trichoderma* and *Penicillium* in cellulolytic aerobic filamentous fungi from subtropical and tropical forests in China, and their use in finding highly efficient β-glucosidase *Biotechnol. Biofuels.* **7** 107
[25] Padmavathi T, Nandy V and Agarwal P 2012 Optimization of the medium for the production of cellulases by *Aspergillus terreus* and *Mucor plumbeu European J. Exp. Biol.* **2** 1161-70
[26] Jahangeer S, Khan N, Jahangeer S, Sohail M, Shahzad S, Ahmad A and Khan Shakeel A 2005 Screening and Characterization of Fungal cellulases isolated from the native environmental source *Pak. J. Bot.* **37**(3) 739-748
[27] Miller GL 1959 Use of dinitrosalicylic acid reagent of determination of reducing sugar *Anal. Chem.* **31** 246-48
[28] Teather RM and Wood PJ 1982 Use of Congo redpolysaccharide interactions in enumeration and characterization of cellulolytic bacteria from the bovine rumen *Appl. Environ. Microbiol.* **43**(4) 777-80

[29] Kar S, Mandal A, Mohapatra PK, Mondal KC and Pati BR 2006 Production of cellulase-free xylanase by *Trichoderma reesei* SAF3 *Brazilian J. Microbiol.* **37**(4) 462-64

[30] Ratnam J, Syed JI, Singh SK 2014 Comparative Study on Production of Penicillin by *Penicillium Notatum* and *Penicillium Chrysogenum* & its Antimicrobial Property *Int. J. of Bio Trends and Techno* **6**(1) 28-30

[31] Zhou X, Chen H and Li Z 2004 CMCase activity assay as a method for cellulase adsorption analysis *Enzyme and Microbial Technology* **35** 455- 59

[32] Sonia NMO and Kusnadi J 2015 Isolasi dan Karakterisasi Parsial Enzim Selulase dari Isolat Bakteri OS-16 Asal Padang Pasir Tengger-Bromo *J. Pangan Agro.* **3** (4) 11-9 (in Indonesian)

[33] Yosmar R, Suharti N and Rasyid R 2013 Isolasi dan Uji Kualitatif Hidrolisat Jamur Penghasil Enzim Selulase dari Tanah Tumpukan Ampas Tebu *J. Farmasi Andalas* **1**(1) 5-12 (in Indonesian)

[34] Vasan PT 2012 Chapter-1 Isolation of Cellulolytic Bacteria http://shodhganga.inflibnet.ac.in/bitstream/10603/56215/7/07_chapter%201.pdf

[35] Florencio C, Couri S and Farinas CS 2012 Correlation between Agar Plate Screening and Solid-State Fermentation for the Prediction of Cellulase Production by Trichoderma Strains *Enzyme Res* Volume 2012 7

[36] Yu SS, Latt ZK, Kyaw EP and Lynn TM 2017 Bioconversion of agricultural cellulose waste to valuable product by cellulolytic nitrogen-fixing bacterial isolates *Sky J. Microbiol. Res.* **5**(2) 010–017

[37] Bintang M 2010 *Biokimia Teknik Penelitian* Penerbit Erlangga (in Indonesian)

[38] Kamble RD and Jadhav AR 2012 Isolation, identification and screening of potential cellulase-free xylanase producing fungi and its production *African J. Biotechnol* **11**(77) 14175-81

[39] Gautam SP, Bundela PS, Pandey AK, Jamaluddin, Awasthi MK and Sarsaiya S 2012 Diversity of cellulolytic microbes and the biodegradation of municipal solid waste by a potential strain *Int. J. Microbiol.* Volume 2012 Article ID 325907 12

[40] Rathore SS, Mannivannan A and Narendhirakannan RT 2014 Screening of cellulase producing microorganisms from lake area containing water hyacinth for enzymatic hydrolysis of cellulose *J. Adv. Sci. Res.* **5**(3) 23-30

[41] Yao G, Wu R, Kan Q, Gao L, Liu M, Yang P, Du J, Li Z and Qu Y 2016 Production of a high-efficiency cellulase complex via β -glucosidase engineering in *Penicillium oxalicum Biotechnology for Biofuels* **9** 78.

[42] Prasanna HN, Ramanjaneyulu G and Reddy BR 2016 Optimization of cellulase production by *Penicillium* sp. *3 Biotech* **6** 162

[43] Nagah M, Abdel-Aziz MS, El-Sherbiny GM, Moghannem SAM and Shawky BT 2016 qualitative and quantitative screening of cellulases from different local egyptian fungal strains middle east *Journal of Applied Sciences* **6**(3) 579-587

[44] Sharma R 2012 Enzyme Inhibition and Bioapplications *IntechOpen*, DOI: 10.5772/39273 https://www.intechopen.com/books/enzyme-inhibition-and-bioapplications/enzyme-inhibition-mechanisms-and-scope

[45] Soliman SA, El-Zawahry YA and El-Mougith AA 2013 *Fungal Biodegradation of Agro-Industrial Waste* **Chapter 4** 75-100

[46] Lu W, Li and Wu Y 2003 Influence of water activity and temperature on xylanase biosynthesis in pilot-scale solid-state fermentation by *Aspergillus sulphureus Enzyme* and *Microbial Technol.* **32**(2) 305–11

[47] Martins LF, Kolling D, Camassola M, Dillon AJP and Ramos LP 2008 Comparison of *Penicillium echinulatum* and *Trichoderma reesei* cellulases in relation to their activity against various cellulosic substrates *Bioresource Technol.* **99**(5) 1417–24

[48] Sethi S and Gupta S 2014 Optimization of Cultural Parameters for Cellulase Enzyme Production from Fungi *Biolife* **2**(3) 989-96

[49] Gusakov AV and Sinitsyn AP 2012 Cellulases from *Penicillium* species for producing fuels from biomass *J. Biofuels* **3**(4) 463-77
[50] Sigres DP and Sutrisno A 2015 Enzim Mananase dan Aplikasi di Bidang Industri: Kajian Pusaka *J. Pangan dan Agroindustri* **3**(3) 899-908
[51] Feller G and Gerday C 2003 Psychrophilic enzymes: hot topics in cold adaptation *Nature Rev. Microbiol.* **1**(3) 200-08
[52] Carrard G, Koivula A, Soderlund H and Beguin P 2000 Cellulose-binding domains promote hydrolysis of different sites on crystalline cellulose *Proceedings of the National Academy of Sciences United State of America* **97**(19) 10342–47
[53] More N, Daniel MR and Petach HH 1995 Research Communication The Effect of low Temperatures on Enzyme Activity *Biochemi. J.* **305** 17-20
[54] Jørgensen H and Olsson L 2006 Production of cellulases by *Penicillium brasilianum* IBT20888—effect of substrate on hydrolytic performance *Enzyme Microbial Tech.* **38** 381–90
[55] Smith, H A. 2014. Production of antimicrobials and antioxidants from filamentous fungi. *A thesis* The National University of Ireland
[56] Solov'eva IV, Okunev ON, Vel'kov VV, Koshelev AV, Bubnova TV, Kondrat'eva EG, Skomarovskiĭ AA and Sinitsyn AP 2005 The selection and properties of *Penicillium verruculosum* mutants with enhanced production of cellulases and xylanases *Mikrobiologiia* **74**(2) 172-178
[57] Camassola M and Dillon AJP 2007 Production of cellulases and hemicellulases by *Penicillium echinulatum* grown on pretreated sugar cane bagasse and wheat bran in solid-state fermentation *J. Appl. Microbiol.* **103**(6) 2196–204
[58] Maeda RN, Barcelos CA, Anna LMMS and Pereira JrN 2013 Cellulase production by *Penicillium funiculosum* and its application in the hydrolysis of sugar cane bagasse for second generation ethanol production by fed batch operation *J. Biotechnol.* **163**(1) 38–44
[59] Nugraha R 2006 Produksi enzim selulase oleh *Penicillium nalgiovense* ss240 pada substrat tandan sawit *Skripsi* Program Studi Biokimia Fakultas Matematika dan Ilmu Pengetahuan Alam Institut Pertanian Bogor, Bogor (in Indonesian)
[60] Krogh KBR, Morkeberg A, Jorgensen H, Frisvad JC and Olsson L 2004 Screening genus *Penicillium* for producers of cellulolytic and xylanolytic enzymes *Appl. Biochemis. Biotechnol.* **113–16** 389–401
[61] Sørensen A, Ahring BK, Lübeck M, Ubhayasekera W, Bruno KS, Culley DE and Lübeck PS 2012 Identifying and characterizing the most significant beta-glucosidase of the novel species *Aspergillus saccharolyticus Canadian J. Microbiol.* **58**(9) 1035–46
[62] Liu D, Zhang R, Yang X, Zhang Z, Song S, Miao Y and Shen Q 2012 Characterization of a thermostable β-glucosidase from *Aspergillus fumigatus* Z5, and its functional expression in *Pichia pastoris* X33 *Microbial Cell Factories* **11** 25
[63] Zhou J, Wang YH, Chu J, Zhuang YP, Zhang SL and Yin P 2008 Identification and purification of the main components of cellulases from a mutant strain of *Trichoderma viride* T 100-14 *Bioresour Tech.* **99**(15) 6826-33
[64] Suryanto D 2009 Prospek Keanekaragaman Hayati Mikroba (Microbial Bioprospecting) Sumatera Utara *Pidato Pengukuhan Jabatan Guru Besar Tetap dalam Bidang Mikrobiologi pada Fakultas Matematika dan Ilmu Pengetahuan Alam*, diucapkan di hadapan Rapat Terbuka Universitas Sumatera Utara Gelanggang Mahasiswa, Kampus USU, Universitas Sumatera Utara Medan (in Indonesian)
[65] Vega K, Sarmiento VR, Ludeña Y, Vera N, Tamariz-Angeles C, Villena GK and Gutiérrez-Correa M 2015 Alkaline Cellulase Production by *Penicillium mallochii* LMB-HP37 Isolated from Soils of a Peruvian Rainforest *British Biotech. J.* **7**(4) 160-8
[66] Hanum AM and Kuswytasari ND 2014 Laju Dekomposisi Serasah Daun Trembesi (*Samanea saman*) dengan Penambahan Inokulum Kapang *J. Sains dan Seni Pomits* **3**(1) 2337-3520 (in Indonesian)

[67] Gandjar I, Syamsuridzal and Oetari A 2006 *Mikologi Dasar dan Terapan* Jakarta: Yayasan Obor Indonesia (in Indonesian)
[68] Pereira J de Cassia, Giese EC, de Souza Moretti MM, dos Santos Gomes AC, Perrone OM, Boscolo M, da Silva R, Gomes E and Martins DAB. 2017 Effect of Metal Ions, Chemical Agents and Organic Compounds on Lignocellulolytic Enzymes Activities **Chapter 6** http://dx.doi.org/10.5772/65934
[69] Pelczar, M and Chan ECS 1986 Dasar-Dasar Mikrobiologi Cetakan I Jakarta: UI-Press
[70] Saskiawan I 2015 Penambahan Inokulan Mikroba Selulolitik pada Pengomposan Jerami Padi untuk Media Tanam Jamur Tiram Putih (*Pleurotus ostreatus J.J.*) *Jurnal Biologi Indonesia* **11**(2) 187-193 (in Indonesian)

IOP Conf. Series: Earth and Environmental Science **166** (2018) 012027 doi:10.1088/1755-1315/166/1/012027

Efficacy of Bacterial Biocontrol Agents and Environmental-Friendly Fine Chemicals for Suppression of *Ganoderma Boninense* Growth

Yuliar[1] and Suciatmih[1]

[1]Research Center for Biology, Indonesian Institute of Sciences (LIPI)
Jl. Raya Bogor Km. 46, Cibinong 16911, Bogor, Indonesia

E-mail: yrivaie@yahoo.com

Abstract. Various means are available to control *Ganoderma boninense*, however none of those methods are efficacious. Biological control agents method is one of promising way to control the pathogen, that has eco-friendly-mannered. The objective of this research was to evaluate the effectiveness of bacterial strains and the application of some environmentally friendly fine chemicals including ammonium, calcium, and salicylic acid in suppressing of mycelia growth of *G. boninense*. *In vitro* results showed that among eight strains, EB13, as identified as *B. amyloliquefaciens*, has reduced *G. boninense* growth diameter significantly (P<0.05%) at the highest percentage (81%). It followed by MB31, MB72, MB58 and MB41 74%, 74%,71% and 63%, respectively. Conversely MB20, EB45 and EB53 did not suppress the fungal pathogen growth. The chemical control test indicated that salicylic acid has inhibited the growth of the pathogen significantly at the highest percentage of 52% (P<0.05%). The lowest growth inhibition was by ammonium nitrate (2%). On the contrary, calcium chloride could not inhibit the *G. boninense* growth. Further assessment on pH in PDA media containing of 200mM ammonium was conducted in order to increase the inhibition activity of the ammonium nitrate. The result exhibited that the medium containing ammonium nitrate with pH 8.5 has reduced *G. boninense* growth significantly at the highest percentage (53%).It followed by pH 8.0 and pH 7.5 at 10% and 2%, respectively.

1. Introduction

Ganoderma species is a soil-borne fungus caused a major basal stem rot (BSR) disease of oil palm (*Elaeisguineensis* Jacq) in Southeast Asia, particularly in Indonesia and Malaysia [1-4]. Incidence of *Ganoderma boninense* BSR in oil palm plantings continues to increase that caused economic losses in the global oil palm industry. Several means have been done in controlling of *Ganoderma* disease including chemical, cultural, and sanitation by the removal of diseased oil palms. Field trials to control *Ganoderma* disease by using systemic and contact fungicides have been performed by various researchers with no conclusive results although *in vitro* they showed to be effective. Application chemical pesticides resulted in the development of insecticide resistance, extensive damage to the environment, and lethal effects on non-target organisms. Therefore, an alternative eco-friendly application is required to increase the production and to generate healthy agricultural products. The possible approach in controlling the diseases caused by *G. boninense* could be through the bacterial biocontrol agents and supplementation of ammonium, calcium, and salicylic acid.

IOP Conf. Series: Earth and Environmental Science **166** (2018) 012027 doi:10.1088/1755-1315/166/1/012027

Interest in biocontrol has increased since public has been concerned over the use of chemical in general [5]. The benefits of biocontrol are 1) potentially self- sustaining, 2) spread on their own after initial establishment, 3) reduced inputs of non-renewable resources and 4) long-term disease suppression in an environmentally manner [6, 7].

Several plant pathogens have been killed at elevated levels of ammonium, such as *Ralstonia solanacearum* [8], *Verticillium dahlia* [9] and *Phytophthora cinnamoni* and *P. parasitica* [10]. Application of calcium (Ca) increased the resistance of tomato to bacterial wilt and decreased the population of *Ralstonia solanacearum* in the xylem [11], suppressed the BSR symptoms on clonal materials [12], decreased disease incidence and delayed the onset of *Phytophtora* stem rot in soybean [13].

Salicylic acid (2-hydroxybenzoic acid, SA) is phenolic compound that induces plant diseases resistance [14]. Induction to pathogen resistance is promising way for controlling of plant diseases. Efficacy of SA in suppression of plant diseases has been studied e.g., *Fusarium gramineamm* in wheat [15], *F. oxysporum* in tomato (Fol) [8], and *Penicillium expansum* in apples [16].

The objective of the present research work was to evaluate the effectiveness of bacterial strains and the application of some environmentally friendly fine chemicals including ammonium, calcium, and salicylic acid in suppressing of mycelia growth of *G. boninense*.

2. Materials and Methods

Pure fungal cultures of *G. boninense* was obtained from Laboratory of Microbiology, Indonesian Research Institute for Biotechnology and Bioindustry, Bogor, Indonesia. Bacterial biocontrol agents ammonium nitrate (NH_4NO_3) (Fisher scientific), salicylic acid ($C_7H_6O_3$) (Merck), calcium chloride dehydrate ($CaCl_2.2H_2O$) (Merck) were taken from the Laboratory of Environmental Microbiology, Research Center for Biology, LIPI.

2.1. Regrowth of G. boninense and bacterial biocontrol agents (BCAs)

The pure culture of the *G. boninense* fungi was sub-culture into a potato dextrose agar (PDA, brand) medium in test tube, and incubated for about 5-7 days. The seven days old of the fungi taken from the test tube was then inoculated on the sterile PDA plates. For the BCAs, eight of the bacterial strains, namely, EB13, EB45, EB53, MB20, MB31, MB41, MB58, and MB72 were regrowth into nutrient agar (NA) medium in test tube, and incubated for about 2 days at room temperature.

2.2. Screening of bacterial biocontrol agents for suppression of G. boninense growth

Eight of bacteria strains (EB13, EB45, EB53, MB20, MB31, MB41, MB58, and MB72) were examined of their ability to inhibit *G. boninense* mycelia growth. A hundred (100) µl of each of the fifth days of the bacterial culture in No. 3 medium (i.e. 10 g polypepton (Becton, Dickinson and Company), 10 g glucose (Merck), 1 g KH2PO4, 0.5 g $MgSO_4 \cdot 7H_2O$ (Merck), pH 6.8, 1 L distilled water) was pipetted into sterile plates. An autoclaved-PDA agar in an erlenmeyer flask was poured into the sterile plates. The tested plates were rotated to mix the bacterial culture and the medium. After the media solidified, in the center of the plates were inoculated with the previously prepared-plug of 5-7 days of *G. boninense* mycelia growth in PDA plates as described in a section 2.1 (the plug diameter size was about 5 mm). For the control were used 100 µl of sterile distilled water. This experiment was performed in a clean bench, in four replicates. Finally, the plates were incubated at room temperature. The diameter of the mycelia growth were measured and recorded at first, fifth, seventh, eleventh, and thirteen days incubation time. Percentage of *G. boninense* inhibition growth was calculated, using the following formula;

$$\text{Inhibition growth} = \frac{\text{ØC} - \text{ØT}}{\text{ØC}} \text{x } 100\%$$

Where is ØC = diameter of control
ØC = diameter of treatments

2.3. Screening of environmentally friendly fine chemical compounds for suppression of G. boninense growth

Three kinds of the chemical compounds: ammonium nitrate (NH_4NO_3), salicylic acid ($C_7H_6O_3$) and calcium chloride dehydrate ($CaCl_2.2H_2O$) were assessed their possibility to suppress the mycelia growth of the fungus. Each compound of 200 µL of 25 mM of NH_4NO_3, 16 mM of $C_7H_6O_3$, and 15 mM of $CaCl_2.2H_2O$ was pipetted into plates. Then PDA medium was poured into the plates and rotated to mix. After the medium was solidified, the mycelia of *G. boninense* sized 5 mm in diameter was inoculated in the center of the tested plates. The plates were incubated at room temperature, the diameter growth of the fungus mycelia was observed and measured on first, seventh, ninth, thirteenth day incubation time. The experiment was performed in four replication.

2.4. Influence of pH of PDA medium containing ammonium nitrate

In order to enhance the ammonium nitrate in suppression of *G. boninense* mycelia growth, the effect of various pH of PDA medium containing ammonium nitrate was investigated. The pH of PDA medium containing 0.2M of NH_4NO_3 was adjusted to 7.5, 8.0, and 8.5 by using 1N NaOH, and 1N HCl. Subsequently, the medium was autoclaved, and pour into sterile plates. The solidified medium was inoculated with a plug of *G. boninense* sized 5mm in diameter, and incubated for about fourteen (14) days. The mycelia growth of the fungus was observed and recorded on first, fourth, seventh, eleventh, and fourteen days. The experiment was performed in four replication.

2.5. Data analysis

Data were statistically calculated using analysis of variance (ANOVA) with Minitab 16 software. The significance of mean differences was determined using the Duncan's test. The responses were judged significant at 5% level.

3. Results and Discussion

3.1. Screening of bacterial biocontrol agents for suppression of G. boninense growth

Significantly, the highest suppression effect of the bacterial biocontrol agents against *G. boninense* mycelia on day seventh was observed in application of EB13 strain, followed by MB31, 72, 58 and 41 strains ($P<0.05$) (**Figure 1, Figure 2,** and **Figure 3**). The suppression effect of EB13 strain increased from 78.43% on day 5 to 90.41% on day 13 (**Figure 2**). The MB72 strain has also increased its suppression effect from 72% on day 5 to 87.14% on day 13 (**Figure 2**). Similarly, the MB31 strain has increased its effect from 78.43% to 86.74% (**Figure 2**). Degrees of *G. boninense* growth inhibition on day 5 to day 13 after application of MB41 and MB58 strains were almost the same. Respectively, they increased from 37% to 81% and from 47% to 82%. In contrary, application of the EB45, EB53 and MB20 strains have not reduced the mycelia growth of fungal pathogen (**Figure 1**). On day eleventh of incubation time, the EB13 strain showed the greatest effect on suppression of the fungal mycelia growth consistently (**Figure 1b**). Our previous research showed that based on 16S rRNA identifications, the EB13 strain was identified as *Bacillus amyloliquefaciens*. *B. amyloliquefaciens* has been reported in increased survivability of tomato against *Ralstonia solanacearum* and *Rhizoctonia solani* significantly [17]. Suppression growth of *G. boninense* due to antifungal compounds produced by *B. amyloliquefaciens* has been reported in Yuliar *et al.* [17]. The author reported that *B. amyloliquefaciens* produced an antifungal of iturin and antibacterial of surfactin. The other bacterial biocontrol reduced *G. boninense* growth has been reported by Azahar *et al.* [2, 18] who found that *Pseudomonas aeruginosa* changed the hyphal tips of *G. boninense*, and caused a great malformation of the hyphae. The culture broth and culture filtrate of *P. aeruginosa* reduced the radial growth of the fungal pathogen by 93% and 96%, respectively.

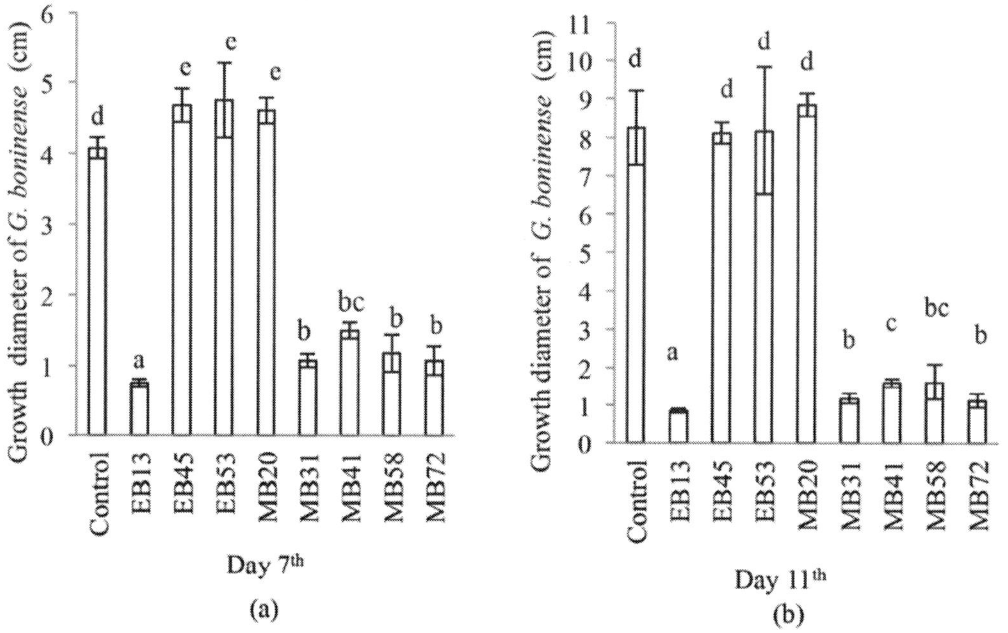

Figure 1. Suppressive ability of the biocontrol agents to inhibit *G. boninense* growth on days seventh (a) and eleventh (b) of incubation time. Means with different letters are significantly different (P<0.05)

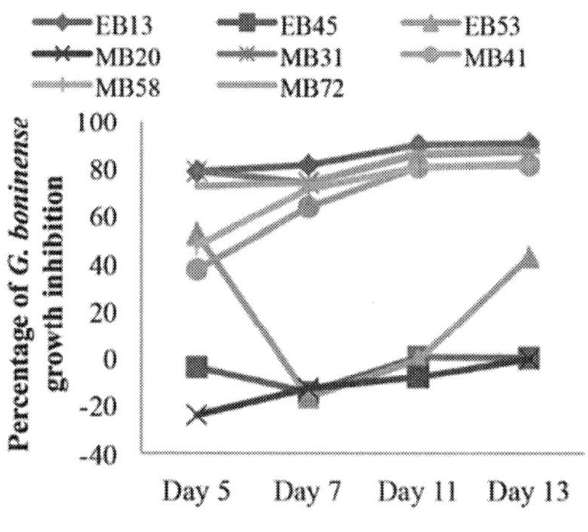

Figure 2. Percentage of the biocontrol agents ability to inhibit *G. boninense* growth during thirteen days of incubation time

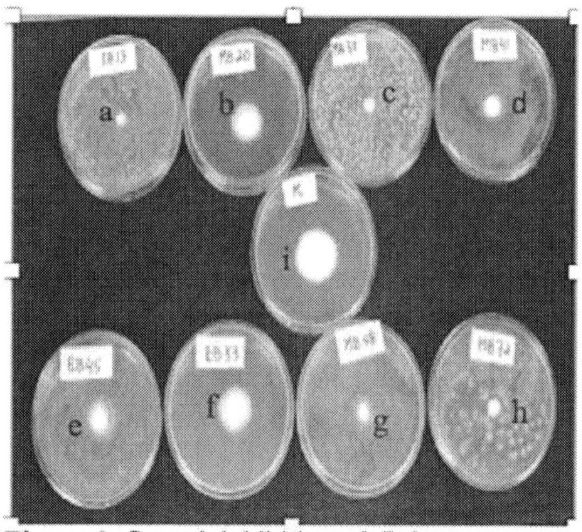

Figure 3. Growth inhibition of *G. boninense* at seventh day incubation by biocontrol agents of EB13 (a), MB20 (b), MB31 (c), MB41 (d), EB45 (e), EB53 (f), MB58 (g), and MB72 (h), Negative control (i)

3.2. Screening of environmentally friendly fine chemical compounds for suppression of G. boninense mycelia growth

G. boninense growth was significantly inhibited by salicylic acid (SA) at the highest rate (48%) compared by ammonium nitrate at 12% (**Figure 4** and **Figure 5**). In contrast, calcium chloride did not inhibit the fungal pathogen growth. Similar results on the effects of SA on plant pathogens were also observed by Saikia *et al.* [19]. The author reported application of *P. fluorescence* pf4-92 with SA has resulted in the highest suppression of chickpea seedling wilting compared to acetyl salicylic acid, DL-norvaline, indole-3-carbinol and lichenan. The presence of increasing concentration of SA in liquid and solid media has significantly inhibited mycelial growth and conidia germination of *Fusarium graminearum*, and finally halted head blight in wheat [15]. Further application of 200 mM SA through root feeding and foliar spray of tomato could induce resistance against *F. oxysporum* f. sp *Lycopersici* (Fol) [14]. Similarly, the germination of *Penicillium expansum* was completely inhibited by the application of 2.5 mM salicylic acid [16].

The possible mechanism of SA in suppressing the plant fungal pathogens was related to its behaviour as dissociating agent. It caused a change in the trans-membrane pH gradient in membranes of organelles and in the plasma membrane, possibly caused cellular energy loss [20]. In addition, SA caused leakage of the pathogen's proteins to the medium, lipid damage, and intracellular disorganization [16,21].

G. boninense was suppressed by ammonium nitrate (**Figure 4**). It was probably because ammonium could kill the fungal pathogen. This finding agrees with the findings of Tenuta and Lazarovit [9] and Amari *et al.* [8] who reported that ammonia are toxic to *Verticillium dahliae*, and *Fusarium oxysporum* and *Ralstonia solanacearum* respectively.

The fungal growth was not inhibited by supplementation of calcium chloride (**Figure 4**). However, different result was obtained by Bivi *et al.* [22] who revealed that the mixture addition of calcium, copper, and salicylic acid had potential to suppress growth of *G. boninense in vitro*. The contradictory of this result was probably due to addition of calcium in different way of application and concentration. In this study, the calcium chloride was applied as a single addition instead of a mixtured addition.

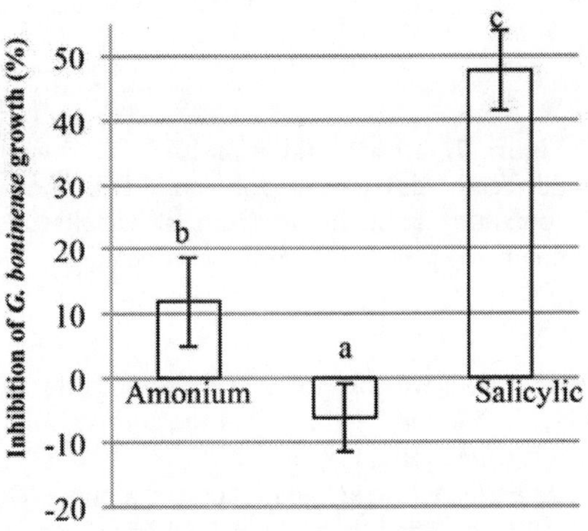

Figure 4. Suppressive ability of ammonium nitrate, calcium chloride, and salicylic acid to inhibit *G. boninense* growth. Means with different letters are significantly different ($P<0.05$)

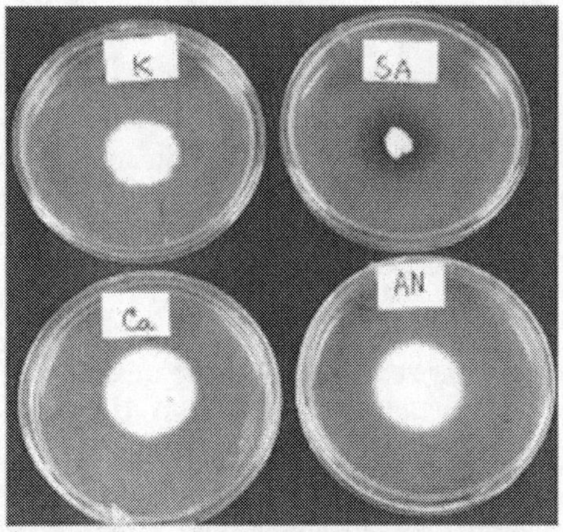

Figure 5. Growth inhibition of *G. boninense*; distilled water as negative control (a), salicylic acid (b),calcium chloride (c), and ammonium nitrate (d)

3.3. Influence of pH of PDA medium containing ammonium nitrate on G. boninense growth

The fungal pathogen grows at pH 8.5 of the medium PDA containing ammonium was highly suppressed from day fourth to day seventeenth consistently **(Figure 6)**. The growth inhibition of *G. boninense* was significantly at the highest at pH8.5 compared to pH8.0 and pH 7.5 in PDA medium containing ammonium nitrate (**Figure** 7). An explanation for this results is probably because of increasing of pH affected the increasing of ammonium uptake rates [9]. Furthermore, the increasing of ammonium uptake leads to increase of the toxicity and finally impacted on the higher reducing of the diameter growth of *G. boninense*.

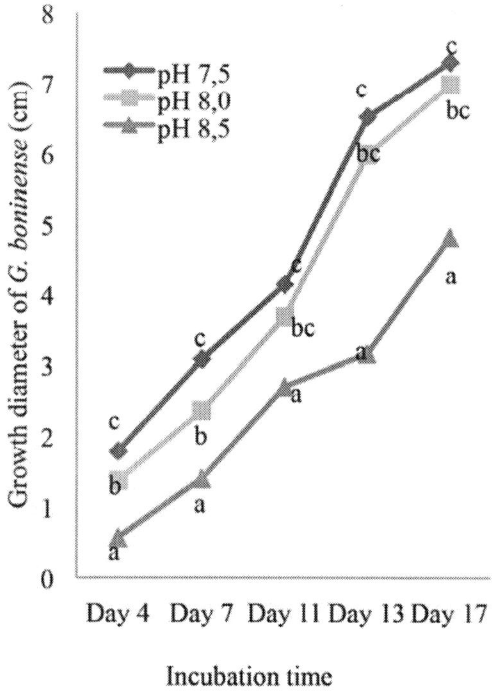

Figure 6. Influence of pH on PDA medium containing ammonium nitrate in suppression of *G. boninense* growth. Means with different letters are significantly different (P<0.05)

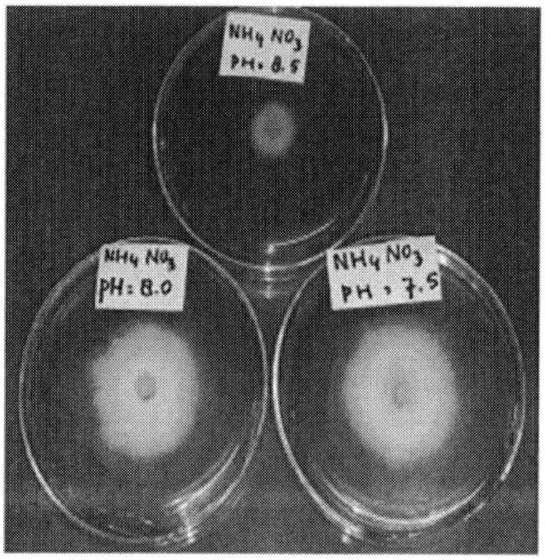

Figure 7. Growth inhibition of *G. boninense* at thirteenth day incubation on PDA medium containing ammonium nitrate at various of pH range. (a) pH8.5, (b) pH 8 and (c) pH 7.5

4. Conclusions

To summarize, the highest suppression of *G. boninense* growth was observed in the application of *B. amyloliquefaciens* (EB13 strain). It followed by MB31, 41, 58, and 72strains. On the contrary, EB45, EB53, and MB29 strains have not suppressed growth of *G. boninense*.

G. boninense growth was inhibited significantly by salicylic acid with the highest percentage (48%) in comparison to ammonium nitrate (12%). Conversely, the growth of fungal pathogen was not inhibited by calcium chloride. Furthermore, the pH of media containing ammonium nitrate influenced its ability in suppressing the growth of *G. boninense*. Its suppressing ability was increased by increasing its pH from 7.5 to 8.5.

Acknowledgment
This research work was partly supported by project of DIPA (Issuance of spending authority) of Research Center for Biology, Indonesian Institute of Sciences. Our gratitude to Mrs. Ety Suryati for her technical assistance during research activities.

5. References

[1] Ariffin D, Idris AS and Singh G 20002 Status of *Ganoderma* in oil palm *Ganoderma diseases of perennial crops*, ed Flood J, Bridge P, Holderness M (Wallingford, UK: CABI publishing) 49-68

[2] Azahar TM, Idris AS, Abu Hassan D and Norazlin 2014 Assessment of basal stem rot disease distribution in palm oil plantation using geographical information system *J. of science and Technol.* **6**(2) 81-92

[3] Susanto A, Purba RY and Sudharto 2005 Enhancing biological control of basal stem rot disease (*Ganoderma boninense*) in oil palm plantations *Mycopathology* **159** 153-157

[4] Susanto A 2009 Basal stem rot in Indonesia biology, economic importance, epidemiology, detection and control *Proceedings of international workshop on awareness, detection and control of oil palm devastating diseases* Kuala Lumpur convention centre Malaysia

[5] Whipps J 2001 Microbial interactions and biocontrol in the rhizosphere *J. Exp. Bot.* **52** 487–511

[6] Quimby FC, King LR and Grey WE 2002 Biological control as a mean of enhancing the suistainability of crop/land management systems *Agric. Ecosyst. Environ.* **88**147-152

[7] Whipps JM and Gerhardson B 2007 Biological pesticides for control of seed- and soil-borne plant pathogens *Modern Soil Microbiology*, ed Van Elsas JD, Jansson JD, and Trevors JT, 2nd edition (Boca Raton, FL, USA: CRC Press) 479–501

[8] Amari K, Koki T, Tajul MI, Kazunari M, Testsuo K and Akira W 2008 Effect of the of anaerobically digested slurry to soil and hydroponics on soil-borne plant diseases *Soil Microorganisms* **62**(2) 106-113

[9] Tenuta M and George L 2002 Ammonia and nitrous acid from nitrogenous amendments kill microsclerotia of *Verticillium dahlia Phytopathology* **92**(3) 255-264

[10] Tsao PH and Oster JJ 1981 Relation of ammonia and nitrous acid to suppression of *Phytophthora* in soil amended with nitrogenous organics substances *Phytophatology* **71** 53-59

[11] Yamazaki H, Kikuchi S, Hoshina T and Kimura T 2000 Calcium uptake and resistance to bacterial wilt of mutually grafted tomato seedlings *Soil Sci. Plant Nutr.* **46** 529-534

[12] Sariah M and Zakaria H 2000 The use of soil amendments for the control of basal stem rot of oil-palm seedlings *Ganoderma diseases of perennial crops*, ed Flood J, Bridge PD, Holderness M (Wallingford,UK:CABI Publishing) 89-99

[13] Sugimoto T, Watanabe K, Yoshida S, Aino M and Irie K 2008 Select calcium compounds reduce the severity of Phytophthora stem rot of soybean *Plant diseases* **92**(11) 1559-1565

[14] Mandal S, Mallick N and Mitra A 2009 Salicylic acid-induced resistance to *Fusarium oxysporum* f. sp. lycopersici in tomato *Plant Physiol. Biochem.* **47**, 642-649

[15] Qi PF, Jhonston A, Balcerzak M, Rocheleau H, Harris LJ, Long XY, Wei YM, Zheng YL and Oulellet T 2012 Effect of salicylic acid on *Fusarium graminearum* the major causal agent of fusarium head blight in wheat *Fungal Biol.* **116**(3) 413-26

[16] Da Rocha Neto AC, Maraschin M and Di Piero RM 2015 Antifungal activity of salicylic acid against *Penicillium expansum* and its possible mechanisms of action *Int. J. Food Microbol.* **215** 64-70

[17] Yuliar, Koki T and Yokota K 2015 Characterization of possible bacterial biocontrol agents, isolated from various plants in Indonesia, against bacterial wilt and damping-off of tomato *Soil Microorganisms* **69**(1) 39-4

[18] Bivi RM, Siti NFM, Khairulmazmi A and Idris A 2010 Control of *Ganoderma boninense*: A causal agent of basal stem rot disease in oil palm with endophyte bacteria *in vitro Int. J. Agric. Biol.* **12** 833-83

[19] Saikia R, Singh T, Kumar R, Srivastava J, Singh K and Arora DK 2003 Role of salicylic acid in systemic resistance induced by *Pseudomonas flourescens* against *Fusarium oxysporum* f. sp. ciceri in chickpea *Microbiological Research* **158** 203-13
[20] Amborabe BE, Essard PFL, Hollet JFC and Roblin G 2002 Antifungal effect of salicylic acid and other benzoic acid derivatives towards *Eutypalata*: structure-activity relationship *Plant Physiol. Biochem.* **40** 1051- 1060
[21] Murphy A M, Chivasa S, Singh D P and Carr J P 1999 Salicylic acid-induced resistance to viruses and other pathogens: a parting of the ways? *Trends Plant Sci.* **4**(4) 155-160
[22] Bivi RM, Siti NFMD, Khairulmazmi A, Idris A, Ahmed OH, Zamri R and Sariah M 2012 *In vitro* effects of salicylic acid, calcium, and copper ions on growth and sporulation of *Ganoderma boninense Afr. J. Biotecnol.* **11**(70) 13477- 13489

IOP Conf. Series: Earth and Environmental Science **166** (2018) 012028 doi:10.1088/1755-1315/166/1/012028

Analysis of Greenbelt in Sibolga for Tsunami Mitigation

A Zaitunah[1], Samsuri[1] and B Slamet[1]

[1]Faculty of Forestry, University of Sumatera Utara
Jalan Tri Darma Ujung No 1, Kampus USU, Medan 20155, Indonesia

E-mail: a_zaitunah@yahoo.com

Abstract. West coast of Sumatra is one of the vulnerable areas to natural disaster, such as tsunami. The prevention of potential destruction has to be done in order to reduce victim and economical loss. Naturally, coastal vegetation can reduce the wave strength, hence reducing the destructive force. The aim of this research was to obtain the potential area for greenbelt to support rehabilitation and prevention of potential destruction. Most of coastal area of Sibolga has very low Normalize Difference Vegetation Index (NDVI) value, indicating that area has less vegetation. Field checking showed that the coastal area has dense settlement with less vegetation. Land system found in the coastal area is Kahayan which is suitable for mangrove. The area is alluvium plain, behind the coastal is mountainous and hilly areas. In 100, 200. and 500 meters of greenbelt area width, highest vegetation density was very low, followed by low vegetation class. It means most of the area has no vegetation or less vegetation, therefore Greenbelt planning is needed to protect the beach from potential destruction of natural disaster, mainly tsunami.

1. Introduction

As an archipelago country, Indonesia has lengthy coastline along its islands. Indonesia has 81,000 km of coastline [1] and considerable amount of people live along the coastline and do many activities for their livelihood. The coastal zone is one of important ecosystem for people who lives in the area. So, the area should be protected from any potential destruction which can harm their lives.

Earthquake and tsunami can create high damage in the coastal area. People needs to prepare themselves to protect their home and lives from any disastrous cases. Tsunami has caused huge loss of lives and materials. Tsunami in Aceh has taken more than 200,000 lives and destroyed the area [2].

Indonesia is divided into six zones based on the relationship between tsunami, earthquake activity and characteristic of seismotectonic, where Sumatra belongs to zone A. Historical record mentioned 16 cases of earthquake in period between 1790 to 2005. Therefore, the repeat period of Zone A is 10-15 years. In 2004, more than 200,000 peoples became victim of Tsunami in Nanggroe Aceh Darussalam and North Sumatera [2]. Tsunami along the southern coast of West Java, Cilacap and Yogyakarta took 378 lives and material destruction with the estimated loss of more than 70 billion rupiahs [3].

Most of Tsunami incident in Indonesia caused by tectonic earthquake along subduction area [2]. Eastern part of Sumatera is one of vulnerable area to tsunami. Its position is in line with subduction line between Eurasian and Indo-Australian plates which have West-East axis in the deep of Indonesian

IOP Conf. Series: Earth and Environmental Science **166** (2018) 012028 doi:10.1088/1755-1315/166/1/012028

Ocean [4]. Earthquake with 9.0 R will create energy bigger than 100,000 times of Hiroshima bomb. Shape of beach and bottom of the sea in the coastal area, incident angle of wave and front shape of tsunami wave which come to the beach will influence to the potential damage [5].

Disaster mitigation is crucial to minimize any potential destruction. According to Republic of Indonesia Constitution Number 27 Year 2007, disaster mitigation is the effort to reduce disaster risk, both structure and physical through development of natural physical and/or man-made, both non structural or non physical, through the improvement of capability in dealing disaster threat in coastal area and small islands. Greenbelt is one of the option to protect the shore by using vegetation to reduce the strength of the wave The aim of this research was to build a spatial model for greenbelt in order to support the rehabilitation effort and prevent the potential destruction of tsunami.

Harada and Imamura (2003) in [2] did a research on effectivity of coastal forest to overcome tsunami. They did the simulation on different width of forest which are 50 m, 100 m, 200 m, and 400 m. The result showed that the coastal forest with width of 200 meters, 30 trees per 100 square meters and diameter of 15 cm can overcome 50% of energy of tsunami with the height of 3 meters. Furthermore, the research of Harada and Imamura showed that the wider the width of the forest the more energy to be reduced. The existence of the forest in the coastal area can reduce the run up distance, height of inundation, current and hydraulic force that reach the beach.

Study on vegetation density in the coastal area is important in order to support protection planning to prevent any potential destruction caused by disaster especially tsunami. Information on land suitability of the area is needed to find out which area is potential to have a greenbelt for coastal protection. The specific aims of the research were to obtain the vegetation density level on the area, to find out the potential area for coastal vegetation and to find out the vegetation density in potential greenbelt area. The specific aims led to the main aim of the research.

This research is conducted to give an insight to stakeholders and decision makers on the potential and condition of the coastal area. Recommendation on greenbelt planning will be given to prevent the possible future damages.

2. Material and Methods

The research area was in the coastal area of Sibolga and Central Tapanuli as the part of eastern area of North Sumatera Province. It was conducted from May to September 2017. There were 3 stages of research including preliminary, fieldwork and spatial analysis. The preliminary stage including proposal compilation, data and information gathering and fieldwork preparation. Fieldwork included field check, vegetation survey/analysis and supportive data compilation. The analysis was including satelite image and map analysis.

Data and equipments of the research were needed for data preocessing, analysis and fieldwork activity. Those were computer, softwares (ArcGis, Erdas, Microsoft Word) for maps and images analysis. Satellite image used in this research was Landsat 8. Maps including administration, land system. For the fieldwork, there were plan map, observation sheets, GPS receiver, compass, measurement tape and camera.

Vegetation density map was obtained from NDVI (Normalized Difference Vegetation Index) analysis. The steps can be seen in Figure 1. NDVI value is the result of image processing using transformation of Normalized Difference Vegetation Index (NDVI) which is the ratio between reflectance of Red Band (R) and Near Infrared Band (NIR). NDVI value can give an information on condition of vegetation and non vegetation such as water, soil and rocks. The value is between -1 to +1. The closer value to 1 shows that the land cover has dense vegetation.

The representation of NDVI value to vegetation and soil relation in land cover is mentioned in [8]. Very low value of NDVI (0.1 and below) correspond to barren areas of rock, sand, or snow. Moderate values represent shrub and grassland (0.2 to 0.3), while high value indicates temperate and tropical rainforests (0.6 to 0.8). Bare soil is represented with NDVI values, which are closest to 0 and water bodies are represented with negative NDVI values.

Buffer analysis was conducted using 3 buffer width that are 100, 200 and 500 meters. The buffer was then combined with NDVI map. Therefore , there is information on NDVI in each buffer area. This will give an insight of how the condition of the vegetation in those areas. It will lead to the need of rehabilitation in terms of planting trees in the area.

Land system map gives an information on which land system are in the coastal area of Sibolga. There will be a potential zone or area that is suitable for greenbelt.

3. Results and Discussion

Sibolga is part of North Sumatera Province (**Figure 1**). The area is in a border of Central Tapanuli and in across with Nias Island. Sibolga coastline is in the western part of Sumatera stretching from north to south. It is in the area of Tapian Nauli bay. Sibolga harbor is a place of transit to Nias Island and other cities in western coast of Sumatera.

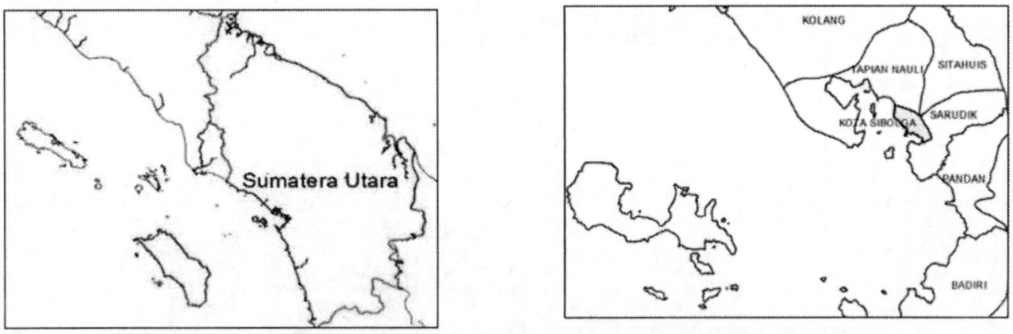

Figure 1. North Sumatera (left) and surrounding cities of Sibolga (right)

Sibolga is in the surrounding of some districts of Central Tapanuli Regency. There are districts of Tapian Nauli, Sitahuis and Sarudik. Other districts include in the bay area are Pandan and Badiri (**Figure 1**).

Islands in the area of Sibolga are Poncan Gadang, Poncan Ketek, Sarudik and Panjang. The Island of Mursala is in across the Sibolga which is in the part of the area of Central Tapanuli. Sibolga is in the border with Cental Tapanuli Regency in the east, south and north part and the west is in the border with Indian Ocean. Coastal area of Sibolga is part of bay area with mainland and small islands in the surrounding.

Shown image gave visual display on coastal area of Sibolga and the surrounding (**Figure 2**). Sibolga is in the bay area and the coast is directed to Nias Island. In the west there is an island and a part of Central Tapanuli called Mursala Island.

Figure 2 also shows that coastal area of Sibolga is an area with dense settlement with only small portion of vegetated area. The area with vegetation is shown by green colour.

Field observation was conducted along the coastline of Sibolga and Central Tapanuli (Figure 3). The field checking showed that the land cover along the coastal area is dominated by settlement and hotel. Only small portion of vegetated area was found (**Figure 4**). Vegetation found in the area are cemara laut, ketapang and coconut. Shrub was also found in the area. There are also bare land and sandy beach

Condition of coastal area of Central Tapanuli is different with Sibolga. Vegetation analysis was conducted in Central Tapanuli. The area were Pandan, Tapian Nauli, Badiri dan Sorkam. In Jago-Jago Village of Badiri District and Tapian Nauli 1 Village of Tapian Nauli District mangrove was found while in other area there were coastal vegetation (**Figure 5**).

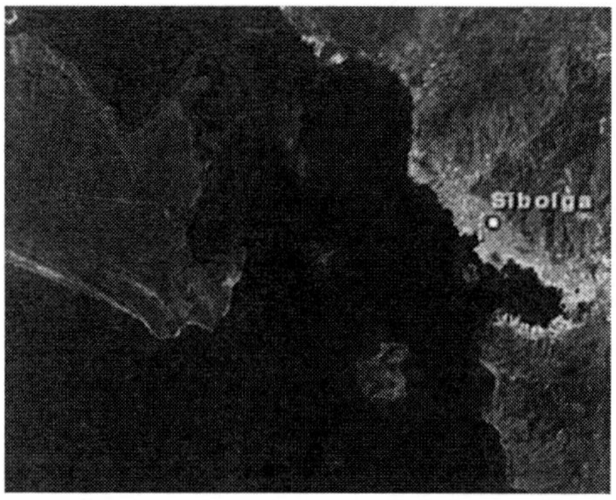

Figure 2. Coastal area of Sibolga

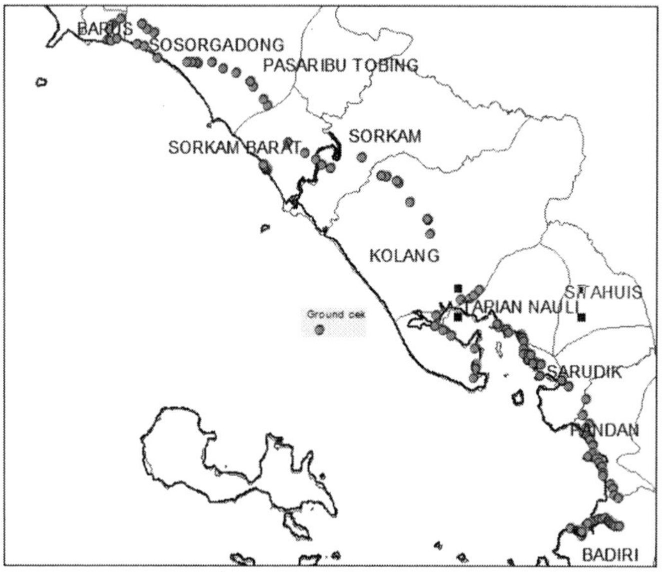

Figure 3. Observation location along the coast of Sibolga and Central Tapanuli (green dots)

Figure 4. Settlements and bare land and sandy field in Sibolga

Figure 5. Mangrove and oil palm in Tapian Nauli

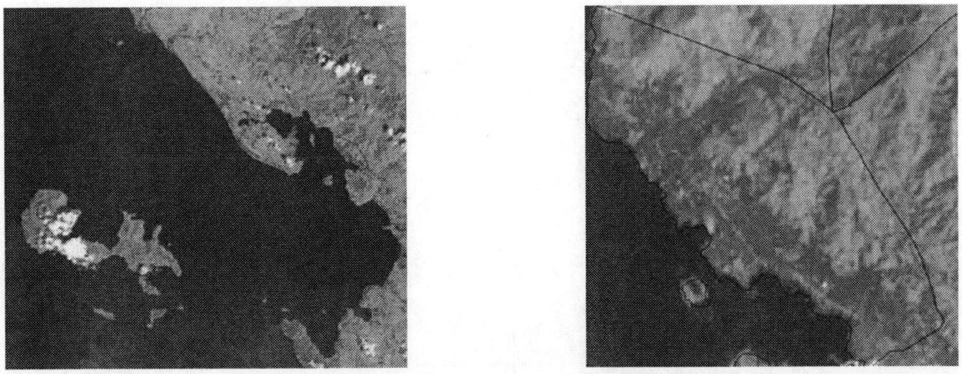

Figure 6. Landsat Image year 2017 of Coastal area of Sibolga and the surrounding

NDVI analysis was conducted using Landsat imagery of year 2017. The result was classified into some classes with 0.1 interval. Figure 6 shows the Landsat year 2017 in coastal Sibolga and Central Tapanuli.

Physical efforts are aimed to decrease the energy of tsunami wave that come to the beach, which can be divided to physical and natural efforts. Those efforts are complementary, depending on the condition of the area. Naturally, it can be done by having greenbelt in the form of coastal forest and mangrove. Mangrove has a strong and unique root system with dense canopy. Mangrove is ideal to protect the beach naturally [7].

Finding in Cuddalore district of India showed that at the river mouth, the tsunami completely destroyed parts of a village and removed a sand spit that formerly blocked the river [7]. The area with mangroves and tree belts were significantly less damaged than other areas. There were different conditions between villages in coast and villages behind the mangrove. The villages on the coast were completely destroyed. Villages behind mangrove did not show any destruction.

Figure 7 shows that most of coastal area of Sibolga has very low value of NDVI which is less than 0.1. It is relevant to the field observation along the coastal area which are covered by building mainly settlement. Higher value can be seen more in the mainland (Figure 15).

Along the area of the bay the NDVI value tends to be lower than the mainland especially in Sibolga which has the lowest NDVI value comparing to other areas (Figure 16). Some part in the mainland shows higher value of NDVI.

Figure 8 shows that the highest vegetation density is in Low class covering 32.5% of the area. Some small part of the area has very dense vegetation which is in the mountainous area of Sibolga.

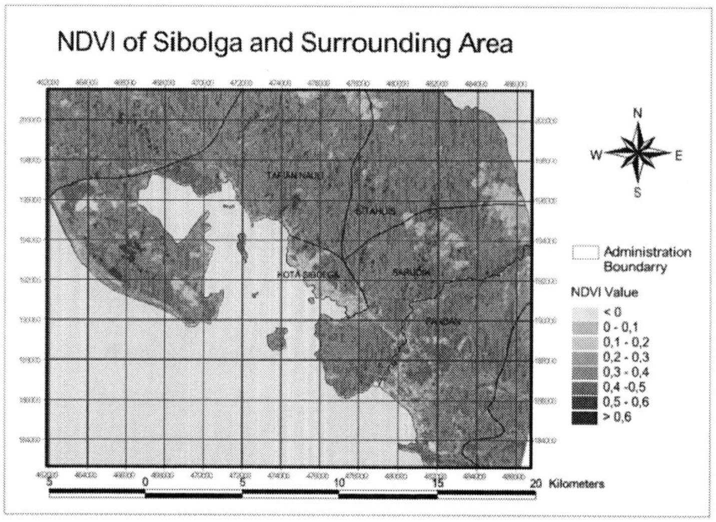

Figure 7. NDVI year 2017 in Sibolga and the surrounding area

Also the dense vegetation area. In the coastal area it can be clearly seen having less vegetation and no vegetation at all.

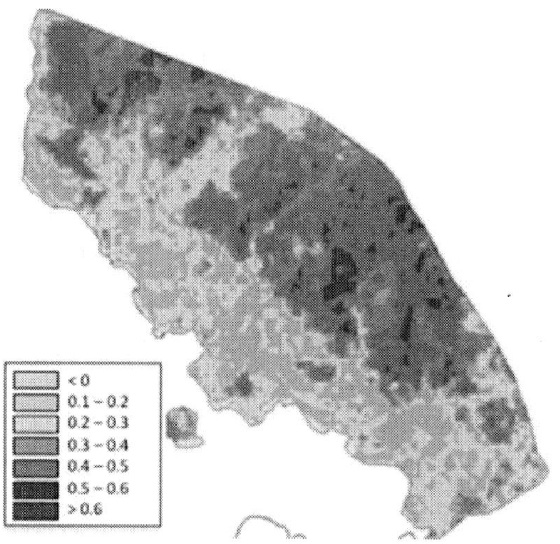

Figure 8. NDVI range in Sibolga

Buffering has been done to find out the condition of vegetation density in some distance for potential greenbelt area of Sibolga. In each buffer, the vegetation density along the coastline can be seen. The vegetation density in 100, 200, and 500 meters buffer can be seen in **Figure 9**. **Figure 9b** gives us an information on percentage of vegetation density in the area of 100 m buffer.

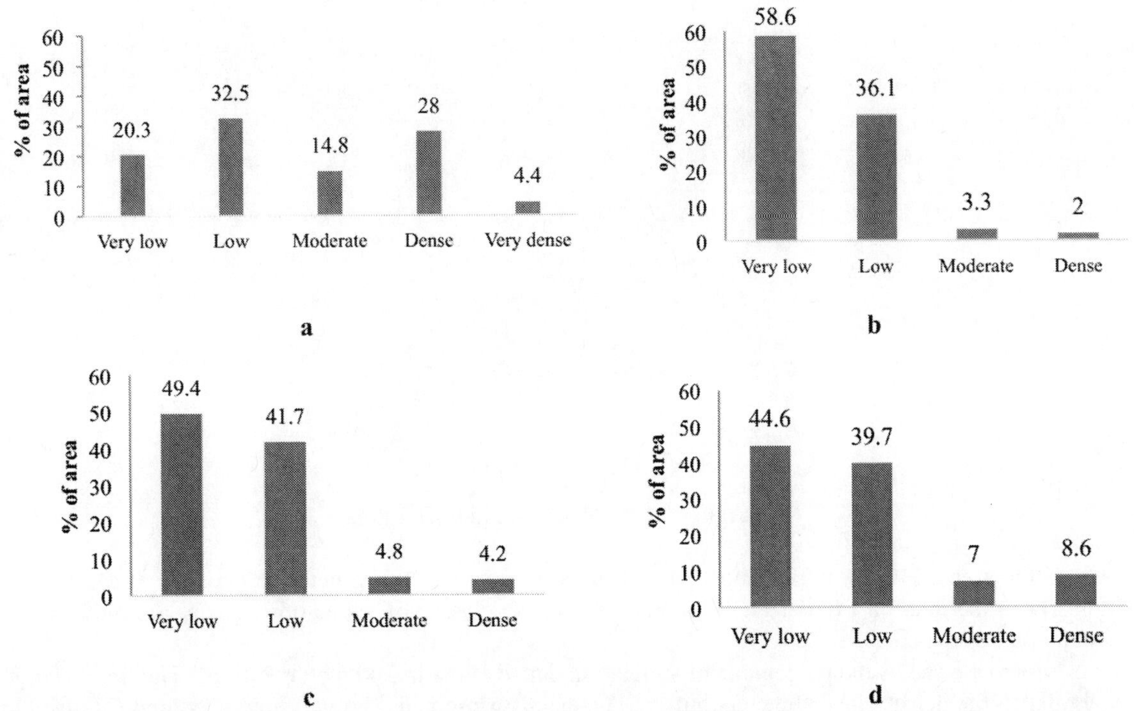

Figure 9. Vegetation density class; a. in Sibolga area, b. In 100 m buffer c. in 200 m buffer, d. In 500 m buffer

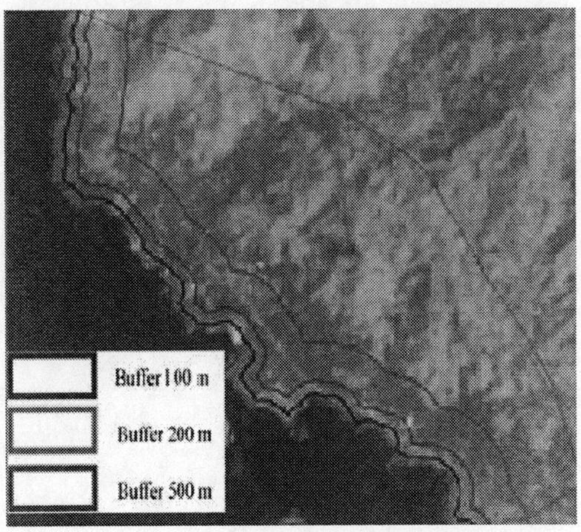

Figure 10. Buffer of 100, 200, 500 m in the coastal area of Sibolga

IOP Conf. Series: Earth and Environmental Science **166** (2018) 012028 doi:10.1088/1755-1315/166/1/012028

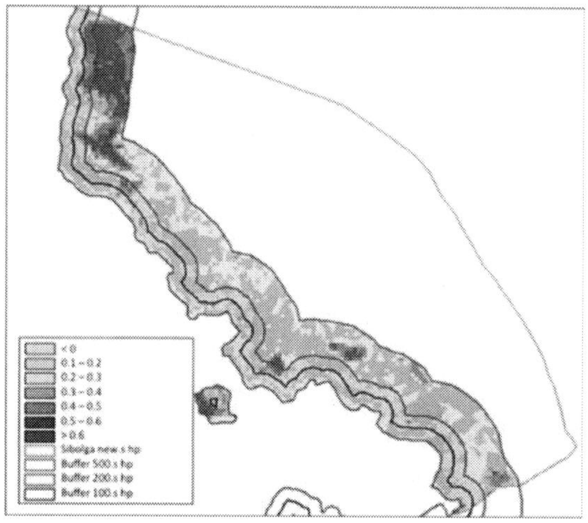

Figure 11. NDVI value in buffer area

In the area of 100 m buffer, most part of the area has very low vegetation density. It can be seen clearly in the map on **Figure 11**. Moderate and dense classes of vegetation density are only in very small part of the area.

Figure 9 c shows the percentage of vegetation density area in 200 meters buffer. The trend is quite similar to the density in 100 meters buffer. The class is dominated by very low class and followed by low class. Moderate and dense vegetation density are still very small portion of the area.

The trends is consistent in the area of 500 meters buffer (**Figure 9d**). The most part of the area has less vegetation and even no vegetation in the area. It means that most part of the coastal area from the coastline to 500 m buffers has small vegetation area.

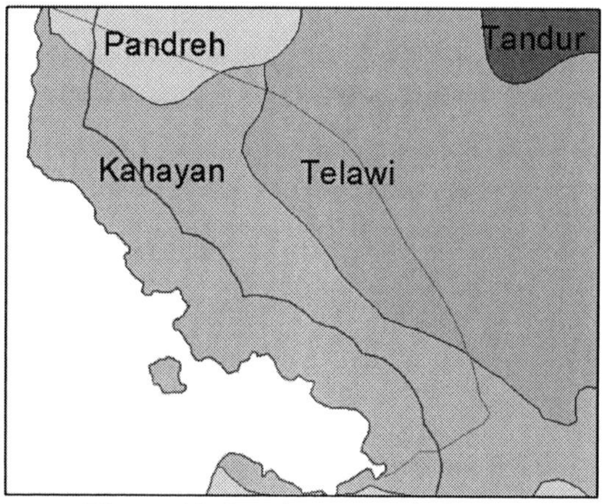

Figure 12. Land system of Sibolga

Land System Map provided the an information that the coastal area of Sibolga has Kahayan as a land system which also mentioned as aluvial plain (**Figure 12**). Regulation of Ministry of Forestry of Republic of Indonesia Number P. 35/Menhut-II/2010 [6] mentioned that Kahayan (KHY) is coastal area with sand sediment in the form of plains which some part is the result of accumulation of sediment in river mouth. The area has slope less than 2% with 2-10 meter relief. Rocks/mineral is alluvium dominat, young estuary and marine mix, river young alluvium and peat. Mangrove vegetation can grow in this type of land system. The area behind the coastal area is a mountainous area.

Mangroves and Casuarina plantations attenuated tsunami-induced waves and protected shorelines against damage. Conserving or replanting coastal mangroves and greenbelts should buffer communities from future tsunami events. Mangroves also enhance fisheries and forestry production [8].

The regulation mentioned about the area along the coastline with important benefit to maintain the coastal function sustainability. The width is proportional to shape and physical condition of beach ini minimum of 100 meters from the highest tide to the land. The rehabilitation of this area is aimed to build a greenbelt considering the community's dependence on the area and forest sustainability. Agroforestry can become part of the rehabilitation effort in this area.

Field observation shows that coastal area of Sibolga and Central Tapanuli do not have dense vegetation. Most part of the area is tourism area with lodging and settlement. In the northern part of coastal area of Central Tapanuli we can see an open coast with no dense vegetation. There is a need to do coastal protection from possible destruction caused by tsunami. Zone planning should put consideration to build greenbelt based on the land system, give consideration to community's benefit and coastal protection.

4. Conclusions

The conclusion of the research are: The highest percentage of the coastal area of Sibolga has low NDVI value. This indicates that the area has less vegetation. The field check shows that the coastal area has dense settlement with less vegetation. Land system found in the coastal area is Kahayan which is suitable for mangrove. The area is an alluvium plain while behind the coastal is mountainous and hilly areas. In 100, 200 and 500 meters width of greenbelt area, the highest vegetation density is very low and followed by low class. It means most of the area has no vegetation or less vegetation, Greenbelt planning is needed to protect the beach from any potential destruction of natural disaster mainly tsunami.

Acknowledgment

Authors would like to say thank you for University of Sumatera Utara for support us in research grant. Thank you to BPKH, USGS (Landsat), ESRI, Hexagon Geospatial (ERDAS) for the assistance of data and software.

5. References

[1] Sukardjo, S 2002 Integrated Coastal Zone Management (ICZM) in Indonesia: A View from a Mangrove Ecologist. Southeast Asian Studies 40/2.

[2] Diposaptono, S dan Budiman 2006 Tsunami Bogor: PT. Sarana Komunikasi Utama.

[3] Pribadi S, Fachrizal, I Gunawan, I Hermawan, Y. Tsuji dan S. S. Han 2006 Gempa Bumi dan tsunami Selatan Jawa Barat 17 Juli 2006 Jakarta: Badan Meteorologi dan Geofisika.

[4] Kastanya, F. J. P., M. H. D. Susilowati dan F. Sitanala. 2000. Daerah Rawan Tsunami di Pesisir Selatan Banyuwangi. Di dalam: Soesilo, I., Widyawati, L. Kusumawati dan S. Wahono, editor. *Ekosistem Pantai Indonesia.* Jakarta: Departemen Kelautan dan Perikanan RI dan Jurusan Geografi Universitas Indonesia.

[5] Sutowijoyo, A. P. 2005. Tsunami, Karakteristiknya dan Pencegahannya. Inovasi 2005; 3/XVII.

[6] Peraturan Menteri Kehutanan Republik Indonesia Nomor P. 35/Menhut-II/2010 . Kehutanan. Jakarta: Departemen Kehutanan Republik Indonesia.

[7] Gandhi, G. M., S. Parthiban, N. Thummalu and A. Christy. Ndvi: Vegetation change detection using remote sensing and GIS – A case study of Vellore District. Procedia Computer Science Volume 57, 2015, Pages 1199-1210

[8] Danielsen, F. M.K. Serensen, M.F. Olwig, V.Selvam, F. Parish, N.D. Burgess, T. Hiraishi, V. M. Karunagaran, M.S. Rasmussen, L. B. Hansen, A. Quarto, N. Suryadiputra. The Asian Tsunami: A Protective Role for Coastal Vegetation. SCIENCE VOL 310 28 OCTOBER 2005

Combination of the Equatorial Atmosphere Radar – Radio Acoustic Sounding System (EAR-RASS) in Determining a Fine Structure of Total Precipitable Water (TPW) over Kototabang, West Sumatera, Indonesia

E Hermawan[1] and F. M. A Wahab[2]

[1]Center of Atmospheric Science and Technology,
Indonesian National Institute of Aeronautics and Space (LAPAN)
Jln. Dr. Djundjunan No. 133, Bandung 40173, Indonesia
[2]Geophysics and Meteorology Department, Faculty of Mathematics and Natural Sciences, Bogor Agricultural University (FMIPA IPB), Jln. Meranti Kampus IPB, Babakan, Dramaga, Bogor, Jawa Barat 16680, Indonesia

E-mail: eddy_lapan@yahoo.com

Abstract. This study is mainly concerned an application the combined of the Equatorial Atmosphere Radar (EAR) and Radio Acoustic Sounding System (RASS) in determining a fine structure of Total Precipitable Water (TPW) in the lower troposphere at Kototabang, West Sumatera (0.20^0 S; 100.32^0 E) during the CPEA I Campaign started from April 10 to May 9, 2004. It looks possible, since EAR-RASS is designed specially to detect the wind at three dimensional direction and virtual temperature with good time and spatial height resolution. By applying the Weisner technique (1970), we found a fine structure of TPW vertically during the above period. While, by applying the Cross-Correlation Function (CCF), we found an a goog agreement between TPW and Optical Rain Gauge (ORG). In the same time, we analyzed the Boundary Layer Radar (BLR) to investigate the updraft and downdraft air mass activities from surface up to 5 km. We found the consistency of BLR, RASS and radiometer data when we investigated the vertical profile of TPW. Futhermore, we found the time variation of TPW, especially on April 24 and May 5, 2005 are about 0.43 and 0.35, respectively. Detailed results of this study are discussed in this paper.

1. Introduction

These major part of the solar radiation reaches the Earth's surface without absorption by the atmosphere. The troposphere is heated by infrared radiation from the Earth's surface, and convection transfers thermal and kinetic energy upward. Especially in the atmospheric boundary layer where the atmosphere undergoes direct infrared heating, active convections often appear and sometimes reach up to the tropopause under low-stability conditions. These deep convections are thought to transport minor constituents such as carbon dioxide between the troposphere and the stratosphere [1]. RASS (Radio Acoustic Sounding System) is a ground based remote sensing technique for measuring a height-profile of atmospheric virtual temperature in the tropoposhere and lower stratosphere with time and height resolution of a few minutes and 150-300 m, respectively.

Acoustic wave emitted from a high-power transmitter on the ground produce sinusoidal variations of atmospheric density, resulting in the radio refractive index perturbations. A radar tracks the acoustic wave fronts to receive strong scattering of radio waves (RASS echo). However, there are two necessary conditions to detect RASS echoes; the wave number vector of incident radiowave must be perpendicular to the acoustic wave fronts, and the acoustic wavelength must be half of the radar wavelength for optimum Bragg resonance.

As one of an MST (mesosphere, stratosphere and troposphere) radar, the EAR (Equatorial Atmosphere Radar) can be used to determine wind velocity fluctuations with good time and spatial height resolution. This is similar to the MU (Middle and Upper atmosphere) radar observation [2].

The construction EAR-RASS capable of detecting fine structure in the vertical profiles of virtual temperature, T_v. From this virtual temperature data, we can obtain a fine structure of Brunt Väisälä frequency squared, N^2 as one of the most important parameters needed to explain the Turbulence Energy Dissipation Rate (TEDR), ε and Vertical Eddy Diffusivity (VED), K [3]. This paper is mainly concerned to describe a brief review of the basic concept of RASS technique, especially on determining a fine structure of Total Precipitable Water (TPW) that represented as W profiles with a good time a spatial height resolution.

2. Materials and Methods

2.1 TPW

TPW is defined as the amount of moisture content in an air vertical column if all the moisture in the column undergoes a precipitated condensation as well as decreased as rain [4], in mm or inches, although no natural process can change the entire moisture content the layer becomes rain. Furthermore, the amount of water that can be converted to rain within an air blob at adequate altitude is calculated by adding pressure or pressure from the surface and observing the temperature, humidity and air pressure above it. The formula that is often used to calculate W in millimeters is:

$$\mathbf{W} = \sum_{i=1}^{N} q\,\Delta p$$

With pressure pa in millibar and q in grams per kilogram, is the average value of specific moisture (moisture mass per moist air mass), in the top and bottom layers of each layer. Specific moisture can be expressed in the formula:

$$\mathbf{q} = 622\frac{e}{p-0.37\,e} \approx 622\frac{e}{p}$$

To estimate the value of this TPW using the formula:

$$W = \int_{z=z_i}^{z=z_i+dz} \rho_v\,dz = \int_{z=z_i}^{z=z_i+dz} q\rho\,dz = -\frac{1}{g}\int_{p=p(z_i)}^{p=p(z_i+dz)} q\,dp$$

where:

- W = *Total Precipitable Water* (cm)
- z_i = height in (m)
- p_{zi} = pressure in (hPa)
- ρ = air density in (kg/m3)

q = specific humidity between two layers p(z) and p(z_i+ dz) in (g/kg)

To process RASS data, because the available data in the form of q, Tv and z ; used the formula:

$$P = \frac{1}{g} \int_{z=p_z}^{z=p_z - dp} q \, dp$$

To complete the above integration by using the technique of integral average value, obtained:

$$\mathbf{W} = \sum_{i=1}^{N} q_i \, \Delta p_i$$

where q_i is the mean value of q between two pressure layer p_z and (p_z - dz)

According to [4] TPW is calculated each isobaric surface height. To determine the height of the isobaric surface for each of the atmospheric layers based on the Geopotential height with the hypsometric equation:

$$z_2 - z_1 = dz = \frac{R_d}{g_0} \overline{T}_v \ln \frac{P_1}{P_2}$$

$$p_n = p_0 \exp\left[(z_2 - z_1) \times \frac{g_0}{R_d \overline{T}_v}\right]$$

where:

T_v = T (1 + 0.608 r) R_d = 287 J/Kg/^{0}K

g_0 = 9.806 m/s^2

$\overline{T}_v$ = The mean value of Virtual Temperature between two layers

While to estimate TPW by using data from Radiometer in accordance with the availability of parameters in the form ρ_v , then using the formula:

$$W = \int_{z=0}^{Z_{top}} \rho_v dz :$$

As well as RASS data, to complete the above integration using the technique of integral integral values, so is obtained:

$$w = \sum_{i=1}^{N} \overline{\rho}_i \, \Delta z_i$$

2.2 Specification of EAR

The EAR has a circular antenna array of approximately 110 m in diameter, which consist of 560 three-element Yagis as presented in Figure 1.

Figure 1. The antenna field of EAR at Kototabang, West Sumatera

It is an active phased array system with each Yagi driven by a solid-state transceiver module. This system configuration makes it possible to direct the antenna beam by electronic control up to 5000 times per second. The EAR transmits an intense radio wave of 47 MHz to the sky, and receives extremely weak echoes scattered back by atmospheric turbulence. It can observe winds and turbulence in the altitude range from 1.5 km to 20 km (troposphere and lower stratosphere). It can also observe echoes from ionospheric irregularities at heights more than 90 km. The explanations more detail is presented at Table 1 which are based on research by [2].

Tabel 1. EAR specification

Location	100.32° E ; 0.20° S ; ± 865 m MSL
Frequency	47 MHz
Output power	100 kW (peak envelope)
Antenna system	Quasi-circular active phased array (110m-diamter, 560 three-element Yagis)
Beam width	3.4° (-3dB, One-way)
Beam direction	Anywhere (within 30° zenith angles)
Observation range	1.5 – 20 km (Atmospheric turbulence) > 90 km (Ionospheric irregularity)

2.3 Radar Equation for RASS Echo

The radar equation for the RASS echo given by [5]. Radiowave echo model of turbulence scattering to the RASS echo was applied by assuming the following conditions :

- The atmospheric temperature is relatively constant over the entire height and the background wind is negligible.
- The radar beamwidth (θ_r) is sufficiently smaller than the acoustic beamwidth.
- The effects of any distortions of the of the acoustic wave fronts caused by turbulence are negligible.

With normal wind profiler radar, successive profiles of three components of the wind velocity, echo power, and the spectral width with certain time and height resolution can be observed. In addition, by applying RASS, we can simultaneously monitor temperature profiles [6,7]. The basic principle and the observational procedure of RASS is shown in Figure. 2.

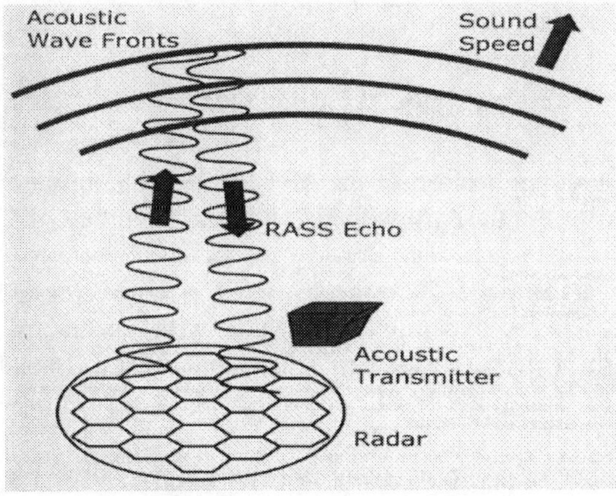

Figure 2. Basic principle of RASS

The basic principle of RASS depends on a relation that sound speed is proportional to the square-root of atmospheric temperature. In a RASS observation, transmitted acoustic pulses from the ground produce refractive-index fluctuations. We can detect the Doppler shift between the transmitted signal and received echoes scattered by acoustic wave-fronts and determine their propagation speed. In the actual atmosphere, the apparent acoustic speed, C_s is the summation of the true acoustic speed, C_a and the background wind velocity, V_r.

$$C_a = C_s + V_r$$

The acoustic speed, C_s measured by RASS then becomes

$$C_s = C_a \cdot n + V_r \cdot n,$$

where n is unit vector parallel to the radar beam direction. The atmosphere temperature can be derived as

$$C_a = K \sqrt{T}$$

where K is a value which depends on the humidity. In a dry atmosphere we can describe K as

$$K = K_d = \sqrt{\{\gamma R / M_r\}}$$

where γ is the ratio of specific heat, R is a gas constant, and M_d is the mean molecular weight. In a moist atmosphere, sound moves slightly faster than in a dry atmosphere. So, the equation of T_v become

$$T_v = \sqrt{\{C_a / K_d\}}$$

where T_v is the atmospheric virtual temperature which is equivalent temperature for a dry atmosphere whose pressure and volume are the same as an actual moist atmosphere with a water vapor mixing ratio of q (kg kg^{-1}). The virtual temperature is defined as

$$T_v = (1 + 0.608\, q)\, T$$

Briefly reviewed here is an application of the MU-RASS technique in monitoring a T_v profile and turbulence echoes from 1.5 to the near tropoppause with an accuracy of T_v within 0.5 K [7, 8, 9, 10, 11].

Three types of acoustic transmitters have been used for the MU-RASS : a pneumatic transducer with a hyperbolic horn, an electro-dynamic loudspeaker with a standard enclosure, and an electro-dynamic loudspeaker with a hyperbolic horn. The pneumatic transducer which can produce transmitted acoustic wave as strong as 147 dB, allowing RASS echoes to be detected up to 22 km was developed some time ago [8].

Ray-tracing results indicated that an acoustic transmitter should be placed windward of the radar antenna. Under disturbed meteorological conditions, the wind velocity tends to fluctuate greatly causing considerable changes in the best position for the acoustic source. To enable continuous observation even under disturbed condition, several acoustic sources composed of loudspeaker until and an amplifier were located around the radar antenna. The output power of a pair of these loudspeaker unit was 132 dB. Because of the lower acoustic power, though, the observation height was limited to below about 8 km.

2.4 Design of Acoustic Transmitter for RASS Measurements with EAR

Since EAR-RASS technique is being constructed now, application of the RASS technique to EAR, in particular a design of a high-power acoustic transmitter are reviewed in this paper.

Figure 3. The hyperbolic horn speaker system at Kototabang. The speaker is mounted on a wooden frame with casters.

Detailed description of these instruments are as follow :

- PC (8 outputs), 1 set :
  - Pentium III (800 MHz), memory 256 MB, H/D 10 GB
  - Digital mixing board : Yamaha DS-2416
  - D/A converter : Yamaha AX-44 (2 sets)
- Acoustic pre-amplifier (4 inputs and 4 outputs), 2 sets :

- Unbalanced input and balanced output
- 4 channels
- UBC-159 (imagenics)

• Power amplifier (2 inputs and 2 outputs), 4 sets :
  - Output power 800 W/channel, 2 channels, total power 1600 W
  - Input terminal should be equiped with a balanced connector
  - Micro-Tech 2401 (Amcron Co.)
  - MPA 750 (JBL)

• Loudspeakers, 16 sets :
  - Electro-voice DL-15 X (diameter :38 cm/15 inch)
  - Nominal impedance : 8 Ohm
  - Maximum continuous input : 400 W
  - Sensitivity : 98 dB at 100 Hz
  - Maximum sound pressure level : 130 dB ((.6 W/m^2)
  - Lower cut-off : 75 Hz

• Acoustic transmitter with a wooden horn with dimension of H850 cm x W1580cm x L2200 cm, 8 sets as shown in **Figure. 3.**

Detailed description of major parameters of RASS-EAR can be shown in **Table 2.**

Table 2. Major parameter of RASS-EAR

	Items
Radar wavelength (EAR)	6.38 m (Freq = 47.0 MHz)
Acoustic power	800 W maximum
Acoust frequency	80 – 120 Hz
Number of speakers	8
Observable height	Up to tropopause
Measurement accuracy	0.5 K (empirical, virtual temperature)
Power amplifier	Yorkville AP 4040
Pre-amplifier	Mackie 1402-VLZ
Speaker box	Wooden buffer box, folded horn (1 m H x 1.5 m W x 2.5 m L)
Speaker	JBL 2226/H
Audio source	Digital mix board and A/D converter controlled with a PC

The example the comparison between RASS and radiosonde in determining the vertical virtual temperature profile at Kototabang on 6 November 2001 are presented in Figure 4.

The thick and thin line indicates radiosonde and RASS data measurement should be noted. The data was taken on 6 November 2001 at 12:57:47 LT from the vertical beam direction. By looking this figure a good agreement between both data below 6 km and between 8-13 km can be detected.

IOP Conf. Series: Earth and Environmental Science **166** (2018) 012029 doi:10.1088/1755-1315/166/1/012029

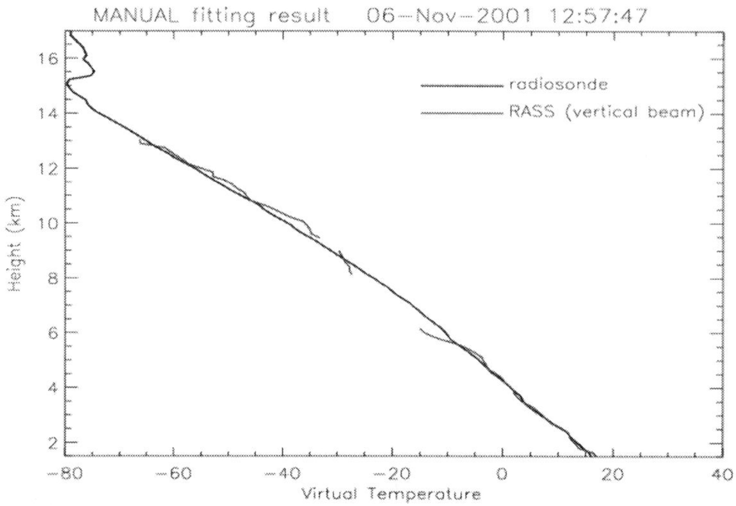

Figure 4. The vertical virtual temperature at Kototabang measured by RASS and radiosonde at 6 November 2001

3. Results and Discussion

The comparison between radiosonde and EAR-RASS techniqusue in determining a fine structure of relative humidity profiles presented in **Figure 5**.

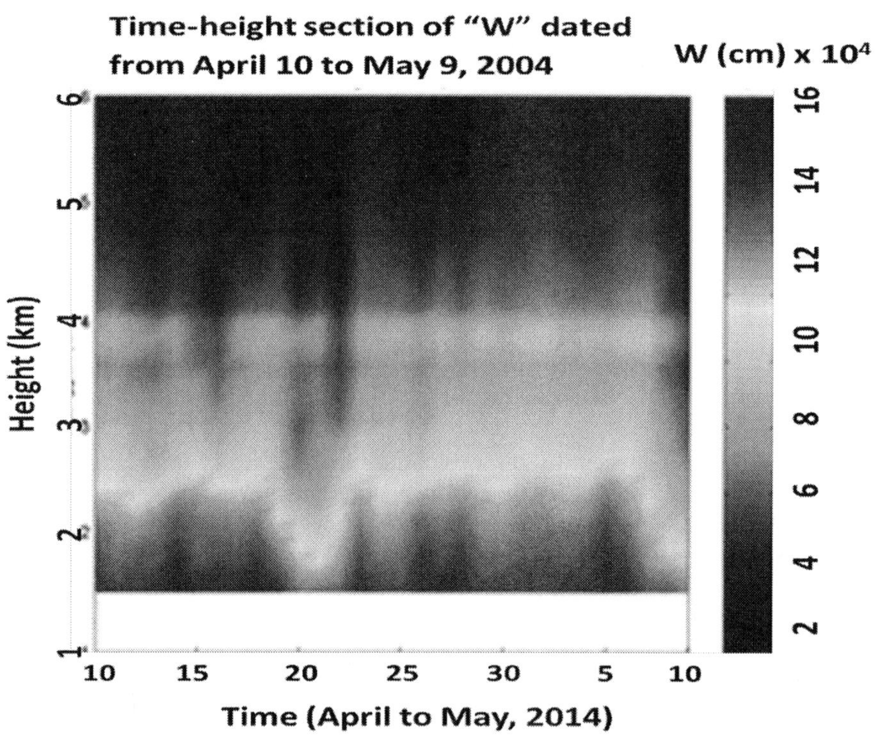

Figure 5. Time-height section of Total Precipitable Water (represented by W) profiles derived from RASS observation from April 10 to May 9, 2004 at Kototabang, West Sumatera.

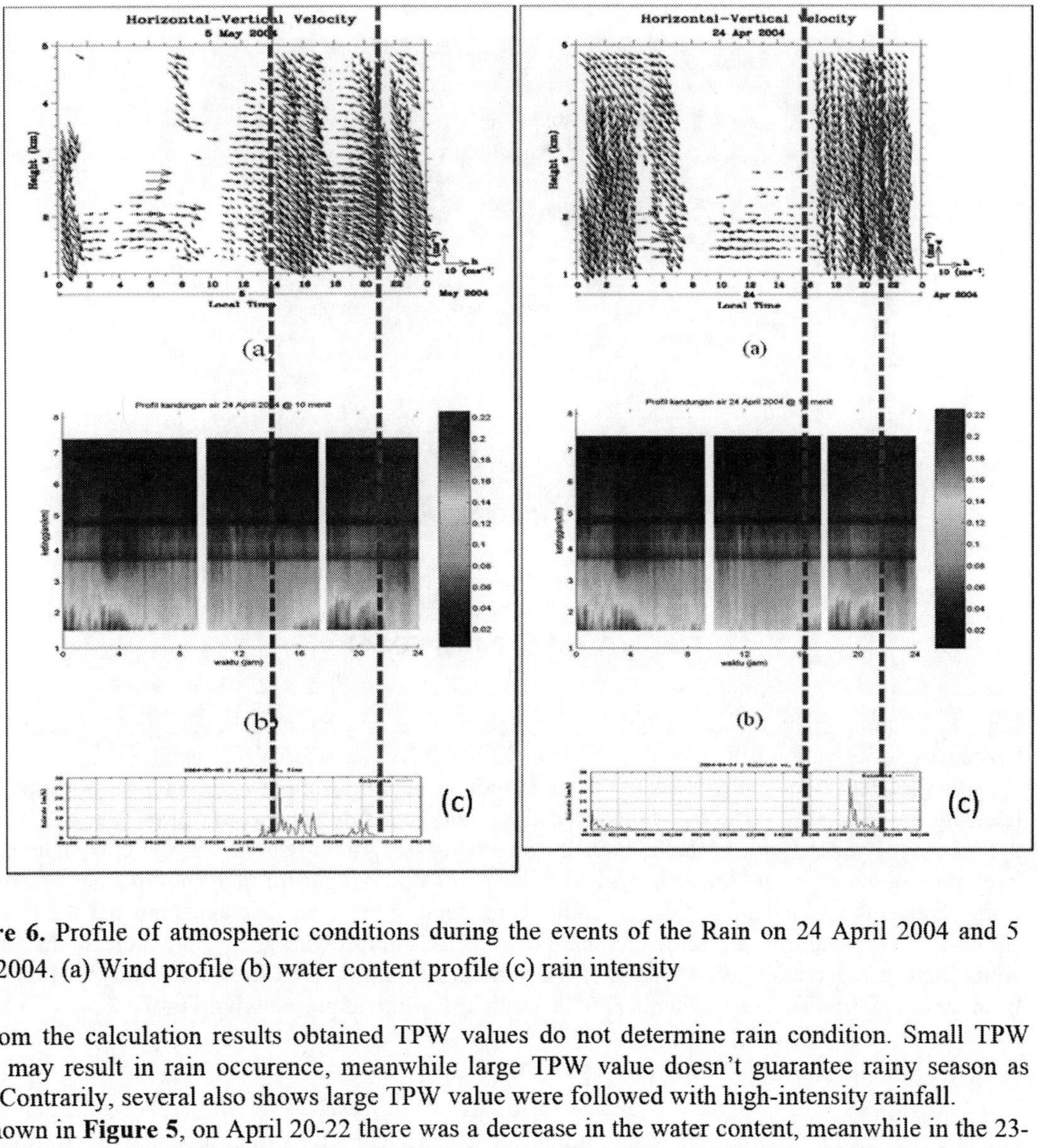

Figure 6. Profile of atmospheric conditions during the events of the Rain on 24 April 2004 and 5 May 2004. (a) Wind profile (b) water content profile (c) rain intensity

From the calculation results obtained TPW values do not determine rain condition. Small TPW value may result in rain occurence, meanwhile large TPW value doesn't guarantee rainy season as well. Contrarily, several also shows large TPW value were followed with high-intensity rainfall.

Shown in **Figure 5**, on April 20-22 there was a decrease in the water content, meanwhile in the 23-24 th there is an increase in water content. Furthermore from **Figure 3** on 21 and 22 there was no record of rainfall intensity and rain fell again on 23 and 24 April. So even when viewed from the data of the wind down in **Figure 6** that on April 24 the activity of the air intake decrease quite intensely at 0-4 and at 17-24. then seen the data contain water content information that at 5-16 there was a decrease in water content and the value rose again after 16 o'clock, other than that from the rainfall data is known at 0-4 and at 17-24 shows there are events of rain.

As on May 5, 2004 when there was an increase in water content of ORG data recorded a large rainfall compared to other days in the period April 10 - May 9, 2004 where the incident rain lasted from 12 to 21 o'clock local time (**Figure 10**). In addition, when the incident rain at that time from the calculation seen the pattern of increase in water content from 12 then the amount of water content is dropped back at 21 o'clock after the rain stopped. As if seen from the BLR data seen intensive air degradation activity at that time.

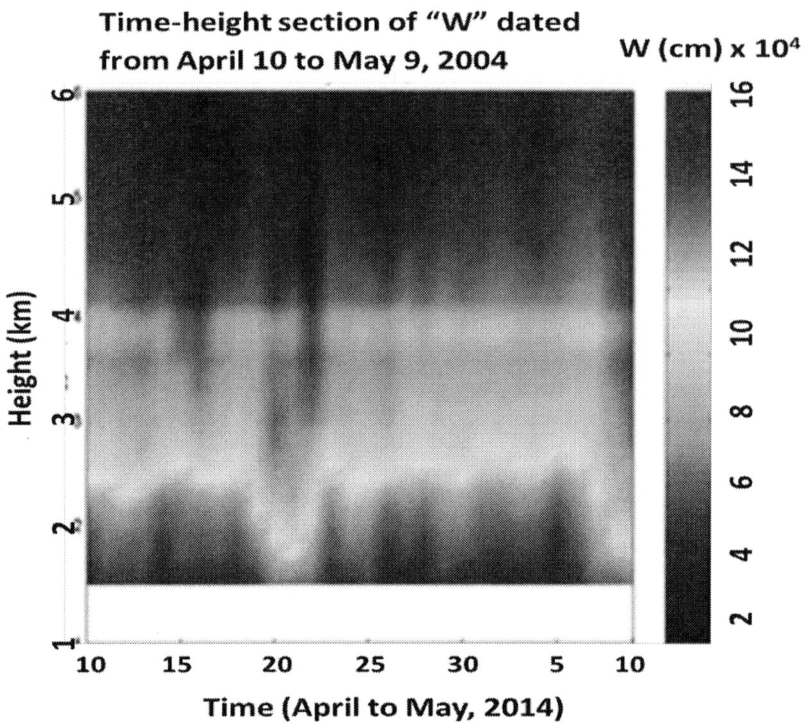

Figure 7. Time-height section of Total Precipitable Water (represented by W) profiles derived from Radiometer observation from April 10 to May 9, 2004 at Kototabang, West Sumatera

This illustrates that the presence of water vapor in the atmosphere is directly proportional to the intensity of rain on the surface and the mass air activity supports the spread of rain. As at 12-16 when the content of water vapor in the column shows a decreasing trend whereas in the ORG data shows a large rainfall event, it could have happened the mass of water vapor formed from the mass air removal from other columns fell in vertical columns Kototabang other than it is apparent that the horizontal winds at 12-21 local time are more dominant than the downdraft winds, this may indicate that at 12-16 hours there is a decreasing vapor mass in the vertical column derived from the removal of air masses from other regions. This result identical with another instrument presented in **Figure 7**.

4. Concluding Remarks

The EAR-RASS offers a great advantage in that it enables user to monitor successive temperature profiles over the entire troposphere into the lower stratosphere with a high time and height resolution. In this paper, the basic principle of RASS especially on design of an acoustic transmitter for RASS measurements with EAR were briefly reviewed. Since the full RASS data campaign at Kototabang has not been obtained. The results of this study indicated the consistency of pattern between BLR, RASS and radiometer data in assessing vertical profile of water vapor content as well as the variation of TPW result from RASS and radiometer data analysis, especially on April 24 and May 5, 2004, 0.43 and 0.35 respectively, besides the mass air degradation activity resulted from BLR data analysis as well as the variation of water content to be an indication of rain.

Acknowledgements

The author wish to thank to the Research Institute for Sustainable Humanosphere (RISH), Kyoto University and Shimane University, Japan for making available the RASS technique at Kototabang station.

5. References

[1] Furumoto, J., Observation of Turbulence Echo Characteristics and Humidity Profiles with the MU Radar-RASS (Radio Acoustic Sounding System), PhD. Thesis, Kyoto University, 2002.

[2] Fukao, S., H. Hashiguchi, M. Yamamoto, T. Tsuda, T. Nakamura, M. K. Yamamoto, T. Sato, M. Hagio, and Y. Yabugaki, The Equatorial Atmosphere Radar (EAR): System description and first results, *submitted to Radio Sci.,* 2002.

[3] Hermawan, E and T. Tsuda, Adachi, T., T. Tsuda, Y. Masuda, T. Takami, S. Kato, and S. Fukao, Effects of the acoustic and radar pulse length ratio on the accuracy of radio acoustic sounding system (RASS) temperature measurements with monochromatic acoustic pulses, *Radio Sci.*, **28**, 571-583, 1993.

[4] Weisner, C.J. 1970. Hydrometeorology. School of Civil Engineering. University of New South Wales Australia.

[5] Marshall, J.M., A.M. Peterson, and A.A. Barnes, Jr., Combined radar-acoustic sounding system, *Appl. Optic,* **11**, 108-112, 1972. May, P.T., K.P. Moran, and R.G. Straauch, The accuracy of RASS temperature measurements, *J Appl. Meteor.*, **28**, 1329-1335, 1989.

[6] Tsuda, T., Y. Masuda, H. Inuki, K. Takahashi, T. Takami, T. Sato, S. Fukao., and S. Kato, High time resolution monitoring of tropospheric temperature with a radio acoustic sounding system (RASS), *Pure and Appl. Geophys.*, **130**, 497-507, 1989.

[7] Matsuura, N., Y. Masuda, H. Inuki, S. Kato, S. Fukao, T. Sato and T. Tsuda, Radio acoustic measurement of temperature profile in the troposphere and stratosphere, *Nature,* **323**, 426-428, 1986.

[8] Adachi, T., T. Tsuda, Y. Masuda, T. Takami, S. Kato, and S. Fukao, Effects of the acoustic and radar pulse length ratio on the accuracy of radio acoustic sounding system (RASS) temperature measurements with monochromatic acoustic pulses, *Radio Sci.*, **28**, 571-583, 1993.

[9] Adachi, T., Detailed temperature structure of meteorological disturbances observed with RASS (Radio Acoustic Sounding System), PhD. Thesis, Kyoto University, 1996.

[10] Tsuda, T., T. Adachi., Y. Masuda, S. Fukao, and S. Kato, Observations of tropospheric temperature fluctuations with*os. Ocean. Technol.,* **11**, 50-62, 1994.

IOP Conf. Series: Earth and Environmental Science **166** (2018) 012030 doi:10.1088/1755-1315/166/1/012030

Influence of ENSO on Deviation of The Season Onset in Java Based on CCAM Downscaling Data

H Satyawardhana[1], Trismidianto[1] and E Yulihastin[1]

[1]Center of Atmospheric Science and Technology, National Institute of Aeronautics and Space, Bandung, Indonesia

E-mail: hariessatha@gmail.com

Abstract. The research of season onset in Java is very important, considering Java is the center of agriculture in Indonesia. The onset of the rainy and drought season in Java are strongly influenced by global atmospheric phenomena, such as monsoon and El Nino Southern Oscillation (ENSO). This study used spatial analysis from previous researches focusing on insitu data. The data used in this current study based on the downscaling output of atmospheric model with ±15km spatial resolution. The result shows that the season onset deviation in Java Island exist when El Nino and La Nina occurred. When the El Nino (La Nina) occurred, the onset of drought season in Java has a tendency to come faster (slower) than neutral conditions. On the other hand, when El Nino (La Nina) occurs, the onset of rainy season in Java comes slower (faster) compared to the neutral condition. This is probably related to disturbance of monsoon in Indonesia when El Nino and La Nina occurred. The study also confirmed that the effects of El Nino and La Nina in the rainy season are stronger than in the dry season.

1. Introduction

Research on the onset of the season in Indonesia, especially in Java is crucial due to its role as the center of agriculture and food production in Indonesia and many Javanese have jobs as farmers. Agriculture is in desperate need the season onset information, both the onset of the rainy season or drought for the determination of planting, fallow and harvest periods.

However, the season onset research in Java is still mostly done with in-situ observation data [1], resulting in less thorough analysis. Related to that, this research using downscaling data from atmospheric and climatic models, it is expected to obtain spatial information with high resolution data. In addition, the use of model data gives in flexibility for choosing the research period, which is not obtained using satellite data, which mostly began to operate in the early 20th century. Research on the season onset taken from Indonesian Agency of Meteorology, Climatology, and Geo-physic (BMKG) based on dasarian rainfall in Indonesia has been done using Tropical Rainfall Measuring Mission (TRMM) satellite data. However, it still has limitations in the analysis period (1998 - 2009) [2, 3].

Dasarian rainfall data that resulted by downscaling using the Conformal Cubic Atmospheric Model (CCAM) has been compared and validated using TRMM data, both values and patterns [4]. Comparisons using satellite rainfall data have similar patterns and values, although there is a difference. CCAM downscaling data can still follow TRMM data on low rainfall, but for high rainfall value, downscaling data is still under-estimate comparing to TRMM rainfall [5]. However, the comparative results still include the sea area around Java, in case the sea part was eliminated, and only use of rainfall in the mainland, then the value of under-estimate can be minimized [4].

The monsoon event greatly influences the pattern of the rainy and dry seasons in Indonesian Maritime Continent (IMC). The monsoon refers to an annual cycle that strictly distinguishes atmospheric conditions during the dry and wet phase. The dry phase is influenced by the winter that occurs in various continents with the air mass in the cold and dry atmosphere [6]. In contrast, the wet phase is influenced by summer with humid air. The monsoon is awakened by the warming differences between the oceans and the mainland due to the pseudo movement of the sun, the shape and topography of the continent, both Asian, European, African, Maritime Continent, American and Australian [7,8,9]. The geographical location of IMC is located between the regional monsoons crossing area of the Asia-Australia monsoon. Besides monsoon, Aldrian et al. (2007) suggested that the monsoon and the El Nino Southern Oscillation (ENSO) are more likely to drive the seasonal variation and inter-annual variations of the rainfall and extreme events in Indonesia [10]. In general, the rainfall pattern in Java Island has a monsoonal pattern where in one year there is one peak of rainfall, which generally occurs in December - February. Research on the influence of ENSO which occurring in the Pacific Ocean on rainfall in Indonesia has been widely studied [11]. But its influence on the onset of the season in Indonesia, especially Java is still need to be study furthermore. Therefore, the purpose of this study was to analyze the influence of ENSO, both in neutral, La Nina and El Nino conditions towards the beginning of the rainy and dry seasons in Java. In addition, this study also analyzes how far the onset of the season in Java deviates (forward or backward), when La Nina and El Nino compared to neutral conditions.

2. Materials and Methods

This research used downscaling data using CCAM, which can give the information about high resolution climate parameter. Downscaling data has an adjustable resolution depending on the research needs, but this study uses data with spatial resolution of 15 km for Java Island in the period 1990-2010. Related to further analysis of the season onset, the research area was split in 2 areas, Western Java (WJ) and Eastern Java (EJ) , as shown in **Figure 1**.

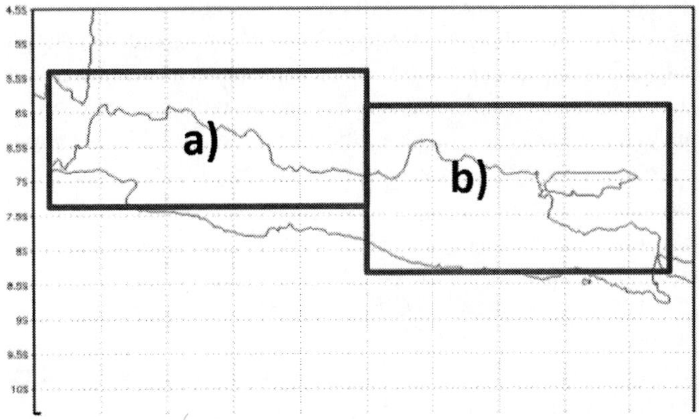

Figure 1. Research areas, where: a) Western part of Java (WJ) (5.8° – 7.8°S and 105.2° – 110°E); and b) Eastern part of Java (EJ) (6.3° – 8.9°S and 110° – 114.7°E)

The criteria for the onset of the season in Indonesia according to BMKG (2009) are based on dasarian rainfall. The onset of the rainy season is occurred if the dasarian rainfall is greater than or equal to 50 mm in one dasarian, followed by the next two dasarians. The opposite is applicable for the onset of the dry season, where the onset of the dry season is occurred if the dasarian rainfall is less than 50 mm in one dasarian and followed by the next two dasarians [12]. The meaning of dasarian (10-days accumulation or decade, decade is not mean 10-years but refer to 10-days, and then will be called dasarian in this paper) is rainfall accumulation in 10 days, but in the end of the month, it can be only 8, 9, 10 or 11 days. Thus in 1 year has 36 dasarians.

This research also used indexes, such as Southern Oscillation Index (SOI) to describing ENSO and Australian Monsoon Index (AUSMI) to describe monsoon and 850mb wind. SOI was one of the methods used by Australian Bureau of Meteorology (BOM) to identification of ENSO, based on calculating the difference of pressure anomaly between Darwin and Tahiti [13]. While AUSMI is the area average of 850mb wind in the area 110-130E, 15S-5S [14].

$$SOI = \frac{[P_{diff} - P_{diffav}]}{SD(P_{diff})} \qquad (1)$$

where:

P_{diff} = (pressure msl)$_{Tahiti}$ – (pressure msl)$_{Darwin}$

P_{diffav} = average of P_{diff} long time periods

SD(P_{diff}) = standart deviation of P_{diff}

$$\text{AUSM Index (AUSMI)} = \text{U850}(110 - 130\text{E}, 15\text{S} - 5\text{S}) \qquad (2)$$

where:

U850 = area average of wind in 850mb, the boundary of the area is 110E-130E,15S-5S

The identification of the El Nino, La Nina and Neutral years is processed by selecting the SOI values as illustrated in **Figure 7**. However, it is done by separating the SOI on JJA for dry season analysis and SOI on DJF for rainy season analysis.

3. Result and Discussion

At the time of the El Nino occurred, rainfall in Java was generally reduced, although there are areas that have high rainfall such as in the south of West Java due to local circulation and topographic influence [11]. In addition to the reduced rainfall, ENSO events also affect the onset of the rainy season as shown in **Figure 2**. The colors in **Figures 2** and **3** show the time of the season onset, both drought and rain, with dasarian units. The sooner the season onset is shown by the color purple - blue, while the yellow - red color indicates the the season onset that comes more slowly.

While ENSO in neutral conditions, the onset of the rainy season in Java on average occurred in 26th – 32nd dasarian. However, the onset of the rainy season in Java came faster than neutral conditions at the time of El Nino occurred. The onset of the rainy season withdraw from 2 to 4 dasarians and become 28th – 35th dasarian, even in eastern part of Java there are areas that have the beginning of rainy season more than 35th dasarian. On the contrary, when La Nina occurred, the onset of rainy season in Java is generally more advanced to 25th - 30th dasarian, even some areas experiencing the beginning of the rainy season is less than 25th dasarian. It is commonly found in western Java. Thus **Figure 2** explains that there was a difference of the season onset in Java according to 3 types of ENSO conditions. While El Nino occurred, the onset of the rainy season in Java tends to occur more slowly compared to La Nina and neutral conditions. In contrast, the onset of the rainy season while La Nina is come faster than neutral conditions.

The onset of the dry season has a contrary impact compared to the onset of the rainy season toward the influence of ENSO. While El Nino occured, the onset of the dry season on the Java island comes faster, while when La Nina slower than neutral conditions. When ENSO in neutral condition, Java had the onset of the dry season on average in dasarian to 10 -18. While at El Nino, the onset of the dry season in Java comes faster than the neutral ENSO conditions. This is marked by the many areas which experiencing the onset of the dry season less than dasarian 10. The interesting phenomenon was, while La Nina with most of areas have slower dry season. The impact in western Java is more affected compared to the eastern part. Western Java had the onset on 15th dasarian until more than 20th dasarian.

The use of satellite rainfall data TRMM explains that the onset of the rainy season on Java Island starts on the 26th dasarian, begin in Banten, then moves to east until comes to East Java in 32nd dasarian. The reverse movement occurs for the onset of the dry season, where the early dry season occurs in East Java on the 11th dasarian then moving westward [2].

The ENSO's effect difference on the season onset in the western and eastern part of Java can be shown in **Figure 2, 3** and **4**. The onset of the rainy season begins from the western area and then moves eastward, as shown in **Figure 4**, the onset value of the rainy season is lower for WJ compared to the EJ. However, in several years such as 2000/2001, 2001/2002 and 2007/2008, there was contrary value which caused by many areas in the west that were not detected in the season onset due to the influence of La Nina. First area that had the onset of the dry season is eastern Java, then moves westward (**Figure 3**), this is also shown by lower value of the EJ area in the season onset than WJ (**Figure 4**).

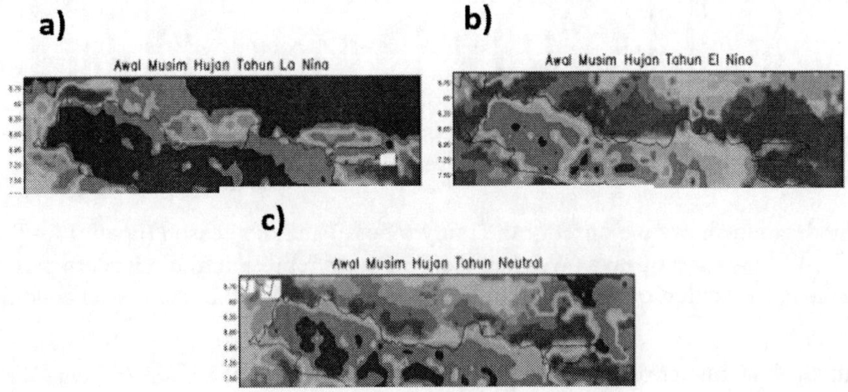

Figure 2. Spatial distribution of the rainy season onset (unit in dasarian) in Java based on CCAM downscaling data of 1990-2010 during (a) La Nina, (b) El Nino, and (c) neutral condition.

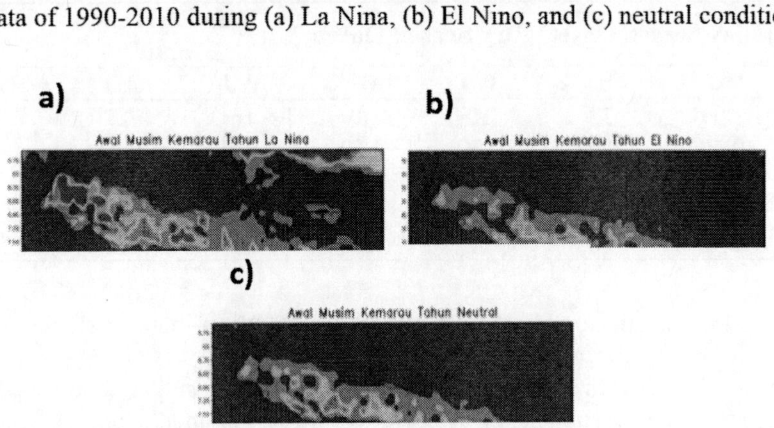

Figure 3. Spatial distribution of the dry season onset (unit in dasarian) in Java based on CCAM downscaling data of 1990-2010 during (a) La Nina, (b) El Nino, and (c) neutral condition.

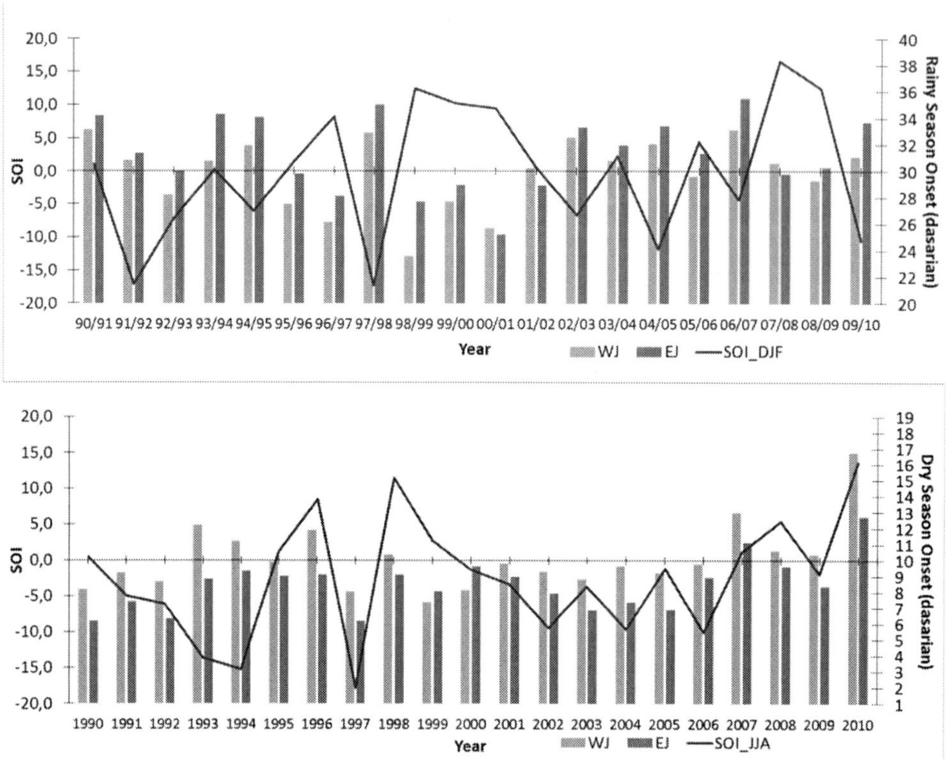

Figure 4. The diagram shows the onset of the rainy (above) and dry season (below) for the light areas represent to the Western part of Java (WJ) and the dark areas represent to the Eastern part of Java (EJ), whereas the line is the Southern Oscillation Index (SOI) value indicating the ENSO condition.

Table 1. Statistical parameter of ENSO to the onset and the length of season in Java (Western part of Java (WJ), Eastern part of Java (EJ)). All parameter have dasarian in unit, except the correlation. Correlation is relation between SOI and the season onset and Standard Deviation (Stdev) show the deviation of the season onset.

Statistic Parameter	Rainy Season Onset		Dry Season Onset		Rainy Season Length		Dry Season Length	
	WJ	EJ	WJ	EJ	WJ	EJ	WJ	EJ
Mean	30	32	10	8	16	13	20	23
Correlation	-0,40	-0,44	0,39	0,50	0,34	0,63	-0,70	-0,64
St dev	3	3	2	2	3	3	3	3
Min	24	25	7	6	12	9	13	16
Max	33	36	17	13	22	20	25	29

Table 1 shows the statistical values of the season onset in Java, both western and eastern during 1990- 2010 period. The average of rainy and dry season onset for WJ is on the 30th and 10th dasarian, therefore the average length of the rainy and dry season for WJ is 16 and 20 dasarians. Whereas for EJ, due to the onset of the rainy season that always comes more slowly and the onset of the dry season that comes faster than WJ, the WJ area has a shorter rainy season and longer dry season, which is 12 and 24 dasarians. The correlation between the season onset, both rain and drought and the value of SOI is also higher for EJ than WJ. Nevertheless, the WJ and EJ areas have the same in the deviation of the rainy season onset during 1990-2010 period, which is ± 3 dasarians, whereas the dry season onset at ± 2 dasarians. The rainy season onset for EJ ever reaches the fastest point to come in 25th dasarian and

for the dry season onset in 6th dasarian, while for the slowest comes in 36th dasarian for the rainy season onset and 13 for dry season. The western part (WJ) has the fastest of the rainy season onset on the 24th dasarian and the slowest in 33rd dasarian. While the onset of the dry season, the WJ region had the fastest coming in 7th dasarian and the most lately in 17th dasarian. This value is the onset of the dry season in 2010, which is the largest deviation from the average.

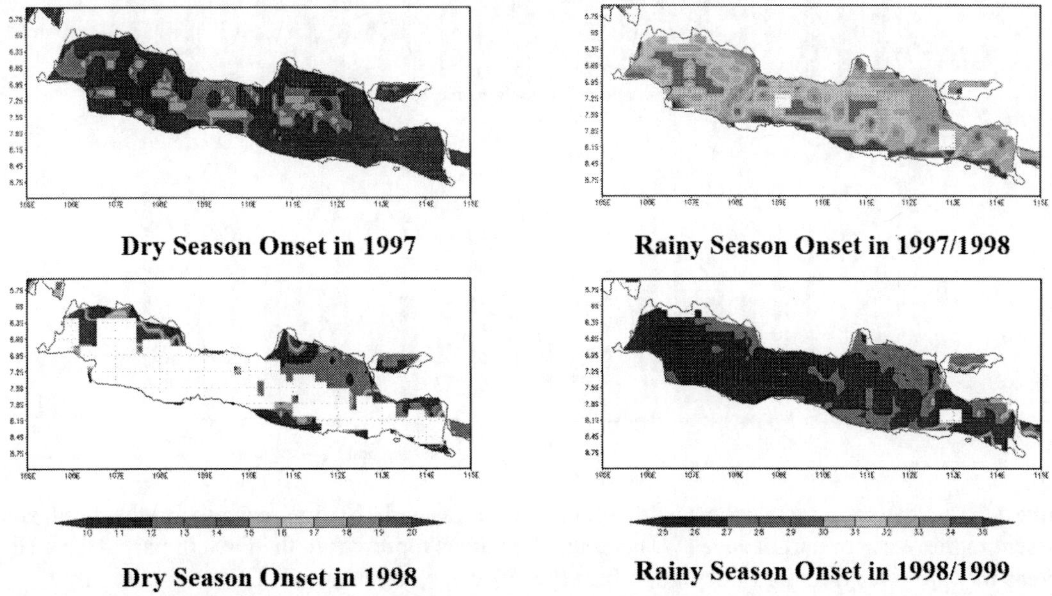

Figure 5. The onset of dry season and rainy season in case of the extreme ENSO's influence to the season onset in 1997 and 1998. White color shows undetectable the season onset.

The extreme El Niño events were depicted in 1997 and examples for the onset of the season while La Nina event were described in 1998. These two years are shown by **Figure 5** examples in studying the influence of ENSO on the season onset. While JJA period in 1997, the SOI value reached -17.8 and consistent at -17.3 until DJF 1997/98, it related to indicates that an extreme El Niño occurred. The SOI values in the middle of 1998 changed drastically to 11.4 and 12.5 in the end of the year, indicating the occurrence of La Nina. The onset of the dry season in 1997 came faster in almost all areas of Java, whereas the beginning of the rainy season was coming too late. In 1998, the onset of the dry season was almost undetectable because it tended to get wet from the beginning of the year, and the rainy season onset was detectable came faster. This shows the extreme El Nino effect that occurred in 1997 and extreme La Nina in 1998 toward the season onset in Java.

The ENSO was also influences the length of the season in Java, where shown by **Figure 6.** The length of the rainy season when El Nino has change becomes shorter, while the length of the dry season is longer than La Nina and the neutral conditions. However, when La Nina is the opposite, the rainy season becomes longer and the dry season becomes shorter. The rainy season at El Nino 1997 was only about 13 dasarians in WJ and 10 dasarians in EJ, whereas during La Nina 1998, the rainy season became longer at 20 and 16 dasarians for each WJ and EJ. The dry season in 1997 (El Nino) became longer, i.e. 29 for WJ and 25 for EJ, and when La Nina (1998) is shorter 7 dasarians for WJ and 4 dasarians for EJ compared to the average **(Table 1)**.

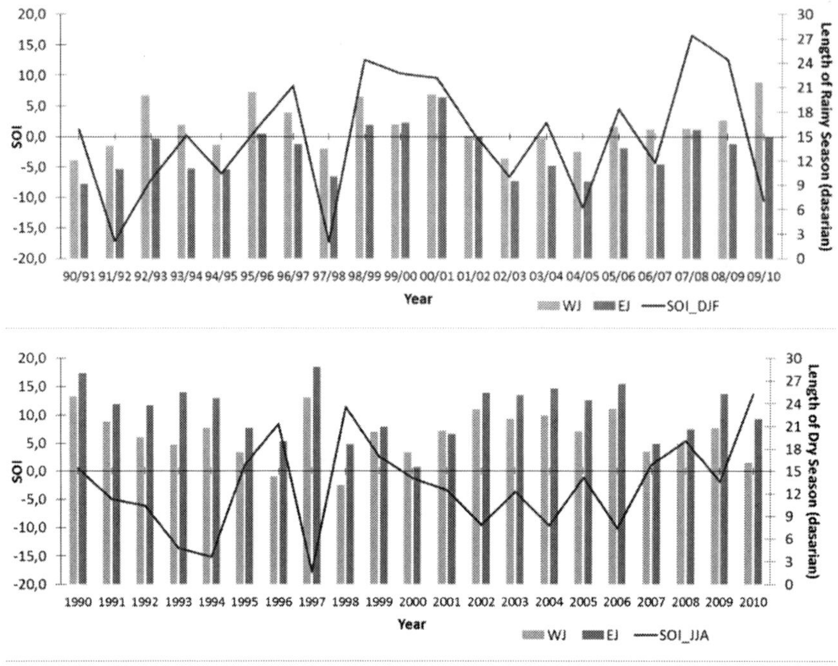

Figure 6. The diagram shows the length of the (a) rainy and (b) dry season, for the light areas represent to the Western part of Java (WJ) and the dark areas represent to the Eastern part of Java (EJ), whereas the line is the Southern Oscillation Index (SOI) value indicating the ENSO condition.

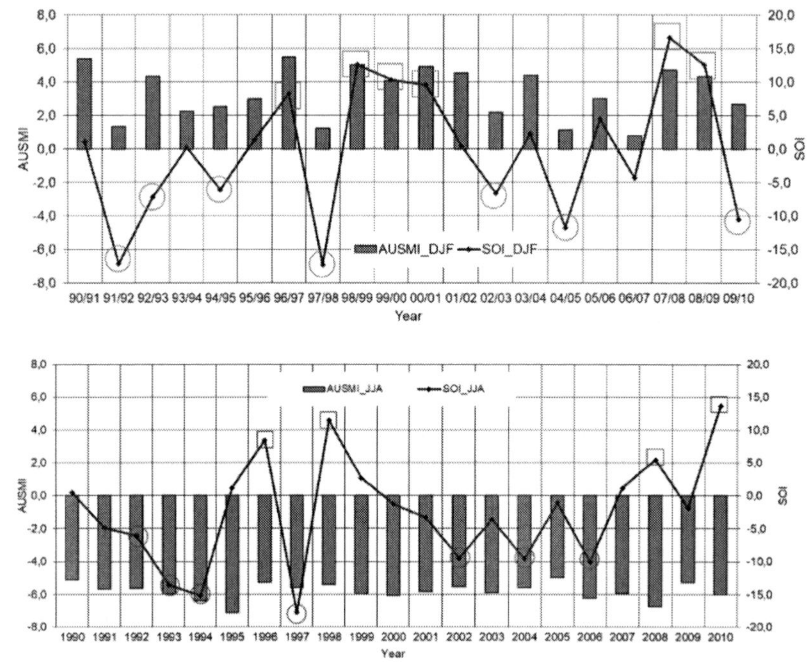

Figure 7. Australian Monsoon Index (AUSMI) and Southern Oscillation Index (SOI) values for a) DJF (December, January and February) and b) JJA (June, July and August), where the circle (○) signs mark El Nino condition and rectangle(□) signs mark La Nina.

The influence of ENSO on the early seasons in Java is due to the disrupted monsoon cycle. Westerlies that occurred during DJF obstructed by an El Nino, and vice versa will be faster during La Nina. This can be explained by the occurrence of a strong El Nino in 1998 resulted DJF westerlies become very weak (**Figure 7**). This condition also causes during DJF westerlies that blow over Java has a shorter duration during the El Nino and long duration when La Nina. In contrast to DJF, the easterly winds over JJA period have longer durations during El Nino, but shorter durations during La Nina, although the intensity is less affected by ENSO. **Figure 7b** explains that AUSMI variation is not affected by ENSO events. Despite the extreme El Nino occurred in 1997, it did not make more slowly the easterly winds.

4. Conclusion

The influence of ENSO during the rainy season *December-January-February (DJF)* and drought *June-July-August (JJA)* results in an early deviation of the rainy and dry seasons in WJ and EJ with different variations. For the DJF, at the time of La Nina occurs, the beginning of the rainy season occurs early between 3-4 dasarians in most WJ and a small part of EJ. While at El Nino, the beginning of the rainy season retreats about 2-3 dasarians in a small portion of WJ and EJ. For JJA, La Nina has postponed the start of the dasarian dry season in most of the WJ and EJ. While El Nino does not really have an impact on the acceleration of the dry season except for a small part of the area south of WJ and EJ. Thus it can be concluded that the beginning of the rainy and dry season in WJ and EJ experienced greater deviations when La Nina compared with El Nino. In addition, although WJ is locally farther from the Pacific Ocean than EJ, the results of this study prove that WJ is more sensitive in responding to ENSO compared to EJ. This is an interesting new finding to further explore the cause through subsequent research. Other new findings also prove that the powerful El Nino DJF 1998 and 2010 have affected the weakening of the western winds so that it impacts slow the beginning of the rainy season and shorten the duration of the rainy season. However, during the time of strong La Nina during the DJF 1999, 2000, 2008, there was no western wind. For the JJA period, winds appear stable from year to year and their variability is not compromised by the ENSO phenomenon.

Acknowledgments

The author would like to thank to Indonesian National Institute of Aeronautics and Space (LAPAN), especially to Center of Atmospheric Science and Technology (PSTA), Bandung for their support in the completion of this research.

5. References

[1] Hidayat, T. 2005. Analisis Perubahan Musim, Kekeringan dan Potensi Waktu Tanam Tanaman Pangan di Provinsi Banten. Tesis IPB. Bogor 2005.

[2] Avia, L.Q., A. Haryanto, N. Cholianawati dan B. Siswanto. 2010. Identifikasi Awal Musim Kemarau dan Musim Hujan Berdasarkan Data Satelit TRMM. Prosiding Seminar Penerbangan dan Antariksa 2010 Sub Seminar Sains Atmosfer dan Iklim. Serpong, 15 Nopember 2010.

[3] Suryantoro A., Krismianto dan Yulihastin E. 2010. Penentuan Onset Monsun Di Jawa Barat, Banten, Dan Dki Jakarta Berbasis Observasi Satelit TRMM. Prosiding Seminar Nasional Fisika 2010. ISBN: 978-979-98010-6-7, hal: 54-66.

[4] Satyawardhana, H., A. Susandi, B. Siswanto, A. S. Muttaqien. 2014. Analisis Curah Hujan Dasarian dan Validasinya Berbasis Data Downscaling CCAM (Cubical Conformal Atmospheric Model) Di Pulau Jawa. Prosiding Seminar Nasional Sains Atmosfer dan Antariksa (SNSAA) 2014. Lembaga Penerbangan dan Antariksa Nasional (LAPAN), Bandung 25 November 2014, hal 252 – 262.

[5] Satyawardhana, H. and A. Susandi. 2015. Proyeksi Awal Musim di Jawa Berbasis Hasil Downscaling Conformal Cubic Atmospheric Model (CCAM) (Season Onset Projection In Java Based on CCAM Downscaling Output), Jurnal Sains Dirgantara **13 No. 1** p :1—14

[6] Webster P. J., V. O. Magaña, T. N. Palmer, J. Shukla, R. A. Tomas, M. Yanai and T. Yasunari (1998): Monsoons: processes, predictability, and the prospects for prediction. Journal of Geophysical Research 103(C7) **14** 451–14 510.

[7] Li, C., and M. Yanai. 1996. The onset and interannual variability of the Asian summer monsoon in relation to land–sea thermal contrast. Journal Climate **9** p: 358–375.

[8] Hung, C.-W., X. Liu, and M. Yanai. 2004: Symmetry and asymmetry of the Asian and Australian summer monsoons. Journal of Climate **17** P: 2413–2426.

[9] Chang, C. P., Z. Wang, J. McBride, and C.-H. Liu. 2005: Annual cycle of Southeast Asia–Maritime Continent Rainfall and Asymmetric Monsoon Transition. Journal of Climate **18** p: 287–301.

[10] Aldrian E., Gates, L.D. and Widodo, F.H. 2007: Seasonal variability of Indonesian rainfall in ECHAM4 simulations and in the reanalyses: The role of ENSO, Theoretical and Applied Climatology **87** p:41–59.

[11] Qian, J. H., A.W. Robertson and V. Moron. 2010: Interactions among ENSO, the Monsoon, and Diurnal Cycle in Rainfall Variability over Java, Indonesia. Journal of Atmospheric Science **67** p:3509-3524.

[12] Badan Meteorologi Klimatologi dan Geofisika. 2009: Prakiraan Musim Hujan 2009/2010 di Indonesia.

[13] Syaifullah, D. 2010. Kajian Sea Surface Temperature (SST), Southern Oscillation Index (SOI) dan Dipole Mode Pada Kegiatan Penerapan Teknologi Modifikasi Cuaca di Propinsi Riau dan Sumatera Barat Juli – Agustus 2009, Jurnal Sains & Teknologi Modifikasi Cuaca, Vol. 11, No. 1, 2010: 1-7

[14] Kajikawa, Y., B. Wang and J. Yang. 2010. A multi-time scale Australian monsoon index, Int. J. Climatol, doi: 10.1002/joc.1955.

Analysis of Rainfall Data based on GSMaP and TRMM towards Observations Data in Yogyakarta

I Sofiati[1] and **L. Q Avia**[1]

[1]Indonesian National Institute of Aeronautics and Space (LAPAN)
Jln. Dr. Djundjunan No. 133, Bandung 40173, Indonesia

E-mail: sofiati07@gmail.com

Abstract. In reality, to obtain a complete and representative rainfall observations data in a particular area is quite difficult. Therefore, to complement the existence of rainfall data, satellite data is used, so the rainfall data can got by us for a region that has limited equipment, then the characteristics of rainfall can knew by us in that region. The purpose of this study is to analyzed the validity of GSMaP (Global Satellite Mapping of Precipitation) and TRMM (Tropical Rainfall Measuring Mission) data with in-situ rainfall data in Yogyakarta during the period (2012-2016). The method used is to process GSMaP and TRMM data into daily, decade, monthly, and seasonal data to compare with in situ data and perform statistical tests. Based on the results it can be seen that during the observation period for DJF (December-January-February), the highest GSMaP and TRMM rainfall correlations occur in decade-1 and decade-3 respectively. In MAM (March-April-May), GSMaP (decade-2) and TRMM (decade-1), while in JJA (June-July-August) for GSMaP (decade-2) and TRMM (decade-3), and in SON (September-October-November) for GSMaP (decade-3) and TRMM (decade-2). Therefore it can be concluded that the correlation between GSMaP and TRMM with in situ data, the best occurs on average in decade-2 and decade-3.

1. Introduction

To obtain a complete and representative rainfall observation data are quite difficult, because of the their variation and incomplete data availability. To overcome this problem, one of them used satellite data that has high spatial and temporal resolution such as GSMaP (Global Satellite Mapping of Precipitation) and TRMM (Tropical Rainfall Measuring Mission), therefore satellite technology can obtain rainfall data for a region with limited equipment and can study the rainfall characteristics in that area [1].

Notable amount of researches using GSMaP and TRMM data were previously conducted [2-6]. Among them, validation research and comparison between satellites for rainfall estimates in Colombia [7] which resulted in good enough validation for daily rainfall in the area. Best results were obtained in the eastern plains of Colombia, and the product performance was relatively low on the Pacific coast. In comparing the various satellites, it appears that PERSIANN and GSMaP-MVK (GSMaP Moving Vector with Kalman filter) show poor performances. While CMORPH (the Climate Precipitation Center Morphing) and GSMaP-MVK + show the best performance among the products evaluated in the study here. Then a study of comparative evaluation of GSMaP rainfall data well with other satellite data, modeling and observational data conducted in America [8], which concludes that overall, GSMaP performance is better compared to other four satellite-based data, with slightly better probability during the season hot compared during winter. But the study in the African river basin [9] compares six satellite performance and in general the performance of GPROF 6.0 and GSMaP-MKV is not very accurate, and RFE 2.0 and TRMM_3B42 are accurate. In the East African region [10], which states that the convective stability of the atmosphere above East Africa is controlled by the relatively cold airflow of the Indian Ocean off the coast. During the rainy season, sea surface temperature (SST) increases (and hottest in the rainy season). As a result, the air flowing into East Africa becomes less stable, and this analysis can be used to help understand the overwhelmingly tropical rainfall of East Africa commonly found in models. Then to conduct flood evaluation that occurred in Kyushu-Japan [11], data model (Generalized Additive Models) for correction bias GSMaP_MVK data was used. The results of this study show that the GSMaP data bias correction is accurate enough with a bias of more than 50%. Furthermore, the TRMM data evaluation study was conducted in Nepal [12]' stating that the use of TRMM data could be to determine the source of water. Then research in China [13] uses data from CMORPH compared with GSMaP that can show variability in diurnal rainfall.

To complement the existence of representative rainfall data, in this research rainfall data from satellite GSMaP and TRMM_3B42 were used and validated further using in situ data. This research are expected to obtain rainfall data for the area without accessible tools and can study rainfall characteristic in a region, especially Yogyakarta and its surroundings.

2. Materials and Methods

Data used in this research are rainfall data generated by GSMaP_NRT, TRMM, (see Table 1), and surface observation data for Yogyakarta area (7.8°S, 110.21°E) in 2012-2016. TRMM data used in the below table is level-3 (3B42RT) which are downloaded from FTP facility [14], with a spatial resolution of 0.25°x 0.25° and temporal resolution every 3 hours. In addition, JAXA (Japan Aerospace Exploration Agency) has also produced TRMM data with better format and quality [15, 16]. As a prototype of GPM (Global Precipitation Measurement), JAXA has developed a near-real-time data processing system and deploys it through the internet site.

Table 1. Data characteristic of GSMaP_NRT and TRMM_ 3B42RT taken from FTP facility

Parameter	GSMaP_NRT	3B42_RT
Product	JAXA	NASA
Spatial Res	0,10° Lat/Lon	0,25° Lat/Lon
Temporal Res	1 hour	3 hours

The algorithm developed based on the project of GSMaP so it has known as GSMaP_NRT (GSMaP near real time). Table 1 shows the characteristic tables of both types of TRMM data from both NASA (3B42RT) and from JAXA (GSMaP_NRT). Viewed from spatial or temporal resolution then GSMaP_NRT is better than 3B42RT. The next analysis is to analyze the correlation between GSMaP, TRMM results, and in situ data with statistical calculations using linear regression equations.

3. Results and discussion

The results of this study support the previous research results expressed by [1, 7, 8] which analyzes two types of NASA TRMM data types (3B42RT) and TRMM Jaxa (GSMaP_NRT) for watersheds. From those studies it is shown that the Jaxa TRMM is closer to observational data than the NASA TRMM. Meanwhile, from this research the general analysis results show that the GSMaP rainfall value has a pattern that follows the in situ rainfall compared to the TRMM although the value tends to be below the estimates, and the results of this study also support the previous research result [9, 11, 13].

The next step of the research we analyzed the temporal distribution of the average daily rainfall that displayed on a seasonal basis. Based on the results it can be seen that in DJF month maximum daily rainfall obtained from GSMaP_NRT data is 46,64 mm, TRMM of 0,2 mm and in situ is 52 mm, as shown in Figure 2 (left). While Figure 2 (right) shows the average daily rainfall seasonally for the MAM month, where the maximum daily rainfall of GSMaP_NRT data is 29.98 mm, TRMM of 4.95 mm and in situ of 18 mm.

Figure 1 below shows the average seasonal spatial distribution of rainfall from GSMaP_NRT (a) and TRMM_3B42RT (b) for December-January-February (DJF) in 2012-2016 in Yogyakarta and surrounding areas. Based on the results seen that rainfall generated from GSMaP reaches 350 mm, then for TRMM only reaches 50 mm, while in situ data is 341.6 mm, thus rainfall generated from GSMaP is closer to in situ compared to TRMM.

To complete the results of this study, analyzed rainfall in MAM, where the spatial distribution of seasonal average rainfall from GSMaP reaches 250 mm, then for TRMM only reaches 100 mm, while from in situ data for 211.2 mm. Meanwhile for JJA, from GSMaP about 50 mm and for TRMM reach 150 mm, while from in situ data equal to 68,5 mm. For SON, from GSMaP is 200 mm, then for TRMM reach 50 mm, while from in situ data is 232,5 mm. Thus based on these results it is seen that the rainfall generated from GSMaP is closer to in situ data than that generated from TRMM.

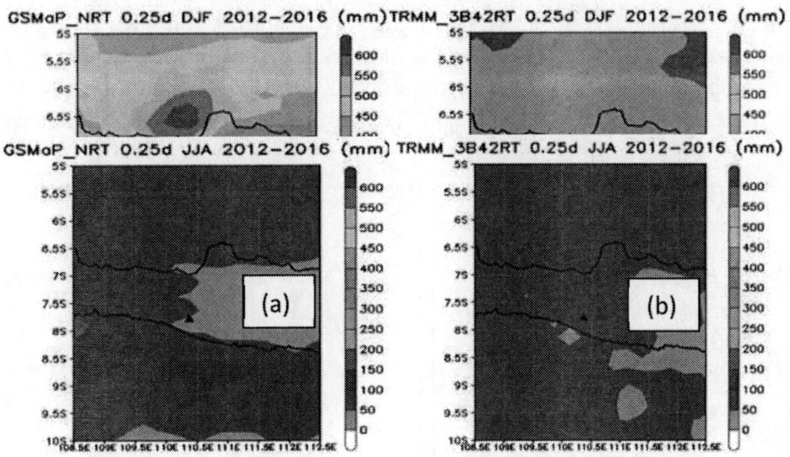

Figure 1. Spatial distribution of seasonal rainfall from *GSMaP* (a) and *TRMM* (b) for *DJF* and *JJA* for 2012-2016 at Yogyakarta. Rainfall with the smallest intensity is marked in blue, and the largest with red.

From the daily average pattern, the distribution of the GSMaP pattern is more similar to the in situ pattern, and it appears that the average daily appearance of the GSMaP pattern is one day earlier than in situ, whereas for TRMM the pattern distribution does not look similar to the in situ pattern, and the intensity always smaller. Meanwhile the daily rainfall and intensity patterns of GSMaP are closer to the in situ rainfall pattern compared to the TRMM.

Other results were analyze for JJA and SON, and it based on the results showed that in the month of precipitation the maximum daily average of GSMaP was 17.69 mm, TRMM 0.5 mm, and in situ

0.25 mm. For the month of maximum daily average daily rainfall of GSMaP of 19.65 mm, TRMM 0.15 mm, and in situ 3.3 mm. Overall, it shows that the magnitude of the GSMaP rainfall is closer to the in situ rainfall value than the TRMM that tends to be below the in situ value.

For comparison, another result mentioned by [17]. **Figure 4** shows monthly mean precipitation averaged over 10 years (December 2000 to November 2010) in Singapore. These studies confirmed that The satellite-borne precipitation can generally capture a temporal variation of rain gauge data in monthly time scale, although GSMaP has discrepancies in winter time. Figure 4b presents long-term annual and seasonal mean precipitation derived from satellite-derived and ground-based rainfall. Observation shows no distinct seasons but relatively low rainfall in summer season. Satellite data are capable of estimating the seasonal- and annual-basis precipitation features appeared in station observation. Overall, satellite-derived precipitation agrees well against the gauge precipitation at coarser time scale.

In further analysis, the first ten days in a month was defined as decade-1, the second ten days as decade-2 and the third ten days as decade-3 and comparing GSMaP and TRMM data with in situ data through statistical analysis as shown in **Figure 5**. GSMaP and TRMM correlations with in situ for decade-1 on DJF of 0.57 and 0.22 respectively.

Another result of GSMaP and TRMM data correlation with in situ data as can be seen in **Figure 7** for GSMaP (left) and TRMM (right). The GSMaP and TRMM correlations with in situ for decade-3 on DJF is 0.57 and 0.22 respectively.

The statistically correlated analyzes were processed for each decade-1, decade-2, and decade-3 for each season (DJF, MAM, JJA, and SON). From ten days rainfall analysis it has seen that during the observation period for DJF, the highest correlation between GSMaP and TRMM with in situ occurs in decade-1 and decade-3 respectively. In MAM, GSMaP (decade-2) and TRMM (decade-1), while in JJA GSMaP (decade-2) and TRMM (decade-3), and on SON GSMaP (decade-3) and TRMM (decade -2). Therefore it can be concluded that the correlation between GSMaP and TRMM with in situ data, the best occurs on average is in decade-2 and decade-3 and the complete results can be seen in **Table 2**.

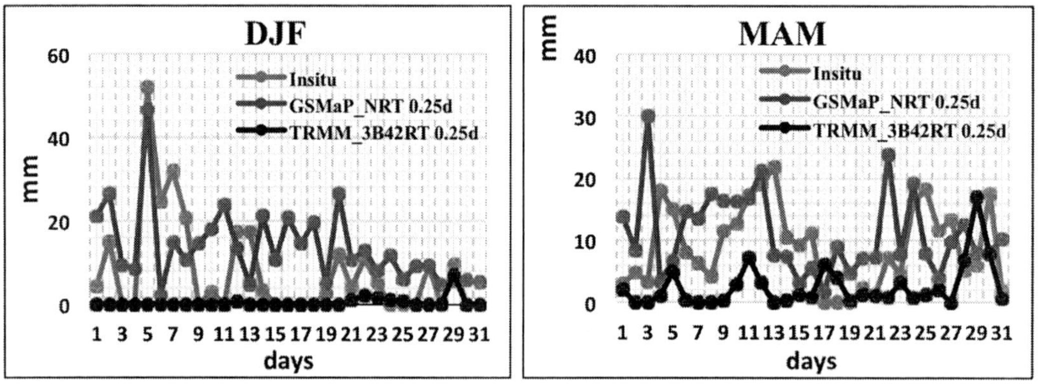

Figure 2. Temporal distribution of average seasonal rainfall from GSMaP, TRMM, and in situ for DJF and MAM for 2012-2016 at Yogyakarta.

IOP Conf. Series: Earth and Environmental Science **166** (2018) 012031 doi:10.1088/1755-1315/166/1/012031

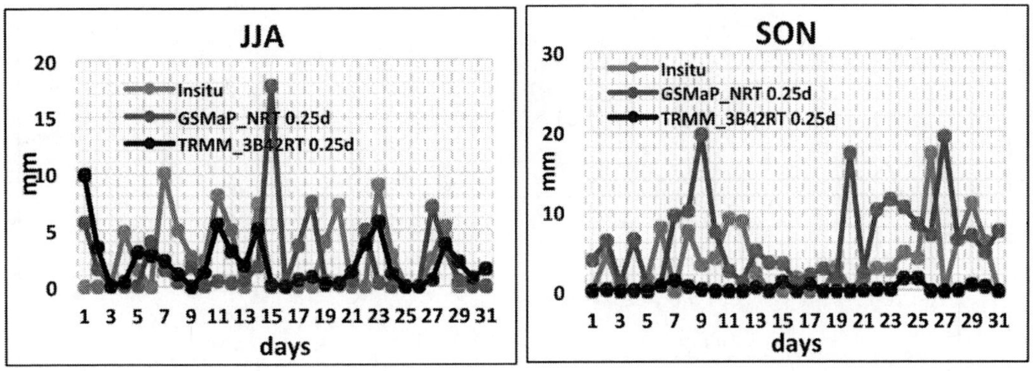

Figure 3. Temporal distribution of average seasonal rainfall from GSMaP, TRMM, and in situ for JJA and SON for 2012-2016 at Yogyakarta.

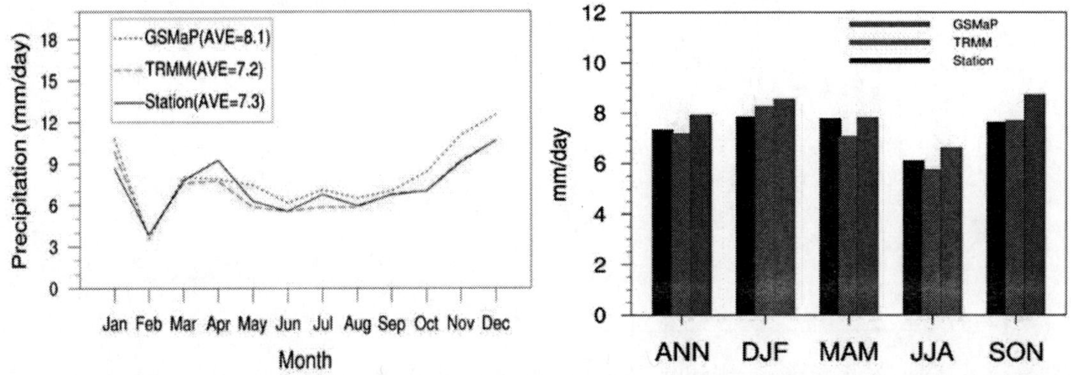

Figure 4. Long-term monthly mean precipitation (mm/day) derived from station (black solid line), TRMM TMAP-3B42 (red line) and GSMaP (blue dotted line) in Singapore (a). Long-term annual and monthly mean precipitation (mm/day) derived from station (black), TRMM TMAP-3B42 (red) and GSMaP (blue) (b). Note : ANN is annual, DJF is December-January-February, MAM is March-April-May, JJA is June-July-August, and SON is September-October-November.

The GSMaP and TRMM data correlation with in situ data as can be seen **in Figure 6** for GSMaP (left) and TRMM (right). The GSMaP and TRMM correlations with in situ for Decade-2 on DJF is 0.25 and 0.56 respectively.

IOP Conf. Series: Earth and Environmental Science **166** (2018) 012031 doi:10.1088/1755-1315/166/1/012031

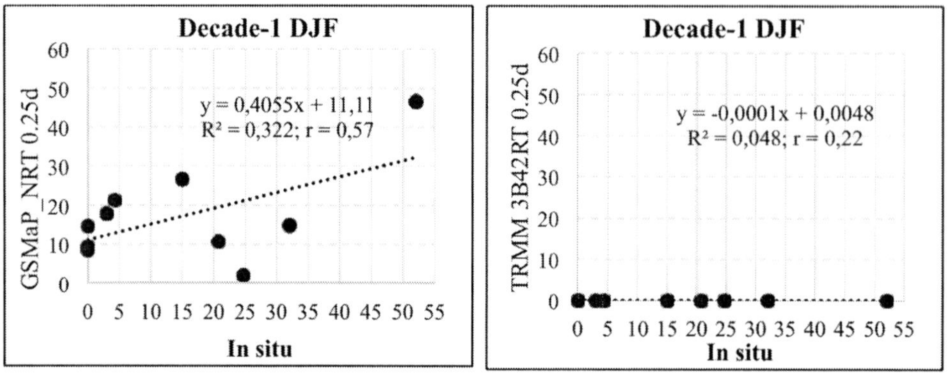

Figure 5. Distribution of correlations for decade-1 on DJF between GSMaP and in situ (left) and TRMM (right) data for 2012-2016 at Yogyakarta.

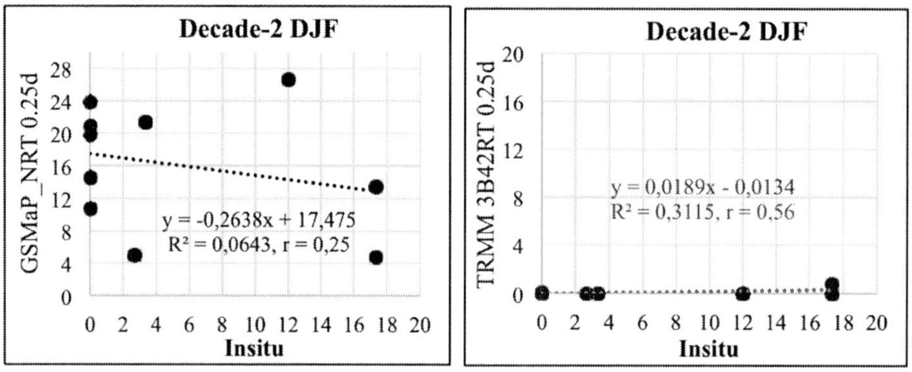

Figure 6. Distribution of correlations for decade-2 on DJF between GSMaP and in situ (left) and TRMM (right) data for 2012-2016 at Yogyakarta.

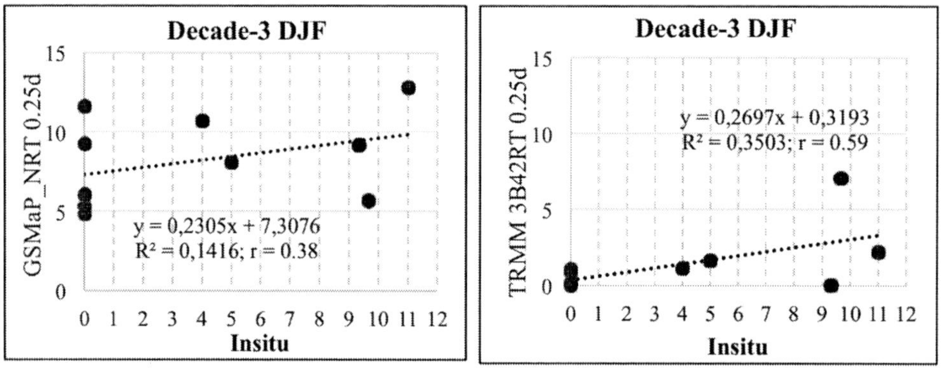

Figure 7. Distribution of correlation for decade-3 on DJF between GSMaP and in situ data (left) and TRMM (right) for 2012-2016 at Yogyakarta.

Tabel 2. Correlation coeffisien of rainfall between GSMaP_NRT, TRMM_3B42RT with in situ data for Decade-1, Decade-2, Decade-3, in 2012-2016 at Yogyakarta.

	DJF		MAM		JJA		SON	
	GSMaP	TRMM	GSMaP	TRMM	GSMaP	TRMM	GSMaP	TRMM
Decade-1	0.57	0.22	0.60	0.53	0.32	0.31	0.10	0.16
Decade-2	0.25	0.56	0.60	0.06	0.46	0.67	0.32	0.37
Decade-3	0.37	0.59	0.03	0.11	0.11	0.72	0.33	0.06

4. Conclusions

From daily rainfall analysis results for each season it was seen that the rainfall generated from GSMaP_NRT is closer to in situ data than that generated from TRMM_3B43RT. While from data analysis of ten days rainfall seen that during observation period for DJF, GSMaP and TRMM highest correlation with in situ data occurs in decade-2 and decade-3, ie 0,56 and 0,59 respectively. On MAM, GSMaP (decade-2) is 0,60 and TRMM (decade-1) is 0,53, while on JJA GSMaP (decade-2) is 0,46 and TRMM (decade-3) 0,72, and in SON GSMaP (decade-3) is 0,33 and TRMM (decade -2) is 0,32. Therefore it can be concluded that the correlation between GSMaP and TRMM with in situ data, the best occurred on average in decade-1 and decade-2.

Acknowledgement

The author would like to thank Miss Rachmawati Syahdiza S. Si, alumnus of Faculty of Earth Science and Technology, Department of Earth Science, Bandung Institute of Technology (ITB) for her help.

5. References

[1] Syaifullah, M D., 2014. Validecadei Data TRMM terhadap Data Curah Hujan Aktual di Tiga DECADE di Indonesia. *Jurnal Meteorologi dan Geofisika* .**Vol. 15.** No. 2: 109-118.

[2] Shrestha M S, Takara K, Kubota T, and Bajracharya S R. 2011. Verification of GSMaP Rainfall Estimates over the Central Himalayas. *Journal of Japan Society of Civil Engineers*, Ser. B1 (Hydraulic Engineering), **Vol. 67**, No. 4, 37-42.

[3] Chen Z, Qin Y. Shen Y, and Zhang S. 2016. Evaluation of Global Satellite Mapping of Precipitation Project Daily Precipitation Estimates over the Chinese Mainland. *Advances in Meteorology*. **Vol. 2016 (2016)**, Article ID 9365294, 15 pages. http://dx.doi.org/10.1155/2016/9365294.

[4] Ning S, Song F, Udmale P, Jin J, RajThapa B, and Hiroshi I. Error Analysis and Evaluation of the Latest GSMaP and IMERG Precipitation Products over Eastern China. Research Articl. *Advances in Meteorology*. **Vol. 2017 (2017)**, Article ID 1803492, 16 pages. https://doi.org/10.1155/2017/1803492.

[5] Mou Leong Tan M L, and Zheng Duan. 2017. Assessment of GPM and TRMM Precipitation Products over Singapore. *Remote Sens*. **Vol. 9**. 720. doi:10.3390/rs9070720.

[6] Sugiartha N, Ogawara K, Tanaka Tc, Mahendra M S. 2017. Article Application of GSMaP Product and Rain Gauge Data for Monitoring Rainfall Condition of Flood Events in Indonesia. *Int. J. Environ. Geosci*. **Vol 1(1)**: 36-47.

[7] Dinku T., Ruiz FYN., Connor SJ., Ceccato P., 2010. Validation and Intercomparison of Satellite Rainfall Estimates over Colombia. *Journal of Applied Meteorology and Climatology*. **Vol 49**. 1004-1014. doi: 10.1175/2009JAMC2260.1.

[8] Tian Y., Lidard CDP., Adler RF., Kubota T., 2010. Evaluation of GSMaP Precipitation Estimates over the Contiguous United States. *Journal of Hydrometeorology*. **Vol 11**. 566-574. doi: 10. 1175/2009JHM1190.1.

[9] Thiemig V., Rojas R. Bigiarini MZ., Levizzani V., Roo AD., 2012. Validation of Satellite-Based Precipitation Products over Sparsely Gauged African River Basins. *Journal of Hydrometeorology*. **Vol 13**. 1760-1783. doi: 10.1175/JHM-D-12-032.1.

[10] Cattani E., Merino A., Levizzani V., 2016. Evaluation of Monthly Satellite-Derived Precipitation Products over East Africa. *Journal of Hydrometeorology*. **Vol. 17**. 2555- 2573. doi: 10.1175/JHM-D-15-0042.1.

[11] Setiawati MD., Miura F., 2016., Evaluation of GSMaP Daily Rainfall Satellite Data for Flood Monitoring: Case Study—Kyushu Japan. *Journal of Geoscience and Environment Protection*, 2016, **Vol. 4**, 101-117. doi: 10.4236/gep.2016.412008.

[12] Krakauer NY., Pradhanang SM., Lakhankar T., Ajay K. Jha AK., 2013. Evaluating Satellite Products for Precipitation Estimation in Mountain Regions: A Case Study for Nepal. *J. Remote Sens.* **Vol 5**, 4107-4123; doi:10.3390/rs5084107.

[13] Shen, Y., P. Zhao, Y. Pan, and J. Yu, 2014. A high spatio temporal gauge-satellite merged precipitation analysis over China, *J. Geophys. Res. Atmos.*, **Vol 119**, 3063–3075, doi:10.1002/2013JD020686.

[14] National Aeronautics and Space Administration (NASA) TRMM Background, (http://trmm.gsfc.nasa.gov/overview_dir/background.html), [Accessed February 2017].

[15] National Space Development Agency of Japan (NASDA), Earth Observation Center, 2001. TRMM Data Users Hand book (*http://www.eorc.jaxa.jp/TRMM/document/text/handbook_e.pdf*), [Accessed February 2017].

[16] Japan Aerospace Exploration Agency Earth Observation Research Center, 2013.*User's Guide for global Rainfall Map in Near-Real-Time by JAXA Global Rainfall Watch (GSMaP_NRT)*, Ver 2.4.

[17] Hur J, Raghavan S V, Nguyen N S, Liong S Y. 2016. Evaluation of High-resolution Satellite Rainfall Data over Singapore. *Procedia Engineering*. **Vol 154**, Pages 158-167.

IOP Conf. Series: Earth and Environmental Science **166** (2018) 012032 doi:10.1088/1755-1315/166/1/012032

Observation of Temperature Profiles of Kototabang (Bukittinggi-West Sumatera Indonesia) Using Radiosonde and Rain Event Identification

I Juaeni[1], G. A Nugroho[1], S. A Rahayu[1], Safrijon[2], Halimurrahman[1]

[1]Center of Atmospheric Science and Technology
Indonesian National Institute of Aeronautics and Space (LAPAN)
Jln. Dr. Djundjunan No. 133, Bandung 40173, Indonesia

[2]Atmosphere and Space Observation Station Agam District, National Institute of Aeronautics and Space (LAPAN), Kototabang, West Sumatera, Indonesia

E-mail: inajuaeni@yahoo.com

Abstract. We recorded temperature profiles of Kototabang (West Sumatera- Indonesia) in a four days observation using Radio Acoustic Sounding System (RASS) and Radiosonde. The activity was carried out in August 2016, on a research cooperation framework of National Institute of Aeronautics and Space (LAPAN), Indonesia and Research Institute for Sustainable Humanosphere (RISH), Kyoto University, Japan. The current article discussed the temperature profile based on 12 times launch of Radiosonde, Convective Available Potential Energy (CAPE) estimation and its relationship with rain events, while the temperature profile based on RASS was discussed in another report. CAPE is the amount of energy an air parcel would have if lifted to a certain distance vertically through the atmosphere. CAPE is effectively the positive buoyancy of an air parcel and is an indicator of atmospheric instability, related to convection clouds. If there is not enough water vapour present, there is no ability for condensation, and therefore, storms, clouds, and rain will not form. CAPE is calculated as the area of positive region in the thermodynamic diagram or through relationships with virtual temperature. On August 30, 2016, at 08 LT, the range of temperatures based on Radiosonde measurements was recorded from 294.9°K at ground level to 186.5° K at an altitude of 26.8 km. The temperature range varied with time. CAPE based on Radiosonde data ranged from 0 to 5075.2 Jkg^{-1} within 30 August 2016 until 2 September 2016. The maximum CAPE value occurred on August 30, 2016 at 15.34 LT. Large CAPE (> 2000 Jkg^{-1}) was coincident with rain events.

1. Introduction

Climate and weather observations are essential for climate change policies and the development of weather services to improve the quality of public life. Progress and reliability in upper air measurements are important for the improvement and the validation of climate model and for numerical weather prediction (NWP).

Some challenges emerge and evolve dynamically, and up to now, an accurate solution regarding those issues is yet to be found. The issues are derived from variable atmosphere of upper air measurements, i.e. climate change issues based on temperature trends, which is important to assess each of atmospheric level, especially the troposphere, the stratosphere, the changes of temperature trends [1-6], troposphere-stratosphere exchange [7-8], and dehydration of the stratosphere [6]. Both single observations with each instrument and a campaign involving many institutions and instruments have been employed in order to answers the problems. Not only research-based observation data, but also the radiosonde data is necessary as the input data and as the main part of the validation of NWP model [4, 9-13].

Recent developments of the technology for conventional (in situ) observation have led to significant improvement in measurement, processing, and transmission while at the same time, it has been able to reduce the expenditure and running cost of ground-based upper air network. Improvements in numerical weather and climate models need observations of atmospheric variables, such as clouds, winds, temperature, and humidity at the increasingly high resolution in space and time. Measurement of the variability of observed quantity at the models' resolved scales allows an optimal assimilation of the information by giving an estimation of the representative error. Interaction of different scales from global to convective scale is important in understanding atmospheric and climate physical process. Information of these interactions needs the data that include high spatial and time resolution from many regions. The future trend of weather forecasting and climate modelling is to optimize operational cost by mixing some existing in situ measurements (balloon-borne radiosondes, aircraft) with integrated ground-based remote sensing systems that provide a continuous sampling of the essential atmospheric variables in complement to satellite observation.

The current paper study discussed the minor part of the afore mentioned atmospheric physics research, through the temperature profiles at Kototabang (West Java-Sumatra, Indonesia) from Vaisala Radiosonde RS-41SG, as Radiosonde new generation has to improve on a level of in-situ observation accuracy and quality. This paper provide additional information on the latest temperature profile in the tropics and its relationship with Radiosonde indices. This research is was carried out in close collaboration between National Institute of Aeronautics and Space (LAPAN), Indonesia and Indonesia and Research Institute for Sustainable Humanosphere (RISH), Kyoto University, -Japan. This paper discusses the temperature profile based on 12 times Radiosonde launch and its relationship with some radiosonde indices, while the temperature profile based on Radio Acoustic Sounding System (RASS) is discussed in another report (in the process of publication).

2. Methods

The temperature profile at Equatorial Atmosphere Radar (EAR) site (Kototabang, West Sumatera, Indonesia) obtained from VAISALA/MW41 Sounding System, RS41-SG, S/N:M2720118 (**Figure 1**) was launched three times a day on the experiments period of August 30th, 2016 to September 2nd, 2016. The surface meteorology information was required for every launch, i.e pressure, temperature, humidity, wind direction, wind speed, latitude, longitude, and altitude of the site. EAR site positional information is shown in **Table 1**.

Table 1. Station position

	Longitude	Latitude	Altitude
EAR site at Koto Tabang (West Sumatera-Indonesia)	0.20°S	100.32°E	857.46 m

Radiosonde provides the profile of pressure, temperature, humidity, as well as wind direction. There are some indices that are evaluated from Radiosonde data. Some of the indices used in this paper are Convective Available Potential Energy (CAPE), a height of Equilibrium Level (EL), a pressure of EL and temperature of EL. CAPE is the amount of energy an air parcel would have if lifted a certain distance vertically through the atmosphere. CAPE is effectively the positive buoyancy of an air parcel and is an indicator of atmospheric instability related to convection clouds. In case there is not enough water vapour present, there is no ability for condensation, storms, clouds, and rain. CAPE is calculated as the positive area in the thermodynamic diagram or through relationships with virtual temperature as in equation (1).

$$CAPE = \int_{z_f}^{z_n} g\left(\frac{T_{v,parcel} - T_{v,env.}}{T_{v,env.}} \right) dz \tag{1}$$

Where:

- z_f is the height of the level of free convection,
- z_n is the height of the equilibrium level (neutral buoyancy),
- $T_{v,parcel}$ is the virtual temperature of the specific parcel,
- $T_{v,env.}$ is the virtual temperature of the environment,
- g is the acceleration due to gravity.

Rainfall data from METEK Micro Rain Radar (MRR-2) and satellite images from HIMAWARI-8, IR-1 channel (http://weather.is.kochi-u.ac.jp/sat/gms.sea/) are used to expand the analysis. Statistical evaluation includes mean, standard deviation, and coefficient of correlation for 12 of data profile is used for common equation. Temperature deviation and coefficient of correlation are shown in equation (2) and equation (3), respectively.

$$\sigma T = T - \overline{T} \tag{2}$$

where:

- σT= temperature deviation
- T= temperature at certain time
- $\overline{T}$ = mean temperature

$$r_{xy} = \frac{\sum_{i=1}^{n}(x_i - \overline{x})(y_i - \overline{y})}{\sqrt{\sum_{i=1}^{n}(x_i - \overline{x})^2 \sum_{i=1}^{n}(y_i - \overline{y})^2}} \tag{3}$$

where:

- Σ is sigma, the symbol for sum up
- $x_i - \overline{x}$ is each x-value minus the mean of x
- $y_i - \overline{y}$ is each y-value minus the mean of y

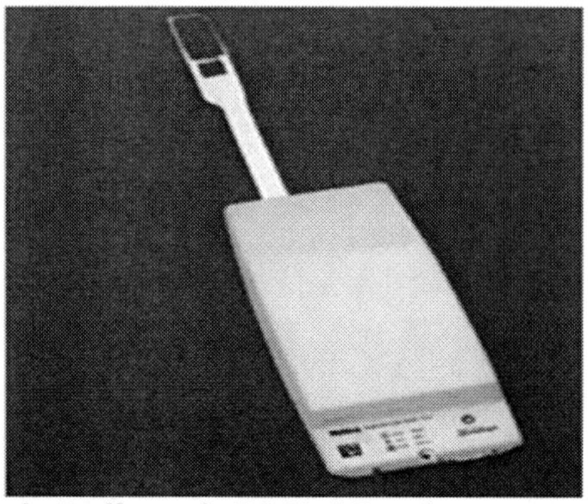

Figure 1. Vaisala Radiosonde RS41-SG and temperature sensor [14]

3. Results

3.1 Diurnal variation and mean temperature

On August 30th, 2016, at 08 LT, temperatures based on Radiosonde measurements ranged from 294.9K at ground level to 186.5 K at an altitude of 26.8 km. The ranges of temperature for all periods are shown in **Table 2**. The Cold Point Tropopause (CPT) maximum occured at evening/night, as shown in column 4 of **Table 2**. The value is quite similar with that was reported by Bhatnagar et al. [8]. CPT fluctuations do not show the trend of heating or cooling, due to limited data. The trend of CPT is important because at troposphere, it could provide an impact on stratosphere state. A positive trend has been found in stratospheric water vapour [15-16]. It was proposed that a warming trend was about 1° K/decade in tropical tropopause temperatures may explain the positive trend in stratospheric water vapour.

Table 2. Temperature range, maximum height and estimated height of tropopause on Radiosonde launch period

Launch time	Temperature range (°K)	Maximum height (km)	Cold point tropopause (CPT) height (km)
August 30th 2016, 08 LT	186.5-294.9	26.8	17.1
August 30th 2016, 11 LT	186.5-299.6	25.9	17.0
August 30th 2016, 15 LT	187.7-297.7	25.6	17.3
August 31th 2016, 07 LT	187.9-294.6	26.1	17.0
August 31th 2016, 11 LT	189.2-299.1	26.9	16.8
August 31th 2016, 19 LT	190.8-294.9	24.6	17.4
September 1st 2016, 07 LT	188.9-293.9	26.6	16.6
September 1st 2016, 10 LT	191.3-298.8	24.3	17.0
September 1st 2016, 15 LT	190.1-294.0	18.3	16.9
September 2nd 2016, 07 LT	190.1-293.6	25.5	17.0
September 2nd 2016, 11 LT	190.1-298.5	24.2	16.6
September 2nd 2016, 18 LT	190.5-294.2	20.7	17.0

On August 30th 2016, small variations (diurnal) occurred from the ground up to 17 km, approximately. Above the layer, there was a relatively large variation. The same feature was showed by temperature data on August 31st, but time variation over 17 km was much greater than that on August 30th. On August 31st, there was a clear difference in the 11:58º LT temperature profile. The temperature profile on September 1st showed similar features with that on August 30, but large variations occurred at 20

IOP Conf. Series: Earth and Environmental Science **166** (2018) 012032 doi:10.1088/1755-1315/166/1/012032

km and above. The temperature variation over time on September 2 is almost non-existent (**Figure 2**). Large variation at low stratosphere could relate to variation in ultraviolet radiation and ozone production.

From those results, we can use the mean of temperature profile as shown on **Figure 3** (left) as a standard deviation of less than <0.7°K at range of 1.8 km to 16.1 km. There was no significant different of standard deviation in relation to the time. The result suggested that the gap of the launch time is small. For accurate estimation, the standard deviation above the height (**Figure 3**, right) should be considered. High standard deviation was also found at the surface up to 1.6 km (unpublished data). These results are appropriate with those reported by Margit & Wolfgang [6] which showed that a standard deviation of the temperature of 0.005° to 0.5° at the tropopause. Homogenization correction did not apply for higher than 1°K error. The high error above 16 km was suggested to come from chemistry process (ozone production) and atmospheric dynamics activities.

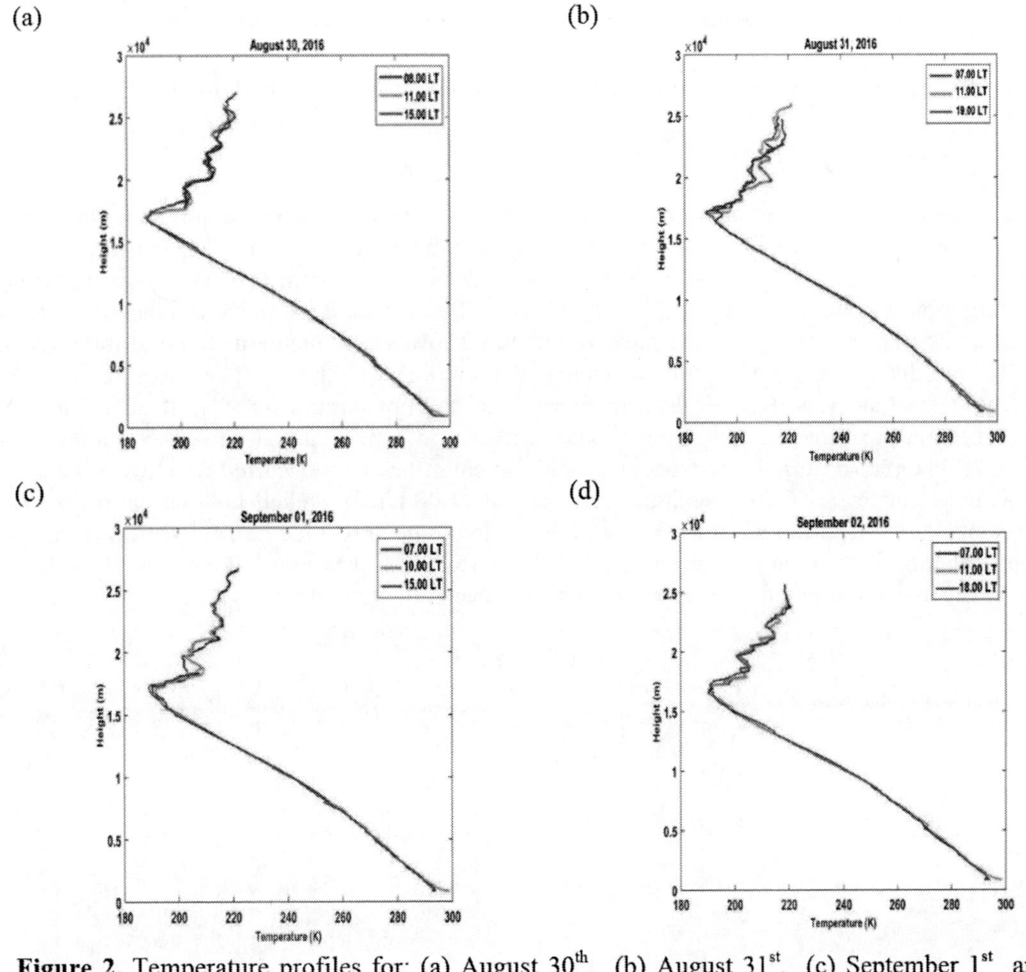

Figure 2. Temperature profiles for: (a) August 30th, (b) August 31st, (c) September 1st and (d) September 2nd 2016

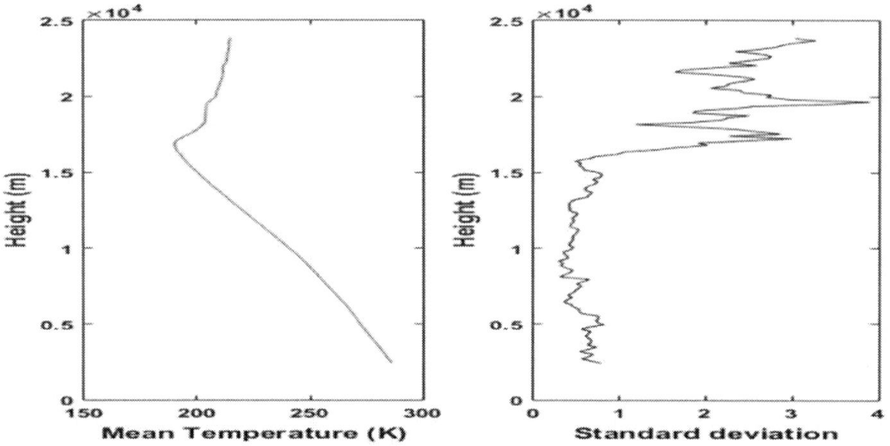

Figure 3. Profiles of mean temperature (left) and standard deviation (right) from 12 Radiosonde launch

3.2 Rainfall

Based on MRR-2, rain occurred on all day observation in different intensity and duration. On August 30th, 2016, rain occurred since 14.00 Local Time (LT) until midnight as shown in **Figure 4**. Maximum intensity occurred at 14:00 LT-14:30 LT. Rain intensity decreased afterward. As shown on **Figure 4**, the rainfall took place on August 30th, August 31st, September 1st and September 2nd, 2016, respectively. The rain intensity and the radar reflectivity showed same pattern. Rainfall intensity on August 31st was lighter than August 30th, as well as the duration was shorter. The maximum of radar reflectivity at the time was over 30 dBz, deviated 10 dBz from rainfall on August 30th. The high intensity of rainfall recurred on September 1st and September 2nd, even though in shorter duration. On September 1st, rainfall occurred since morning, but maximum intensity occurred in between 12:00 to 13:00 LT. Rain stopped at 16:30LT and then fell again at 21:00 LT for an half-hour. Radar reflectivity had the same pattern with rainfall intensity and the maximum occurred at 12:30 LT was over than 40 dBz. On September 2nd, rainfall occurred since 14:00 to 15:30 LT. The rain fell again at 18:30 LT to 20:00 LT. The maximum of radar reflectivity maximum occurred at night.

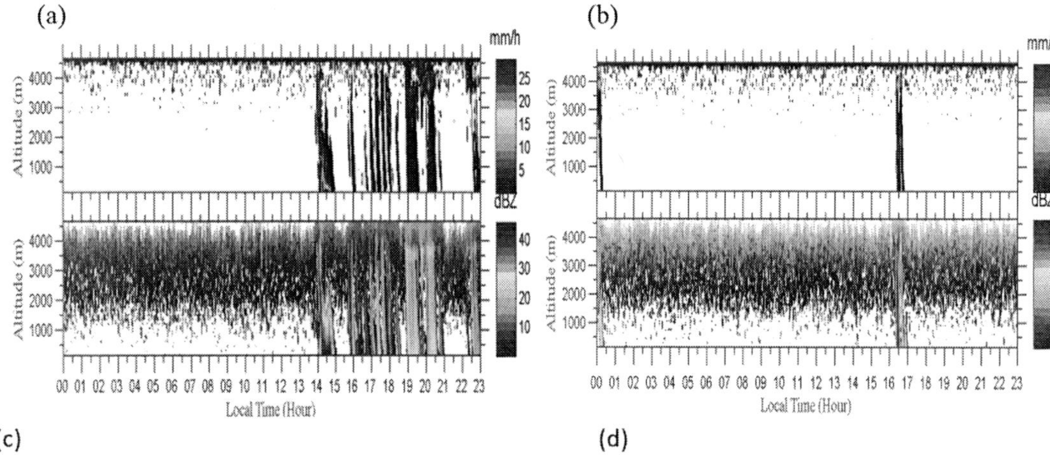

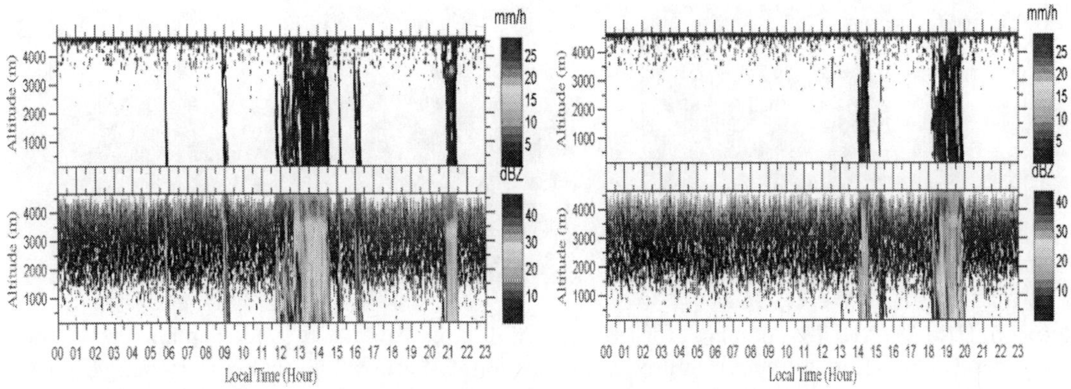

Figure 4. Rainfall intensity (above panel) and radar reflectivity (bottom panel) from MRR-2 for (a) August 30th, (b) August 31st, (c) September 1st, and (d) September 2nd 2016

Rainfall events that was detected by MRR-2 corresponded to HIMAWARI satellite images, as shown on **Figure 5**. The images time was adjusted to Radiosonde launch time and it turned out to coincide about two hours before the rain fell. The cloud covered the station during 4 days observation (**Figure 5**).

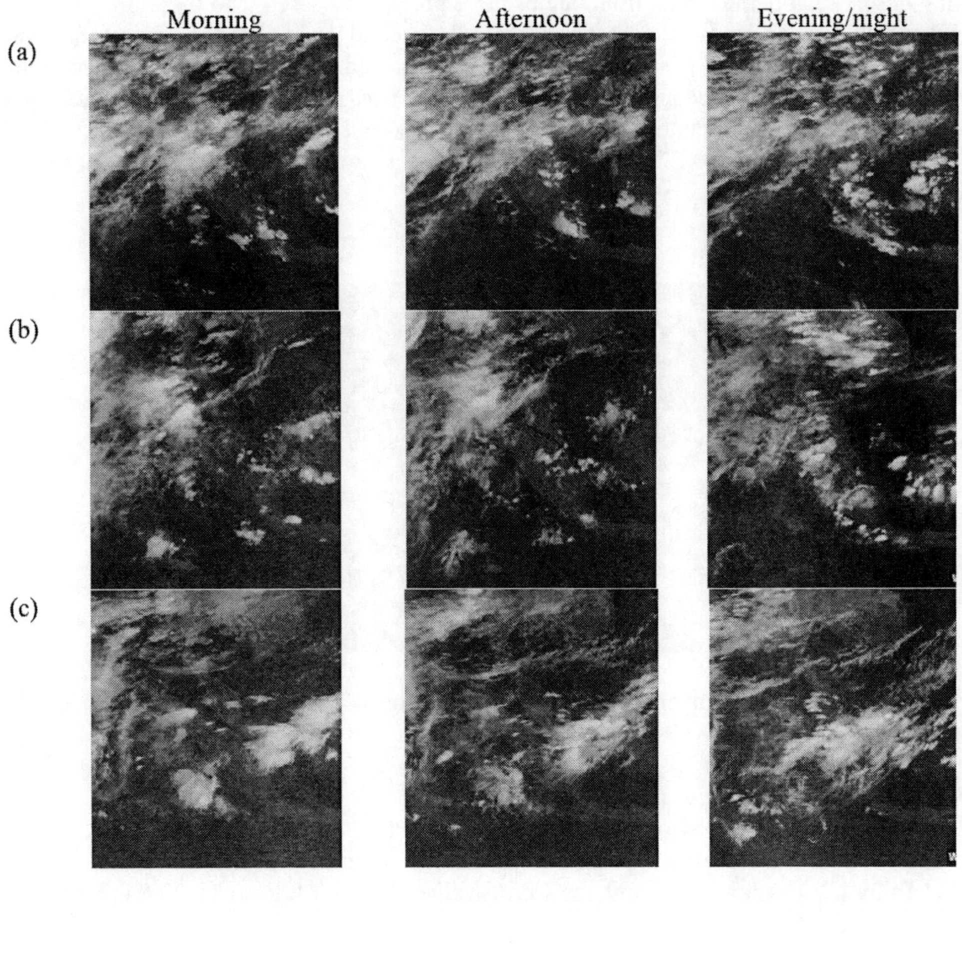

(d)

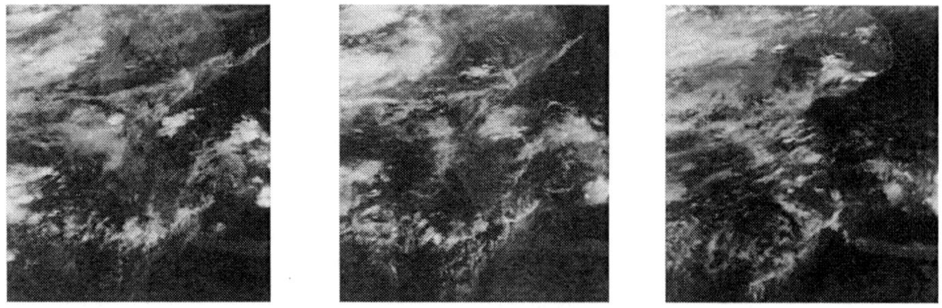

Figure 5. HIMAWARI IR-1 images for (a) August 30th, (b) August 31st, (c) September 1st and (d) September 2nd 2016 (left column: morning, middle column: afternoon, left column: evening/night)

3.3 Rainfall identification

3.3.1 Temperature deviation

Temperature deviation on **Figure 6** showed that the heating was mainly happened on August 30th and August 31st, while the cooling was on September 1st and September 2nd, 2016 from the surface up to 16 km. Above 16 km, the heating and the cooling had been quickly changing in higher intensity. The heating occurred when heat was released from the cloud or rain, while the cooling occurred when heat was absorbed from the environment. Radiosonde data could not be identified all the process, depending on the time resolution or the number of launch. Higher resolution data should catch the heating and cooling during each rain event. One thing can be deduced from the present study that rain during observation period came from one or more precipitation cell, and therefore, it indicated that the cloud and its developments were continuing into the second event after the first rainfall- and so on.

When the number of data is sufficient, it can be extracted to the stratosphere-troposphere coupling. Stratosphere-troposphere coupling at the tropopause was reported to be crucially important to the exchange of energy, momentum, and constituents such as H_2O and O_3 between the two regions [17]. Moreover, it plays a key role in the influence of stratospheric circulation on the QBO in the troposphere [18] PDO [19].

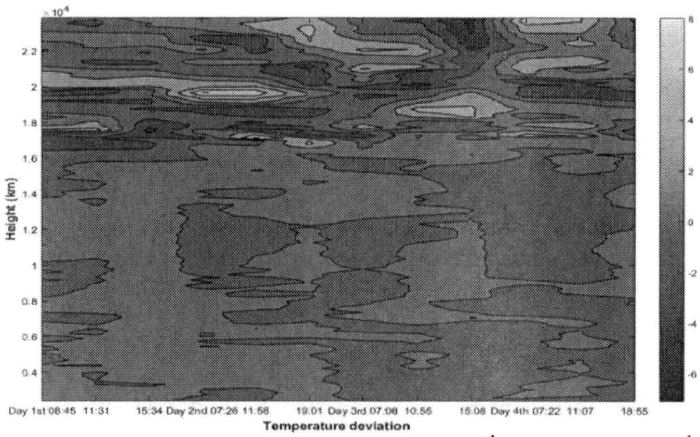

Figure 6. Temperature deviation from August 30th to September 2nd, 2016

3.3.2 CAPE and other Radiosonde indices

Rain identification was also performed using CAPE. Fluctuations in an intensity of rainfall as previously discussed correspond to CAPE fluctuations (**Figure 7**). The rain that occurred on August 30th at 14:00 LT was coincident with the CAPE value of >2000 Jkg^{-1} at 11:00 LT. Rainfall at 16:00 LT was identified by the CAPE value of >5000 Jkg^{-1} at 15:00 LT. A low rain intensity on August 31st was identified by small CAPE (<2000 Jkg^{-1}). The same track can be applied for rainfall intensity on September 1st and September 2nd.

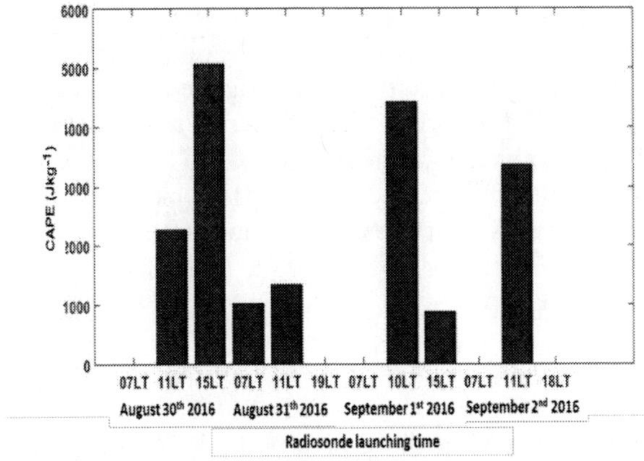

Figure 7. CAPE estimation from Radiosonde data

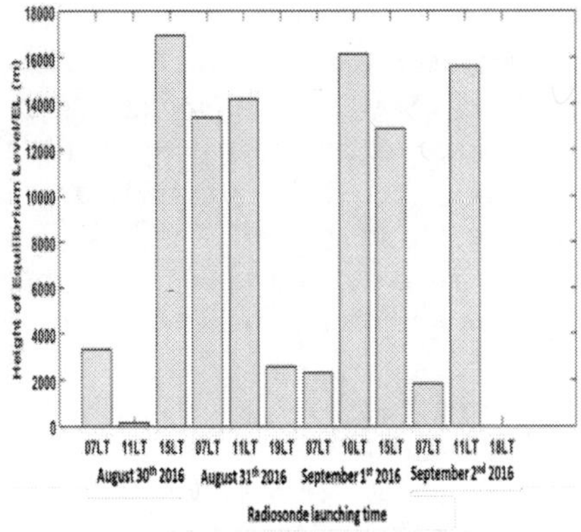

Figure 8. Equilibrium level height

A strong relationship between CAPE (1-2 hours before the rain) and rainfall supported by the coefficient of correlation as shown in **Table 3**. The first row of **Table 3** indicates four of height classification of rainfall from MRR-2 data and the second row indicates coefficient of correlation between CAPE and MRR-2 rainfall for height classification. The maximum coefficient of correlation (0.85) appears at the third of height classification, from 3.2 to 4.5 km.

Table 3. Coefficient of correlation MRR-2 rainfall for different height and CAPE

Height (m)	Coefficient of correlation
150-1500	0.63
1650-3000	0.71
3150-4500	0.85
150-4500	0.80

Equilibrium level (EL) is the level above the level of free convection (LFC) at which the temperature of a rising air parcel again equals the temperature of the environment. The height of the EL is the height at which thunderstorm updrafts no longer accelerate upward. From Radiosonde estimation, the increasing of EQ altitude (**Figure 8**) is related to CAPE's values (**Figure 7**), as indicated by pattern correlation on 0.69 (**Table 4**). The coefficient of correlation of CAPE with EL temperature and EL pressure are also presented in **Table 4**.

Table 4. Pattern correlation between CAPE and EL

	Pattern correlation
CAPE vs EL height	0.69
CAPE vs EL temperature	-0.83
CAPE vs EL pressure	-0.76

4. Conclusions

This paper provides the latest temperature profile in the tropics and its relationship with Radiosonde indices based on 12 times Radiosonde launch. Radiosonde can reach the maximum height well enough on average over 25 km. The temperature profile varies over time with a standard deviation of 0.7.

Minimum temperature occurs in the morning while maximum temperature occurs in the afternoon. The relationship between rain and CAPE indicated the strong correlation. On the other hand, CAPE has a significant correlation with equilibrium level (EL), CAPE has a strong positive correlation with the height of EL and negative correlation with temperature and pressure of EL.

Acknowledgement
The EAR observatory is operated jointly by the Research Institute for Sustainable Humanosphere (RISH), Kyoto University, Japan and the Indonesian National Institute of Aeronautics and Space (LAPAN). This study is supported by RISH, Kyoto University, and PSTA-LAPAN. The authors would like to thank Prof. Toshitaka Tsuda, Prof. Mamoru Yamamoto, Prof. Hiroyuki Hashiguchi from RISH-Kyoto-University for their support. Also to thank the colleagues at the EAR site who have helped on Radiosonde launched.

5. References

[1] Xue-Long Z H, Marvin A. G and Minghua Z H 2001 Cooling trend of the tropical cold point tropopause temperatures and its implications *J. of Geophy. Res* 106 D2 1511-2
[2] Peter W T, David E P, Simon F B T, Phil D J, Mark M, Holly C, and Philip B 2005 Revisiting Radiosonde upper air temperatures from 1958 to 2002 *J. of Geophy. Res* 110 D18105 doi:10.1029/2004JD005753
[3] Melissa F and Dian J S 2005 Causes of differing temperature trends in radiosonde upper air data sets *J. of Geophy. Res.* 110 D07101 doi:10.1029/2004JD005481
[4] David H D, John R C, Benjamin D P and S. Fred Singer A 2008 comparison of tropical temperature trends with model predictions *Int. J. Climatol.* 28 1693–01
[5] William J R, Keith P S, John A, John B, Chantal C, Nathan P G, Philippe K, Ulrike L, Roger L, Craig L, Carl M, Alvin M, John N, Dian J S, David W J T, Fei W and Shigeo Y 2009 An update of observed stratospheric temperature trends *J. of Geophy, Res.* 114 D02107 doi:10.1029/2008JD010421
[6] Margit PÁ and Wolfgang S 2015 Temperature Trends over Germany from Homogenized Radiosonde Data *J. Climate* 28 5699-15
[7] Shoichiro F 2006 Coupling Processes in the Equatorial Atmosphere (CPEA):A Project Overview *J. of the Met. Soc. of Jpn* 84A 1-18
[8] Bhatnagar, Panwar , Shibagaki , Hashiguchi, Fukao, Takahashi, Dhaka 2013 *Indian Journal of Radio & Space Physics* 42 277-91
[9] Zapotocny, Jung, Le Marshall and Treadon 2008 A Two-Season Impact Study of Four Satellite Data Types and Rawinsonde Data in the NCEP Global Data Assimilation System *Weather and Forecasting, Volume* 23 80-100.
[10] Gelaro and Zhu 2009 Examination of observation impacts derived from observing system experiments (OSEs) and adjoint models *Tellus A* 61 179-193
[11] Laroche and Sarrazin 2010 Impact study with observations assimilated over North America and the North Pacific Ocean on the MSC global forecast system. Part I: Contribution of radiosonde, aircraft and satellite data *Atmosphere-Ocean* 48 10-25
[12] Singh, Ojha, Kishtawal and Pal 2014 Impact of various observing systems on weather analysis and forecast over the Indian region *J. of Geophy. Res.: Atmospheres* 119 232-246
[13] Kutty and Wang 2015 A Comparison of the Impacts of Radiosonde and AMSU Radiance Observations in GSI Based 3DEnsVar and 3DVar Data Assimilation Systems for NCEP GFS *Ad. in Met.* 2015 1-17
[14] Michael P J , Donna J H, Petteri S, Raisa L, Shannon B, Tami T, and Karen L J 2016 Comparison of Vaisala radiosondes RS41 and RS92 at the ARM Southern Great Plains site *Atmos. Meas. Tech.* 9 3115–29.
[15] Oltmans and Holmann 1995 Increase in lowerstratospheric water vapor at a mid-latidude northern hemisphere site from 1981 to 1994 *Nature* 374 146-9

[16] Nedoluha, Bevilacqua, Gomez, Siskind, Hicks, Russell, and Conner 1998 Increases in middle atmospheric water vapor observed by the Halogen Occultation Experiment and the ground-based water vapor millimeter-wave spectrometer from 1991 to 1997 *J. Geophys. Res.* 103 3531-43.
[17] James R H, Peter H H, Michael E M, Anne R D and Richard B R, and Leonhard P 1995 Stratosphere-Troposphere Exchange *Rev. of Geophy.* 33 403-39
[18] Yasunari 1990 Impact of Indian monsoon on the coupled atmosphere ocean system in the tropical Pacific *Meteor. Atmos. Phys.* 44 29–41
[19] Wuke W, Katja M, Nour-Eddine O and Mojib L 2016 Decadal variability of tropical tropopause temperature and its relationship to the Pacific Decadal Oscillation *Sci. Rep.* 6 29537 doi: 10.1038/srep29537

IOP Conf. Series: Earth and Environmental Science **166** (2018) 012033 doi:10.1088/1755-1315/166/1/012033

Simple Design of Transmitter Circuit and Optimization Design of Receiver Circuit for Wireless Power Transfer

I Sukma[1], S. W Hidayat[1], A. R Hidayat[1] and I Supono[1]

[1]Research Center for Quality and Testing Technology,
Indonesian Institute of Sciences (LIPI),
Puspiptek District. Building 417, South Tangerang, Indonesia 15314

E-mail: irawan.sukma@lipi.go.id

Abstract. Two important parts of wireless power transfer are transmitter circuit and receiver circuit. Previous research has been completed which use CMOS H-Bridge in transmitter and full wave rectifier in receiver. This paper presented simple design of transmitter circuit and optimization of receiver circuit in rectifier circuit with transient analysis and Fast Fourier Transform (FFT) from LT spice software. The simple transmitter circuit used an integrated circuit (IC) NE555 with push-pull as oscillator circuit and LC resonant, while the receiver circuit used LC resonant, rectifier circuit, CLC filter for smoothing direct current (DC) output and 500 Ohm as load. The circuit design had resonant frequency of 250 kHz on both of transmitter and receiver, which comply with Power Matters Alliance (PMA) standard. The optimization of rectifier circuit had been done by comparing output of power load when it used basic rectifier (full wave), voltage doubler (dickson circuit and Villard circuit) and voltage multiplier (cockcroft-walton circuit). The results showed the receiver with Dickson circuit as rectifier has the highest efficiency compared to other rectifier circuits with efficiency of 10.93 % when it used coupling coefficient of 0.99, because Dickson rectifier has resonant frequency that is closest to 250 kHz and the highest level power, compared to other rectifier circuits.

1. Introduction

Research topic about Wireless Power Transfer (WPT) is highly increased recently. WPT can be used in several functions depending on the power, such as the use on low power charging-battery of implantable medical devices and the use in high power charging-battery of electric vehicles [1]. WPT has two categories: (1) Near-field transfer which well-known as non-radiative technique, and (2) far-field transfer which well-known as radiative technique [2,3]. The near-field transfer happens when the resonating frequency of the electromagnetic field is comparatively low (Hz - MHz) and the transfer range is comparatively short, inversely to far-field transfer which transferring power by a beam of electromagnetic source [2,3].

The resonant frequency from WPT circuit regulated in different three standards, that is the Alliance for Wireless Power (A4WP), the Wireless Power Consortium (WPC), and the Power Matters Alliance (PMA) or Rezence [4]. WPC and PMA use frequency ranges 110~205 kHz and 205~300 kHz, respectively, while A4WP use frequency 6.78 MHz ± 15 kHz carrier frequency [5,6].

A lot of researches was already done in WPT circuit with near-field categorized, it can be seen in **Table 1** [7–10]. The common WPT circuit consists of two important parts, transmitter and receiver. The transmitter circuit consists of oscillator circuit such as power amplifier (PA) or royer oscillator and resonant tank (LC tank). The receiver consists of resonant tank (LC tank), rectifier circuit (full wave or basic rectifier, voltage doubler or voltage multiplier) and load.

Table 1. Recent oscillator and rectifier circuit for research in WTP circuit

	Transmitter Circuit	Receiver Circuit	Resonant Frequency
Aldhaher et al	Class E Power Amplifier	Class E Rectifier	800 kHz
Irawan et al	Royer Oscillator	Full Wave	145 kHz
Ranum el al	Power Generator	Voltage Doubler	10 MHz
Schuetz et al	Direct Digital Synthesis (DDS)	Full Wave	13.56 MHz

State of the art of this paper, we proposed a simple design for transmitter circuit part which using an integrated circuit (IC) NE555 timer with astable multi-vibrator and push-pull transistor circuit and optimization in the receiver circuit. The transmitter circuit is different from previous research which was used CMOS H-Bridge and the receiver circuit was used full wave rectifier [11]. In the receiver circuit, the optimization of rectifier was used to define which rectifier that can give the highest efficiency. Furthermore, WPT circuit design had fixed resonant frequency at 250 kHz, which meet the requirement of PMA standard. In this paper, multifunction and free software namely LT spice software simulation tool was used for circuit design with command K1 Lt Lr .99, which had function to set coupling coefficient (k) between inductor LC tank of transmitter (L_t) and inductor LC tank of receiver (L_r) from 0 to 0.99 with increment 0.05. In addition, there was air gap between L_t and L_r. Furthermore, FFT was used to get resonant frequency of circuit design. the coupling coefficient similar to the fraction of magnetic flux resulted by L_t which flows through L_r and ranges from 0 to 1. Maximum coupling is obtain for transmission efficiency [12].

2. Circuit Design

2.1. Transmitter Circuit

The NE555 IC timer is famous and low-cost IC which can generate various waveforms, such as integrated square waveform and triangular waveform [12]. The NE555 IC timer also can be used in astable and mono stable mode [13].

The astable multivibrator mode, named astable a result of the output of NE555 IC timer is regularly inconstant between low and high. The transmitter circuit using NE555 IC timer in astable multivibrator mode to make square wave with sharp transition between low and high voltage. Furthermore, push-pull transistor was used to make the square waveform to sinusoidal waveform, so the LC tank resonate in sinusoidal waveform. The transmitter circuit design can be seen on **Figure 1**.

Input voltage of NE555 IC timer and push-pull transistor were using direct current (DC) with value 5 volt, as shown on **Figure 1**. The value of component in astable multivibrator mode was calculated by period and frequency (number of cycles per second) formula (equation 1 & 2) [13]. Furthermore, the value of LC tank was generated from resonant frequency (equation 3 & 4) [14]. The output frequency of NE555 IC timer was same as the resonant frequency of LC tank, with value frequency close to 250 kHz.

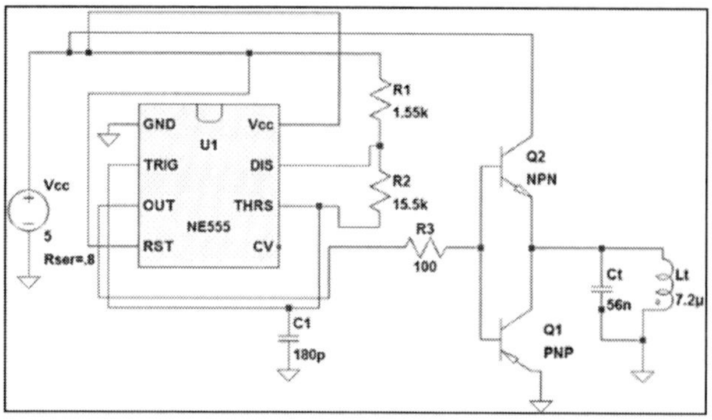

Figure 1. Proposed transmitter circuit using LTspice

$$\mathrm{f} = \frac{1.44}{(R_1 + 2R_2)\times C_1} \tag{1}$$

$$\mathrm{T} = 0.693\times(R_1 + 2R_2)\times C_1 \tag{2}$$

$$\omega_\mathrm{t} = \frac{1}{(L_t \times C_t)^{1/2}} \tag{3}$$

$$\mathrm{f_t} = \frac{1}{2\pi\omega_t} \tag{4}$$

Where,
f = frequency of NE555 IC timer (hertz)
R = resistor (ohm)
C = capacitor (farad)
T = period of NE555 IC timer (second)
ω_t= angular frequency transmitter of LC tank (radiant per second)
f_t = frequency transmitter of LC tank (hertz)
L_t= inductor transmitter of LC tank (henry)
C_t= capacitor transmitter of LC tank (farad)

2.2. The Receiver Circuit

The receiver circuit consists of LC tank, rectifier, CLC filter and load. The LC tank resonant frequency 250 kHz and the value component of LC tank calculated by equation (5 & 6) [14].

$$\omega_\mathrm{r} = \frac{1}{(L_r \times C_r)^{1/2}} \tag{5}$$

$$\mathrm{f_r} = \frac{1}{2\pi\omega_r} \tag{6}$$

Where,
ω_r= angular frequency receiver of LC tank (radiant per second)
f_r = frequency receiver of LC tank (hertz)
L_r = inductor receiver of LC tank (henry)
C_r = capacitor receiver of LC tank (farad)

If the alternating current (AC) voltage from LC tank would be used in charging direct current (DC) battery, it must be converted to direct current (DC) after LC tank part by using rectifier circuit. A lot of research has interest about rectifier circuit in energy harvester, but small has interest of the use of

WPT. Therefore, this paper describes the optimization of rectifier part, in order to get highest output of DC for WPT application. The rectifier circuit has two categories, basic rectifier (full wave rectifier) and voltage doubler (cockroft-walton, dickson charge pump and villard). The optimization of using 4 different circuits from basic rectifier and voltage doubler circuit is presented on **Figure 2**.

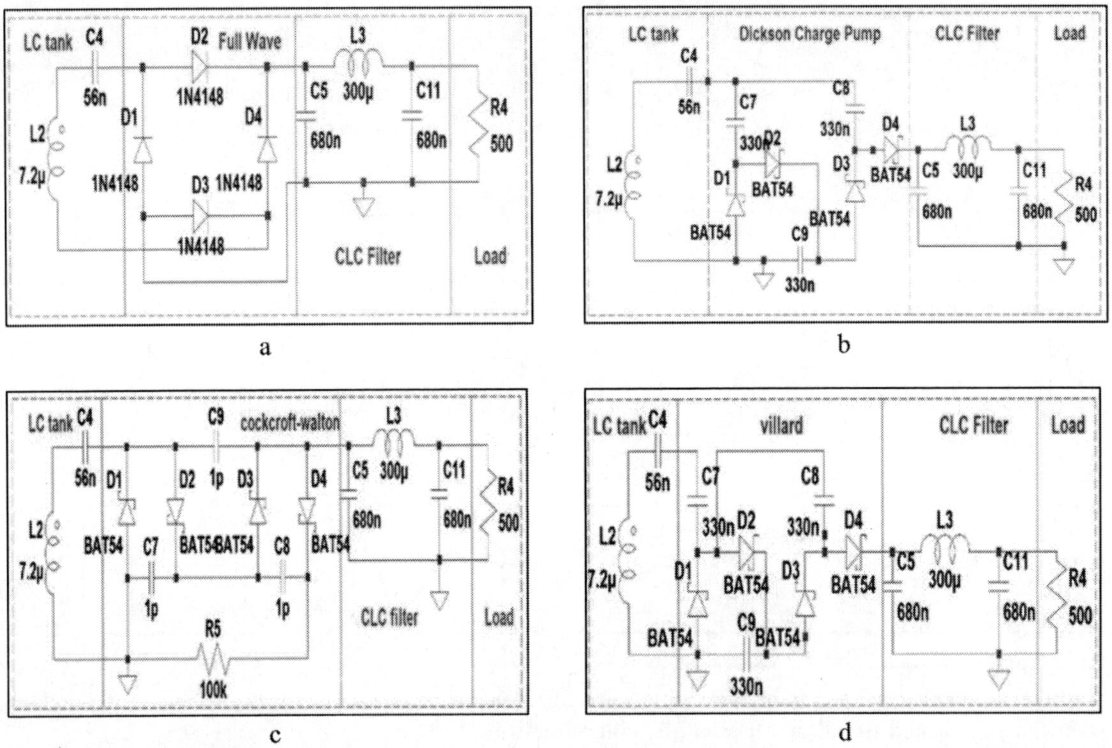

Figure 2. The receiver circuit for wireless power transfer use different rectifier circuit : a) full wave rectifier ; b) dickson rectifier ; c) cockroft-walton rectifier ; d) villard rectifier

3. Results And Discussions

3.1. Result of the Simulation

The simulation were conducted by using transmitter with IC NE555 timer and receiver circuit with full wave, Dickson, cockroft-walton and villard rectifier circuit, which resulting resonant frequency, voltage and current both of Lt and L_r.

Figure 3 show the closest and the farthest resonant frequency of circuit design in L_t and L_r when k of 0.99. The red and blue line on those images refer to f_t and f_r, respectively. The rectifier circuit with Dickson rectifier has resonant frequency of 254.9 kHz, which is the closest to 250 kHz and the highest-level of power in dB unit among rectifier circuits. Furthermore, full wave rectifier has resonant frequency of 257 kHz, which is the farthest resonant frequency and the lowest power in dB unit among rectifier circuits. The low-level power from full wave rectifier has an effect to the efficiency of WPT circuit.

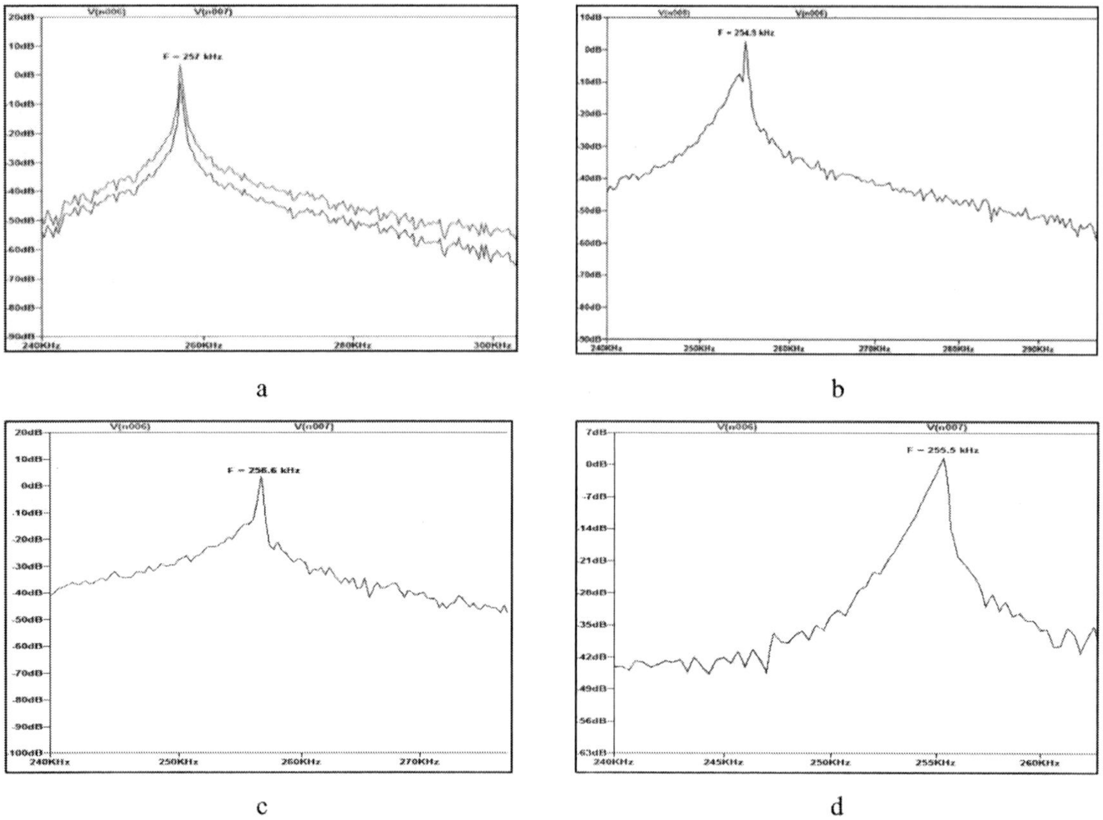

Figure 3. Result resonant frequency which use different rectifier : a) full wave rectifier ; b) dickson rectifier ; c) cockcroft-walton rectifier ; d) villard rectifier

The resonant frequency and level of the power can effect to voltage and current measured in load. The red, blue, pink, green lines on **Figure 4** refer to voltage in Lt, current in L_t, voltage in load, and current in load, respectively **Figure 4** show that the full wave rectifier only generated 0.7 mV in load, while Dickson rectifier resulted 4.2 V in load.

3.2. Efficiency (η)

Efficiency (η) of the circuit design can be calculated by using voltage and current from simulation both of Lt and of load. Efficiency (η) can be calculated by equation 7 [14]. Efficiency (η) of the circuit design can be seen in **Figure 5**. The result of η was based on different rectifier, as Dickson rectifier had the highest result of $\mathbf{P_r}$ and $\mathbf{P_t}$. **Figure 5** shows the maximum of circuit design that used Dickson rectifier could generate when it was using k of 0.95.

$$\eta = \frac{P_r}{P_t} = \frac{V_r i_r}{V_t i_t} \quad (7)$$

Where,

$\boldsymbol{\eta}$ = efficiency (%).
$\mathbf{P_r}$ = power receiver (watt).
$\mathbf{P_t}$ = power transmitter (watt).
$\mathbf{V_r}$ = voltage receiver (volt).
$\mathbf{i_r}$ = current receiver (ampere).

$\mathbf{V_t}$ = voltage transmitter (volt).
$\mathbf{i_t}$ = current transmitter (volt)

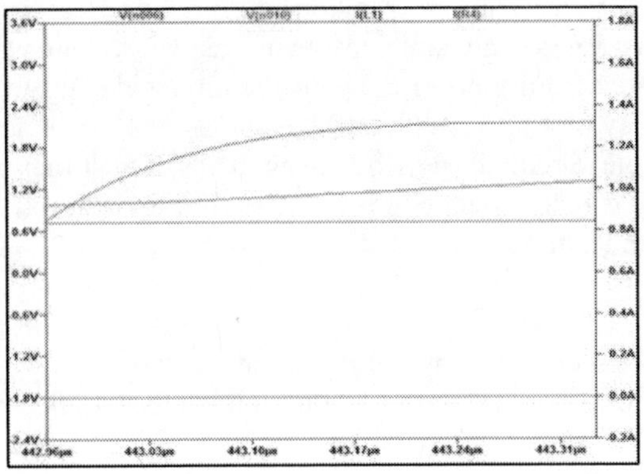

a

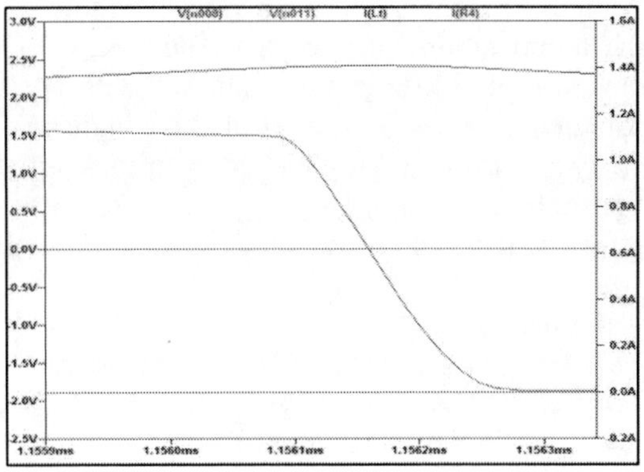

b

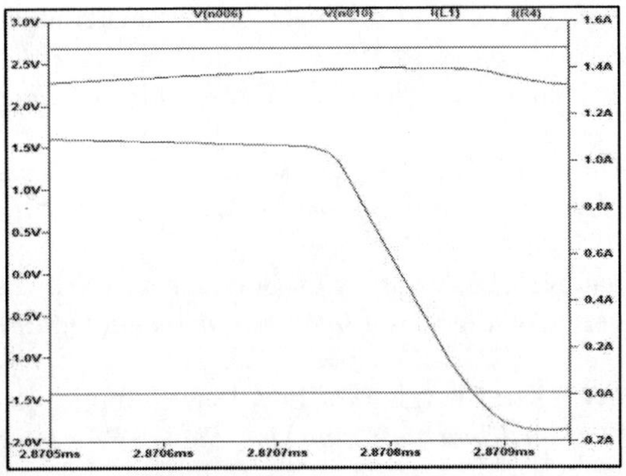

c

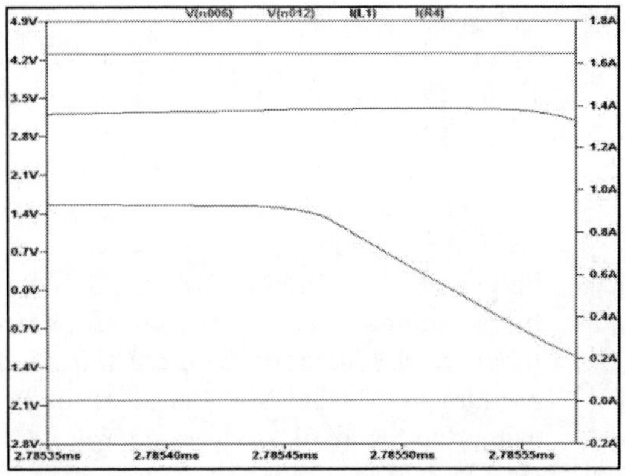

d

Figure 4. Result V_t, i_t,V_r and i_r which use different rectifier : a) full wave rectifier ; b) dickson rectifier ; c) cockcroft-walton rectifier ; d) villard rectifier

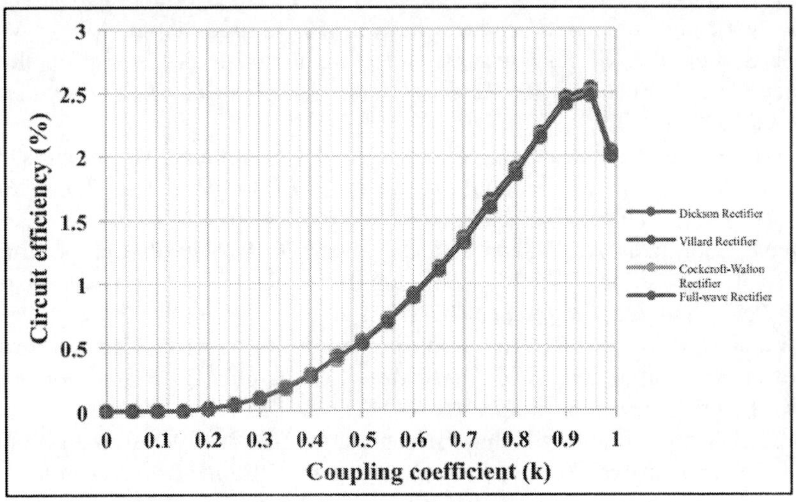

Figure 5. Efficiency of design circuit when using different rectifier

4. Conclusion and Future Work

The simple design circuit with optimization of receiver circuit for WPT was successfully designed. Although the circuit design had resulted lower efficiency than those of previously reported researches, the optimization resulted that Dickson rectifier was able to generate resonant frequency that was closest to 250 kHz, which complied with PMA standard. Furthermore, when the circuit design was using Dickson rectifier, it resulted the highest efficiency of 2.53 %and k of 0.95.

The aim of this research was to make real simple circuit design by using IC NE555 timer in transmitter and to optimize the receiver circuit, so the real distance can be measured and the other effects which can disturb the efficiency (η) will be understood.

Acknowledgments

This research was performed under the research program endorsed by the Research Center for Testing Technology and Quality System (P2SMTP), Deputy for Scientific Services, and funded by Indonesian Institute of Sciences, Indonesia.

5. References

[1] Wang B, Yerazunis W and Teo KH 2016 Wireless Power Transfer Based on Metamaterials *In: Nikoletseas S, Yang Y, Georgiadis A, editors. Wireless Power Transfer Algorithms Technologies and Applications in Ad Hoc Communication Networks Springer International Publishing AG* 31–53

[2] Challa VR, Mur-miranda JO and Arnold DP 2012 Wireless power transmission to an electromechanical receiver using low-frequency magnetic field *Smart Mater Struct IOP Publishing Ltd* **21**(11)

[3] Perera TDP, Jayakody DNK, De S and M A Ivanov 2016 A Survey on Simultaneous Wireless Information and Power Transfer *In: International Conference on Information Technologies in Business and Industry IOP Publishing Ltd*

[4] Kiruthiga G, Jayant MY and Sharmila A 2016 Wireless Charging for Low Power Applications using Qi Standard *In: International Conference on Communication and Signal Processing (ICCSP) IEEE* 1180–4

[5] Ahn D and Mercier PP Wireless Power Transfer with Concurrent 200 kHz and 6 . 78 MHz Operation in a Single Transmitter Device *In: IEEE Transactions on Power Electronics* 5018–29

[6] *Richtek Technologies Corporation. Introduction to RT1650 Wireless Power Receiver [Internet]* https://www.richtek.com/assets/product_file/RT1650/DS1650-00.pdf

[7] Aldhaher S, Luk PC, El K, Drissi K and Whidborne JF 2015 High-Input-Voltage High-Frequency Class E Rectifiers for Resonant Inductive Links *In: IEEE Transactions On Power Electronics* 1328–35

[8] Ranum BT, Wayan N, Eka D and Munir A 2014 Development of Wireless Power Transfer Receiver for Mobile Device Charging *In: The 2nd IEEE Conference on Power Engineering and Renewable Energy (ICPERE)* 48–51

[9] Schuetz M, Georgiadis A, Collado A and Fischer G. A 2015 Particle swarm optimizer for tuning a software-defined , highly configurable wireless power transfer platform *In: IEEE Wireless Power Transfer Conference (WPTC)* 1–4

[10] Sukma I and Supono I 2017 Design and Simulation of 145 kHz Wireless Power Transfer for Low Power Application *In: IEEE International Conference on Radar, Antenna, Microwave, Electronics, and Telecommunications (ICRAMET)* 79–82

[11] Nataraj C, Khan S, Eniola FF and Selvaperumal SK 2017 Design of Simple DC-to-DC Wireless Power Transfer Via Inductive Coupling *In: 2017 Third International Conference on Advances in Electrical, Electronics, Information, Communication and Bio-Informatics (AEEICB)* 50–5

[12] Ali H, Ieee SM, Ahmad TJ and Khan SA 2009 Inductive Link Design for Medical Implants.

In: IEEE Symposium on Industrial Electronics and Applications (ISIEA) 694–9

[13] Senani R, Bhaskar DR, Singh VK and Sharma RK 2016 Basic Sinusoidal Oscillators and Waveform Generators Using IC Building Blocks *In: Glaser C, editor. Sinusoidal Oscillators and Waveform Generators using Modern Electronic Circuit Building Blocks* Springer Cham Heidelberg 3–70

[14] *Texas Instruments. xx555 Precision Timers package* http://www.ti.com/lit/ds/symlink/ne555.pdf

[15] Li H, Li J, Wang K, Chen W and Yang XA 2015 Maximum Efficiency Point Tracking Control Scheme for Wireless Power Transfer Systems Using Magnetic Resonant Coupling. *In: IEEE Transactions on Power Electronics* 3998–4008

Analysis of El Niño and IOD Phenomenon 2015/2016 and Their Impact on Rainfall Variability in Indonesia

L. Q Avia[1] and I Sofiati[1]

[1]Center of Atmospheric Science and Technology
Indonesian National Institute of Aeronautics and Space (LAPAN)
Jl. Dr. Djundjunan 133, Bandung 40173, Indonesia

E-mail: qodrita@yahoo.com

Abstract. The El Niño phenomenon that occurs in 2015/2016 fell into category of strong intensity. The El Niño phenomenon had a very large impact on meteorological droughts in Indonesia. The aim of this study is to determine the dynamics of the atmosphere and the sea during the phenomenon of El Nino 2015/2016 and its impact on rainfall in the Indonesia region. The research methods consisted of several steps, in order to obtain an overview of the spatial conditions of rainfall during the El Niño period, data processing of both accumulation and anomaly of rainfall per-season (MAM, JJA, SON, DJF) were evaluated. The results showed that strong intensity of the El Niño occurred from early July 2015 to early March 2016. The shift of the convection center occurred in the Pacific Ocean during this period where the sea surface temperature anomaly in the eastern Pacific appeared warmer 2-3°C than normal temperature (1981-2010). In 2015, the positive Indian Ocean Dipole (IOD) also occurred in the Indian Ocean from August to November. These phenomena resulted wider drought impacts throughout Indonesia especially during the transition from dry season to wet season period (September-November).

1. Introduction

Indonesia is an archipelago country and its territorial position lies between the Pacific and Indian Oceans. This unique geographical profile has lead to rainfall variability in Indonesia, which is strongly influenced by the interaction between the atmosphere and the oceans around Indonesia. El Niño - Southern Oscillation (ENSO) is a phenomenon that consists of two phases, i.e, heat phase (El Niño) and cold phase (La Nina). Southern Oscillation is the changes of wind circulation caused by the difference of sea surface temperature (SST) between eastern and western Pacific Ocean [1]. This difference causes reversal of atmospheric circulation over the ocean located near the equator [2], and causes a global climate anomaly [3]. El Niño occurs when there is an increase of SST in the eastern and central Pacific Ocean equatorial areas, whereas La Nina is the opposite condition where there is cooling temperature. Several previous studies have shown that when the El Niño phenomenon in the equatorial Pacific Ocean appears to be associated with decreased rainfall in Indonesian territory [4-9].

El Niño with strong intensity had occurred in 2015/2016. In the El Niño period, there was also an ENSO-like phenomenon in the Indian Ocean commonly called the Indian Ocean Dipole (IOD). IOD is the difference in temperature conditions of the east coast of Africa and the west coast of Sumatra [10]. In the case of positive phase of IOD, the sea surface anomaly in the western Indian Ocean is warmer, whereas in the eastern part it is colder than normal. Positive phase of IOD has been reported to have relationship with rainfall reduction in part of Indonesia [6, 8-11].The relation of ENSO with rainfall variability over Indonesia has been studied by many researchers. The variability of SST in the Pacific Ocean especially the location of El Niño 3.4 affects 50% of the variation in rainfall throughout Indonesia while the variability of ocean surface temperature in the ocean affects only 10-15% [12]. Several other studies have shown that the positive El Niño and IOD impacts have resulted in the reduction of rainfall in Indonesia, thus the dry season tends to be longer and is getting dry [8 - 9]. Research on the influence of ENSO on rainfall in Java based on RegCM3 modeling and image interpretation has been done [13]. The results showed a strong correlation of ENSO and rainfall in Java Island compared to other islands in Indonesia, This is because Java Island is the center of Asia-Australia monsoon region. The impact of El Niño on reducing rainfall appears significant in the transition of the dry season to the rainy season (SON period), and also significant in the rainy season (DJF). This El Niño can be predicted based on an indicator of SST conditions in the Pacific Ocean as well as on the difference in surface pressure between Tahiti and Darwin better known as Southern Oscillation Index (SOI).How the dynamics of the atmosphere and oceans during the ongoing El Niño 2015/2916 and how their impact was toward the intensity and distribution of rainfall in Indonesia by using a database of Global Satellite Mapping of Precipitation (GSMaP) will be discussed in this study. This research is very important to assess the impact of El Niño as an evaluation in an effort to raise awareness of any areas that have the potential drought. Currently, the satellite-based remote sensing is the only practical way to measure rainfall on a global scale. Satellite sensors can produce high-resolution rainfall measurements and cover a wide and diverse area on the Earth's surface such as oceans, deserts and high mountains, which are not feasible to be monitored by conventional rain gauges or meteorological radar [14]. The use of satellite data GSMaP in this study is a good alternative considering the limited amount of rainfall observation stations in Indonesia or the unsustainability of available data. The use of weather satellites in measuring rainfall intensity is very useful for monitoring weather conditions that continue to develop. To generate global rainfall maps with high-resolution and high-precision retrieval. The GSMaP incorporates rainfall from TRMM and other polar orbit satellites [15]. The GSMaP is a project of the Japan space agency to approach the value of rainfall using space satellite media. The goal of the project is mapping global rain with a spatial resolution and high temporal and has a high accuracy using satellite with MWR sensor (microwave radiometer) [16]. GSMaP is managed by the Earth Observation Research Center (EORC) of Japan Aerospace Exploration Agency (JAXA) [15].

2. Data and Methods

In this study used several indexes global phenomenon to determine the atmosphere and sea dynamics that occurred in the Pacific Ocean and the Indian Ocean associated with El Niño 2015/2016 and IOD positive phase. These data are the Sea Level Pressure (SLP), Zonal Wind, Outgoing Longwave Radiation (OLR) and SST for El Niño 3.4 location. The data were obtained from NOAA [17]. While the Southern Oscillation Index (SOI) and Indian Ocean Dipole (IOD) data was obtained from the Australian Bureau of Meteorology [18]. Furthermore, to determine the pattern of rainfall in Indonesia during the 2015/2016 El Niño and its distribution in zonal and meridional used rainfall data obtained near real time from the GSMaP satellite [19]. The GSMaP data that are used in this study have a spatial resolution is 0.1 x 0.1 degree latitude/longitude and temporal resolution is daily. The impact on rainfall in Indonesia region was spatially observed by comparing the data of rainfall anomaly to its normal condition (1979-2000), based on IRI data [20]. The IRI data has a spatial resolution is 2.5 x 2.5 degrees latitude/longitude and temporal resolution is seasonally (3 monthly). Two types of data resolution is used in this research because of the limitations of new high-resolution GSMaP data that was available in 2014. The boundaries of this research area include the territory of Indonesia and its surroundings that are bounded between 12N-12S and 90-150E. The

IOP Conf. Series: Earth and Environmental Science **166** (2018) 012034 doi:10.1088/1755-1315/166/1/012034

research methods consisted of several steps. First of all was to collect main data and supporting data during period of 2015/2016. The second step was plotting the index of SST and IOD for analysis of period and intensity level of the phenomenon. The third step was to extract the satellite data that covering Indonesia for rainfall and OLR parameters. In order to obtain an overview of the spatial conditions of rainfall during the El Niño period, data processing of both accumulation and anomaly of rainfall per-season (MAM, JJA, SON, DJF) were evaluated. Based on GSMaP rainfall data, we processed the mean of latitude and mean longitude versus time to know the spatial distribution of rainfall in zonal and meridional. Data processing and display of rainfall spatial pattern in this research was proceeded by using Grid Analysis and Display System (GrADS) tools. The final step was analysis and interpretation of the results.

3. Results and Discussions

As an indicator of El Niño activity in the Pacific Ocean, the study used SST in the Pacific Ocean, especially in the Niño 3.4 region (5N-5S and 120-170W). The results indicated that the Niño 3.4 index was the strongest index that can illustrate the relationship of SST anomaly and indicator of drought or wetness in Indonesian region. The anomaly of SST of El Niño 3.4 during the period of January 2015 to June 2016 was indicated by the red color graph on **Figure 1**. **Figure** 1 shows that in mid-March 2015 to late June 2015, the phenomenon of El Niño started with weak intensity (SST anomaly 0.57°C), and continued to increase with moderate intensity (an anomaly of SST 1.37°C). The El Niño phenomenon arrived in strong intensity in early July 2015, where SST anomaly was over 1.5°C. Strong intensity lasted until early March 2016 and then gradually weakened. The El Niño peak with strong intensity occurred in November 2015 where SST anomaly reached 2.48°C. The current results were in accordance with the monitoring on the atmospheric dynamics conducted by Meteorology, Climatology and Geophysics Agency (BMKG), which suggested that El Niño was at a Moderate status and was predicted to decline slowly to Neutral in April-May 2016 [21].

Apart of the analysis of El Niño which occurred in the Pacific Ocean, the analysis of anomalies in the Indian Ocean, i.e. IOD, was important as an indicator of drought in some parts of Indonesia. IOD index is calculated from the difference of SST anomalies in the Indian Ocean near Africa (50-70E ; 10N-10S) and the SST anomalies in the Indian Ocean east of the island of Sumatra (90-110E ; 10 N-10S) [10,11]. In the period of the 2015/2016 El Niño, positive phase of IOD also took place in the Indian Ocean since mid-August 2015 to mid-November 2015. The peak condition occurred at the end of September 2015 with an intensity of 1.17°C as shown in the yellow graph on **Figure 1**.

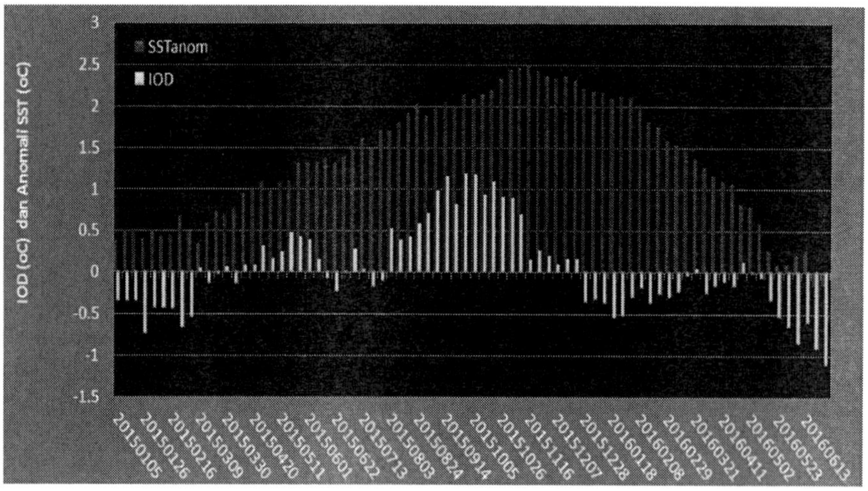

Figure 1. Niño3.4 SST index (red color chart) and IOD (yellow color chart) in the period January 2015 - June 2016.

The positive phase of IOD index becomes a drought indicator for the west region of Indonesia because the convergence centers were formed near Africa thus blocking the supply of clouds from the Indian Ocean to Indonesia. Normally, in the western Indian Ocean, the temperature of the sea surface is cooling, but warm in parts of the eastern hemisphere which is characterized by a fairly equal distribution of sea surface temperature around the equator [10, 11].

Figure 2 (left) shows the variation of SST anomalies in the equatorial Pacific Ocean. It suggested that since April 2015, the tendency of sea surface temperature in eastern Pacific Ocean and central parts always seemed warmer than the normal temperature (climatological, 1981-2010). The warm temperatures continued to rise, the dominance of positive SST anomalies 2°C -3°C means that warmer sea surface temperature 2°C -3°C normally occurred around July 2015 to February 2016. Even at around position 100-160W looks around November-December 2015 was a peak with SST more than 3°C from the normal temperature. **Figure 2** (right) shows the variations of SLP anomalies in the Pacific Ocean. Since the period of July 2015 to April 2016, the pressure above sea surface in the western Pacific Ocean continued to increase. The highest SLP was about 140-160E, to reach over 2hPa which was higher than normal conditions (1981-2010). While in the center and eastern of the Pacific Ocean, the surface pressure is lower than normal (up to 2hPa).

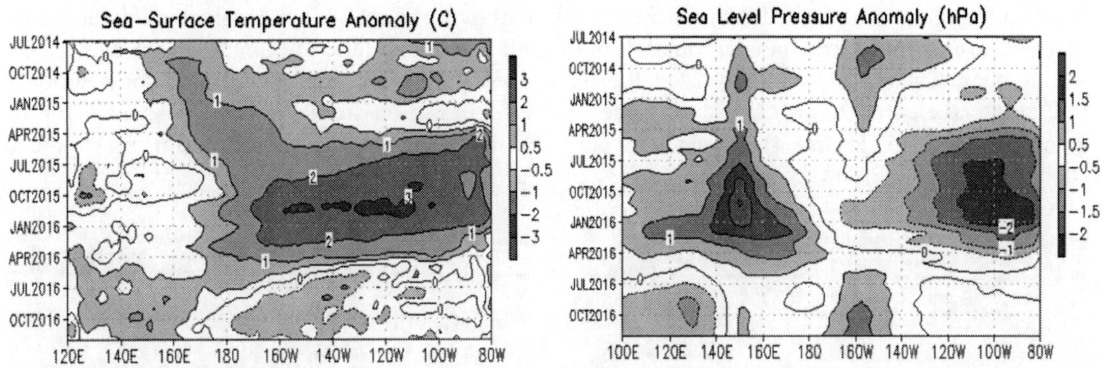

Figure 2. SST anomalies (left) and SLP anomalies (right) in the Pacific Ocean during the period July 2014 to December 2016.

The pressure difference will then cause a zonal wind movement, due to the nature of the air to move from high to low air pressure. The condition caused air from the western Pacific Ocean moved towards the central and eastern Pacific Ocean. It was also the cause of convective clouds over the Indonesian territory shifted to the middle and east pacific that reduced rainfall in the territory of Indonesia. **Figure 3** (left) shows the zonal wind anomaly pattern in the Pacific Ocean. It can be seen on **Figure 3** (left) around March 2015 to March 2016, in the central and eastern Pacific Ocean around 140-120W in the dominance of a positive anomaly of 2-6 m/s, which means winds are blowing hard around 2-6 m/s stronger than normal.

The shift of the convection center from its normal conditions in the Pacific Ocean caused an increase in water vapor that will lead to high rainfall in the central and eastern Pacific Ocean region. This was supported by the fact that the lower OLR data was about 20-40 W / m2 compared to normal (1981-2010) in the region, as shown on **Figure 3** (right).

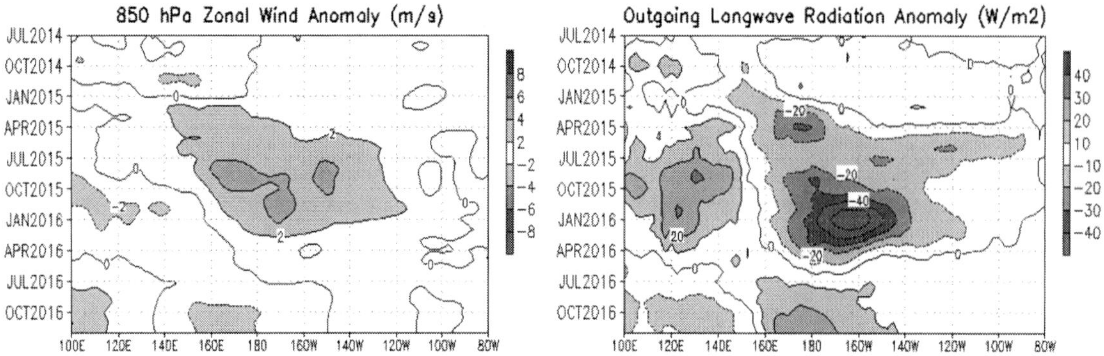

Figure 3. Zonal wind anomalies (left) and a nomali OLR (right) in the Pacific Ocean during the periode July 2014 to December 2016.

Rainfall data *near real time* with the new high-resolution GSMaP is available since March 2014, so there is a long period of limited data for the determination of normal / climatological precipitation that is needed to determine how much reduction in rainfall that occurs due to climatic anomalies such as El Niño them. Therefore, this study used the rainfall data from IRI from the climatology period of 1979-2000. Based on those data, during the period of transition from the rainy season to the dry season (MAM 2015, **Figure 4**), in which El Niño was in weak intensity, the rainfall in most parts of Indonesia are still in normal condition.

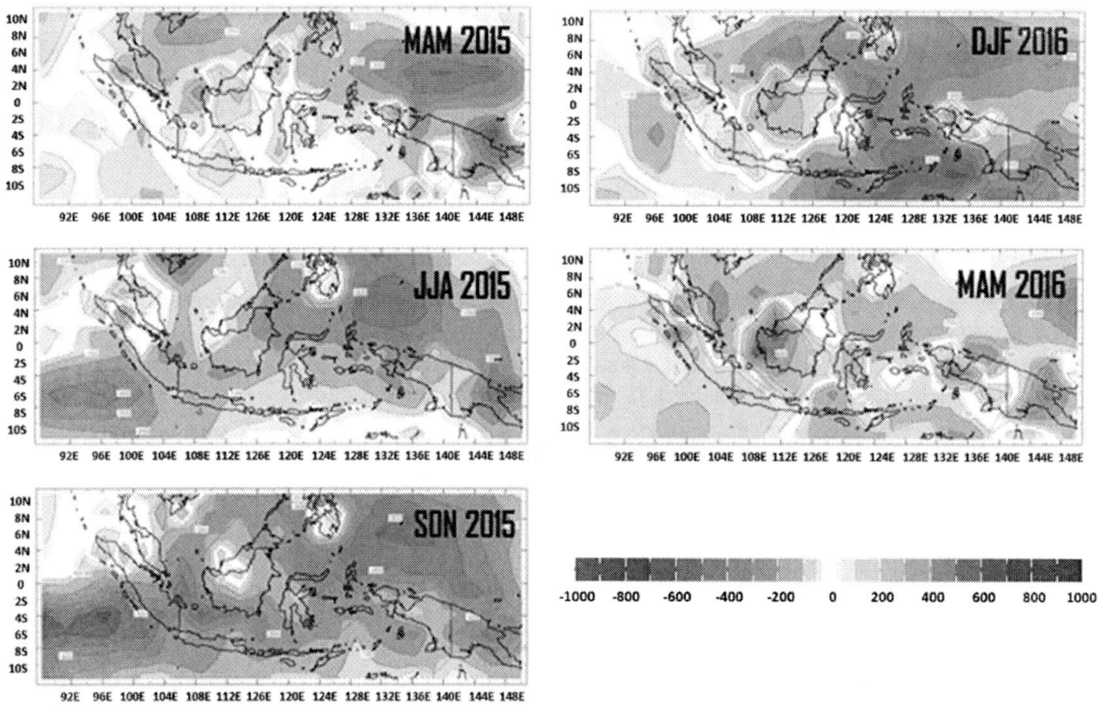

Figure 4. The spatial pattern of precipitation anomalies (mm/season) in El Niño 2015-2016.

Even in southern part of Sumatra, Java, Kalimantan and southern parts of western and central Sulawesi, the rainfall tends to be greater about 10-30 mm than the normal average, as shown on **Figure 4**. While the strong El Niño intensity underway in the Pacific Ocean, it seems to have a significant impact on the drought in almost uniformly in all parts of Indonesia. This condition was called the dry season (JJA 2015, **Figure 4**), in which the negative anomalies reached 200-400 mm of rainfall index. As for the transition period from the dry season to the wet season (SON 2015, **Figure 4**), it suggested the increasingly dry conditions and widespread in region of Indonesia. At the same time, the rainfall decreased to reach about 400-600 mm compared to normal. During this period, the El Niño reached the peak where SST anomalies also reached 2:48 ° C in November 2015, in conjunction with the positive phase of IOD phenomenon that occurred in the Indian Ocean.

In the rainy period (DJF 2016, **Figure 4**), the visible pattern of rainfall in most parts of Indonesia were back to normal even tend to be higher than the average, except in East Nusa Tenggara, North Sulawesi and Maluku which seemed still dry. The transition from the rainy season to the dry season (MAM 2016, **Figure 4**) was coinciding with the period of moderate El Niño intensity with SST anomalies amounted to 1, 44 ˚ C in mid-March 2016. The El Nino continued to get weaker in intensity with SST anomaly of 0.59 and then ended up being neutral by mid-May 2016. During the 2016 MAM period, the rainfall in Indonesia was generally around 20-40 mm higher than normal, except in NusaTenggara, North Sulawesi and parts of Papua, which was still in dry as suggested by rainfall was lower than normal, about 200 mm on average. In some parts of Sumatra, Bali, East Kalimantan and part of Papua, rainfall was in normal category. The results supported previous report [22] based on GPCP data finding the Asian monsoon rainfall anomalies associated with the 2015/2016 El Niño. The station and satellite records confirm the extremely deficient rainfall over Indonesia and the Philippines. During the period JJA 2015 to DJF 2016, it was generally a negative rainfall anomaly between 45 to 90 mm/month in the region.

Based on GSMaP rainfall data for the period 2015/2016, the spatial patterns of El Niño accumulation per-season rainfall occurred in Indonesia were obtained (**Figure 5**). **Figure 5** shows the transition from rainy season to dry season (MAM 2015) where during that period, a weak El Niño intensity, the cumulative rainfall in most parts of Indonesia ranging from 00-1500 mm 3, lasted ,. The Asian monsoon still gives influence as seen in the transition. Even though its influence which brings a little mass of water vapour, Asian monsoon was also visible in the south in which small rainfall less than 300 mm was around Nusa Tenggara. The accumulation of heavy rainfall over 1800 mm happened in the waters west of Sumatra, south Palopo (Central Sulawesi), the waters north of Nabire (Papua Barat) and partly on the mainland of center Papua.

During the dry season in the period of El Niño with strong intensity, rainfall reduced significantly throughout the territory of Indonesia. The accumulated rainfall ranging from 300-900 mm that is less than 300 mm/season occurred in the whole areas south of the equator, except Papua. Areas in the northern part of the equator have an accumulated rainfall ranging from 300-600 mm. It can be seen that the reduced rainfall occurred not only because of the Asian monsoon influence but also because of stronger El Niño anomalies. El Niño anomalies are caused by the decline of STT in most of the Indonesian waters, and the shifting center of low pressure in the Pacific Ocean due to the weakening easterlies winds.. The results supported previous report suggesting a strong relationship between drought and ENSO during the dry season (JJA) because the ENSO phenomenon affects the condition of SST in Indonesian waters [23]. Likewise, the studies of SST on some parts of Indonesia have a negative correlation with the ENSO events during the dry season (JJA) [12]. At the time, SST Pacific Ocean warms the central part of SST conditions in Indonesian waters so that it is cooler than the normal conditions. This condition resulted in the weakening of the southeast monsoon winds and east-west zonal winds that are the source of convection in Indonesian territory. Besides, SST addition also inhibits the evapotranspiration process which is the source of water vapour for the process of cloud occurrence.

During the transition from dry season to wet season (SON 2015), there normally should be an increase in rainfall before the rainy season due to the influence of south-west monsoon. However, the rain did not occur in the waters north of Bali and Nusa Tenggara because of the dry conditions in the Indonesian territory south of the equator.. It shows the result of south easterly winds influence is still dominant in the area. However, in Papua, part of Sumatraand Kalimantan and Papua (between 300-

600 mm) are still higher than the highlands in Papua and Aceh reaching 1500 mm. Accumulated rainfall is generally less in the south of equator during this period due to the combined effects of El Niño anomaly in the Pacific Ocean and IOD positive phase in the Indian Ocean.

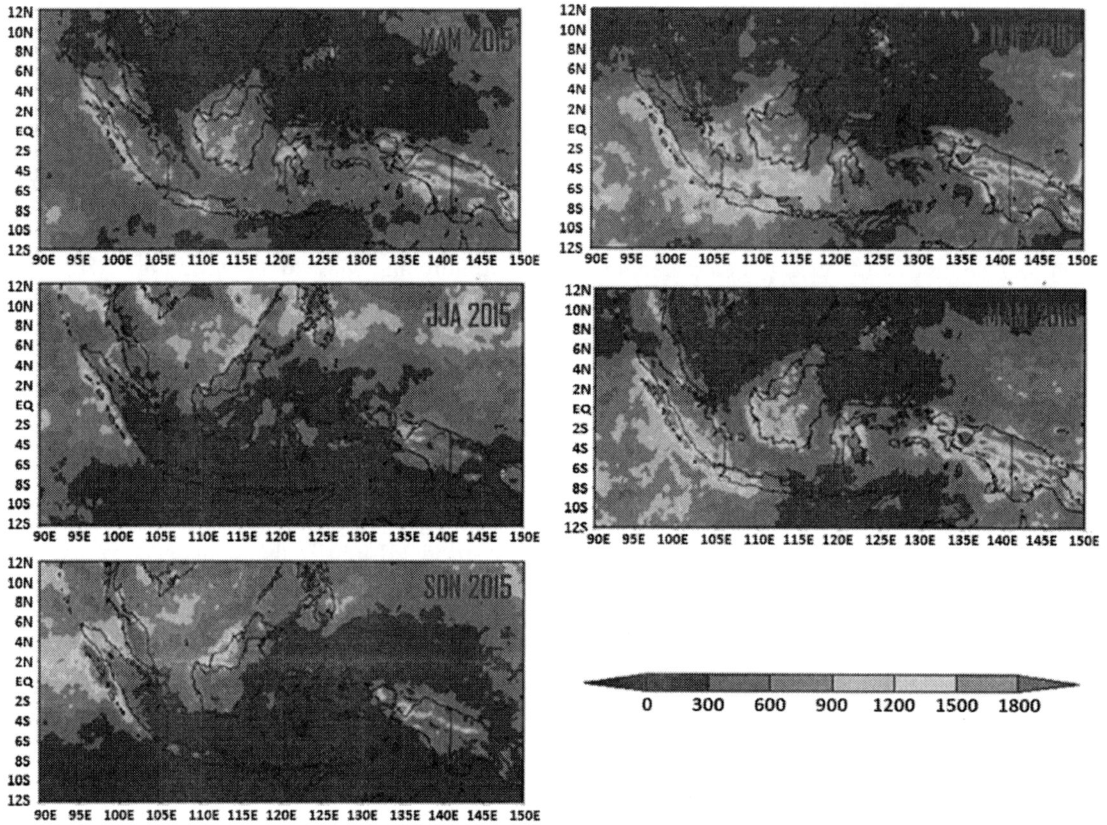

Figure 5. The spatial pattern of accumulated precipitation (mm) in El Niño 2015-2016 in Indonesia.

In the rainy period (DJF 2016), the visible pattern of rainfall in most parts of Indonesia was back to normal where the accumulation ranging from 600-1500 mm, except in East Nusa Tenggara, North Sulawesi, Maluku, was still dry with accumulated rainfall of less than 300 mm. In that period, El Niño still had a strong intensity but IOD had normal phase. The strengthening of monsoon wind Asia was more dominant in parts of Indonesia so that it would support the process of convection in most parts of Indonesia. The presence of wet air masses was brought into the territory of Indonesia. These results are in line with previous studies [13] which reported that the geographical aspects had an effect on the rainfall anomaly pattern. Based on data from the model output, the peak period of the rainy season (DJF) has relationship with El Niño and positive rainfall anomalies in Java.

However, during the transition from the rainy season to the dry season (MAM 2016) the moderate intensity of El Niño continued to weaken and became neutral. During the period of MAM 2016, the rainfall in all parts of Indonesia ranged between 600-1800 mm except in Papua highlands (over 1800 mm), Bali, Nusa Tenggara which were dry less than 300 mm as shown in **Figure 5**.

This research also analyzed the distribution of rainfall and OLR in zonal (E-W) and meridional (N-S) in Indonesian region. Zonal distribution was done by daily average rainfall / OLR data processing based on latitude vs time, and the distribution was done meridional with daily data processing of rainfall / OLR daily based on longitude vs time. The results of data processing for daily rainfall distribution and daily OLR are zonal and meridional as shown in **Figures 6** and **Figures 7**, respectively.

Figure 6 shows the pattern of zonal distribution of rainfall indicating that in the El Niño period with weak intensity to moderate while IOD is still in neutral condition (early March - early June 2015). Indonesian territory is dominated by daily rainfall with very light intensity, less than 5 mm especially the area around 102.5E-130E, while in the west (90-102,5E) and east (130-150E) visible rainfall intensity varies from very light to light (5-20 mm). However, in mid to late June mild the mild and moderate intensity of rainfall predominates in Indonesia especially in the east (125-150E).

During a strong intensity of El Niño accompanied by positive phase IOD, from early July to October, the drought-stricken Indonesian region was expanding and its longer period was indicated by the dominance of very mild daily rainfall intensity of less than 5 mm, except in the west (90-100E) that varied from very light to light (5-20 mm). At the peak of the strong El Niño around mid-November 2015 and subsequently entering moderate and weakening around the end of March 2016, the areas experienced severe drought where the decreaseing of rainfall intensity less than 5 mm mainly occurred around 115E -130E because IOD entered the neutral phase.

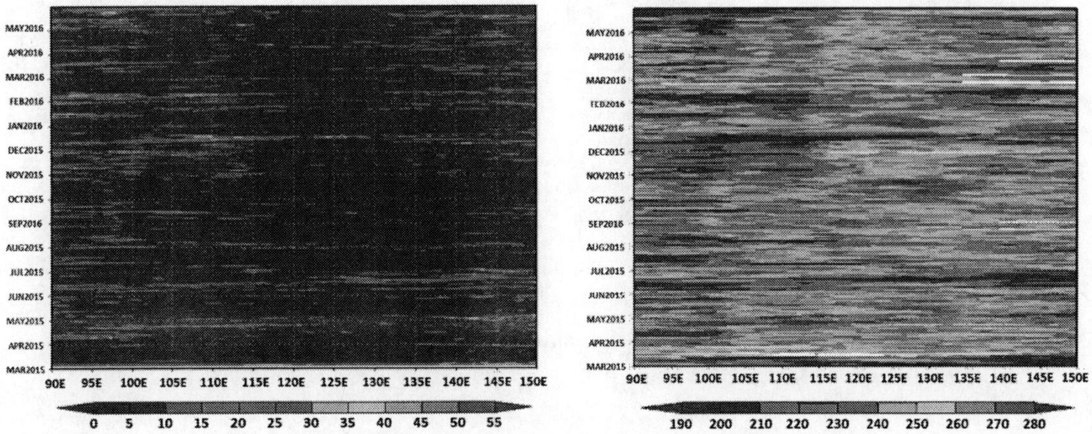

Figure 6. The zonal distribution of daily rainfall (left) and daily OLR (right) in the area of research in the period March 2015-May 2016.

While in the west (90-115E) and eastern (130-150E), the intensity of rainfall dominated by light intensity is 5-20 mm as shown in **Figure 6** (left). Based on the OLR data, during the study period the relationship is close where the obtained OLR pattern is very similar to the rainfall pattern which has opposite value. **Figure 6** (right) shows that during the period of rainfall with a very light intensity of less than 5 mm in the location, the high OLR value was between 230-280 W/m2. While the areas with greater rainfall or with light intensity of 5-20 mm rainfall had smaller OLR value dominated less than 210 W/m2.

Based on the results of meridional distribution, data processing for rainfall (**Figure 7**, left) during the weak and moderate El Niño period (March-June), the dominant pattern of daily rainfall with a very light intensity of less than 5 mm occurred in the north of the equator, especially 2N-12N and south of equator especially 8S-12S. However, when El Niño took place with strong intensity and with positive phase IOD (July-December), the southern areas of the equator sufferred from severe drought where the dominance of daily rainfall intensity was very light less than 5 mm. As for the area around 6N-2S, the rainfall intensity varied from very light (<5 mm) to light (2-20 mm), but the daily rainfall to the north (6N-12N) with moderate intensity (20 -50 mm) was higher. Furthermore, during the period of El Niño to moderate and weaker with IOD entering the neutral phase, the rainfall with a very light intensity of less than 5 mm was predominantly in the north of the equator while in the south of the equator it varied from very mild to light intensity. The daily distribution of rainfall occurred during El Niño 2015/2016 in Indonesian territory which was in line with the distribution of OLR over Indonesian territory. In the meridional distribution, the smaller the rainfall in the region, the greater the value of OLR (**Figure 7** right).

Based on the analysis of zonal and meridional distribution of rainfall and OLR, the the relationship (both negative correlation) between the two parameters was significant, because OLR can describe the distribution of convective areas indicating their thickness and the number of cloud cover. In this study, a pattern that shows smaller OLR and greater rainfall occurred in the area was obtained. This result is line with research [24] that OLR is negatively correlated with rainfall, and also research [25] suggests that OLR is negatively correlated with the ITCZ (Inter Tropical Convergence Zone).

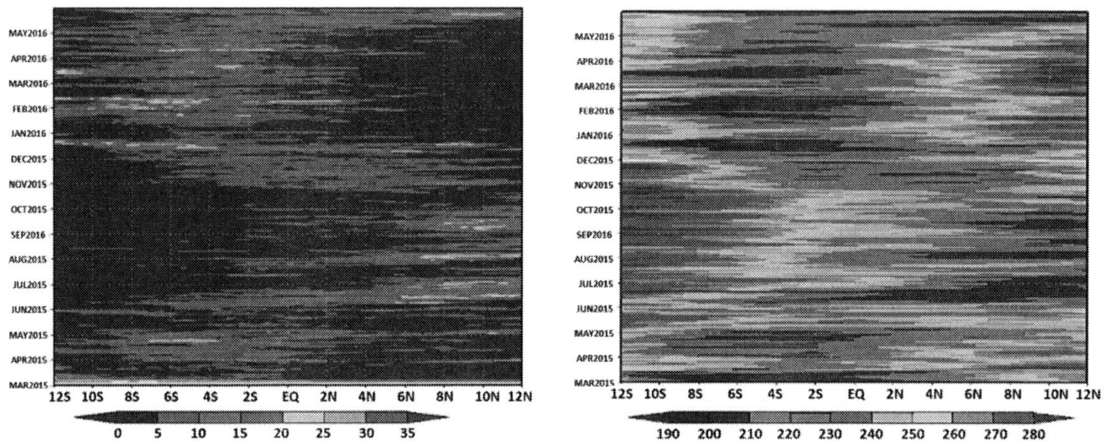

Figure 7. Meridional distribution of daily rainfall (left) and daily OLR (right) in the area of research in the period March 2015-May 2016.

4. Conclusion

Based on the analysis of SST anomalies in the Niño 3.5, it shows that El Niño phenomenon occurred in the Pacific Ocean and El Niño intensity was strong (positive anomalies> 1.5°C) since the beginning of July 2015 until early March 2016, and reached its peak with a positive anomaly 2.48 ° C in mid-November 2015. In mid-August to mid-November 2015, the data analysis shows that the IOD phenomenon in the Indian Ocean also has a positive phase. The analysis shows that SST and winds in the Pacific Ocean and the Indian Ocean significantly affect rainfall patterns in Indonesia. During El Niño 2015/2016 with strong intensity, especially the period of 2015, JJA precipitation declined and reached 200-400 mm / season in parts of Indonesia except for Aceh and North Sumatra which are partially in normal conditions. In the transitional period, (SON) coincided with the lasting period of positive IOD phase. It has led to increasingly severe drought and widespread in Indonesia. The worst drought areas occurred mainly in the southern part of the equator. The rainfall of 400-600 mm / season is lower than normal conditions. The merger of two global phenomena in the Pacific Ocean and the Indian Ocean exacerbated drought in Indonesia. The rainfall data have good GSMaP which can be used as an alternative in the study on the impact of El Niño in Indonesia spatially.

Acknowledgments

We are grateful to Center of Atmospheric Science and Technology, Indonesian National Aeronautics and Space Agency (LAPAN) to support this research. We also thank to JST and team GSMaP Project, IRI, BoM, Australia, and CPC/NOAA for their data used in this study.

References

[1] McGregor G R and Nieuwolt S 1998 *Tropical Climatology*. John Wiley & Sons Ltd. United Kindom
[2] Shelton M L 2009 *Hydroclimatology: Perspectives dan Applications*. Cambridge University Press. United Kingdom
[3] Trenberth K E and Caron J M 2000 The Southern Oscillation Revisited: Sea Level Pressures, Surface Temperatures and Precipitation *Journal of Climate* **13** 4358 – 4365
[4] Kane R P 1997 Relationship of El Niño-Southern Oscillation and Pacific Sea Surface Temperature with Rainfall in Various Region of the Globe *Mon. Wea. Rev.* **125** 1792-1800
[5] Ropelweski C F and Halpert M S 1987 Global and Regional Scale Precipitation Pattern Associated with The El Niño /Southern Oscillation *Mon. Wea. Rev.* **115** 1606-1626.
[6] Luo J J, Zhang R, Behera S K, Masumoto Y, Jin F F, Lukas R, and Yamagata T 2010 Interaction between El Niño and extreme Indian Ocean Dipole *Journal of Climate* **23** 726–742
[7] Aldrian E, Gates L D, and Widodo F H 2007 Seasonal Variability of Indonesian Rainfall in ECHAM4 Simulations and in the Reanalyses: The Role of ENSO. *Theoretical and Applied Climatology*, **87** 41–59
[8] As-syakur A R 2012 Analisis Pola Spasial Hubungan Curah Hujan dengan ENSO dan IOD di Indonesia Melalui Observasi Data TRMM 3B43 ed. Wikantika K. dan Fajri L *Bunga Rampai Penginderaan Jauh Indonesia* Pusat Penginderaan Jauh Institut Teknologi Bandung (ITB) Cetakan Pertama 92-107
[9] As-syakur A R, Adnyana I W S, Mahendra M S, Arthana I W, Merit I N, Kasa I W, Ekayanti N W, Nuarsa I W and Sunarta I N 2014 Observation of Spatial Patterns on the Rainfall Response to ENSO and IOD over Indonesia Using TRMM Multisatellite Precipitation Analysis (TMPA) *International Journal of Climatology* **34** 3825–3839
[10] Saji N H, Goswami B N, Vinayachandran P N and Yamagata T 1999 A Dipole Mode in the Tropical Indian Ocean *Nature* **401** 360-363
[11] Yamagata T, Behera S K, Luo J J, Masson S, Jury M R and Rao S A 2004 *Coupled Ocean-Atmosphere Variability in the Tropical Indian Ocean*, AGU Book Ocean-Atmosphere Interaction and Climate Variability, ed. Wang C, Xie S P and Carton J A Geophys. Monogr. **147** AGU Washington D C 189-212
[12] Hendon H H 2003 Indonesian Rainfall Variability: Impacts of ENSO and Local Air–Sea Interaction *Journal of Climate* **16** 1775–1790
[13] Qian J H, Robertson A W, Moron V 2010 Interaction Among ENSO, the Monsoon, dan Diurnal Cycle in Rainfall Variability Over Java, Indonesia. *Journal of the Atmospheric Sciences* **67** 3509-3524
[14] Ushio T, Sasashige K, Kubota T, Shige S, Okamoto K, Aonashi K, Inoue T, Takahashi N, Iguchi T, Kachi M, Oki R, Morimoto T, Kawasaki Z, 2009 A Kalman filter approach to the Global Satellite Mapping of Precipitation (GSMaP) from combined passive microwave and infrared radiometric data. *J. Meteor. Soc. Japan* **87A** 137-151
[15] Tian Y, and Peters-Lidard C D, 2010. A global map of uncertainties in satellite-based precipitation measurements *Geophys. Res. Lett.* **37** L24407 doi:10.1029/2010GL046008
[16] Kubota T, Shige S, Hashizume H, Aonashi K, Takahashi N, Seto S, Takayabu YN, Ushio T, Nakagawa K, Iwanami K 2007 Global precipitation map using satellite-borne microwave radiometers by the GSMaP Project: production and validation *IEEE Transactions on Geoscience Remote Sensing* **45** 2259–2275
[17] http://www.cpc.ncep.noaa.gov. (accessed on December 17, 2016)
[18] http://www.bom.gov. (accessed on January 7, 2017)
[19] GSMaP Earth Observation Research Center accessed on April 27, 2017 available online on: http://sharaku.eorc.jaxa.jp/GSMaP_crest/ (accessed on April 27, 2016)

[20] IRI, International Research Institute for Climate and Society/Earth Institute/ Columbia University, tersedia online https://iridl.ldeo.columbia.edu (accessed on April 27, 2017)
[21] http://www.bmkg.go.id/BMKG_Pusat/Sestama/Humas/MEI_JUNI_sebagian_besar_ wilayah_ Indonesia_masuk_awal_musim_kemarau.bmkg (accessed on January 7, 2017)
[22] Wang B, Li J, and He Q 2017 Variable and robust East Asian monsoon rainfall response to El Niño over the past 60 years (1957-2016). *Advances in Atmospheric Sciences*, **34** (10), 1235-1248, doi:10.1007/s00376-017-7016-3. IPRC-1259.
[23] Aldrian E, and Susanto R D 2003 Identification of three dominant rainfall regions within Indonesia and their relationship to sea surface temperature *International Journal of Climatology* **23** 1435–1452
[24] Motell C E and Weare B C 1987 Estimating Tropical Pacific Rainfall Using Digital Satellite Data *J. Climate Appl. Meteor.* **41** 604-613
[25] Murakami T and Matsumoto J 1994 Summer Monsoon over the Asian Continent and Western North Pacific, *Journal of the Meteorological Society of Japan* **72** 719-745

IOP Conf. Series: Earth and Environmental Science **166** (2018) 012035 doi:10.1088/1755-1315/166/1/012035

Development of Flash Flood Hazard Map in Bima City (NTB) using Analytical Hierarchy Process

R. P Romdani[1], M Tamamadin[1], A Susandi[1], A Pratama[1] and A. R Wijaya[1]

[1]Faculty of Earth Sciences and Technology, Institut Teknologi Bandung, Ganesa 10 Bandung 40132, West Java, Indonesia

E-mail: mamadtama@yahoo.com

Abstract. Flash flood is one of the hydro-meteorological hazards that often occurs in Indonesia. Some major factors affecting this hazard are high topography and weak soil structure. Every incident of flash flood has caused losses in the form of damage to infrastructure, victims, and the other losses. Therefore, flash flood hazard areas should be mapped so that community and local government can anticipate the event during heavy rainfall. This study developed a flash flood hazard index using a decision-making model, with approach of the Analytic Hierarchy Process (AHP). With this method, four parameters were calculated, namely the distance to the stream channels, slope, cover and soil types. The weight and rank values were assigned to the layers and classes of each layer respectively. A hazard map was obtained using an algorithm that combines factors in weighted linear combinations. The index of flash flood hazard zone with rainfall factor was analysed with sensitivity analysis shows better result since 13 of 16 flash flood points (81.25%) are in high vulnerability to flooding.

1. Introduction

Flash flood is a flood occurred in a short time caused by heavy rainfall or broken dam, characterized by rapid increase of river water level [1]. Flash flood is noticed as the most dangerous disasters since it combines destructive hazards with high speed and unpredictable flow [2]. The factors contributing on this disaster consist of intensity of rainfall, rainfall duration, land use, topography, and slope of river.

The disaster of flash flood has always occurred in Bima City, Nusa Tenggara Barat (NTB). It has occurred 4 times until 2016. The hugest flash flood was on 21 December 2016 and 23 December 2016 [3]. The occurrence impacted on mortality, damage of shelters, damage of buildings, failure of agricultures, and breakage of other infrastructures. Based on the final report of Bappenas (2017), the damage of building because of flash flood on December 2016 reached 439 buildings that consist of houses, offices, schools, healthy units building, stores, bridges and farm land. The total of financial damage was about Rp. 984,4 billion.

To solve the problem, central government has possessed agency on disaster management to address the disaster occurrence coordinating with local level government. In operational action to cope the flash flood, government has developed index of flash flood in form of spatial mapping for Indonesia coverage. People could visit this URL at www.inarisk.bnpb.go.id to access [4]. Index of flash flood in every district can be identified with pixel resolution up to 100 meter. One of the factors that is not included in this index of flash flood is rainfall parameter.

The development of flash flood hazard index was ever done by Kim (2011). He develop a flash flood index through the average of the same scale relative severity factors quantifying characteristics of hydrograps generated from a rainfall-runoff model for the long term observed rainfall data, and presents regression equations between rainfall characteristics and flash flood index. Overall, the flash flood index shows a much stronger relation to some rainfall data with relatively high coefficient of determination with the value is 0,860.

There is a need to input rainfall factor in the index of flash flood since the disaster usually occurs if there is occurrence of rain. One method in building flash flood index that can be rainfall as one of inputs is Multi Attribute Utility Theory (MAUT) supported with Analytical Hierarchy Process (AHP). The method was used by Kazakis (2015) [5] in mapping the flash flood vulnerability zone with the rainfall as one of inputs. Through this paper, the research is to analyze and calculate flash flood in Bima city as one of the biggest flash floods once occurred and determines the biggest influence of parameter that cause the flash flood.

2. Data and Methods

The data used in this research consist of Digital Elevation Model (DEM) from Shuttle Radar Topography Mission (SRTM) in 30 meters of resolution (https://earthexplorer.usgs.gov/), land cover, watershed map, and rainfall from Global Precipitation Measurement (GPM) (https://pmm.nasa.gov/data-access/downloads/gpm). The whole data was resampling to be 30 meters of resolution as similar to DEM data.

Analytical Hierarchy Process (AHP) was used to produce various weights of each parameter in order to develop index of flash flood vulnerability zone in Bima City. In this method, the complex condition is structured into the certain parts. Then, the parts are arranged in a hierarchical arrangement and provide numerical values based on the subjective of important variables. The AHP also synthesizes the highest priority and acts to affect to final results [6]. In case of flood, the mapping evaluation and emergency management could be identified by the AHP effectively [7].

Equation of flash flood index mapping was made based on the flood index conducted by Kazakis (2015) as shown on equation (1).

$$IBB = \sum_{i=1}^{n} P_{ri} P_{wi} = SLP_r SLP_w + FA_r FA_w + SG_r SG_r + LC_r LC_w + CH_r CH_w \quad (1)$$

equation explained:

IBB : Flash flood index
P : Parameters
ri : rate of parameter
W_i : Weight of parameter
N : Amount of parameter
CH : Rainfall
LC : Land Cover
FA : Flow Accumulation
SLP : Slope
SG : Distance to River

In determining the most priority from the various parameters effecting to flash flood, pairwise comparison was used to compare each pair of parameters based on the determined criteria. Comparison of the pairs was represented in a matrix. The scale used in this case study to fill the matrix is 1 to 9. The value 1 means same dominant, 3 means a bit more dominant, 5 means more dominant, 7 means strongly dominant, and 9 means absolute dominant. Value 2, 4, 6, and 9 are as intermediate values. The weight was computed using method used by Saaty (2004) [8] to obtain the most priority parameters in a matrix of pairwise comparison.

Then, the sensitivity analysis was used to evaluate the impact of each parameters to index of flash flood zone map. This attempt could make us understand the role of each parameters in identifying the flash flood since it provides information on effect or weight value given to the criteria. This method has been also implemented by Kazakis (2015) in analyzing the change of parameters in the evaluation of flood risk area using index from AHP in Rhodope-Evros, Yunani.
In the sensitivity analysis, the initial weight obtained from the AHP process was changed to effective weight obtained from the Equation 2.

$$W = \frac{P_r . P_W}{V} . 100 \qquad (2)$$

Equation explained:
W : Effective weight for parameter
P_r : Parameter rate
P_w : Parameter weight
V : Aggregate value of index

The effective weight was obtained from the calculation then it was used to make and calculate the flash flood zone index revised by sensitivity analysis.

In this research, the rainfall data are divided into 2 conditions, those are unprocessed condition (normal condition) and processed condition (distributed evenly in the watershed). The parameter data is reclassified according to the reference and then the weight was searched with the AHP method. The weight gain was then given into Eq. (1). The index obtained from the calculation was then analyzed using Sensitivity Analysis method.

3. Result and Discussion

The selection of parameter sequences in pairwise comparison matrices is based on the research of [9] which states that the slope of the slope is an important factor because flash flood usually occurs in the lower slope. After that, the flow accumulation is as the second most important parameter because this parameter could identify the high flow accumulation area that will occur.

The distance of the territory to the river network becomes the third most important parameter, because, in the case of flash flood, it is mentioned that the most affected area by flash flood is near the river, as a consequence of river overflow. Furthermore, land cover becomes the fourth most important parameter, since land cover reduces infiltration capacity and runoff from paved areas that can add substantially to total runoff. Urbanization usually leads to decrease in time lag, increase of peak discharge, and increase of total discharge for certain floods.

Table 1. Normalization of flash flood parameters to generate weights (W_i)

Parameter	Slope	Flow acc	River	Land cover	Rainfall	Mean	wi
Slope	0,3900	0,4000	0,3100	0,3300	0,4200	0,3900	3,9000
Flow acc	0,1900	0,2000	0,3100	0,2200	0,2100	0,1950	1,9500
River	0,1900	0,1000	0,1500	0,1100	0,2100	0,1467	1,4700
Land cover	0,1300	0,2000	0,1500	0,1100	0,0500	0,1475	1,4700
Rainfall	0,1000	0,1000	0,0800	0,2200	0,1100	0,1220	1,2200

IOP Conf. Series: Earth and Environmental Science **166** (2018) 012035 doi:10.1088/1755-1315/166/1/012035

Rainfall becomes the fifth important parameter because the study of rainfall in flash flood events is still not much discussed and the limitations of rainfall data due to spatial resolution are still low enough for a small study area. So as to know the weight of each parameter, the result of pairwise comparison matrix has been normalized. The results of parameter weights can be seen in **Table 1**.

In the calculation of the index of flash flood zone (*Indek Bahaya Banjir*, IBB) from the weight of AHP as seen in the condition of unreacted rainfall shows much less vulnerable area compared to IBB with rainfall processed. This is shown by the red mark in **Figure 4**.

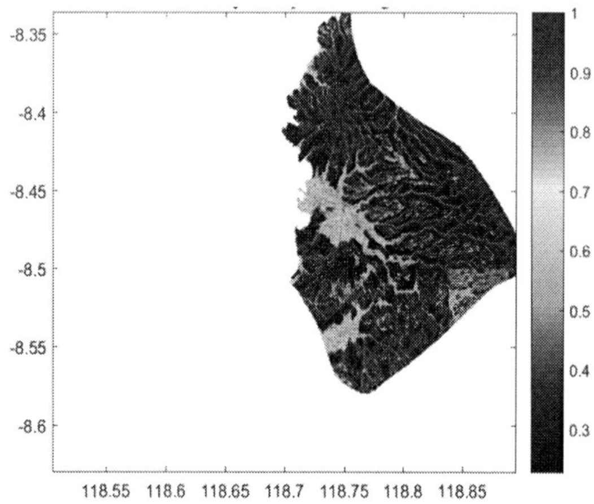

(a)

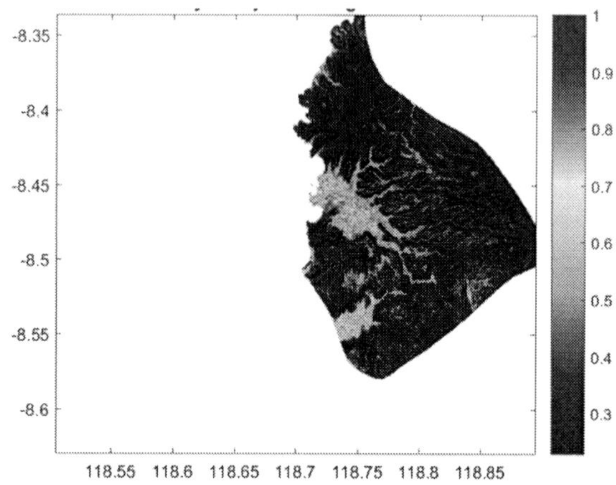

(b)

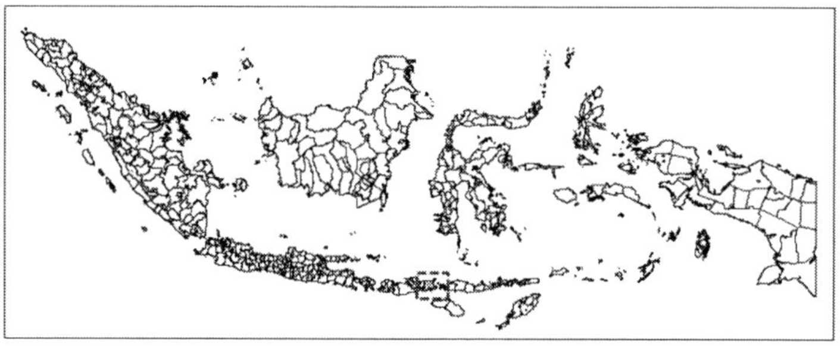

(c)

Figure 1. (a) IBB without rainfall factor, (b) IBB with rainfall in the process, (c) Bima city position in Indonesia map.

The author has paired IBB with a sensitivity analysis process that evaluates the impact of each criterion in the method used. This helps to understand the role of each parameter in the risk of flash flood disaster, since sensitivity analysis is used to clarify the significance of subjective assessments of the various criteria used. The results of the sensitivity analysis can be seen in **Table 2**.

From the sensitivity analysis, it can be seen that almost of all IBB parameters assumed to use AHP is underestimated. This results can be compared with the weight of the theory with the effective weight of the sensitivity analysis. In the study area, it has been identified that the slope, land cover and accumulation have considerable influence. Need to be noticed that the influence of rainfall is processed larger than the unprocessed.

The index of flash flood hazard is classified into 3 zone classes, namely low vulnerability (0-0.33), middle vulnerability (0.33-0.66) and high vulnerability (0.66-1). Then the level of compliance with the flash floods was analyzed for index of flash flood in Bima City on occurrence on December 21, 2016.

Table 3 shows that there are 16 flash flood points in the flash flood incident in Bima City. 8 of them are classified into high vulnerability class at IBB. Whereas in IBBS, 3 flash flood points are classified into middle vulnerability class and 13 points are in high vulnerability class. This suggests that IBBS is better in representing the actual conditions. From the analysis above, map of flash flood index zone is suitable using map considering rainfall factor (**Figure 2**). The map uses equation of IBBS = 1,7 slope + 0,37 flowacc + 0,23 river + 0.75 lc + 0.15 ch.

Table 2. Sensitivity analysis with rainfall and no rainfall factor

Parameter	Weight	Effective weight with rainfall factor	Effective weight with rainfall factor
Slope	3.9	18.24	17.55
Flow accumulation	1.95	3.87	3.72
Distance to river	1.47	2.45	2.38
Land cover	1.47	7.88	7.58
Rainfall	1.22	1.23	1.54

Table 3. The suitability of index class of vulnerable zone with flash flood point on the occurrence of flash flood in Bima City on December 21, 2016

Class	IBB		IBBS	
	Area (%)	Point of flash flood	Area (%)	Point of flash flood
Low (0-0.33)	81,48%	0	62,62%	0
Middle (0.33-0.66)	17,01%	8	30,59%	3
High (0.66-1)	1,51%	8	6,79%	13

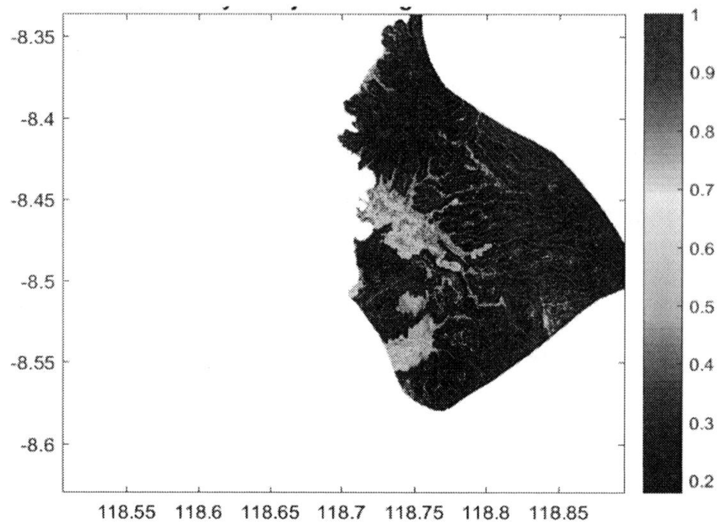

Figure 2. The index of flash flood vulnerability map with rainfall factor in flash flood occurrence on December 21, 2016

4. Conclusions

This research has produced a map of flash flood hazard map in Bima City of West Nusa Tenggara Province using Analytical Hierarchy Process (AHP). Using this method, it can be seen in fact that the weight for the calculation of flash flood hazard zone index from weighting result is still underestimated. The equation for determining the index of flash flood hazard zone in Bima City uses equation which parameter weights were analyzed using sensitivity analysis. However, since the index of flash flood prone zone considering rainfall parameter analyzed combined with sensitivity analysis shows better result. The map shows that 13 of the 16 flash flood locations (81.25%) are in high vulnerability to flooding. Therefore, the AHP method should be consideration in mapping of flash flood index combined with rainfall parameter.

Acknowledgment

This paper describes the result of research that is under the joint cooperation of Bandung Institute of Technology and National Disaster Management Authority. The publication of this paper is also supported by Bandung Institute of Technology and Ministry of Research, Technology and Higher Education of the Republic of Indonesia.

5. References

[1] Adi S 2013 Characterization of Flash Flood Disaster in Indonesia *Journal of Sains dan Teknologi Indonesia* **15** (1) 42-51
[2] Shahirah 2012 A Study on Causes of Flood at Sungai Isap and Jalan Tun Ismail, Kuatan, Pahang. Undergraduates Project Papers, University Malaysia Pahang
[3] http://bimakota.go.id/ accessed on 27 August 2017
[4] http://dibi.bnpb.go.id/ accessed on 6 March 2017
[5] Kazakis N, Kougias I, Patsialis T 2015 Assesment of Flood Hazard Areas at Regional Scale Using an Index-based Approach and Analytical Hierarchy Process: Application in Rhodope-Evros Region, Greece. *Science of the Total Environment* **538** 555-563
[6] Saaty T L and Forman E H 1993 The Hierarchon - A Dictionary of Hierarchies. Pittsburg: RWS Publications
[7] Kim S and Choi H 2011 Assessment of Vulnerability to Extreme Flash Floods in Design Storms *International Journal of Environmental Research and Public Health* Vols 2 to 15
[8] Tanggap Bencana team of PPRPT 2017 *Kajian Singkat Pasca-Bencana Bandang Kota Bima 2016* Kementerian Agraria dan Tata Ruang

IOP Conf. Series: Earth and Environmental Science **166** (2018) 012035 doi:10.1088/1755-1315/166/1/012035

[9] Levy J K 2005 Multiple Criteria Decision Making and Decision Support System for Flood Risk Management *Stochastic Environment Research and Risk Assessment* **19** (6) 438-447

Land Cover and Groundwater Recharge Changes Study Using Conservation Index in the Highland Ciburial Village, Bandung Regency

M. H. Z Nurrachman[1*], I Sumardi[1] and T Lastini[1]

[1]Forestry Engineering Department, School of Life Sciences and Technology, Bandung Institute of Technology, Ganesa 10, Bandung, West Java 40132, Indonesia

E-mail: mhafizh.zhafran@outlook.com

Abstract. Ciburial Village, Cimenyan Region, is a potential location for Bandung Basin groundwater recharge area. Based on West Java Local Regulation No. 1 Year 2008 (Perda No.1/2008) about Northern Bandung Region Management, Ciburial Village is classified as protection area. As a matter of fact, constructed area in Ciburial Village has been increasing. Thus,it causes decreasingin water protection function. Hence, study on degraded land impact estimation to water conservation in Ciburial Village is needed. To estimate how much impact had been done by degraded land, spatial use in Ciburial Village was analyzed. To analyze spatial use, land cover of Ciburial Village isdivided into four types, settlements, forests, agricultures and grasslands. Each land cover's water conservation ability was approached by conservation index. The spatial analysis between 2006 and 2015 shows the decreased area of good level and normal level water conservation index by 1.45 hectares and 1.25 hectares respectively and shows the increased area of critical level water conservation index by 2.7 hectares. Land cover changes also impactonactual conservation index, whichdecreased from 0.610 on 2007 to 0.606 in 2015.

1. Introduction

Groundwater is one of primary raw water sources for municipal activity in Bandung City. Nowadays groundwater in Bandung has become primary source for domestic use instead of surfacewater which has been polluted [1,2]. Bandung Public Waterworks uses groundwater 38 million m^3 annually and industrial activity uses it up to 50 million m^3 annually. The consequence of high groundwater abstraction in Bandung City was groundwater level in Bandung City dropped from +4 (upward) m on 1920 to -60 (downward) m in 1990[3]. The amount is still increasing, recorded in 2004, groundwater level in Bandung City dropped into -100 m below thegroundand it keeps running until today [3].

One of groundwater recharge zones for Bandung City's groundwater is Northern Bandung Region. However, the development in Northern Bandung Region is inevitable [4]. Cimenyan District, one of the districts of Bandung Regency in Northern Bandung Region, has area of 5,308.00 hectares and it has constructed area of 1,130.85 hectares (21.3% of Cimenyan's area) [5]. Development in Cimenyan District shows erosion potential of 57.59 tons/hectare/year [6]. The erosion results in the increased runoff from 331.1 mm out of 1,474.3 mm precipitation (22%) in 2006 to 457.4 mm out of 1,800.3 mm precipitation (25%) in 2015 [6]. Because of the increased runoff, it causes decreasing in infiltration. In order to conserve more water for Bandung City's groundwater, we need to calculate or at least estimate the water entering Bandung City's aquifer. Hence, in order toestimate the conserved water, we conduct research in water conserving model using conservation index.

2. Method and materials

2.1 Study site

The research was conducted in Ciburial Village, Cimenyan District, Bandung Regency. Ciburial Village has an area of 5.99 km^2. The area of the study site is 1.65 km^2 and located in the upper stream of Ciburial Village, 6.835-6.852 °S and 107.654-107.675 °E (**Figure 1**). Plot elevation was on 1,175-1,375 m asl. Ciburial Village has climate condition with average annual precipitation of 1,500 mm, average humidity of 77%, and average air temperature of 24-25 °C [7].

Study site's land coveris divided into four types, namely constructed area, forest area, agriculture area, and grassland area. To estimate the conserved water, we used conservation index by spatial analysis.

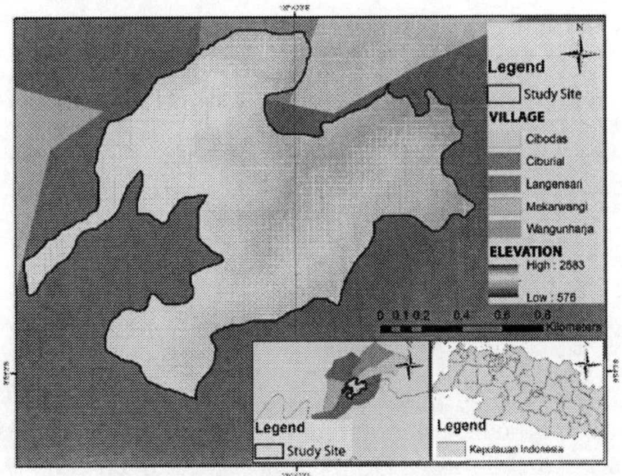

Figure 1. Study site area.

2.2 Spatial analysis

*2.2.1 Types and data source.*The spatial data source was secondary data. The spatial data used was spatial physical characteristic data. The detail of data types and data source is presented in **Table 1**.

Table 1. Data types and source.

Data Types	Data Source
1. Administrative Map	Geospatial Information Agency, scale 1:25.000, year 2015
2. Contour and River Map	Geospatial Information Agency, scale 1:25.000, year 2015
3. Soil Texture Map	Settlement and Housing Service West Java Prov., year 2005
4. Satellite Imagery	Digital Globe (Quick Bird year 2007, WorldView-02 year 2012, and WorldView-03 year 2015) and Airbus (SPOT 6 year 2014)

2.2.2 Spatial processing. To get useful data from the raw data, we processed every data uniquely. Contour and river map data were processed to be a slope/topography data. Satellite imagery was digitized to be a land cover data. Ground truth was conducted to validate digitization process.

*2.2.3 Conservation index analysis.*Conservation index is a parameter of hydrological ability in an areato conserve water. The index has value ranging from 0 to 1, in which 0 is totally runoff and 1 is totally infiltrated.The conservation indexcomes from hydrological balance, which precipitated water is equal to infiltrated water plus runoff water and evaporated water. Assuming that there is not enough energy to evaporate water, the balance becomes precipitation (P) and it is equal to infiltration (I)and runoff (R). Then, the equation was divided by precipitation. The result is 1 equal to infiltration per precipitation and runoff per precipitation.The infiltrated water per precipitated water is the conservation index (*IK*)[8]. Since *R/P* is equal to runoff coefficient (*C*), and we need to estimate conservation index,the runoff coefficient is calculated. The runoff coefficient uses five parameters, namely rainfall, soil texture, slope, elevation, and land cover. However, with a small catchments of study site, rainfall and elevation parameters can be neglected[8].

$$P = I + R \tag{1}$$

$$1 = (I/P) + (R/P) \tag{2}$$

$$IK = 1 - C \tag{3}$$

To calculate the runoff coefficient, we used modified Hassing Runoff Coefficient[8].Hassing Runoff Coefficient is a runoff coefficient based on topography, soil texture, and land cover[9].The detail of Hassing Runoff Coefficient is presented in **Table 2**.

Table 2. Modified Hassing Runoff Coefficient Classification.[11]

Runoff coefficient $C=C_T+C_S+C_V$					
C_T (Topography)		C_S (Soils)		C_V (Land Cover)	
Very flat (<1%)	0,03	Sand and gravel	0,04	Forest	0,04
Undulating (1-10%)	0,08	Sandy clays	0,08	Farmland	0,11
Hilly (10-20%)	0,16	Clay and loam	0,16	Grassland	0,21
Mountainous (>20%)	0,26	Sheet rock	0,26	Settlement	0,40

Conservation index can be divided into two groups, natural conservation index (IK_A) and actual conservation index (IK_C). Natural conservation index is a parameter which shows ability of an area to conserve water naturally or assumed forest. On the other side, actual conservation index is a parameter which shows ability of an area to conserve water on actual/existing land use. Then, water conservation status can be estimated by actual conservation index which has normal level value ranging from 0.46-0.69. If it is higher than 0.69 it means good level and if it is lower than 0.46 it means critical level[8].

3. Results and Discussion

3.1 Topography

Based on contour map data from Geospatial Information Agency, the study site was dominated by high slope/mountainous area (>20% slope). The mountainous area consists of 72.11 hectares or 44% of study site area (**Table 3**).Figure 2ashows almost every part of study site which has high slope. Part of study site that has flat slope is scattered nearby streams or ridges.

3.2 Soil texture

Based on soil texture map data from Settlement and Housing Services West Java Province year 2005, the soil texture was sandy clays throughout the study site (**Table 3** and **Figure2b**).

Table 3. Distribution of topography and soil texture in study site.

Physical Condition	Area (ha)	Percentage from Total (%)
	Topography	
Very flat (<1%)	0,46	0,3
Undulating (1-10%)	28,50	17,4
Hilly (10-20%)	62,82	38,3
Mountainous (>20%)	72,11	44,0
	Soil Texture	
Sandy clays	163,89	100

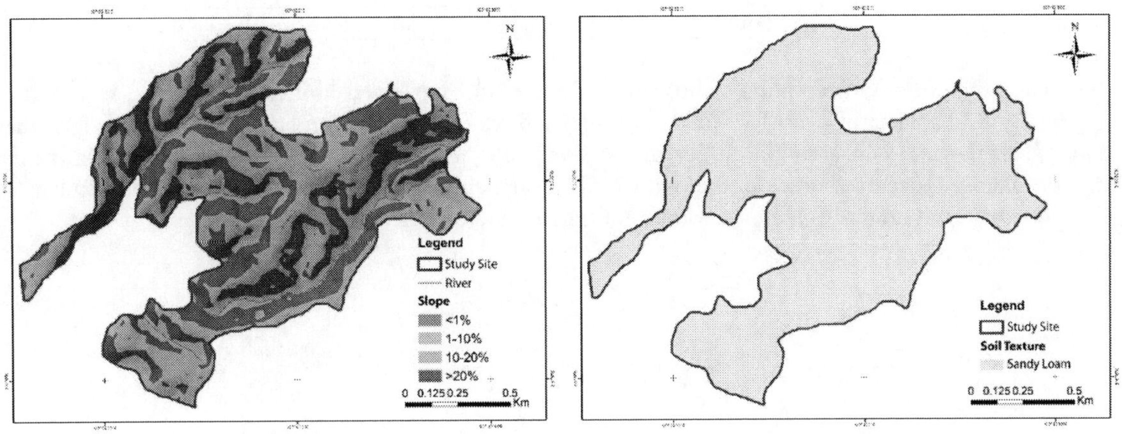

(a) (b)

Figure 2. Map distribution of (a) Topography (b) Soil texture.

3.3 Land cover

Data from satellite imagery shows changing area of settlements, forests and agricultures (**Table 4**). The settlements seem increasing over the time, the forests and agricultures are both increasing and decreasing. Population growth in upper Ciburial Village has pushed local communities to expand settlements[4]. Local activities affect land use. Between 2007 and 2012, *Eucalyptus* spp. and *Gmelinaar borea*were planted in part of the study site, it was shown by satellite imagery and ground truthing. On the other side, the grassland has no changes.According to the interviews with local community, the grassland was used as grazing area which has been established since 2000.

Table 4. Land cover changes in study site between 2007-2015.

No	Land Cover	2007		2012		2014		2015	
		Area (ha)	%	Area (ha)	%	Area (ha)	%	Area (ha)	%
1	Settlements	7,21	4,4	9,34	5,7	9,45	5,8	9,92	6,1
2	Forests	6,71	4,1	10,29	6,3	10,06	6,1	9,66	5,9
3	Agricultures	149,59	91,3	143,89	87,8	144,01	87,9	143,94	87,8
4	Grasslands	0,37	0,2	0,37	0,2	0,37	0,2	0,37	0,2
	Total	163,89	100	163,90	100	163,90	100	163,90	100

After collecting the needed data, we could estimate average actual conservation index in study site as in **Table 5**. After estimating actual conservation index, then we valued water conservation status. Water conservation status in study site is shown in **Table 6**.

Table 5. Average actual conservation index changes in study site between 2007-2015.

IK_C 2007	IK_C 2012	IK_C 2014	IK_C 2015
0,61	0,608	0,607	0,606

Table 6. Water conservation status in study site between 2007-2015.

No	Water Conservation Status	Area (ha)			
		2007	2012	2014	2015
1	Good	88,66	87,89	87,63	87,17
2	Normal	68,01	66,66	66,81	66,80

3	Critical	7,21	9,33	9,44	9,91
	Total	163,88	163,88	163,88	163,88

Generally, water conservation ability in study site still has good ability to conserve water. It is shown by 87.17 hectares (>50% of total area) of good water conservation status in 2015. On the other side, critical water conservation ability in the study site has been increasing over years (**Figure 3**). Major changes occurred due to the increased settlements. Settlements has the lowest ability to infiltrate water. It affects to increase runoff coefficient and decrease conservation index.

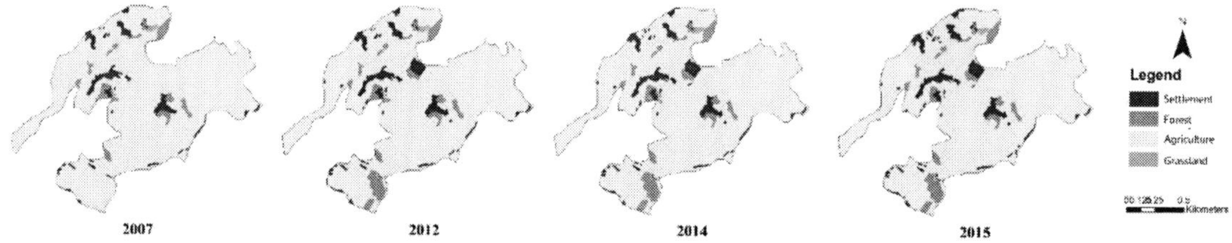

(a)

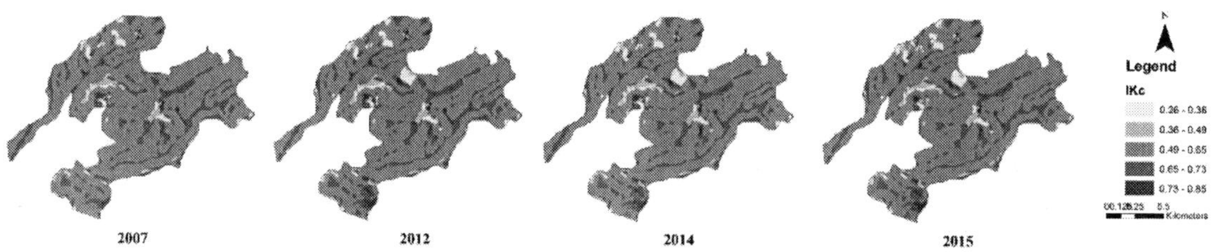

(b)

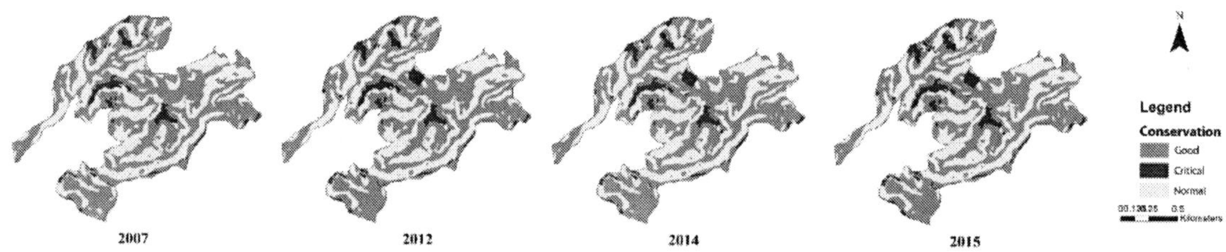

(c)

Figure 3. Map of changes in study site between 2007-2015 (a) Land cover, (b) Distribution of actual conservation index, and (c) Water conservation status.

4. Conclusion

During the period between 2007-2015 land use changes occurred in study site. Major changes cover land use changing into settlements because of population growth. Spatial analysis between 2006 and 2015 shows the decreased area of good level and normal level water conservation index by 1,45 hectares and 1,25 hectares respectively and theincreased area of critical level water conservation index by 2,7 hectares. Land cover changes also impact on actual conservation index which decreased from 0,610 on 2007 to 0,606 in 2015.

Overall, water conservation ability in the highland Ciburial Village is still at good level despite the land degradation. However, we have to be wary of land degradation because settlements growth has increased relatively stable over the years. We need to keep it balance in order to conserve water availability.

5. References

[1] Delinom R M 2009 Structural geology controls on groundwater flow: Lembang Fault case study, West Java Indonesia*Hydrogeology Journal* **17(4)** 1011-1023

[2] Surtikanti H 2005 Kesehatan lingkungan di daerahaliran Sungai Cikapundung akibat pencemaran air *Jurnal Pengajaran MIPA* **6(2)**38-46

[3] Wangsaatmadja S, Sutadian A D, and Prasetiati MAN 2006 *Groundwater resources management in Metropolitan Bandung*(Kanagawa: Institute for Global Environment Studies (IGES))

[4] RasyidiM S 2009 *Menujuketersediaan air yang berkelanjutan di DAS Cikapundung Hulu: Suatupendekatan SYSTEM DYNAMICS*Master Thesis(Bandung:InstitutTeknologi Bandung)

[5] Statistics of Bandung Regency 2017 *CimenyanSubdistrics in Figures* (Bandung: Statistics of Bandung Regency)

[6] Rigine M,Ardiadika W, NurachmanR, and Nurrachman M H 2016*RancanganAgroforestriuntukPengendalianErosi di DesaCimenyandanDesaCiburial* Bachelor Thesis(Bandung: InstitutTeknologi Bandung)

[7] Desa Ciburial 2015 *ProfilDesaCiburial.* <https://ciburial.desa.id/profil-desa-ciburial/> (Accessed on 22 September 2016)

[8] Sabar A 1999 *IndeksKonservasiSebagaiInstrumenPengendalianPemanfaatanRuang di KawasanBopunjurdalamRangkaRancanganKeppres*MakalahBahanDiskusi di Bappenas. (Bandung: InstitutTeknologiBandung)

[9] Hassing J M 1995Hydrology In: B. Thagesen, ed*Highway and Traffic Engineering in Developing Countries*(London: E&FN Spon) p198-210.

IOP Conf. Series: Earth and Environmental Science **166** (2018) 012037 doi:10.1088/1755-1315/166/1/012037

Seasonal Variability Of Ozone Vertical Profiles in Indonesia Based on AQUA-AIRS Data

N Komala and N Ambarsari

Center for Atmospheric Science and Technology of LAPAN
Jl. Dr Djundjunan 133, Bandung 40173, Indonesia

E-mail: ninongk@yahoo.com

Abstract. We examined the seasonal vertical profiles of ozone mixing ratios in Indonesia based on ozone monthly data of AQUA-AIRS from 2002 to 2016,analysed and discussed the characteristic of its seasonal profiles of ozone mixing ratios. The method used is by processing the ozone monthly data to be the seasonal profiles for December, January, February (DJF), March, April and May (MAM), June, July and August (JJA) and September, October and September (SON) from the surface to upper stratosphere (1000 hPa to 1 hPa) and comparing the time series, average and deviation of each ozone seasonal profile. The results indicate good agreement in shape of ozone seasonal profiles but the bias shows in the peak of ozone which was found at 10 hPa. The peak of ozone ranged between 8.58 ppm to 9.56 ppm in DJF, and 8.98 ppm to 9.87 in MAM ppm. In JJA, ozone peak ranged between 8.55 ppm to 9.44 ppm and the peak of ozone in SON showed 8.29 ppm to 9.48 ppm.

1. Introduction

The amount of ozone in the atmosphere can be measured by instruments on the ground level (sparsely distributed measurement sites). These instruments are thereafter set aloft on balloons, aircrafts, and/or satellites. Through ground and aircraft measurements, observations are only limited to the surface. As such, measurements are more sensitive to sources and sinks. However, these same instruments are unable to make evaluations based on daily global variations. The satellite remote sensing has very good global coverage to increase our capability to access the influence of human activities on the chemical composition of the atmosphere and on the climate changes [1]. The satellite also can provide the quantitatively data with high spatial or temporal resolution. In addition, the free download satellite data are provided by infrared measurements from the Atmospheric Infra Red Sounder (AIRS) on the NASA AQUA satellite to establish and produce the high quality spectrally resolved radiance climatologic for observing and detecting climate change, to understand the distribution of trace gases, sources and sinks, and to validate weather and climate models [2].

Atmospheric ozone is one of the highly variable trace gas constituents of the atmosphere. The concentration of ozone can vary in short to long-term scales which can profoundly influence the vertical thermal structure of the atmosphere [3]. Deviation in ozone concentration can change the radiative properties of the atmosphere inflicting a negative radiative forcing due to the absorption of shortwave radiation and a positive radiative forcing in the troposphere due to the absorption of long wave radiation. The change in ozone amount over a region can be due to natural and anthropogenic factors. Human activities can influence the ozone amounts in the atmosphere through the emission of certain harmful substances that can destroy the ozone layer and also by the release of certain chemicals that can cause the production of ozone in the troposphere [4].

Although chemically identical, ozone in different layers in the atmosphere exerts different influences on the well-being of Man. In the stratosphere, where approximately 90% of the atmosphere's ozone is found, ozone resides in the so-called ozone layer and prevents the sun's ultra-violet radiation from reaching the surface of the Earth. In the troposphere, ozone is a direct greenhouse gas. In the boundary layer, ozone is a pollutant which harms the health of both Man and his crops [5].

Launched on board NASA's Aqua satellite on 4 May 2002, Aqua's cross-track scanning Advanced Microwave Sounding Unit (AMSU) with AIRS cross-track scanning grating spectrometer coupled, provides vertical profiles of the atmosphere with a nadir 45 km field-of-regard (FOR) across a 1650 km swath. AIRS broad spectral coverage (3.7 to 16 μm with 2378 channels) includes spectral features of CH_4, CO, O_3, and CO_2. The objectives of the AIRS is to determine the factors that control the global energy and water cycles, investigate the atmosphere-surface interactions, improve numerical weather prediction, assess climate variations and feedbacks and detect the effects of increased carbon dioxide, methane, ozone, and other greenhouse gases. The term "sounder" in the instrument's name refers to the fact that water vapour and temperature are measured as functions of height. AIRS measures CO total column by (36) channels with uncertainty estimate 15-20% at 500mb, vertical coverage 1000 - 1 mb, global non-polar, mid-tropospheric total, (7 - 9) layers. Troposphere carbon monoxide (CO) abundance is retrieved in the 4.58-4.50 μm (2180-2220 cm^{-1}) region from the AIRS measured radiances of the IR spectrum. AIRS radiance data in the 9.6-μm band are used to retrieve column ozone and ozone profiles for both day and night [6-8].

A study by Bian et al., [9]. has compared the AIRS ozone profiles with ozonesonde data from Beijing using a new type of ozonesonde called the Global Positioning System Ozone sensor (GPSO3). The ability to measure ozone variability has widespread application [10] and it also has used AIRS to show how transient convective events, associated with the Asian summer monsoon, lead to vertical transport of low ozone and high-water vapor into the UTLS region. AIRS-retrieved ozone has been evaluated against several datasets. Initial validation studies have shown that IR-retrieved ozone profiles using AIRS measurements capture the variability in the extra tropical upper troposphere and lower stratosphere (Ex UTLS) region observed in the ozonesonde data [9, 11].

While providing the near-real-time AIRS-retrieved ozone at several pressure levels in the Upper troposphere and Lower Stratosphere (UTLS) during the 2005 and 2008 Stratosphere–Troposphere Analyses of Regional Transport Experiments (START05 and START08), significant agreement between the AIRS-retrieved ozone near the tropopause and aircraft ozone measurements was also found [12]. A detailed description of the AIRS retrieval method is presented in [13]. AIRS also provides information on several greenhouse gases, CO_2, CH_4, CO and O_3 [14].

2. Materials and methods

AIRS ozone profiles are available from 1000 hPa to 1 hPa. This study used the version 006 Level 3 ozone profile data which are available from the Goddard Earth Sciences Distributed Archive Center (http://disc.sci.gsfc. nasa.gov/data/dataset/AIRS/). The analysis of ozone volume mixing ratio was investigated for the period 2002-2016 in Indonesia (12N-12S, 94E-146 E) using the retrieved AIRS monthly product (AIR3STM) Version 006 data. The ozone mixing ratios from 1000 hpa to 1 hPa Satellite data were evaluated over Indonesia in 14 years study period. The ozone mixing ratios data were gridded monthly at 1°×1° spatial resolution (latitude x longitude).

3. Results and Discussions

Figure 1 shows the results of the retrieved monthly analysis (AIRX3STM). The Atmosphere infrared sounder (AIRS) data were utilized to analyze vertical profile of ozone mixing ratio in Indonesia for the period 2002-2016. The analysis of monthly ozone profile for Indonesia region shows that maximum values occur in 2004, 2006, and 2008, with minima in 2005, 2007 and 2010. Ozone concentration at 10 hPa (`~ 26 km) or in the stratosphere was between 8.3 ppm ~10.39 ppm and it seemed high in January-May. However, it decreased during June-July and increased again from August to December. This condition relates to the cycles of the sun's movement which is away from the equator during June-July. The previous study using data derived from AURA-MLS [15] shows that the peak of ozone mixing ratio profile occurred at a pressure of 10 hPa with a concentration ranged from 9.04 to 10.24 ppm. Ozone from surface to 100 hPa was 0.021 ppm to 0.010 ppm, and the stratospheric ozone average was 9.15 ppm with standard deviation of 0.44 ppm.

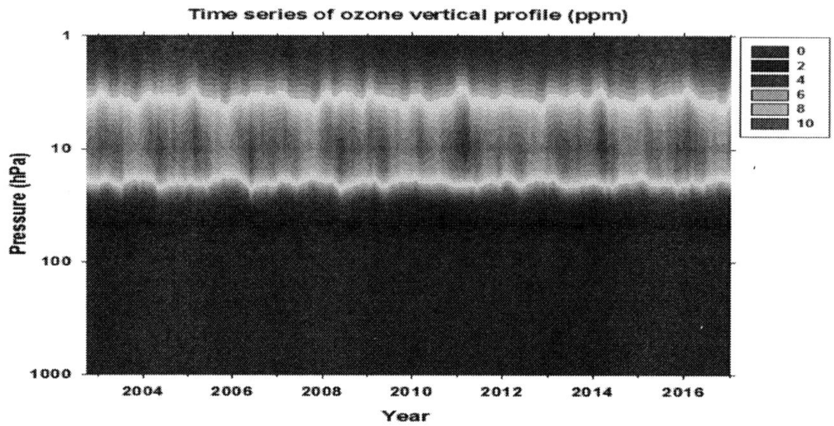

Figure 1. The time series of monthly vertical profile of ozone in Indonesia from 1000 hPa to 1 hPa from 2002 to 2016.

Figure 2 shows the time series of seasonal vertical profile of ozone in Indonesia from 1000 hPa to 1 hPa for 2002 to 2016. The seasonal profile of DJF shows that the O_3 mixing ratios in the stratosphere were higher in 2005, 2007 and 2011 and lower in 2012, 2015 and 2010. The seasonal profile of MAM shows that the O_3 mixing ratios in the stratosphere were higher in 2004, 2008 and lower in 2010. The JJA profile shows that the O_3 mixing ratios in the stratosphere were higher in 2008, 2006 and 2011 and lower in 2005, 2007 and 2010, while the SON profile shows that the mixing ratios of ozone were higher in 2015, 2006 and 2004, and lower in 2005, 2007, 2009 and 2012. In the lower troposphere, the mixing ratios of O_3 were 0.02 ppm to 0.05 ppm. While the mixing ratios of ozone in tropopause (100 hPa) shows 0.100 ppm to 0.150 ppm, and the peak of ozone in 10 hPa (~ 26 km) ranged between 8.3 ppm to 9.87 ppm.

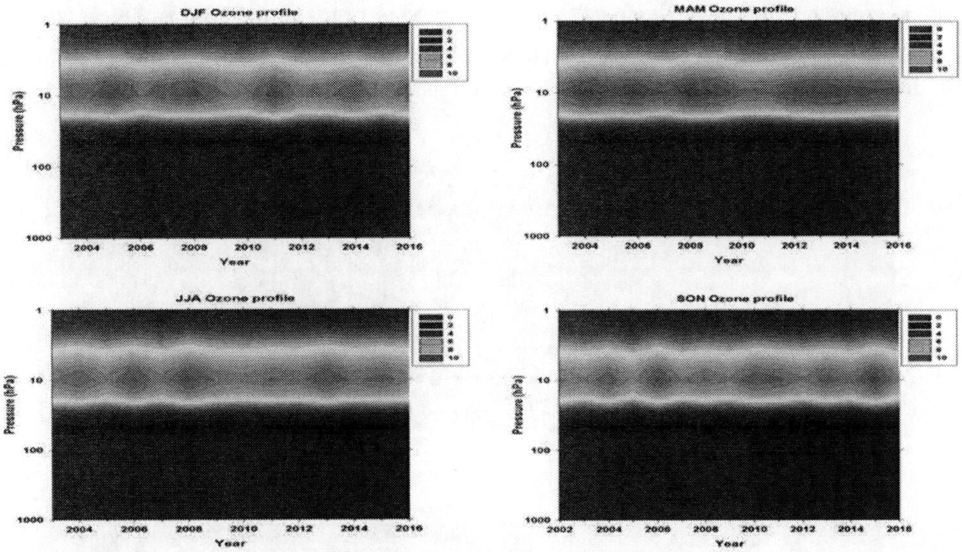

Figure 2. The time series of seasonal vertical profile of ozone in Indonesia from 1000 hPa to 1 hPa from 2002 to 2016.

Figure 3 describes the ozone average profile of DJF and its deviation. It shows that the ozone mixing ratios in the stratosphere were higher in 2005, 2007 and 2011 and lower in 2012, 2015 and 2010. The deviation of stratospheric ozone was between -600 ppb to 600 ppb. The average profile of MAM shows that the O_3 mixing ratios were higher in 2004, 2008 and lower in 2010. The deviation also shows -600 ppb to 600 ppb with the highest minus deviation of ozone in 2010.

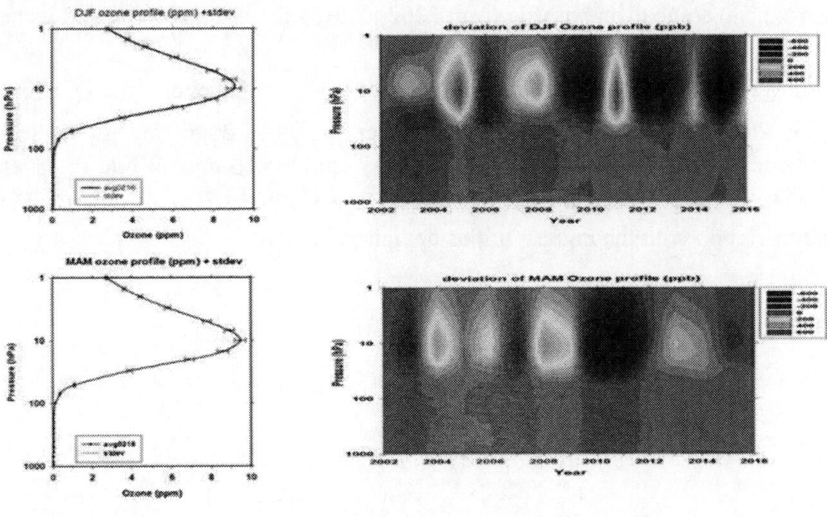

Figure 3. The average ozone profile from 2002 to 2016 for DJF and MAM and its deviation.

Figure 4 describes the ozone average profile of JJA and its deviation. The ozone mixing ratios in the stratosphere were higher in 2008, 2006 and 2011 and lower in 2005, 2007 and 2010. The deviation of stratospheric ozone was between -600 ppb to 600 ppb and the highest minimum deviation was in

2010. The average profile of SON shows that the O_3 mixing ratios in the stratosphere were higher in 2015, 2006 and 2004, lower in 2005, 2007, 2009 and 2012. The deviation also shows -600 ppb to 600 ppb with the highest minus deviation in 2005.

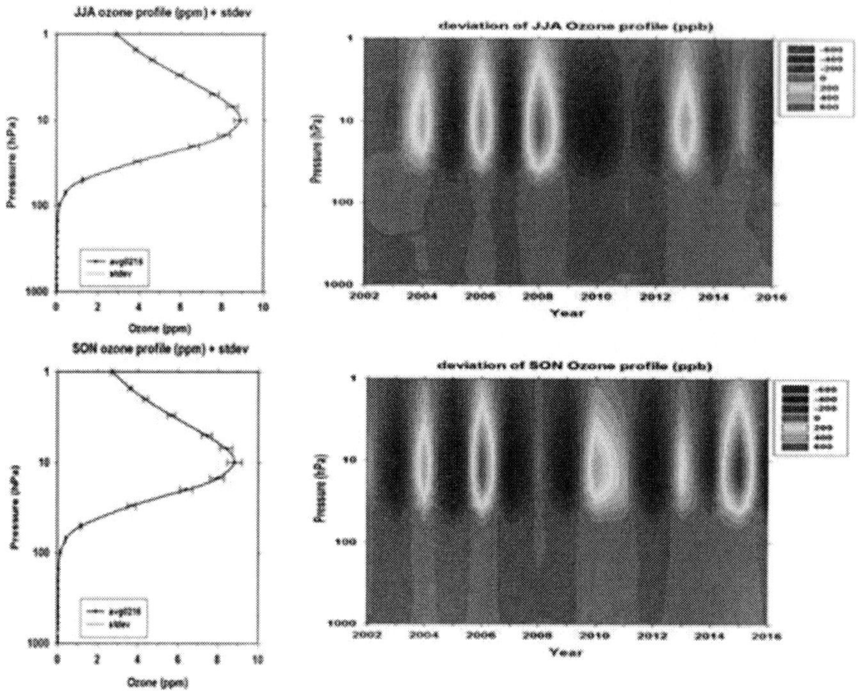

Figure 4. The average ozone profile from 2002 to 2016 for JJA and SON and its deviation.

Figure 5 shows the 1000 to100 hPa profile of DJF. In the troposphere, the O_3 mixing ratios were higher in 2011, 2014 and 2016 and 2011 and lower in 2003, 2006 and 2008. The deviation of tropospheric ozone between in DJF ranged between -10 ppb to 10 ppb. While the average profile of MAM shows that the O_3 mixing ratios were higher in 2015, and lower in 2011. The deviation also shows -10 ppb to 10 ppb with the highest minus deviation in 2010.

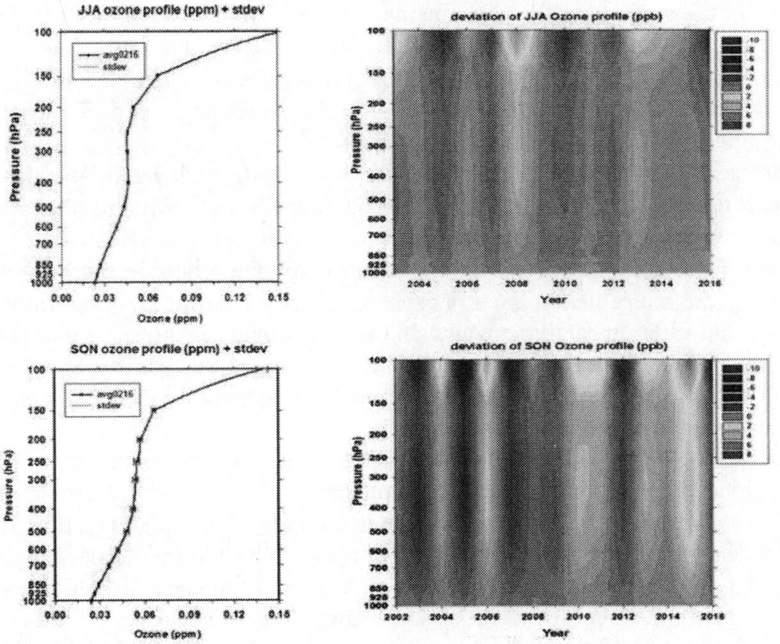

Figure 5. The average ozone profile from 2002 to 2016 in 1000 hPa to 100 hPa for DJF and MAM and its deviation.

In **Figure 6**, the 1000 hPa to 100 hPa ozone profile of JJA shows that in the troposphere the O_3 mixing ratios were higher in 2008 and 2013, and lower in 2005, 2007 and 2010. The ozone deviation of tropospheric ozone is between -10 ppb to 10 ppb. While average profile of SON shows that the O_3 mixing ratios were higher in 2004, 2006, 2010 and 2015, and lower in 2005, 2007 and 2012. The ozone deviation also shows -10 ppb to 10 ppb with the highest minus deviation of ozone in 2005.

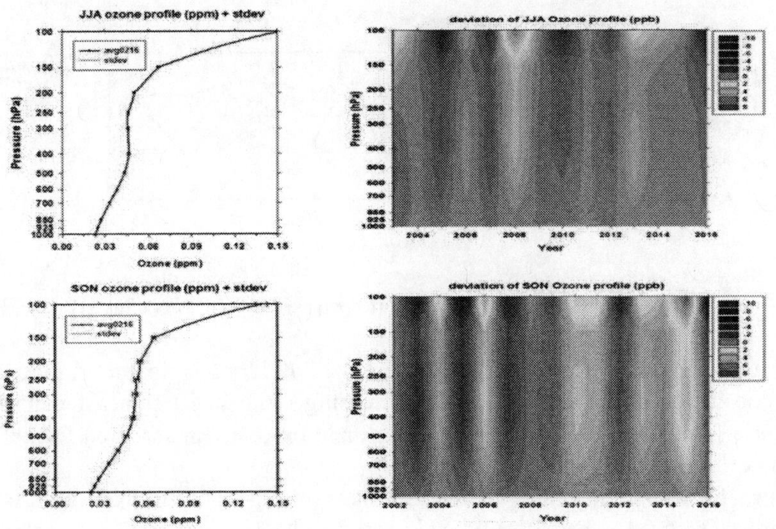

Figure 6. The average ozone profile from 2002 to 2016 in 1000 hPa to 100 hPa for JJA and SON and its deviation.

Measurements of the vertical profile of ozone have been carried out over the last few decades by a large number of instruments, operating in situ or from remote vantage points on the ground and in space [16]. These indisputably show globally declining ozone levels during the 1980s and a large part of the 1990s in the lower and upper stratosphere (~5% – 7% decade^{-1}), and to a lesser extent also in the middle stratosphere (1% – 2 % decade^{-1}) [17, 18].

For many decades, it has been known that the stratospheric ozone layer shields the Earth's surface from harmful solar ultraviolet radiation (UV), thus enabling life on Earth and protecting humans and the biosphere against adverse effects. This protective layer could be depleted by anthropogenic emission of chlorofluorocarbons (CFCs) to the atmosphere. The photodecomposition of CFCs and other long-lived organic molecules in the stratosphere releases chlorine (Cl) and bromine (Br) atoms that destroy ozone molecules in catalytic cycles. In the last Scientific Assessment of Ozone Depletion: 2014 of the World Meteorological Organization [17], it is stated that global ozone levels decreased through the 1980s and early 1990s while stratospheric abundances of ozone-depleting substances (ODSs) were increasing.

The presence of declining ozone in the stratosphere and the increasing of tropospheric ozone occurrence in Indonesia were analyzed by investigating the long-term variation of ozone in the stratosphere (at ozone peak level ~ 26km) and ozone in the lower troposphere (at 700 hPa ~ 3 km).
In Figure 7, the time series monthly ozone peak at 10 hPa (~26 km) in Indonesia show almost constant trend. It can explain that the declining of stratospheric ozone in Indonesia does not occur. In seasonal variation of ozone peak, the trend of each season shows a small negative variation. In DJF season trend of ozone peak show coefficient of determination of 0.113, in MAM 0.016, in JJA 0.028 and in SON 0.005.

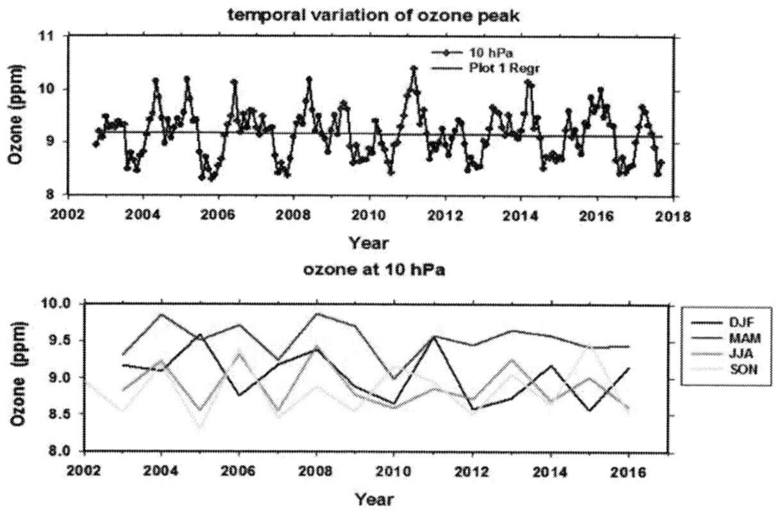

Figure 7. The long term variation of ozone at stratosphere (at peak level) in Indonesia.

Figure 8 presents the time series of monthly ozone at 700 hPa (~3 km) in Indonesia. The trend of tropospheric ozone in Indonesia shows a little increasing, with coefficient of determination 0.007. Lower tropospheric ozone shows coefficient of determination 0.367 in seasonal DJF, 0.225 in MAM , 0.058 in JJA and 0.115 in SON .

The increasing of lower tropospheric ozone in Indonesia especially in SON (end of dry season) is affected by anthropogenic activities. Ozone levels will also be affected by the expected anthropogenic increases in abundances of other ozone-relevant gases (carbon dioxide (CO_2), methane (CH_4), and nitrous oxide (N_2O) as well as by the natural influences of volcanic eruptions, solar activity, and the natural variability in the Earth's climate [17].

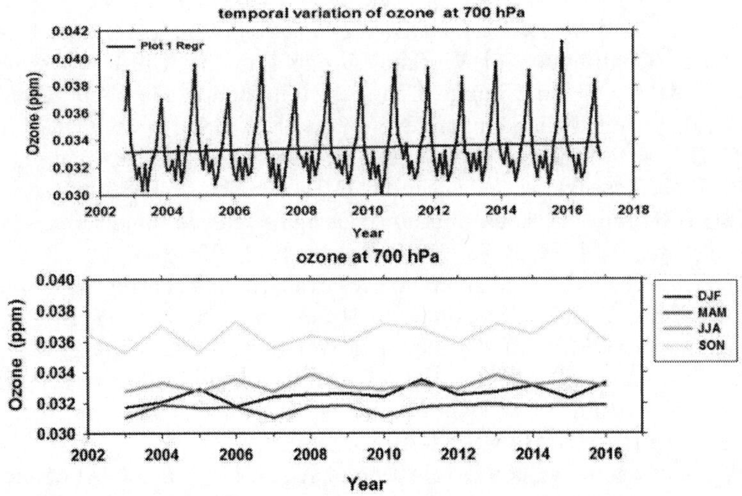

Figure 8. The long term variation of ozone at lower troposphere (~ 3km) in Indonesia.

4. Conclusion

The seasonal profile of ozone mixing ratio in Indonesia based on AQUA-AIRS data for the period of 2002 to 2016 indicates good agreement in shape of ozone seasonal profiles and the bias shows in peak of ozone at 10 hPa. The peak of ozone in DJF ranged between 8.58 ppm to 9.56 ppm, the MAM ozone peak showed 8.98 ppm to 9.87 ppm. The JJA ozone peak ranged between 8.55 ppm to 9.44 ppm and the peak of ozone in SON showed 8.29 ppm to 9.48 ppm.

Time series of monthly ozone peak at 10 hPa (~26 km) in Indonesia are almost constant. It can explain that the declining of stratospheric ozone in Indonesia does not occur. In seasonal variation of ozone peak, the trend of each season shows a small negative variation (negative trend). The time series of monthly ozone at 700 hPa (~3 km) in Indonesia trend of tropospheric ozone in Indonesia show a little increased. The increasing of lower tropospheric ozone in Indonesia especially in SON (end of dry season) is affected by anthropogenic activities.

Acknowledgements.

This work is supported by Center for Atmospheric Science and Technology, the National Institute of Aeronautics and Space of LAPAN. The authors would like to thank for the NASA scientists and the PI who have prepared the AQUA satellite data with AIRS sensor that could be accessed through the facilities of MIRADOR, and GIOVANNI so that they could be used in this study.

5. References

[1] Clerbaux C, Hadji-Lazaro J, Turquety S, Mégie G, and Coheur P –F 2003 Trace Gas Measurements From Infrared Satellite For Chemistry and Climate Applications *Atmospheric Chemistry and Physics*, **3** 1495-1508

[2] Dousset B and Gourmelon F 2003 Satellite Multi-sensor Data Analysis of Urban Surface Temperatures and Landcover *ISPRS Journal of Photogrammetry and Remote Sensing* **58** 43-54

[3] Langematz U, Kunze M, Kru¨ger K, Labitzke K, and Roff G L 2003 Thermal and dynamical changes of the stratosphere since 1979 and their link to ozone and CO_2 changes *J. Geophys. Res.* **108** (D1) 4027, doi:10.1029/2002JD002069

[4] Forster P M F and Tourpali K 2001 Effect of tropopause height changes on the calculation of ozone trends and their radiative forcing *J. Geophys. Res.* **106** (D11) 12241–12251

[5] Prather M J 2001 *Atmospheric chemistry and greenhouse gases* In: Climate Change The Scientific Basis. Contribution of Working Group I to the Third Assessment Report of the Intergovernmental Panel on Climate Change [Houghton J T (eds.)]. Cambridge University Press, Cambridge, United Kingdom and New York, NY, USA, pp 239–287

[6] Aumann H H, M T Chahine, C Gautier, M D Goldberg, E Kalnay, L M McMillin, H Revercomb, P W Rosenkranz, W L Smith, D H Staelin, L L Strow, and J Susskind 2003 AIRS/AMSU/HSB on the aqua mission: Design, science objectives, data products, and processing systems, IEEE Trans *Geosci. Remote Sens.* **41** 253-264

[7] Chahine M T, T S Pagano, H H Aumann, M Divakarla, E J Fetzer, M Goldberg, C Gautier, S Granger, F W Irion, R Kakar, E Kalnay, B H Lambrigtsen, Sung-Yung Lee, J L Marshall, W W McMillan, Larry McMillin, E T Olsen, H Revercomb, P Rosenkranz, W L Smith, D Staelin, L L Strow. J Susskind, D Tobin, W Wolf, and L Zhou 2006 AIRS: Improving Weather Forecasting and Providing New Data on Greenhouse Gases American *Meteorological Society* **87** 911- 926 DOI: 10.1175/ BAMS-87-7-911

[8] Pittman J V and Coauthors 2009 Evaluation of AIRS, IASI, and OMI ozone profile Retrievals in the extratropical tropopause region using in situ aircraft measurements *J. Geophys. Res.* **114** D24109 doi:10.1029/2009JD012493

[9] Bian J, A Gettelman, H Chen, and L L Pan 2007 Validation of satellite ozone profile retrievals using Beijing ozonesonde data *J. Geophys. Res.* **112** D06305 doi:10.1029/2006JD007502

[10] Randel W J and M. Park 2006 Deep convective influence on the Asian summer monsoon anticyclone and associated tracer variability observed with Atmospheric Infrared Sounder (AIRS) *J. Geophys. Res.* **111** D12314 doi:10.1029/2005JD006490

[11] Monahan K P, L L Pan, A J McDonald, G E Bodeker, J Wei, S E George, C D Barnet, and E Maddy 2007 Validation of AIRS v4 ozone profiles in the UTLS using ozonesondes from Lauder, NZ and Boulder, USA *J. Geophys. Res.* **112** D17304 doi:10.1029/2006JD008181

[12] [12] Pittman J V and Coauthors 2009 Evaluation of AIRS, IASI, and OMI ozone profile Retrievals in the extratropical tropopause region using in situ aircraft measurements *J. Geophys. Res.* **114** D24109 doi:10.1029/2009JD012493

[13] Susskind J, C D Barnet, J M Blaisdell, L Iredell, F Keita, L Kouvaris, G Molnar, and M T Chahine 2006 Accuracy of geophysical parameters derived from Atmospheric Infrared Sounder/Advanced Microwave Sounding Unit as a function of fractional cloud cover *J. Geophys. Res.* **111** D09S17 doi:10.1029/2005JD006272

[14] Tian B, D E Waliser, E Fetzer, B Lambrigtsen, Y Yung and B Wang 2006 Vertical moist thermodynamic structure and spatial-temporal evolution of the Madden-Julian oscillation in AIRS observations *J. Atmos. Sci.* **63** 2462-2485

[15] Komala N and N Ambarsari 2015 *International Journal of Remote Sensing and Earth Sciences* **12 (2)** 125 – 134

[16] Hassler B, Petropavlovskikh I, Staehelin J, August T, Bhartia P K., Clerbaux C, Degenstein D, Mazière M De, Dinelli B M, Dudhia A, Dufour G, Frith S M, Froidevaux L, GodinBeekmann S, Granville J, Harris N R P, Hoppel K, Hubert D, Kasai Y, Kurylo M J, Kyrölä E, Lambert J-C, Levelt P F, McElroy C T, McPeters R D, Munro R, Nakajima H, Parrish A, Raspollini P, Remsberg E E, Rosenlof K H, Rozanov A, Sano T, Sasano Y, Shiotani M, Smit H G J, Stiller G, Tamminen J, Tarasick D W, Urban J, van der A R J, Veefkind J P, Vigouroux C, von Clarmann T, von Savigny C, Walker K A, Weber M, Wild J, and Zawodny J M 2014 Past changes in the vertical distribution of ozone – Part 1: Measurement techniques, uncertainties and availability *Atmos.Meas.Tech.***7** 1395–1427doi:10.5194/amt-7-1395-2014.

[17] WMO 2014 *Scientific Assessment of Ozone Depletion Global Ozone* Research and Monitoring Project–Report No. 55. http://ozone.unep.org/Assessment_Panels/SAP/ Scientific_Assessment_2014/ Scientific_Assessment_Report_2014.pdf.

[18] Harris N R P, Hassler B, Tummon F, Bodeker G E, Hubert D, Petropavlovskikh I, Steinbrecht W, Anderson J, Bhartia P K, Boone C D, Bourassa A, Davis S M, Degenstein D, Delcloo A, Frith S M, Froidevaux L, Godin-Beekmann S, Jones N, Kurylo M J, Kyrölä E, Laine M, Leblanc S T, Lambert J-C, Liley B, Mahieu E, Maycock A, de Mazière M, Parrish A, Querel R, Rosenlof K H, Roth C, Sioris C, Staehelin J, Stolarski R S, Stübi R, Tamminen J, Vigouroux C, Walker K A, Wang H J, Wild J, and Zawodny J M 2015 Past changes in the vertical distribution of ozone – Part 3: Analysis and interpretation of trends *Atmos. Chem. Phys.* **15** 9965–9982 doi:10.5194/acp-15-9965-2015, 2015

IOP Conf. Series: Earth and Environmental Science **166** (2018) 012038 doi:10.1088/1755-1315/166/1/012038

Vertical profile variations of ozone in lower stratosphere in Indonesia and influence to upper troposphere ozone based on satellite

N Ambarsari[1] and N Komala[1]

Center for Atmospheric Science and Technology, LAPAN
Jl. Dr. Djunjunan No. 133 Bandung, Indonesia

E-mail: novitaambar@yahoo.com

Abstract. This study examined the vertical profile and ozone variation in the lower stratosphere in Indonesia using AQUA/AIRS and AURA/MLS satellite data. Ozone in the stratosphere as an important component in atmospheric chemistry as well as radiation and climate becomes important to learn. Seasonal vertical profiles, time-varying vertical profiles, ozone seasonal variations at pressures of 100 hPa and 70 hPa, ozone time series at pressures of 100 hPa and 70 hPa from 2002 to 2016, average zonal variation (influence latitude) ozone concentrations in Indonesia, and the comparison of the vertical profiles of ozone, HCl, and N_2O in the lower stratosphere layer in Indonesia from MLS/AURA satellite data have been done to understand the variation of ozone in the lower stratosphere in Indonesia. The results show that the vertical ozone profile increased from the tropopause then the lower stratosphere to the middle stratosphere with the highest concentration of up to 9 ppmv. The monthly variation of ozone climatology at 70 hPa shows a pattern with a maximum peak in August while at a pressure of 100 hPa it indicates the presence of two maximum peaks in June and September. Time series of ozone at 100 hPa and 70 hPa from 2002 to 2016 show that ozone concentration tended to be constant with very small concentration increase occurring at 70 hPa and 100 hPa. Zonal variation shows the lowest ozone concentration at zero degrees latitude compared to other latitudes, and the vertical profile of ozone has the same pattern as HCl while it is different from N_2O. Vertical-time series cross section of ozone and Potential Vorticity (PV) in upper troposphere show influenced of stratospheric air into upper tropospheric air which showed by higher ozone and PV value and can reach until pressure 150 hPa occurred in the same season.

1. Introduction

Ozone (O_3) in the stratosphere is known to shield the surface from harmful ultraviolet radiation, while in the middle and high troposphere ozone it is the third most important greenhouse gas after carbon dioxide (CO_2) and methane (CH_4). Scientific efforts to understand the trends and variation in ozone observed over the past few decades have demonstrated the role of both photochemical and meteorological factors in driving stratospheric ozone changes. It has been known that these stratospheric changes have altered the tropospheric ozone burden over the past few decades and will continue to affect it in the future [1]. The abundance of stratospheric ozone is sensitive to both chemical and dynamical factors and has been altered by anthropogenic activity [2].

The basic overturning of the stratosphere described by Brewer [3] more than fifty years ago, with low-ozone tropospheric air entering the stratosphere in the tropics and high-ozone air returning to the troposphere in the mid and high latitudes [3,4]. The characterization of troposphere and stratosphere had been done by different dynamical and chemical properties with strong gradients of Potential Vorticity (PV) and ozone at the tropopause. Some definitions of chemical tropopause are popular, such as ozone tropopause criterion which defines the tropopause as the altitude at which ozone mixing ratio is greater than 80 ppbv, at which ozone gradient exceeds 60 ppbv/km in a depth of 200 m, and the ozone above the tropopause exceeds 110 ppbv [5]. Lower stratospheric and upper tropospheric ozone air arc highly correlated [6,7]. The stratosphere is characterised by high value of PV and ozone concentration, so intrusion of stratospheric air is expected to bring PV and ozone rich air into troposphere [8,9]. Stratospheric intrusions along the subtropical jet stream occur outside the tropics but can bring extratropical lower stratosphere air into the tropical middle and upper troposphere [10,11]. For ozone in the tropical lower stratosphere, the chemistry is dominated by ozone production from photolysis which contributes to annual cycle, and in lower stratospheric ozone concentrations has been linked to annual cycle in upwelling of the stratospheric residual circulation [12,13].

Various chemical and dynamical process that control concentration of upper tropospheric and lower stratospheric ozone such as photochemical production within the troposphere, downward/upward transport from the stratosphere/troposphere through Stratosphere-Troposphere-Exchange (STE) [14], deep convection [15], and transport process within the troposphere [16]. Long term trends in tropospheric and lower stratospheric ozone show a complex pattern over the globe and in some cases are driven by regional influences. There are few focused studies reporting ozone trends in the lower stratosphere over the tropical region. Randel and Thompson [17] reported statistically significant negative trends between 0,2-0,4% per year in the tropical lower stratosphere (17-21 km) using Stratospheric Aerosol and Gas Experiment (SAGE II) data (1984-2005) and Southern Hemisphere Additional Ozonesondes (SHADOZ) measurements (1998-2009). Another result about decreasing ozone trends in the tropics (-10% below 20 km) has been reported by Randel and Wu [18]. Another research about seasonal variation of ozone in the lower stratosphere has been done by Stolarski [19]. The research has shown that the seasonal cycle of ozone is stronger in the northern tropical lower stratosphere than that in southern tropical lower stratosphere. These differences in the seasonal variations between hemispheres imply hemispheric differences in the transport in the tropical lower stratosphere [19].

This study focuses on analyses variation of vertical profile ozone in lower stratosphere and the influence of ozone from lower stratosphere to upper troposphere in Indonesia using AQUA/AIRS satellite data. This paper also reports the vertical profile of ozone in lower stratosphere compared with another trace gases that are nitrous oxide (N_2O) and hydrochloride acid (HCl) from AURA/MLS satellite data. To determine the influence of ozone from lower stratosphere to upper troposphere, we compare the vertical profile-time series and seasonal for ozone and PV data from Modern Era Retrospective-analysis for Research and Application (MERRA) reanalysis in the upper troposphere.

2. Data and Methodology

We examined the behaviour of vertical profile lower stratospheric ozone over Indonesia using Atmospheric Infra-Red Sounders (AIRS) sensors on AQUA satellite data from 2002 to 2016. We chose vertical profile ozone area average from 90° E – 150 °E and 15 °South – 15 °North at pressure 100 hPa to 10 hPa for analysis of lower stratospheric ozone profile over Indonesia. We compared behaviour of ozone at 70 hPa as the initial pressure of the lower stratosphere layer after the tropopause (100 hPa) measured by AIRS/AQUA sensor, with behaviour of ozone in the tropopause layer. The change of ozone concentration in lower stratosphere compared with tropopause was analysed by time series (trend) ozone concentration at 70 hPa and 100 hPa. The influence of geographic position (latitude) on changes of ozone concentration was analysed by zonal variation of ozone concentration at 70 hPa and 100 hPa. We also compared the vertical profile of ozone with vertical profile of another trace gases that are Nitrous oxide (N_2O) and Hydrochlorid Acid (HCl) from Microwave Limb Sounder

(MLS) sensor data on AURA satellite to know the differences of their behaviour in lower stratosphere layer. Influences of lower stratospheric ozone concentration into upper tropospheric ozone were analysed by comparing vertical profile-time series and seasonal of ozone in upper troposphere at pressure 250 hPa until 100 hPa from AQUA/AIRS data with Potential Vorticity (PV) variation from MERRA reanalysis data as the tracer of stratospheric air influence with the higher PV value.

AIRS was launched on May 4, 2002 installed on the Aqua satellite with the primary objective was to accurately measure the vertical profile of temperature and water vapor in the Earth's atmosphere. The AIRS scientific team then developed methods for measuring the physical and chemical parameters of temperature, water vapor, ozone, and CO. The ozone data from AIRS used in this paper are version 6 level 3 gridded with 1° x 1 spatial resolution [20]. Vertical profile ozone and temperature data from AIRS have been validated using ozonesonde data in Watukosek Jawa Timur. The comparison result show good agreement with the AQUA-AIRS satellite overpass data with correlation coefficient of 0,98 for ozone mixing ratio and 0,99 for temperature [21].

The AURA MLS sensor has the ability to measure atmospheric chemical parameters including HNO_3, N_2O, and ozone with global coverage area and with good enough data resolution with 3 km vertical resolution and 200 km spatial resolution [22]. MLS observes thermal radiation on Earth-emitted micro waves and records data from surface to a height of 90 km. MLS measures vertical profiles at 3500 locations in the world every 24 hours [22]. The MLS instrumentation is equipped with four types of radiometers operated at ambient temperatures in five spectral regions which are capable of detecting the O_2 spectrum including measuring temperature and pressure, H_2O, HNO_3, O_3 in the upper troposphere, CO, HCl, ClO, BrO, HO_2, N_2O, and OH [23].

The Modern Era Retrospective-Analysis for Research and Applications (MERRA) was undertaken by NASA's Global Modeling and Assimilation Office with two primary objectives: to place observations from NASA's Earth Observing System satellites into a climate context and to improve upon the hydrologic cycle represented in earlier generations of reanalyses. MERRA was generated with version 5.2.0 of the Goddard Earth Observing System (GEOS) atmospheric model and data assimilation system (DAS), and has covered the modern satellite era from 1979 to the present [24]. In this study, we use Potential Vorticity (PV) data from MERRA monthly averages in upper troposphere with spatial resolution 0.5° x 0.625°.

3. Result and Discussion

3.1 Lower stratosphere ozone vertical profile variation

Time series-vertical cross section variation and seasonal ozone vertical profile at lower stratosphere for December-January-February (DJF), March-April-May (MAM), June-July-August (JJA) and September- October-November (SON) are shown in **Figure 1**. Ozone concentrations show higher value at higher altitude or lower pressure with maximum ozone shows at pressure 10 hPa, where the ozone layer exists, the values reached 9000 ppbv or 9 ppmv (**Figure 1a**). Vertical profile of ozone concentrations shows small differences in tropopause (pressure 100 hPa) while higher differences are shown in upper part of lower stratosphere (pressure 10 hPa). At pressure 10 hPa the minimum value of ozone in JJA reached 8891 ppbv or 8,891 ppmv and the maximum value of ozone in MAM reached 9479 ppbv or 9,479 ppmv (**Figure 1b**). Ozone in lower stratosphere over Indonesia also shows annual and semi-annual variation due to the annual variation of Brewer Dobson Circulation (BDC) and Quasi Biennial Oscillation (QBO) [12,25].

Monthly variation of ozone and temperature in tropopause (pressure 100 hPa **Figure 1a**) and in lower stratosphere (pressure 70 hPa **Figure 1b**) is shown in Figure 2 and Figure 3. Ozone in tropopause as a boundary layer between troposphere and stratosphere is influenced by tropospheric and stratospheric air characteristics [26]. While in the lower stratosphere, the characteristics of the troposphere air are diminished with the stratospheric air characteristics, so comparing the monthly variations of ozone in the tropopause and in the lower stratosphere is important to recognize differences in ozone characteristics in both regions. Due to both radiative and dynamical processes, there is a strong positive correlation between ozone and temperature in lower stratosphere [27]. We can see from **Figure 2** and **Figure 3** that monthly ozone variation in the tropopause has 2 maximum peak in June and September with value around 0,16 ppmv which is also consistent with monthly variation of tropopause temperature (**Figure 3**). Monthly variation of ozone in the lower stratosphere at pressure 70 hPa (**Figure 2b**) also shows similar feature with temperature at the same level pressure (**Figure 3b**) which has one maximum peak value in Augustus with value around 0,5 ppmv.

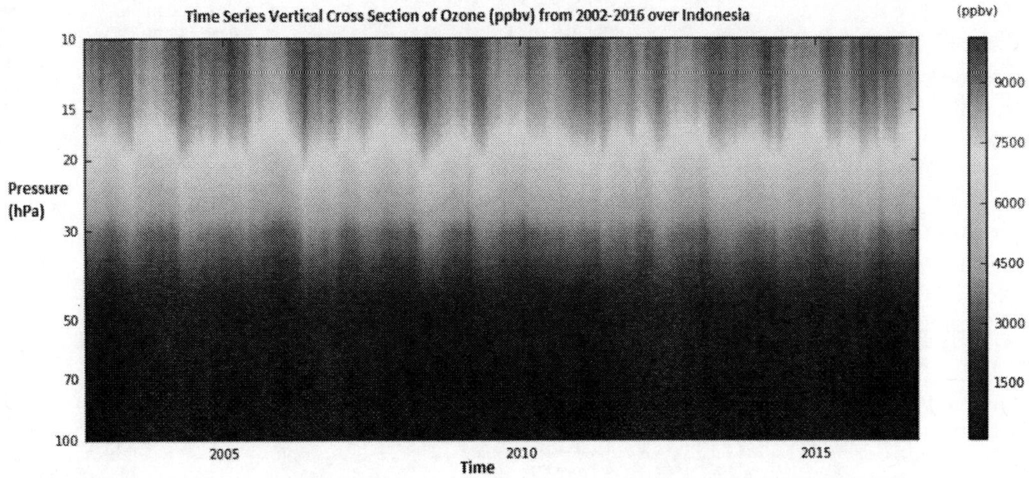

(a)

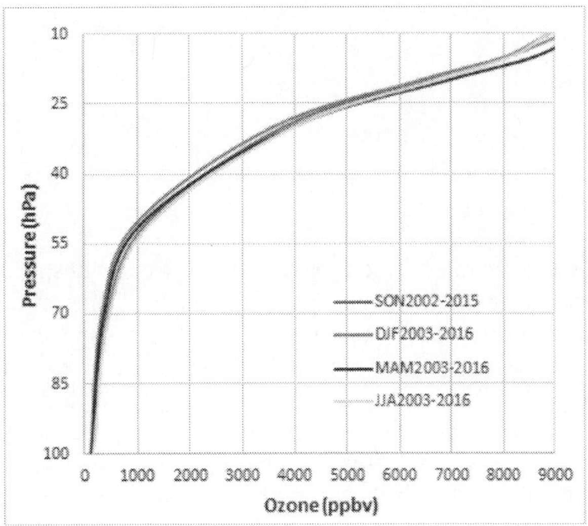

(b)

Figure 1. (a) Time series-vertical cross section and (b) seasonal vertical profile ozone concentration (ppbv) at lower stratosfer (pressure 100 hPa until 10 hPa) over Indonesia from 2002 to 2016.

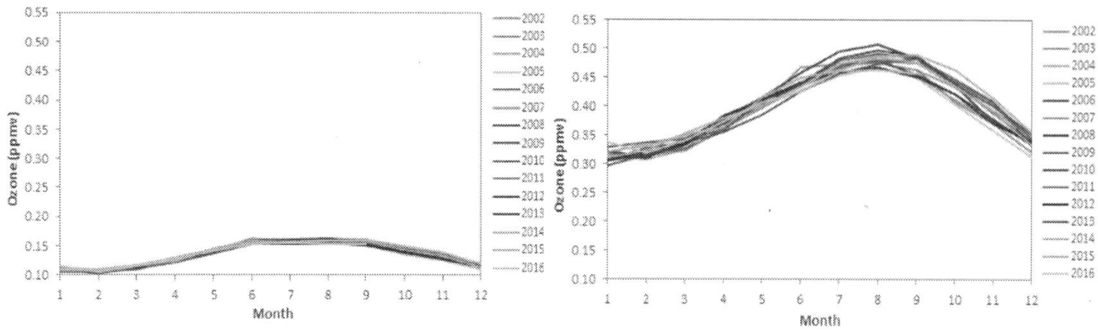

(a) (b)

Figure 2. Monthly average climatology variation of ozone concentration (ppmv) (2002-2016) over Indonesia at pressure (a) 100 hPa and (b) 70 hPa.

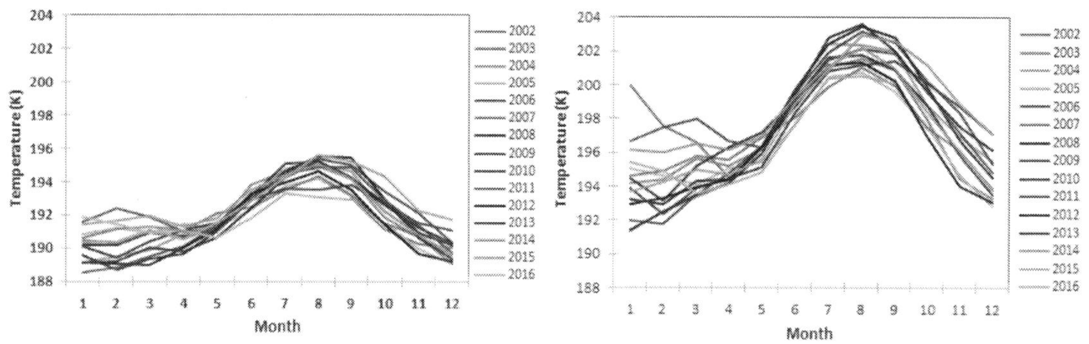

(a) (b)

Figure 3. Monthly average climatology variation of temperature (K) (2002-2016) over Indonesia at pressure (a) 100 hPa and (b) 70 hPa.

The changes of ozone concentration in tropopause and lower stratosphere are shown in **Figure 4**. Ozone concentration at two different level pressure shows higher value of ozone at lower stratosphere (pressure 70 hPa) than at tropopause (pressure 100 hPa). It happened because ozone in mid-stratosphere is actually very rich on pressure 10 hPa that makes ozone becomes higher in lower stratosphere than in tropopause. Ozone concentration in lower stratosphere shows the value around 0,3 ppmv until 0,5 ppmv with linear trend shows very small increase around 0,01 ppmv for 14 years of measurement. Comparing to ozone in lower stratosphere, ozone in tropopause (pressure 100 hPa) shows a lower value around 0,1 ppmv until 0,165 ppmv with a smaller increase only 0,0035 ppmv for 14 years measurement. The increase of ozone concentration in the stratosphere was estimated because of several factors, which are the decreased amount of ozone depleting substances (ODS) and the decrease of stratospheric temperature due to the increased amount of greenhouse gases in troposphere [28]. Ozone in lower stratosphere and tropopause shows similar pattern to annual and semi-annual variation with peak value occurs around Augustus-September and minimum value occurs around February-March.

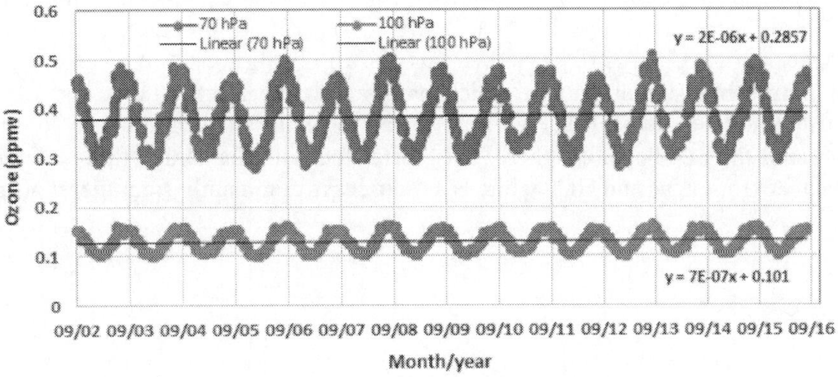

Figure 4. Time series ozone concentration (ppmv) at pressure 100 hPa and 70 hPa from 2002 to 2016 over Indonesia.

The differences of ozone variation also depend on geographic positions or latitude as shown in **Figure 5**. Ozone over Indonesia on latitude -15 °S to 15 °N shows that the concentration of ozone is higher in Nothern part than in Southern part of Indonesia, while over the equator on latitude 0 degree ozone shows minimum peak value. Stolarski [19] shows the same result on ozone concentration which is higher in the Northern hemisphere than Southern hemisphere due to the interhemispheric differences in both magnitude and phase of variation in the upwelling and mixing.

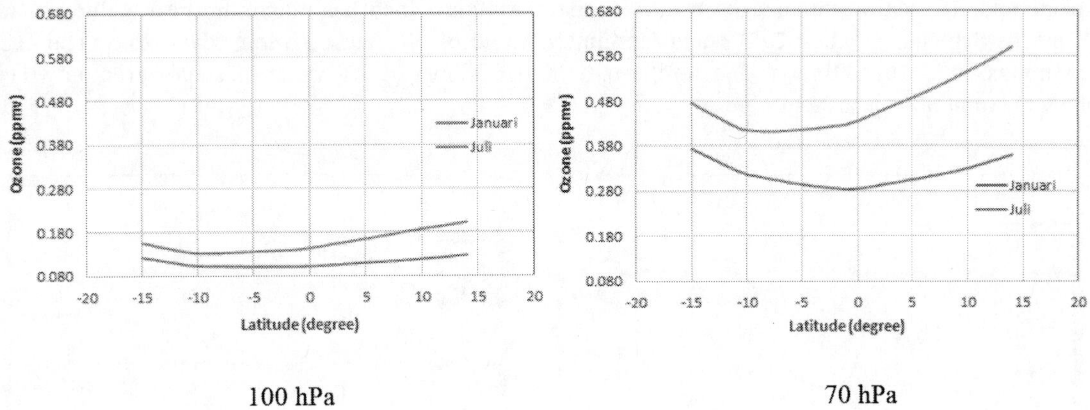

Figure 5. Zonal mean (based on latitude) of ozone concentration (ppmv) in 2002-2016 at pressure 100 hPa and 70 hPa over Indonesia.

The same phenomenon occurs at two different level pressure in the tropopause (100 hPa) and in the lower stratosphere (70 hPa) with ozone concentration higher in lower stratosphere compared to tropopause. Based on zonal mean variation, ozone concentration in lower stratosphere (70 hPa) shows value around 0,26 ppmv until 0,61 ppmv, while ozone concentration in tropopause (100 hPa) shows lower value around 0,1 ppmv until 0,2 ppmv. The concentration of ozone also shows higher value on July compared to January due to the rainy season over Indonesia occurs on January that makes ozone destructed by water vapour.

Ozone is a stratospheric tracer which is dominantly formed in the middle to upper stratosphere and then transported to the rest of the stratosphere. This feature also similar with HCl which is a long lived reservoir fo chlorine in the stratosphere, while N_2O has different feature that is released in the troposphere and transported up to the stratosphere [19]. To compare ozone at lower stratosphere with another trace gases which are HCl and N_2O is shown on vertical profile of ozone, HCl, and N_2O in the lower stratosphere from AURA/MLS satellite data shown on **Figure 6**. We can see that ozone and HCl vertical profile has similar pattern which has lower concentration at lower stratosphere then increase until the middle stratosphere, while N_2O shows higher concentration in lower stratosphere then decrease until the middle stratosphere. It is correlated with the source that is dominantly in the middle stratosphere for ozone and HCl, while N_2O sources is dominantly from the troposphere.

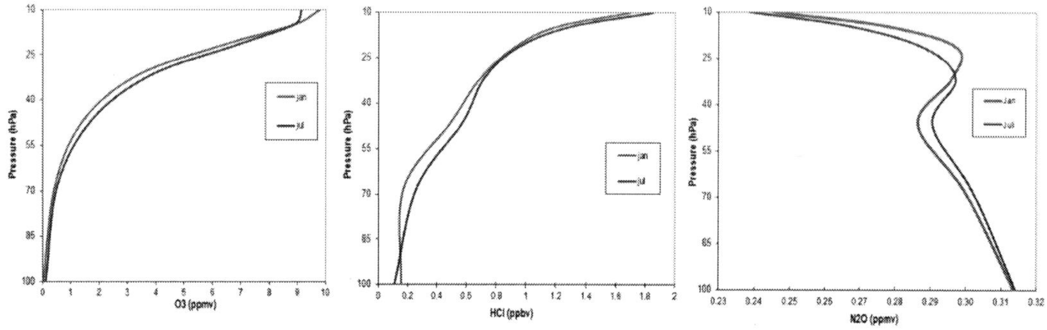

Figure 6. Vertical profile mean of ozone, HCl, and N_2O concentration at lower stratosphere for January and July at pressure 100 hPa until 10 hPa from MLS/AURA satellite.

The differences between ozone, HCl, and N_2O also shown on seasonal variation on **Figure 7**. Ozone and HCl show similar pattern on seasonal variation which has maximum peak value around Augustus-September, while N_2O occurs minimum value on the same season when ozone and HCl reach maximum concentration. The similar pattern also shown by PV value seasonal variation which show maximum value around June.

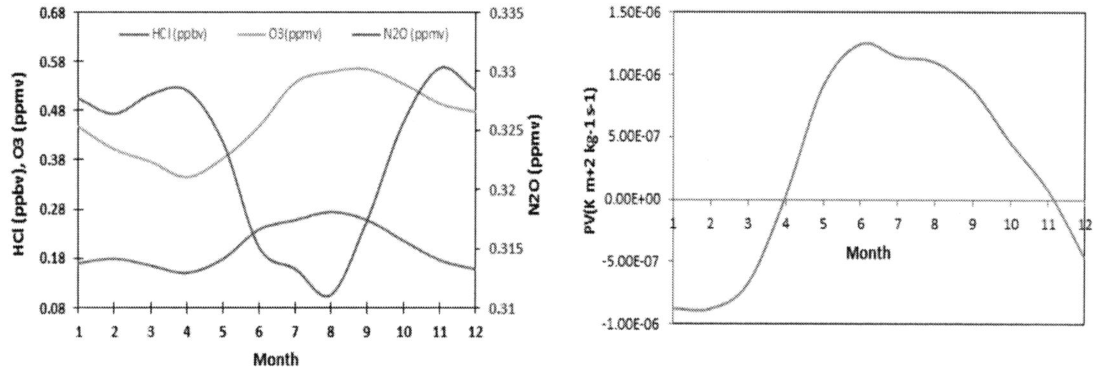

Figure 7. Seasonal variation of ozone, HCl, and N_2O concentration and PV value in lower stratosphere at pressure 68 hPa from MLS/AURA satellite and MERRA data reanalysis.

This different behaviour between ozone, HCl, and N_2O in the tropical lower stratosphere over Indonesia is also shown in the result by Stolarski's study [19]. It happens due to seasonality of transport operating on tracer gradients. Transport in tropical lower stratosphere is primarily due to a combination of vertical advection and mixing between the tropics and extra tropics. The phase and relative amplitude of the tracer annual cycles will depend on the seasonality of the upwelling and mixing between the tropics and extra tropics as well as the vertical and horizontal tracer gradients [19].

3.2 Influence of lower stratospheric ozone into upper tropospheric ozone in Indonesia

Time series-vertical cross section of ozone and potential vorticity in the upper troposphere are shown in **Figure 8**. It can be seen that ozone in the tropopause at pressure 100 hPa shows higher value and it can reach until upper tropopause at pressure 150 hPa. This feature also shows the variation of PV value in the tropopause which also shows higher value and it reaches until the upper troposphere at pressure 150 hPa. It indicates that ozone from the lower stratosphere can also influence ozone in the upper troposphere. Higher ozone and PV value indicate stratospheric air characteristics which can be transported into upper troposphere through stratosphere-troposphere transport.

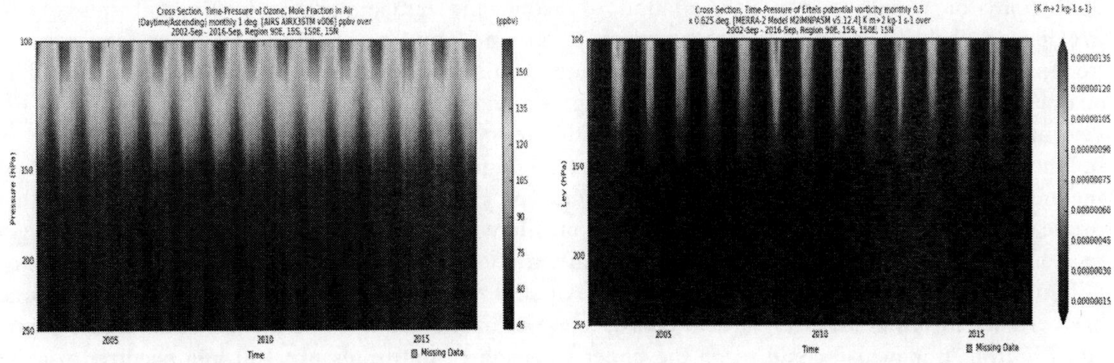

(a) (b)

Figure 8. Time series-vertical cross section for ozone concentration (a) and Potential Vorticity (PV) (b) in the upper troposphere from pressure 250 hPa until 100 hPa.

Similar feature also can be seen in the seasonal variation in time series-vertical cross section of ozone and PV in 2014-2015 as shown in **Figure 9**. It can be seen clearly that higher concentration of ozone ~150 ppbv in the tropopause can crosses and enters the upper troposphere until pressure 150 hPa.

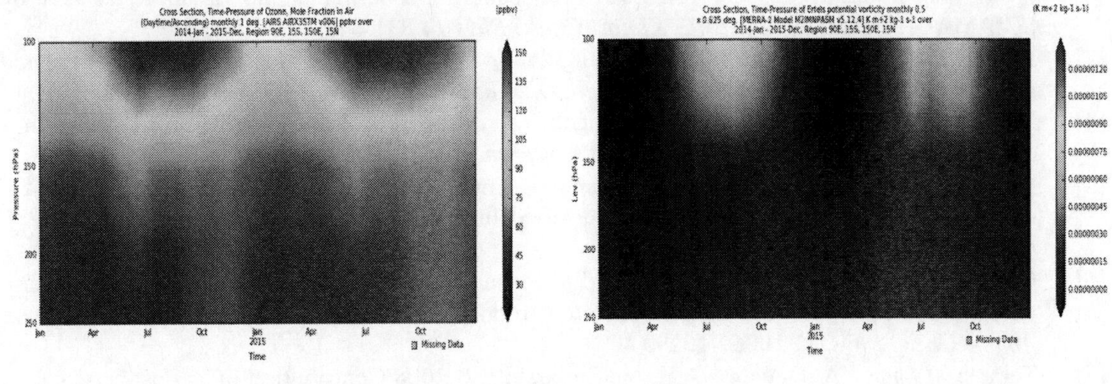

(a) (b)

Figure 9. Seasonal vertical cross section of ozone concentration (a) and Potential Vorticity (PV) (b) from year 2014-2015 in the upper troposphere from pressure 250 hPa until 100 hPa.

The pattern also happens to variations of PV in which higher value was reached until upper troposphere and consistent with ozone variations. **Figure 9** shows the same season when higher concentration of ozone reaches the upper troposphere from Mei to October when high PV also reached the upper troposphere. It indicates that some of stratospheric air has been influenced to upper tropospheric air.

4. Conclusions

Vertical profile variation of ozone in the lower stratosphere Indonesia shows higher value at higher altitude with the maximum peak occurred at pressure 10 hPa (middle stratosphere), the value reached 9000 ppbv or 9 ppmv. Monthly variation of ozone and temperature at lower stratosphere and tropopause shows consistently 2 maximum peaks in June and September which occurred in tropopause, while 1 maximum peak occurred in Augustus-September in lower stratosphere. The changes of lower stratosphere ozone concentration show a very small value around 0,01 ppmv for 14-year measurement, smaller changes occurred in the tropopause. Ozone variation in lower stratosphere Indonesia is also influenced by geographic position which is shown that ozone in Northern part of Indonesia is higher than in equatorial region. Ozone and another trace gases, HCl and N_2O, shows that ozone and HCl have the same pattern in vertical profile with lower value in lower stratosphere and get increased until middle stratosphere, and seasonal variation shows maximum value around Augustus. Different pattern occurred to N_2O. The influence of stratospheric ozone into upper tropospheric ozone has been seen from time series vertical cross section of ozone and PV which shows that higher ozone and PV from tropopause could reach the upper troposphere until pressure 150 hPa occurred for the same season.

Acknowledgments

The authors would like to thank to the Center for Atmospheric Science and Technology, LAPAN for financing assistance for ISSH 2017 activities, and to NASA as a provider of atmospheric composition data used in this study.

5. References

[1] Hsu J and Prather M J 2009 Stratospheric variability and tropospheric ozone *J. Geopyhsical Research* **114** D06102

[2] Gebhart C, Rozanov A, Hommel R, Weber M, Bovensmann H, Burrows J P, Degensteins D, Froidevaux L and Thompson A M 2014 Stratospheric ozone trends and variability as seen by SCIAMACHY from 2002 to 2012 *Atmos. Chem. Phys* **14** 831-846

[3] Brewer A M 1949 Evidence for a world circulation provided by the measurement of helium and water vapour distribution in the stratosphere *Q. J. R. Meteorol. Soc* **75** 351-363

[4] Hsu J and Prather M J 2005 Diagnosing the stratosphere-to-troposphere flux of ozone in a chemistry transport model *J. Geopyhsical Research* **110** D19305

[5] Kuang S, Newchurch M J, Burris J, Wang L, Knupp K and Huang G 2012 Stratosphere-to-troposphere transport revealed by ground based lidar and ozonesonde at a midlatitude site *J. Geopyhsical Research* **117** D18305

[6] Greenslade J W, Alexander S P, Schofield R, Fisher J A and Klekociuk K A 2017 Stratospheric ozone intrusion events and their impacts on tropospheric ozone in the Southern Hemisphere *Atmos. Chem. Phys.* **17** 10269-10290

[7] Terao Y, Logan J A, Douglass A R and Stolarski R S 2008 Contribution of stratospheric ozone to the interannual variability of tropospheric ozone in the northern extratropics *J. Geophys. Res.* **113** doi:10.1029/2008jd009854

[8] Holton J R, Haynes P H, McIntyre M E, Douglass A R, Rood R B and Pfiser L 1995 Stratosphere-troposphere exchange *Rev. Geophys.* **4** 403-439

[9] Barre J, Peuch V H, Attie J L, Amraoui L E, Lahoz W A, Josse B, Claeyman M and Nedelec P 2012 Stratosphere-troposphere ozone exchange from high resolution MLS ozone analyses *Atmos. Chem. Phys* **12** 6129-6144

[10] Waugh D W and Polvani L M 2000 Climatology of intrusions into the tropical upper troposphere *Geophys. Res. Lett* **27** 3857-3860

[11] Boothe A C and Homeyer C R 2017 Global large-scale stratosphere-troposphere exchange in modern reanalyses *Atmos. Chem. Phys* **17** 5537-5559

[12] Randel W J and Wu F 2007 A stratospheric ozone profile data set for 1979-2005: variability, trends, and comparisons with coloumn ozone data *J. Geopyhs. Res* **112** D06313

[13] Ploeger F, Fueglistaler S, Groob J U, Gunther G, Konopka P, Liu Y S, Muller R, Ravegnani F, Schiller C, Ulanovski A and Riese M 2011 Insigth from ozone and water vapour on transport in the tropical tropopause layer (TTL) *Atmos. Chem. Phys.* **11** 407-419

[14] Fadnavis S, Chakraborty T and Beig G 2010 Seasonal stratospheric intrusion of ozone in the upper troposphere over India *Ann. Geophys.* **28** 2149-2159

[15] Randel W J, Park M, Emmons L, Kinnison D, Bernath P, Walker K A, Boone C and Pumphrey H 2010 Asian monsoon transport of pollution to the stratosphere *Science* **328** 611-613

[16] Ghude S D, Kulkarni S H, Kulkarni P S, Kanawade V P, Fadnavis S, Pokhrel S, Jena C, Beig G and Bortoli D 2011 Anomalous low tropospheric coloumn ozone over Eastern India during severe drought event of monsoon 2002: a case study *Environ Sci Poll Res* **18** 1442-1455

[17] Randel W J and Thompson A M 2011 Interannual variability and trends in tropical ozone derived from SAGE II sateliite data and SHADZ ozonesondes *J. Geophys. Res* **116** D07303

[18] Fadnavis S, Dhomse S and Ghude S 2013 Ozone trends in the vertical structure of upper troposphere and lower stratosphere over the Indian monsoon region Int. J. Environ. Sci. Technol. Doi 10.1007/s13762-013-0258-4

[19] Stolarski R S, Waugh D W, Wang L, Oman L D, Douglass A R and Newman P 2014 Seasonal variation of ozone in the tropical lower stratosphere: Southern tropics are different from northern tropics *J. Geophys Res: Atmospheres* 10.1002/2013JD021294

[20] Tian B, Manning E, Fetzer E, Olsen E, Wong S 2014 AIRS Version 6 L3 User Guide *Jet Propulsion Laboratory California*

[21] Komala N 2016 Watukosek tropospheric ozone and temperature profiles from ozonesondes: Comparison with AQUA-AIRS *Quadrennial Ozone Symposium 2016*

[22] Ahmad S P, Waters J W, Johnson J E, Gerasimov I V, Leptoukh G G and Kempler S J 2006 Atmospheric Composition Data Products from the EOS Aura MLS *Proc. Amer. Meteorological Soc. Eighth Conf. on Atmospheric Chemistry, Atlanta, Georgia*

[23] Waters J W, Froideuvaux L, Harwood R S 2006 The Earth Observing System Microwave Limb Sounder (EOS MLS) on The Aura Satellite *IEEE Transactions on Geoscience and Remote Sensing* **44** 5

[24] UCAR 2017 Climate data guide NASA MERRA https://climatedataguide.ucar.edu/climate-data/nasa-merra

[25] Sioris C E, McLinden C A, Fioletov V E, Adams C, Zawodny J M, Bourassa A E and Degenstein D A 2013 Trend and variability in ozone in the tropical lower stratosphere over 2.5 solar cycles observed by SAGE II and OSIRIS *Atmos. Chem. Phys. Discuss* **13** 16661

[26] Henz, D 2010 A modelling study of the tropical tropopause layer *Thesis University of Wisconsin-Madison*

[27] Chandra S, Ziemke J R, Bhartia P K and Martin R 1996 The mid-latitude total ozone trends in the northern hemisphere *Geophys. Res. Lett* **23** 555-558

[28] Steinbrecht W, Froidevaux L, Fuller R, Wang R, Anderson J, Roth C, Bourassa A, Degenstein D, Damadeo R, Zawodny J, Frith S, McPeters R, Bhartia P, Wild J, Long C, Davis S, Rosenlof K, Sofieva V, Walker K, Rahpoe N, Rozanov A, Weber M, Laeng A, Clarmann T V, Stiller G,

Kramarova N, Beekman S G, Leblanc T, Querel R Nedoluha G, Vigouroux C, Blumenstock T, Schneider M, Garcia O, Jones N, Mahieu E, Smale D, Kotkamp M, Robinson J, Pertopavlovskikh I, Harris N, Hassler B, Hubert D and Tummon F 2017 An update on ozone profile trends for the period 2000 to 2016 *Atmos. Chem. Phys* **17** 10675-10690

Analysis of Heavy Rainfall with Hail using Reflectivity and Radial Velocity Derived from X-Band Radar in Bandung (Study Case: March 17, 2017)

T Sinatra[1] and G. A Nugroho[1]

[1]Center of Science and Technology Atmosphere,
Indonesian National Institute of Aeronautics and Space (LAPAN)
Jln. Dr. Djundjunan No. 133, Bandung 40173, Indonesia

E-mail: tiim.sinatra@lapan.go.id

Abstract. A local heavy rainfall along with hail and strong wind occurred in Bandung basin on March 17th, 2017. This event was documented using an X-band Doppler radar and supported by rain scanner, and C-band radar. Based on three types of radar, it was concluded that this was a local heavy rainfall, started at 10.50 LT. The location of the first reflectivity was detected from 25 km northward (107.6°W, 6.92°S) of X-band Doppler radar location. Reflectivity pattern showed a V-shaped with duration of about 30 minutes and reached its peak at about 11.30 local time (LT). Maximum horizontal wind speeds from Automatic Weather Station (AWS) took place during those periods. Based on vertically integrated liquid (VIL), the potential of hail was located at the right flank of the V-shaped. Three vertical cores with maximum reflectivity were also looked at in the right flank at altitude of 2-6 km with echo top reached up to 9 km above ground surface.

1. Introduction

Severe convective events occur most frequently in location where warm, moist air in the lower troposphere is overlain by cooler, drier air [1]. It is usually accompanied by heavy rain, strong wind, lightning, and hail. This phenomenon has a direct impact on environment, building, and of course, human and their activities. Severe convective storms, in Bandung, occur characteristically in the peak of rainy season and in transition period.

The convective events, like hail storms, have high variability in time and space, so that these events are not always captured accurately by single ground weather station. The occurrence of hail usually occurs only about a few minutes. So that, the use of weather radar is a feasible option because it gives information in spatial and temporal resolution well. Several studies have used reflectivity data and Doppler's velocity based on single polarization radar for observing evolution of an updraft and downdraft cell [2] [3] [4]. The two splitting (formed V-shaped) storms have typical situation where the right member is the dominant one [3]. This echo V-shaped generally indicates a separation process from an object (generally a storm) due to a new updraft formed in the storm [3]. The storm development can be divided into 3 stages, i.e. the initialization stage, the mature stage, and the decaying stage [5].

Bandung is a basin located at ~800 m above sea level (asl) and surrounded by mountains with the peak of northern is higher than the southern. The peak of accumulation of monthly rainfall of Bandung

IOP Conf. Series: Earth and Environmental Science **166** (2018) 012039 doi:10.1088/1755-1315/166/1/012039

from ground observation data from 1981-2010 occurred in March [6].Water vapour within the bottom of basin during morning and low-level wind convergence induced by valley winds are important factors of convection in Bandung [7]. Indonesian Agency for Meteorological, Climatological and Geophysics (BMKG) gets report from society that on March 17, 2017 heavy rain occurred and it was accompanied by hail from 11.50 until 12.10 LT over Bandung city. The initial condition of this event was recorded by transportable radar operated during campaign. This study wants to investigate the hail event based on reflectivity and radial velocity data on March 17, 2017.

2. Data

The data used in this study were obtained during campaign in Bandung on March 17, 2017. Transportable radar is an x-band Doppler weather radar with single-polarization. It was installed at Rancakasumba Substation (107.6 °E, 6.92 °S) located in Majalaya, south of Bandung city. The area coverage of radar is 100 km radii. Because of power failure, radar was off around 11.32 LT. Even so, the weather radar captured the initial condition of the event. Rain scanner located in Pasteur was used to complement radar data. Rain scanner is a marine radar development by LAPAN to observe rain since 2012. This rain scanner has 44 km radii with temporal resolution 2 minutes. It can detect heavy rainfall well [8] [9] and hail event [10]. The location of transportable radar and rain scanner is shown by A and B in **Figure 1**, respectively.

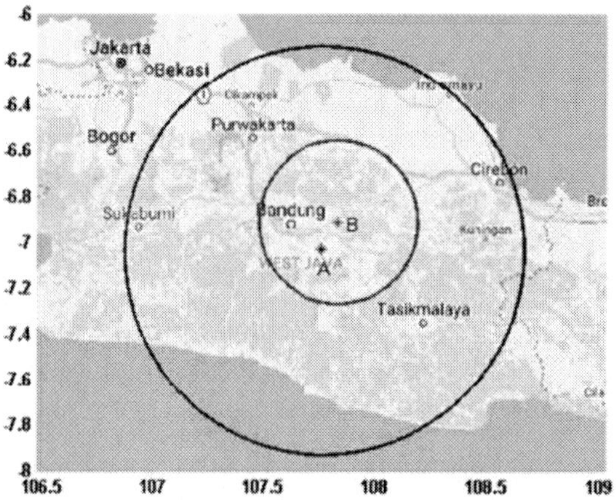

Figure 1. Map of location of the transportable radar (A) and rain scanner (B). The two circles show the area coverage of radar, respectively.

Reflectivity and radial velocity CAPPI product from transportable radar were utilized to analyse the rain. Reflectivity from rain scanner can give the information of the rain overall, from growth until decayed. Radial velocity is in addition to reflectivity data, which is component of movement either toward or away from radar. Doppler velocity can give the information about divergent/convergent signature

To describe the condition of environment during the event in the affected area, the automatic weather station (AWS) data are used. AWS are located in Pasteur (107.59°E, 6.89°S) and Antapani (107.65°E, 6.92°S).

3. Method

The weather radar completes one volume scan in 10 minutes. Each volume scan consists of 20 elevations that can detect cloud rain until 15 km height. The radar products were derived from volume scan.

In this study, we computed the vertically integrated liquid (VIL) from CAPPI product. The VIL is the total amount of concentration of liquid water in a cloud. It can be a useful tool for estimating the

IOP Conf. Series: Earth and Environmental Science **166** (2018) 012039 doi:10.1088/1755-1315/166/1/012039

severity of a thunderstorm [11]. CAPPI product from 1-15 km height were calculated to get the VIL value using Equation (1):

$$VIL = \sum 3.44 \times 10^{-6} [Z_i + Z_{i+1}/2]^{4/7} \Delta h \tag{1}$$

where VIL has unit of kg m^{-2}, Z_i and Z_{i+1} are bottom and top of the sample layer of reflectivity value in mm^6 m^{-3}, respectively and Δh is the layer thickness in m unit.
Amburn and Wolf (1994) show that VIL density (VILD) is a useful method for assessing hail potential in thunderstorms. The VILD can be used to identify the convective storms with high reflectivity relative to their height. The VILD was computed by using Equation 2:

$$VILD = \frac{VIL}{ET} \times 10^3 \tag{2}$$

where VILD is in g m^{-3} and ET is echo top in m. As VILD value increased, the hail core should become deeper and more intense, and increase the hail size [12].

4. Case study

Figure 2 shows the reflectivity distribution in case study on March 17, 2017. The **Figure** describes the pixel echo shown in the 25 km at northwest (300°) from radar location with no contribution from other echo around it. This echo started to develop at 3.50 UTC and grown rapidly in vertical in the next 1 hour. CAPPI in several heights (3, 5, 7 km) shows different growth in every level of height with different echo top. CAPPI 1 km height shows near surface level of the convective activity. From 3.50 LT until 4.00 LT, the convective cell shows growth but no major changes in shape. After that in 4.10 LT the cell started to expand horizontally to the left forming a V-shaped with an angle of around 80° lookalike until 4.30 LT.

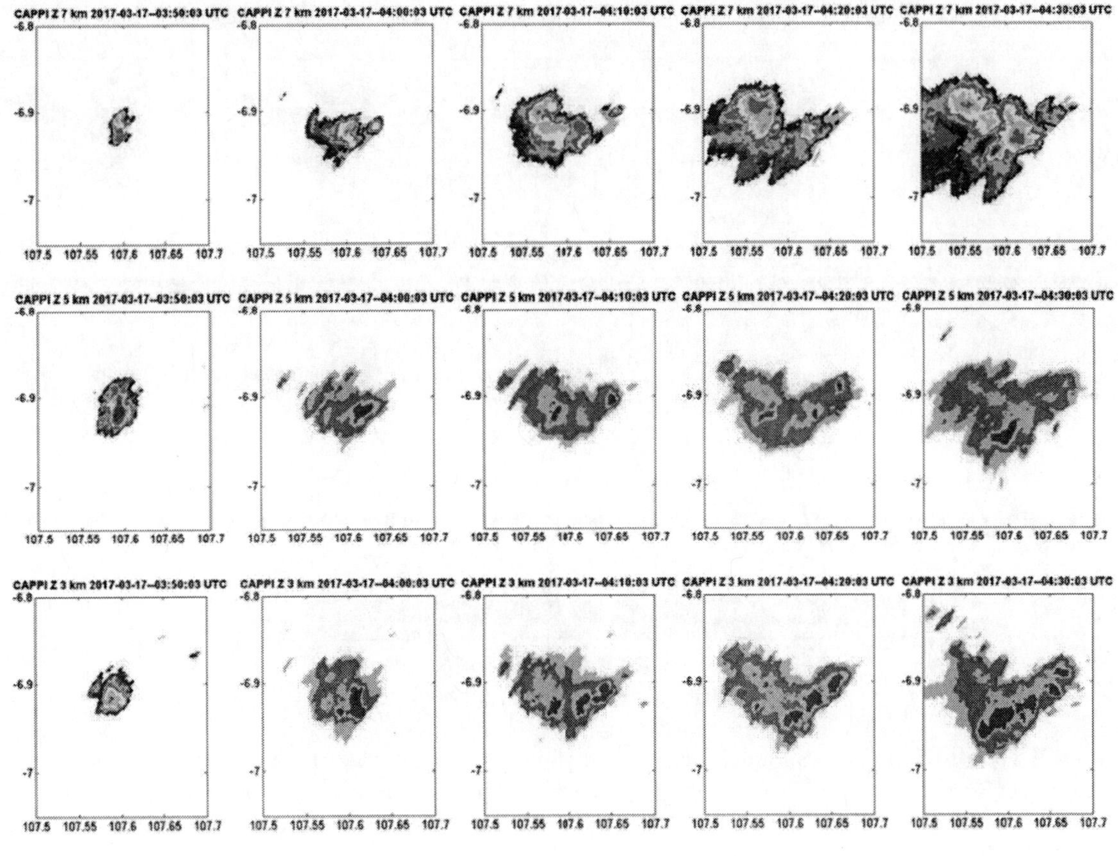

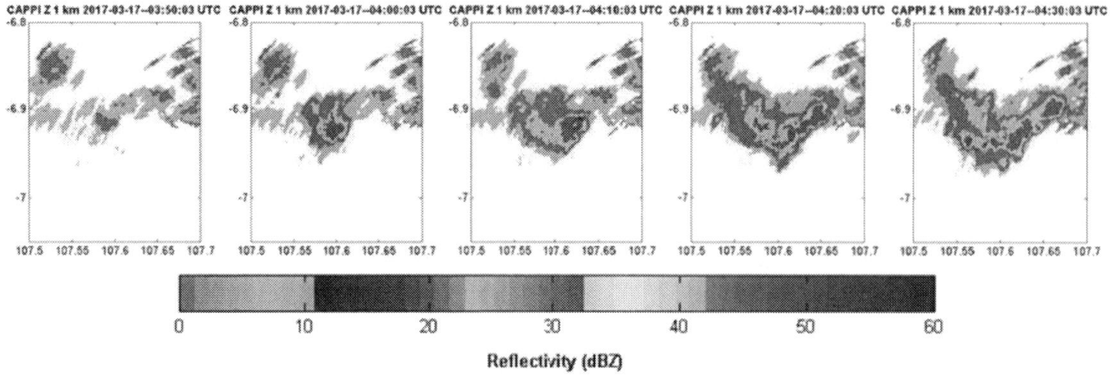

Figure 2. Spatial distribution of CAPPI product at 3, 5, and 7 km height.

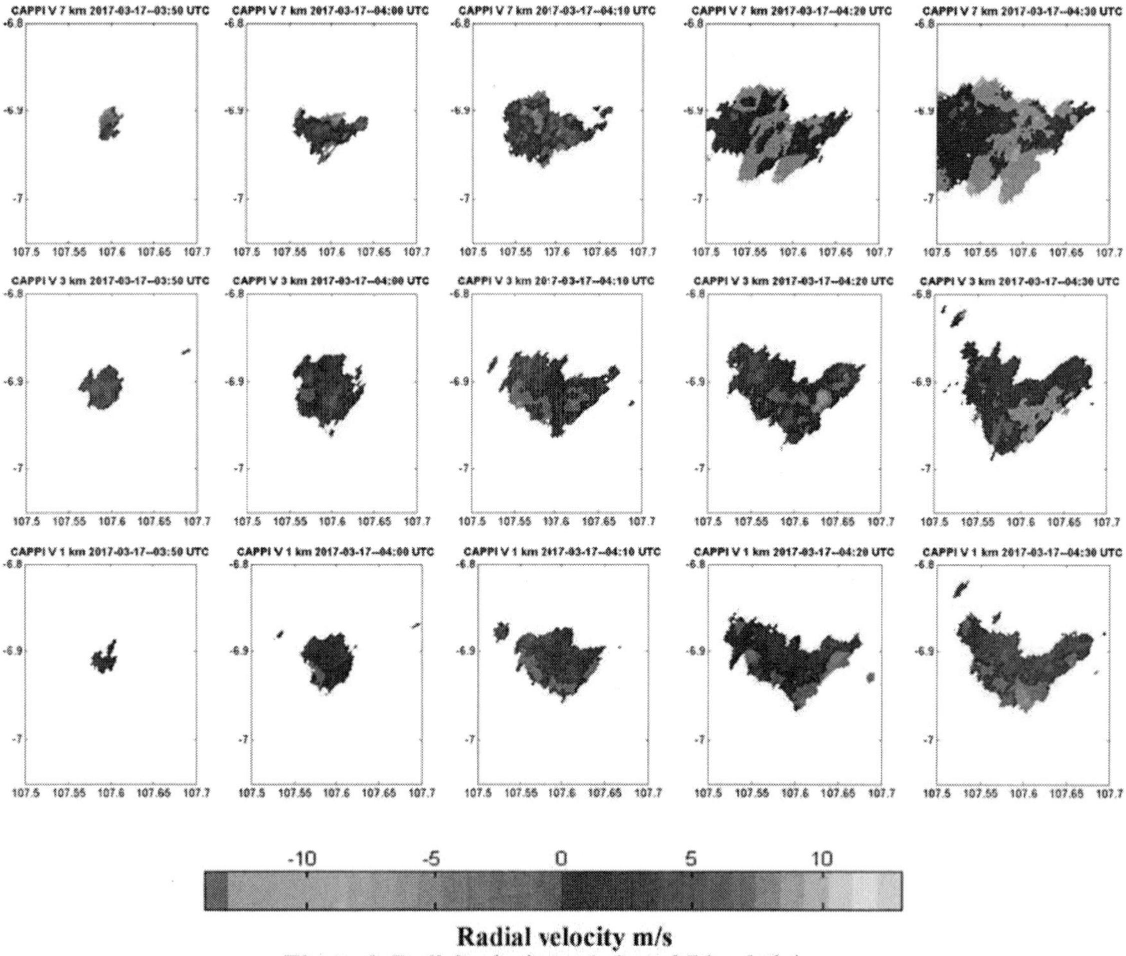

Figure 3. Radial velocity at 1, 3, and 7 km height.

CAPPI 1 km to 5 km shows the middle cross section reflectivity area of the V-shaped. At 11.00 LT (4.00 UTC) convection cell became larger and higher with the reflectivity value was more than 50 dBZ. From 11.10 LT until 11.20 LT, the V-shaped still existed with two distinct cell appeared. Then, at 11.30 LT, the left flank was started to dissipate while the right flank still increased rapidly. The position of convective cores was at right-flank with the largest core was in the middle. V-shaped radar reflectivity pattern was formed at 11.10 LT and clearly seen at 11.20 LT at 3 and 5 km height. CAPPI 7 km shows the upper part of the echo. This **Figure** shows a wider spread reflectivity from 11.00 LT - 11.30 LT. The recorded data show that the echo top still existed in 9 km.

The radial velocity (V) of CAPPI product for 1, 3, and 7 km height is shown in **Figure 3**. Based on the orientation of the radar location, there were convergent signatures at bottom layer (1 and 3 km height) until 11.20 LT. At the upper layer, the divergence signatures was seen. This convergence-divergence processed in different level is an indication of the presence of an updraft [3]. The divergence process on the CAPPI V 7 km at 11.20 LT until 11.30 LT was also an indication of the dissipated process of the left flank of the V-shaped.

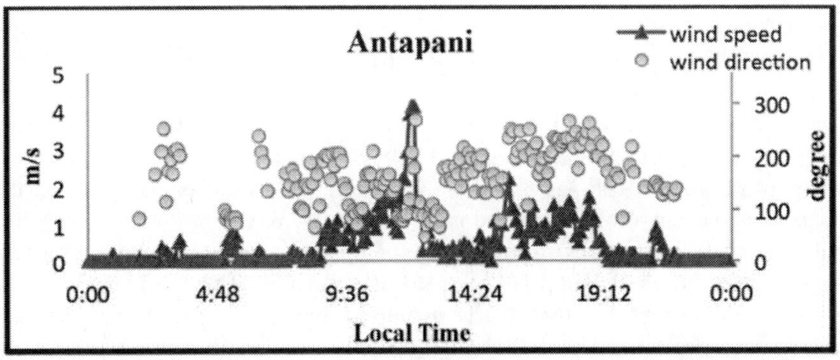

(a)

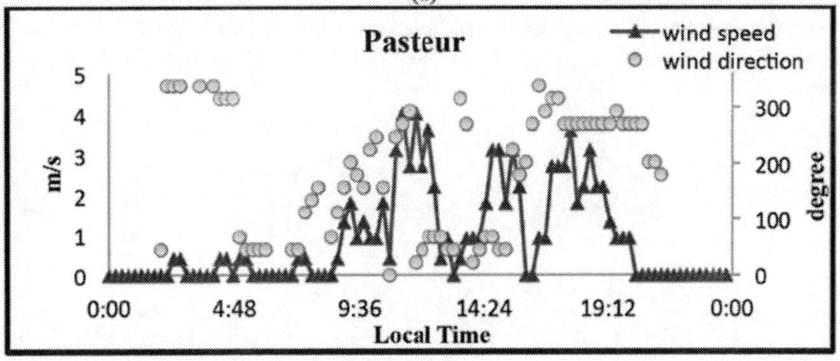

(b)

Figure 4. Environmental situation from AWS data located in Antapani (a) and Pasteur (b) on March 17, 2017.

Pasteur and Antapani were affected area of this event where Pasteur area is on the left and Antapani is on the right. The wind speed at Antapani area started to blow at 9.00 LT. It became stronger around 11.40 LT and reached the strongest at 12.00 LT, and moved southeastward. In Pasteur, the wind speed started to blow at 9.00 LT and then got stronger at 11.00-12.30 LT westward. The wind speed soared up when the echo increased and formed the V-shaped. The left-flank propagated roughly 40°-50° to the right of the mean wind at the time of -V-shaped. While, the right-flank propagated roughly 90°-100° to the left of the mean wind at the time of V-shaped. The right-flank moved away to the southeast.

IOP Conf. Series: Earth and Environmental Science **166** (2018) 012039 doi:10.1088/1755-1315/166/1/012039

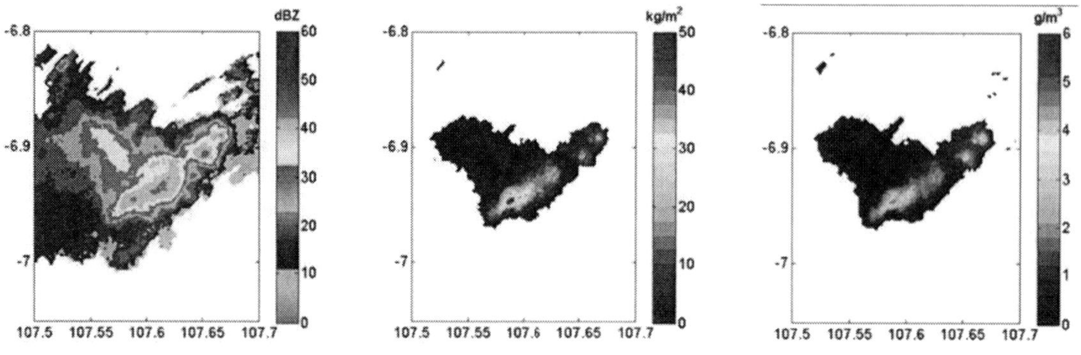

(a) (b) (c)

Figure 5. Composite reflectivity a), VIL b), VILD c) at 11.30 LT (4.30 UTC).

The maximum composite reflectivity distribution (**Figure 5a**) at 11.30 LT revealed the maximum value of reflectivity inside the V-shaped area. The right-flank had larger reflectivity (reached >50 dBZ) than left-flank. **Figure 5b** shows that there is a large VIL value > 30 kg m^{-2} at the right-flank. The VIL increased as the rain cloud extended more vertically. Based on the echo top at this level, the VIL value indicates that there is still strong energy of updraft to maintain reflectivity value in upper level [13]. Spatial distribution of VILD is presented in **Figure 5c**. The VILD shows value >3.5 g m^{-3}. Amburn and Wolf (1994) reported that VILD increased above 3.5 g m-3 in a thunderstorm would potentially produce severe hail. The VILD increased due to the increase in target size. If VILD increases, the hail cores tend to be deeper and more intense, and hail size tends to be larger [12]. This VILD value was also used to show the hail core location compared to VIL. VILD value shows that the hail event occurred at the right-flank, with the largest hail even in the middle of V-shaped.

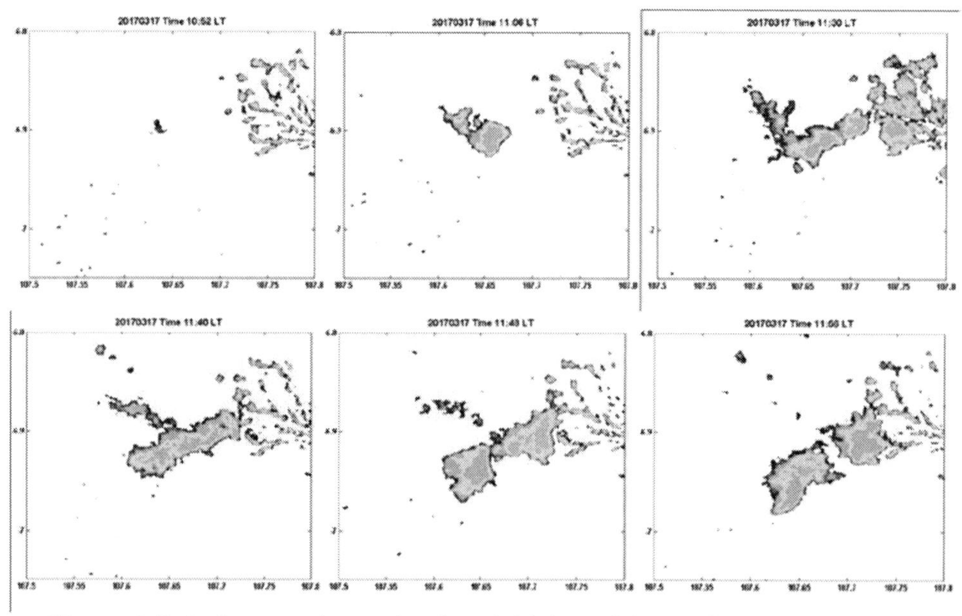

Figure 6. Rain Scanner observation from initial condition until V-shape decayed.

SANTANU rain scanner was used to observe the lost data from transportable radar. **Figure 6** describes the event from rain scanner observation. There is suitability between transportable radar and rain scanner results. The rain started at 10.52 LT and then the rain spread into 2 regions at 11.06 LT, the left region elongates northwestward and the right rain expand northeastward and formed the V-shape. The intensity of right-flank rain is higher than the left-one. At 11.30 LT the left-echo decayed, but the intensity of the right-echo became higher. The lifetime of the left-rain is about 45 minutes. On the other hand, the right rain spread to eastward and southeastward and at 12.48 the rain disappeared. Weather radar C-Band located in Tangerang also detected the event (Figure not shown). The initial echo was seen at 11.02 LT and the V-shaped was seen from 11.18 LT -until 11.34 LT.

From all those data, it can be divided into 3 stages i.e. the initial phase of convective storm lasted roughly from 10.50 LT until 11.10 LT, the mature phase took place at from 11.20 LT until 11.40 LT when reflectivity got maximum value, and the decaying phase.

5. Conclusion

Case study on March 17, 2017 showed a local convective activity in the Bandung area. This local convective activity was recorded by 2 X-band radars. From CAPPI products in certain height showed a distinct V-shaped echo with the region of 1 km until 5 km while the upper part of echo showed wider spread area. The V-shaped possesses two flanks and the right-flank of V-shape has more convective activity than the left-flank. The V-shaped possesses high updraft activity indicated from the convergence and divergence of the radial velocity in the lower and upper level. VIL and VILD show there is a strong potential of hail occurred in the right flank of the v shape. The echo was then gradually dissipated from the left flank after one and a half hour and moved towards southeast that influenced by the dominant wind.

Acknowledgments
This work is fully supported by Center of Science and Technology Atmosphere of Indonesian National Institute of Aeronautics and Space. The authors thank to all staff in Rancakasumba substation of State Electricity Company (PLN) for their cooperation and for letting us campaign at Majalaya's area. The authors also thank to Weather and Climate Prediction Laboratory (WCPL) for providing and allowing us to use the AWS data.

6. References

[1] Barnes G M 2010 Meteorological hazards in the tropics: severe convective storms and flash floods. Tropical Meteorology, Encyclopedia of Life Support Systems (EOLSS) www.eolss.net (sponsored by the UNESCO, 109 pp)

[2] Boe B A, Stith J L, Smith P L, Hirsch J H, Helsdon Jr J H, Detwiler A G, Orville H D, Martner B E, Reinking R F, Meitin R J, and Brown R A 1992 The North Dakota thunderstorm project: A cooperative study of high plains thunderstorms *Bull. Amer. Meteor. Soc.* **73** 145-160

[3] Brown R A and Meitín R J 1994 Evolution and morphology of two splitting thunderstorms with dominant left-moving members *Mon. Weath. Rev.* **122** 2052-2067

[4] Brown R A and Torgerson K L 2003 Interpretation of single-Doppler radar signatures in a V-shaped hailstorm: Part 1-Evolution of reflectivity-based features *Natl. Wea. Digest.* **27** 3-14

[5] Hóller H, Bringi V N, Hubbert J, Hagen M, and Meischner 1994 Life cycle and precipitation formation in a hybrid-type hailstorm revealed by polarimetric and Doppler radar measurements *J. Atmos. Sci.* **51** 2500-2522

[6] Suryantoro A 2014 Analisis Sifat Hujan di Wilayah Bandung dan Sekitarnya Berbasis Observasi Permukaan dan Satelit TRMM *Prosiding Seminar Nasional Fisika Terapan IV* ISSN: 2407-2281

[7] Oigawa M, Matsuda T, Tsuda T, and Noersomadi Coordinated observation and numerical study on a diurnal cycle of tropical convection over complex topography in West Java, Indonesia *J. Meteor. Soc. Japan* **95** 261-281

[8] Nugoho G A, Awaludin A, and Rahayu S A 2014 Hasil pengamatan scanner hujan dan AWS pada kejadian hujan tanggal 2-3 Maret 2014 di daerah Bandung dan sekitarnya *Buku Variabilitas Cuaca dan Iklim di Indonesia* ISBN 978-979-1458-81-8

[9] Nugroho G A and Awaludin A 2015 Analisis distribusi dan pertumbuhan area hujan hasil observasi RDH (Radar Deteksi Hujan) pada tanggal 19 februari 2014 di daerah bandung dan

sekitarnya berdasarkan metode pengolahan citra *Buku Pengembangan Teknologi Atmosfer dan pemanfaatannya* ISBN :978-979-2458-92-4

[10] Nugroho G A, Sinatra T, Awaludin A, and Handiana D 2016 Hail detection by a low cost local weather radar operated for disaster early warning system *Proceeding of Internasional IJJSS (Indonesia Japan Join Scientific Symposium)* ISSN: 978-4-901404-15-0 p.33-39

[11] Clark R A and Greene D R 1972 Vertically integrated liquid water- A new analysis tool *Mon. Wea. Rev.* **100** 548-552

[12] Amburns S A and Wolf P L 1994 VIL density as a hail indicator *Wea. Forecasting* **12** 473-478

[13] Martín F, San Ambrosio I, and Carretero O 2002 Supercélula severa en el area mediterránea *Nota técnica del STAP núm. 37* Instituto Nacional de Meteorología 174pp

IOP Conf. Series: Earth and Environmental Science **166** (2018) 012040 doi:10.1088/1755-1315/166/1/012040

The Global Population of Mesoscale Convective Complexes (MCCs) over Indonesian Maritime Continent during 15 Years

Trismidianto[1]

[1]Center of Atmospheric Science and Technology
Indonesian National Institute of Aeronautics and Space (LAPAN)
Jln. Dr. Djundjunan No. 133, Bandung 40173, Indonesia

E-mail: trismidianto@lapan.go.id

Abstract. Mesoscale convective complexes (MCCs) were identified and tracked during 15-years over Indonesian Maritime Continent (IMC) by infrared satellite imagery using an algorithm that combines criteria of cloud coverage, eccentricity, and cloud lifetime. Infrared satellite imagery is obtained from the Himawari generation satellite data. This study results that there are several characteristics of MCC over the IMC, among others; Most of these MCCs were over the continental area, mainly near the mountains and the high elevation areas which have similar characteristic with the global population of MCCs in the world. The peak of MCC occurrence over the IMC is during MAM with around 33.56% due to likely linked to the synoptic-scale environment, which is more baroclinic in nature than in the late summer months. Around of 26,26% and 25,29%, MCC occurrences are found in DJF and SON, respectively. The season with fewer events is JJA, around of 14.79%. The average of cloud shield area size of MCCs is around 315,000 km^2. The average duration of the MCCs over IMC was approximately ~9.5 hours and the maximum duration was 10 hours. Those observed in this study were usually nocturnal and reached a maximum at midnight.

1. Introduction

According to Maddox [1], mesoscale convective complexes (MCCs) are a special case of mesoscale convective systems (MCSs), where MCS is portrayed as an organized ensemble of convective elements, whose lifecycle is longer than that of the individual convective elements, and the largest of the convective storms. He has divided MCS into two types as linear and circular type, and MCC includes as the circular type found in midlatitude and tropics region. Laing and Fritsch [2] have studied the characteristics of MCCs globally. They are of the view that, more than 400 MCCs occur at various locations around the globe each year, but mostly in tropical and mid-latitude locations, with approximately 66 percent of the occurrences in the Northern Hemisphere. They also found that 91% of MCCs occur over continents even though they can occur over oceans. The average duration of these systems is 10 hours with a cloud shield size of 354,000 km^2 at the time of maximum extent, but this varies according to location. In general, the systems are long lasting in the Southern Hemisphere rather than in the Northern Hemisphere, and MCCs originating over the oceans are more long lasting than those originating over land. The typical MCC develops in the late afternoon, the maximum cloud-shield reaches its maximum size in the early morning hours of the following day, and the system dissipates just after sunrise.

Trismidianto et all [3,4] have reported that MCC in the Indian Ocean near of Sumatra Island influences weather pattern in several regions over Sumatra Island. The result has been posited that there is a role of cold pools during MCC in influencing the rainfall, but they only analyzed the present of new convective systems using convective index by spatial analysis, not only over the Indian Ocean but also over Indonesian Maritime Continent (IMC). Such that, the enthusiasm gives a deep understanding of the mechanism of MCC that relates to the land-sea breeze influencing the rainfall. For 10-years (i.e, 2000 - 2009), Trismidianto [3] also realized the greatest frequency of MCC occurrence in the transition season March, April, and May, and duration of the life cycle of MCC in the Indian Ocean about 12 to 15 hours.

The existence and characteristic of MCCs over IMC have been studied using infrared (IR) satellite imagery from geostationary meteorological satellite (GMS) observations of 5-years (i.e., 2001-2005) period cited by Ismanto [5]. He has reported that 36% of MCC was found in the Eastern Indian Ocean (including Sumatra Island), 15 % in Papua Island, 13% in Northern Papua, 13% in Kalimantan Island, 9.5% in Northern Australian and 13.5% in the widespread area except those regions. He found that MCCs appearances are predominantly nocturnal with an average lifetime of 12.2 hours, but, Ismanto [5] only used short data to explain the characteristics of the MCC over the IMC. However, the climatology of MCCs has not been given attention in detail over IMC by the previous research. So that, the focus of this study is to provide a detailed description of the MCC characteristics in IMC using long period data for 15-years in the period of 2001-2015.

2. Data and Method

The previous study, such as Maddox [1], Anderson and Arrit [6], Durkee et al. [7] and others have utilized IR satellite image from Geostationary Operational Environmental Satellites (GOES). Laing and Fritsch [8,9], Morel and Senesi [10] and several others have utilized the METEOSAT IR images from the International Satellite Cloud Climatology Project (ISCCP) B3 stage radiance data. The METEOSAT is a geostationary meteorological satellite, operated by the European Organization for the Exploitation of Meteorological Satellites (EUMETSAT), with multiple series. IR data provide the variable brightness temperature (T_{BB}) that gives information about the cloud temperature and altitude, where, as a general rule, the colder the object, the larger the altitude. T_{BB} is uniquely related to the (sensed) wavelength.

This study utilized T_{BB} from the Himawari geostationary satellites, provided by the Japan Meteorological Agency (JMA). Since the launch of GMS-1 (Himawari-1) in 1977, there have been three generations, including GMS (Geostationary Meteorological Satellite), MTSAT (Multi-functional Satellite imagery Transport SATellite), and Himawari 8/9. However, this research utilized data from Himawari-5 until Himawari-8 and added the data GOES-9 because from May 2003 to June 2005, JMA rented the GOES-9 satellite from the National Oceanic and Atmospheric Administration (NOAA) for observation of the western Pacific region to fill the gap between the operational periods of Himawari-5 and Himawari-6. In detail, the data consist of Himawari-5/GMS-5 for data from January 2001-April 2003, Pacific GOES/GOES-9 for data from May 2003- June 2005, Himawari-6/MTSAT-1R for data from July 2005 - June 2010, Himawari-7/MTSAT-2 for data from July 2010 - June 2015, and Himawari-8 for data from July 2015 - December 2015.

MCC was identified by inputting the temperature, latitude, and longitude values for each cloud shield pixel obtained from IR to a modified version of a computerized MCC program using MATLAB based on characteristics of MCCs [1]. This method adapts the method from Ismanto [5] and Trismidianto [3] based on the maximum spatial correlation tracking technique (MASCOTTE) method by Carvalho and Jones [11]. Maddox [1] stated that MCC has cloud shield with continuously $T_{BB} \leq -32^{o}C$ (in this study converted to 241 K) and must have an area $\geq$ 100,000 km^2. The interior cold cloud has $T_{BB} \leq -52^{o}C$ (in this study converted to 221 K) and must have an area $\geq$ 50,000 km^2 for over six hours, although the cloud shield does not have to maintain an eccentricity of ≥ 0.7 for its entire life cycle. The eccentricity is also important to distinguish MCC from the other MCS type, for example; squall line, bow echo, cloud cluster and the others. The areal extent and eccentricity values were used to determine the duration and life cycle of the MCCs, while the centroid position indicates the path of propagation of the systems. The explanation of MCC based on Maddox [1] criteria over IMC is still rarely found, so it becomes interesting to explore MCC over IMC based on Maddox [1].

3. Result and Discussion

3.1 Global distribution of MCC occurrences

A climatological study of MCCs during 2001 - 2015 over IMC has been identified using IR satellite imagery with creating an algorithm that combines criteria of cloud coverage, eccentricity, and cloud lifetime. A total of 1028 MCCs were identified and tracked during fifteen years for the 2001 – 2015 period as shown in **Figure 1**. The MCCs were developed in several concentrated areas; some of them were over the ocean and coastal areas, and others were over inland and high elevation areas. By examining the location of the interior cloud centroids for all MCCs, one can see that MCCs mostly frequently occurred in nine regions as shown in red dotted line in **Figure 1**. That nine regions are the Indian Ocean near Sumatra Island, along the western coastal of Sumatra Island, the South China Sea near Kalimantan Island, the Central Kalimantan around 3°S-1°N; 111° -115°E, the East Kalimantan around 1° - 4°N; 116° - 120°E, Makassar Strait near coasts of Kalimantan Island, the Central Sulawesi around 4° - 2°S; 120° - 123°E, Merauke over Papua Island in location around of 8° - 5°S and 135.5° - 140°E, and Cendrawasih Bay near coasts of Papua Island in location around of 132.5° - 136.5°E and 4° - 1°S.

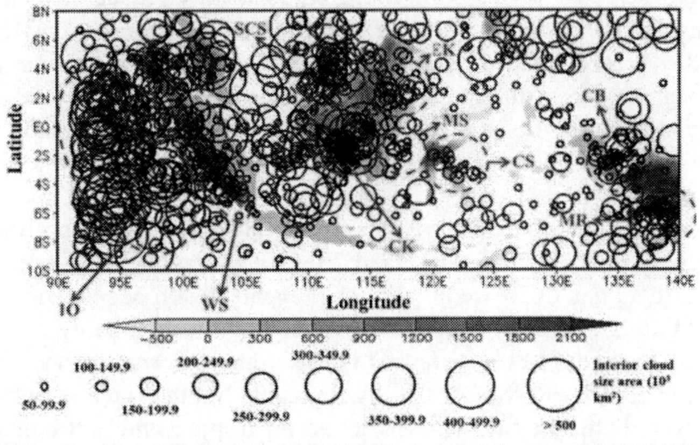

Figure 1. Geographical distribution of the MCCs in the IMC during 15-years (2001-2015). Locations are for the MCC at the time of maximum extent of the interior cloud size area when the mature stage. The circles represent the interior cloud size area (10^3 km^2). The red dotted line indicates the MCCs concentration region, where IO,

WS, SCS, CK, EK, MS, CS, CB and MR refer to Indian Ocean, Western Sumatra, South China Sea, Central Kalimantan, East Kalimantan, Makassar Strait, Central Sulawesi, Cendrawasih Bay and Merauke, respectively. The color shaded refers to the elevation of the unit in meters (m).

Figure 1 shows that the MCC occurrences was frequently found in the near-equatorial regions in the range of latitude 5°S – 5°N. It is associated with the surplus of radiative energy in the tropics and a net deficit in middle and in high latitudes, requiring on average a poleward transport of energy by the atmospheric circulation. The fact is that this surplus energy heats the ocean and land surfaces and evaporates moisture. In turn, some of this heat lands its way into the atmosphere in the form of sensible and latent heat, and it is the energy of this type that is transported polewards by the atmospheric circulation. Cloud systems directly affect the surface energy budget through their modification of the net radiative fluxes. Clouds are also a key factor in the coupling between the atmosphere and the ocean through their influence in the surface radiative, heat, and momentum fluxes [12]. The energy associated with the equatorial maximum in solar radiation released through vigorous atmospheric convection is the ultimate driver of the mean position of the Hadley circulation [13]. Atmospheric convection cell warm air rises at the equator. As it rises, it cools and generates large amounts of precipitation. The development of the MCC over near-equatorial is also possibly related to the ITCZ as a belt of low pressure which circles the Earth generally near the equator where the trade winds of the Northern and Southern Hemispheres come together. It is characterized by convective activity which often generates vigorous thunderstorms over large areas. It is most active over continental land masses by day and relatively less active over the oceans.

Most of MCCs concentrated near the mountainous and high elevation areas. This result is similar to Ashley [14] and Laing and Fritsch [15] who reported that the mean location for the development of the MCCs in America was found on the lee side of the mountain. Morel and Senesi [10] also reported that a large number of MCS over a large part of Europe using satellite infrared imagery was found at mainly continental, strongly related to orography and in their majority are in phase with the diurnal radiative heating. MCC was frequently found near mountain because the effect of topography is important for the low level jet formation. The jet appears to contribute to the formation of the MCS (including MCC) structure in the northern part of the convective system and the formation of low-level intense wind. Additionally, the effect of the convective system on the formation of the jet could also be substantial because the convective system may hydrostatically reduce the low-level pressure behind it, thereby enhancing the horizontal pressure gradient force across the mountain range [16]. A mountain breeze and a valley breeze are two related; localized winds that occur one after the other on a daily cycle which also gives a contribution to convective activity over the IMC due to IMC is a unique geographical region composed of a complex system of mountainous islands. The interaction between the topography with the westerly wave propagation is assisted in the development of convective storm or MCC in Africa [17].

Variations in actual MCC totals per year as shown in Figure 2 have been observed with an obvious peak of MCC activity in 2004, 2005, 2011 and 2013 when more than 80 events occur each year. 2001, 2007, 2010, 2012, 2014 and 2015 are all above 70 events which occur each year. A total of 53 and 62 events occurred in 2003 and 2008, respectively. The only years notably fewer events are 2002, 2006 and 2009 with 45, 45, and 44 events, respectively. Numerous studies have demonstrated that ENSO can modulate the synoptic and mesoscale environments South America, which in turn can influence convective activity and precipitation patterns across the tropical and subtropical regions [e.g.,18,19,20]. Specifically, Velasco, and Fritsch [20] found that the number of MCCs doubled during the 1982/83 El Nino event. However, this research shows no apparent relationship between MCC frequency and

ENSO that is similar to Durkee and Mote' analysis [21] that ENSO and MCC cloud-top characteristics, longevity, and the distribution of MCC rainfall show no apparent relationship.

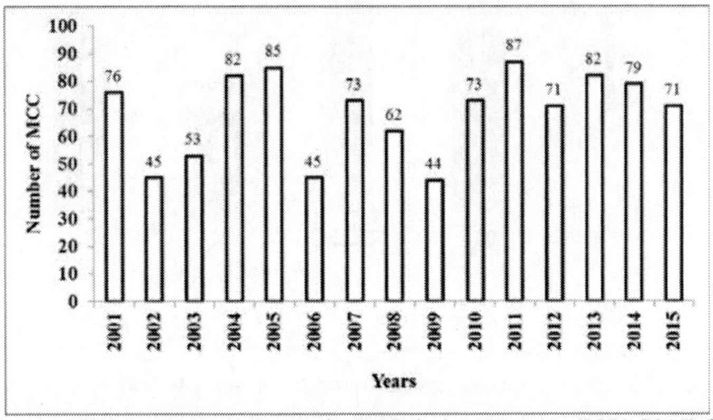

Figure 2. A total of MCC occurrences over the IMC during 15-years (2001-2015) for each year.

This result is also supported by previous research, among other; Trismidianto [3] stated that there are many MCC occurrences over the Indian Ocean near Sumatra. In addition to that, Ismanto [5] stated that one of concentration areas for oceanic MCC is the Indian Ocean. Moreover, Mori et al. [22] stated that IMC consists of many large/small islands with very long coastlines, many narrow straits control the global (Pacific to Indian) ocean circulation and the deep convection is the major contributor to changes in tropical rainfall [23].

3.2 Seasonal distribution of MCC occurrences

The season in IMC is divided into four seasons, namely in December, January and February (DJF) as the wet season, March, April and May (MAM) as a transitional season from wet season to the dry season. June, July, and August (JJA)is as dry season and September, October, and November (SON) is a season of transition from dry season to the wet season. **Figure 3** shows that the peak of MCC occurrence over the IMC is during MAM with around 33.56% MCC occur in this season. This result is similar to Tyson and Preston-Whyte [24] that stated the peak of MCC activity in southern Africa mostly occur during the early before summer months due to likely linked to the synoptic-scale environment, which is more baroclinic in nature than in the late summer months. Velasco and Fritsch [20] suggest that the longer MCC season maybe due to the influence of the oceans on the relatively smaller landmass. However, some of the previous paper stated that there is an effect of ENSO on MCC activity in Africa [25] and in America [7]. 26,26 % and 25,29% MCC occurrences were found in DJF and SON, respectively. The season with fewer events is JJA, around of 14.79%.

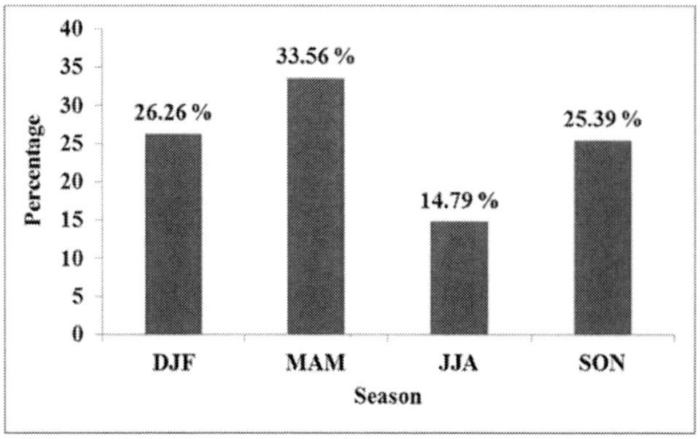

Figure 3. Percentage of the MCC occurrences over the IMC during 15-years (2001-2015) for each season.

The location of the interior cloud centroid for each season was examined. In DJF, one can see MCCs are the most frequent concentrated in the continent in Central Kalimantan, the South China Sea near northern Kalimantan, Central Sulawesi and Merauke as shown in **Figure 4(a).** However, the MCCs that occurred in the continent have an interior cloud in small size around 50,000 - 149,999 km^2. In contrast, most of the MCC with large size was frequently found in the Indian Ocean near Sumatra Island. Most of them have an interior cloud in size more than 300,000 km^2. The MCC in medium size was also concentrated in the Pacific Ocean. During MAM as shown in **Figure 4(b)**, the frequency of MCCs was almost similar to DJF, but MCCs over the Central Kalimantan became more widespread throughout to Kalimantan. Small-sized MCCs dominate MCCs that occurred over the Central Kalimantan during MAM, but several large-sized MCCs were also found. MCCs over Merauke also became more widespread until the Arafura Sea. Several MCCs with medium size around 200,000 - 300,000 km^2 were found over Sumatra Island and western coastal of Sumatra. The MCCs with large-sized were still found in the Indian Ocean. Several MCC in small-sized also occurred over the Java Island.

During JJA as shown in **Figure 4(c)**, the MCCs occurred in the continent were just found in several areas, among others; Central Kalimantan and northern Sumatra in small frequency. The MCCs are more concentrated over the ocean with the greatest concentration are over the Indian Ocean. MCC never occurred along the Java Island until East Nusa Tenggara during JJA. There are five concentration regions of MCC events during SON as shown in **Figure 4(d)**. i.e., over the Indian Ocean, the South China Sea near northern Kalimantan, Central Kalimantan, Cendrawasih Bay and Merauke. The MCC with large size was still concentrated over the Indian Ocean, while small size was concentrated over Kalimantan, especially central Kalimantan. Overall, the MCCs with large size more than 300,000 km^2 most frequently occurred over the Indian Ocean each season with the greatest frequency in SON season. The Central Kalimantan is one favored region for MCC occurrences each season, but most of the MCCs in this area have a small and medium size less than 300,000 km^2. MCCs are rarely found along of the Java Island to the East Nusa Tenggara. **Figure 4** also shows that MCC most frequently occurs in region nearly equator around 2°S to 2°N. It is interesting to note that MCC is possible related with ITCZ.

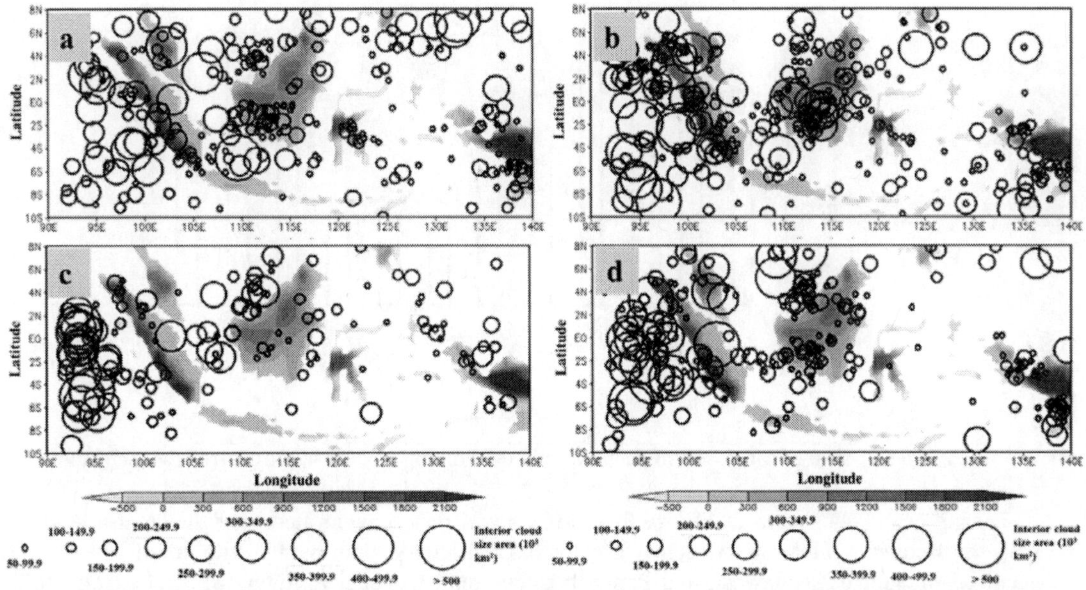

Figure 4. Geographical distribution of MCCs in the IMC during 15-years (2001-2015) for each season; (a) DJF, (b) MAM, (c) JJA and (d) SON. Locations are for MCC at the time of maximum extent of the interior cloud size area when the mature stage. The circles represents the interior cloud size area (10^3 km^2). The color shaded refers to the elevation of the unit in meters (m).

3.3 *Monthly distribution of MCC occurrences*

Variations in actual MCC totals per month as shown in **Figure 5** are reported with an obvious peak of MCC activity in April and November when more than 120 events occur each month. The frequency of MCC occurrences in January, March, and May is also great with 107, 111 and 105 events, respectively. A total of 77, 66, 94 and 86 events occurred in February, June, October, and December, respectively. The fewer events were found in July, August and September that have frequency 41, 45 and 40 events, respectively.

Figure 6 shows the monthly analysis of MCCs over the IMC during 15-years by plotting the location of the interior cloud centroid of MCCs for January until June. In January as shown in **Figure 6(a)**, the MCCs are concentrated in the Central Kalimantan and Central Sulawesi but in small size less than 100,000 km^2. The MCCs with large size are predominantly the MCC that are most frequently found in the Indian Ocean. The MCC with small size were concentrated over the coastal of Merauke and the coastal of northern Kalimantan over the South China Sea. The frequency of MCCs over the Central Kalimantan during February as shown in **Figure 6(b)** became widespread until western Kalimantan. MCCs were also concentrated in Lampung, southern Sumatra, and several MCCs were found in the coastal of North Sumatra. The frequency of MCC over the Indian Ocean and the other ocean decreased, just occurred in a few events.

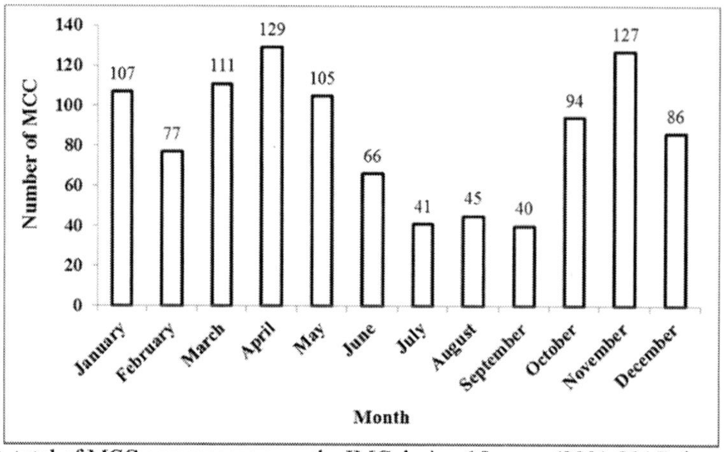

Figure 5. A total of MCC occurrences over the IMC during 15-years (2001-2015) for each month.

During March as shown in **Figure 6(c)**, MCCs was just concentrated over the Central Kalimantan with the frequency of MCCs which is greater than February, dominantly with small size. MCC also occurred in Sumatra, Sulawesi, and Papua but the small frequency. A few of large-sized MCC was found in the Indian Ocean. The MCC with small size increased over the Central Kalimantan in April as shown in **Figure 6(d)**. Most of MCCs were also located in along western of Sumatra Island. Most of them have medium size around 200,000 km^2. The MCC with medium size also occurred in the Indian Ocean near the coast of northern Sumatra. Several small-sized MCCs were also found over Papua and Java Island. Most of the MCCs that occurred in May as shown in **Figure 6(e)** has a large size and were located in the Indian Ocean. MCC was rarely found over Sulawesi, Papua, and Java. In June as shown in **Figure 6(f)**, most of MCCs were concentrated over the ocean with the greatest frequency in the Indian Ocean. Only a few were found in the continent, such as in the Central Kalimantan and North Sumatra. Like May, MCC does not occur over Sulawesi, Java, and Papua.

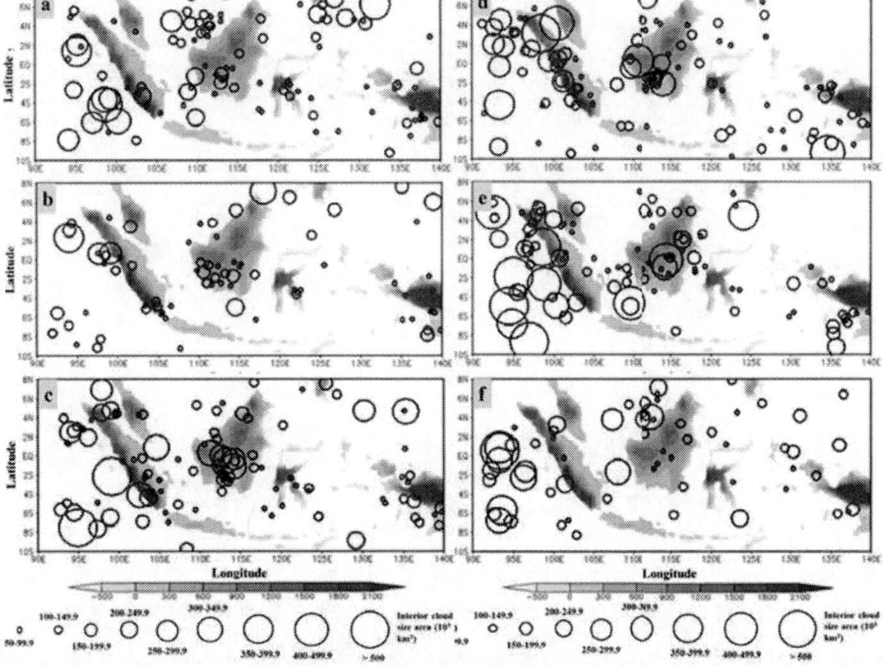

Figure 6. Geographical distribution of MCCs in Indonesian Maritime Continent during 15-years (2001-2015) for each month; (a) January, (b) February, (c) March, (d) April, (e) May and (f) June. Locations are for the MCC at the time of maximum extent of the interior cloud size area when the mature stage. The circles represents the interior cloud size area (10^3 km^2). The color shaded refers to the elevation of the unit in meters (m).

The frequency of MCC in July as shown in **Figure 7(a)** was similar to June, where only found in the Indian Ocean in longitude 91° - 97°E. The MCC rarely occurred in the continent during this month. The frequency of the MCC with large size over the Indian Ocean increased in October and November as shown in **Figure 7(d)** and **(e)**, while in August it was slightly same as in July as shown in **Figure 7(b)**. However, this frequency decreased in September and December as shown in **Figure 7(c)** and **(f)**. In August and September, the frequency of the MCC was slightly same as in July, only a few were found in the large island of MCC. In fact, almost never occur along of the Java Island to East Nusa Tenggara. The frequency of MCC over the Central Kalimantan increased in October and November as shown in **Figure 7(d)** and **(e)**. However, this frequency decreased in August, September, and December as shown in **Figure 7(b)**, **(c)** and **(f)**. The MCC was more concentrated in coastal of Merauke from October until December. Overall, it is similar to seasonal that the MCCs with large size more than 300,000 km^2 most frequently occurred over the Indian Ocean each month with the greatest frequency in October and November. Central Kalimantan is one favored region for MCC occurrences each month except June until September, but most of the MCCs in this area have a small and medium size less than 300,000 km^2. Along Java Island to East Nusa Tenggara is the region where MCC events were rarely found.

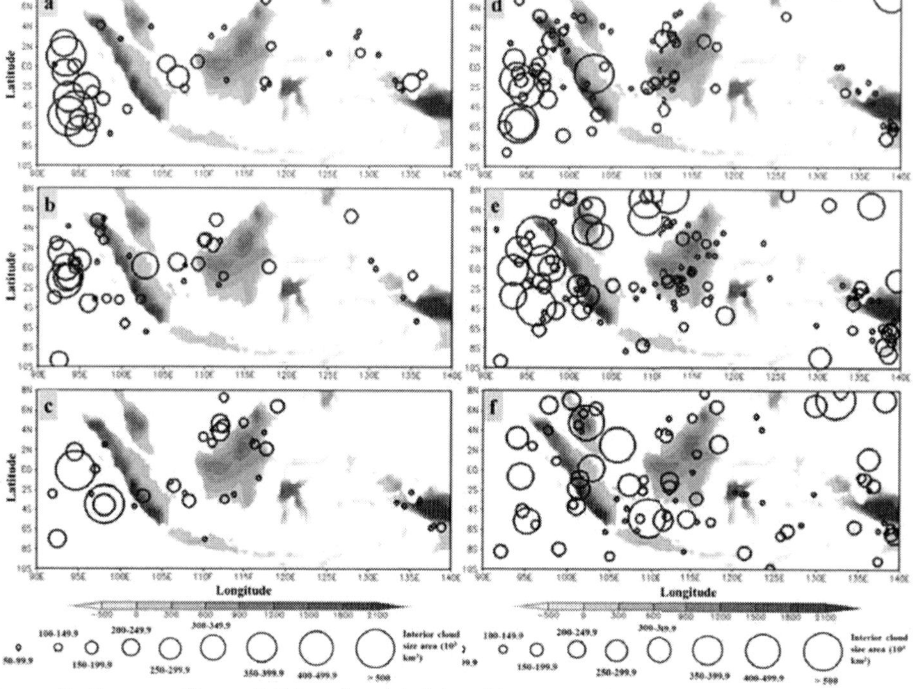

Figure 7. Same as Figure 3.7 but for; (a) July, (b) August, (c) September, (d) October, (e) November and (f) December. The color shaded refers to the elevation of the unit in meters (m).

3.4 *Diurnal distribution of MCC occurrences*

In the diurnal analysis, most of MCCs that occurred in the continent reached maximum extent or mature stage in the midnight until morning from 20.00 - 0700 LT as shown in **Figure 8(f)**, **(a)** and **(b)** with the greatest frequency of the MCC was concentrated over the Central Kalimantan. The MCCs which occurred in the ocean reached the most frequent maximum extent in 0800 - 1100 LT as shown in **Figure 8(c).** During the daytime at 1200 - 1500 LT and the evening at 1600 - 1900 LT, the frequency of MCC occurrences was concentrated most frequent over the Indian Ocean as shown in **Figure 8(d)** and **(e).** MCCs rarely reached maximum extent at night in 2000 - 2300 LT.

Most of previous studies indicate that diurnal cycles of deep convection over land and sea show a distinct contrast. Convection over continents and large islands is most intense in late afternoon (around 19 LT). In contrast, convection over the seas near continents and large islands is most intense in the morning (around 03 LT); diurnal cycle over continents and large islands are likely forced by strong surface heating during the day, whereas diurnal cycles over adjacent maritime areas arise from interactions between sea-land breeze circulations and large-scale environmental flows [26, 27]. The other previous researcher, Sui et al. [28] found the average diurnal variation of rainfall from satellite infrared over an extensive tropical region that is in the central and eastern Pacific. It shows the maximum value in the afternoon (mid-afternoon) and the second peak approaching dawn. They also stated that the strongest convection on the surface of the continent occurred towards evening or afternoon depending on the dominance of diurnal cycle of heating surface, except in areas where the difference between land and sea is very real or orographic very powerful force. On top of the ocean area that is free from the influence of the mainland, the coldest clouds and the maximum rainfall is often observed in the morning.

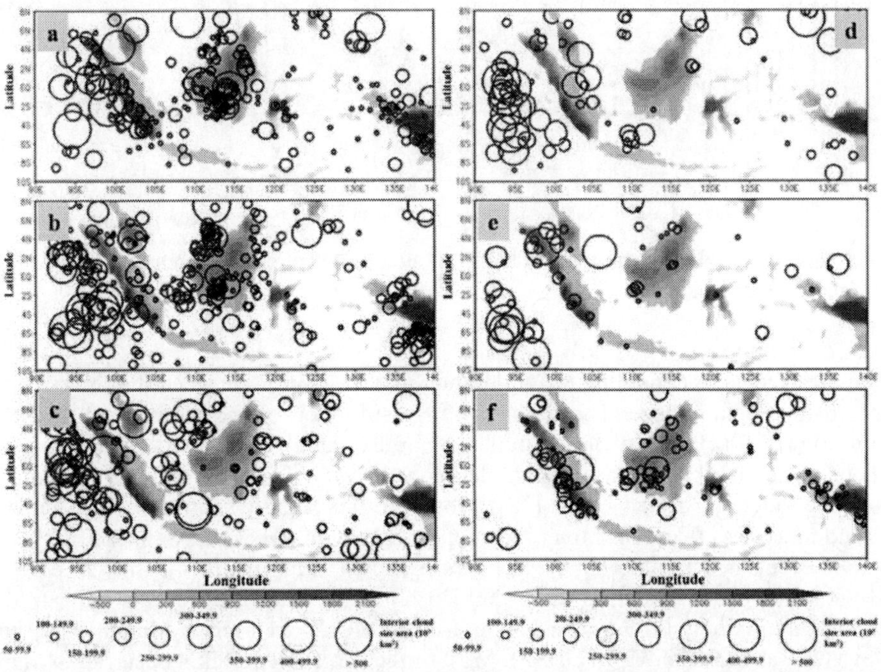

Figure 8. Diurnal distribution of MCCs in Indonesian Maritime Continent during 15-years (2001-2015); (a) 0000 - 0300 LT, (b) 0400 - 0700 LT, (c) 0800 - 1100 LT, (d) 1200 - 1500 LT, (e) 1600 - 1900 LT and (f) 2000 - 2300 LT. Locations are for the MCC at the time of maximum extent of the interior cloud size area when the mature stage. The circles represents the interior cloud size area (10^3 km^2).

3.5 MCC maximum-extent distribution

The area size of cloud shield and an interior cloud of MCCs during the mature stage are shown in **Figure 9**. Most of the MCCs have cloud shield around 100 - 200 x 10^3 km^2 in a total of 433 events as shown in **Figure 3.10(a)** and interior cloud less than 200 x 10^3 km^2 in a total of 834 systems as shown in **Figure 3.10(b).** It indicates that most of the MCCs over the IMC are included in the category of small-sized MCCs. However, the large-sized MCCs with cloud size area more than 600 x 10^3 km^2 also occurred over the IMC. In fact, a total of 29 events have cloud shield more than 1000 x 10^3 km^2, and 20 events have interior cloud more than 500 x 10^3 km^2.

The average of cloud shield area size of MCCs is around 315,000 km^2. It is slightly less than the global MCC by Laing and Fritsch [2] who stated the mean cloud-top area for the global population is 354,000 km^2. However, the MCCs in IMC are more than from MCC in the western Pacific region that has the greatest frequency of MCC between 2 x 10^5 km^2 until 3 x 10^5 km^2 as reported by Miller and Fritsch [29]. This MCC is also more than MCC over subtropical South America that has cloud size area 256,500 km^2 [21]. Although some MCCs produce cold cloud shields that extend over a million km^2, the size, and duration of MCSs and MCCs are positively correlated, i.e., larger systems tend to last longer [30, 31, 2].

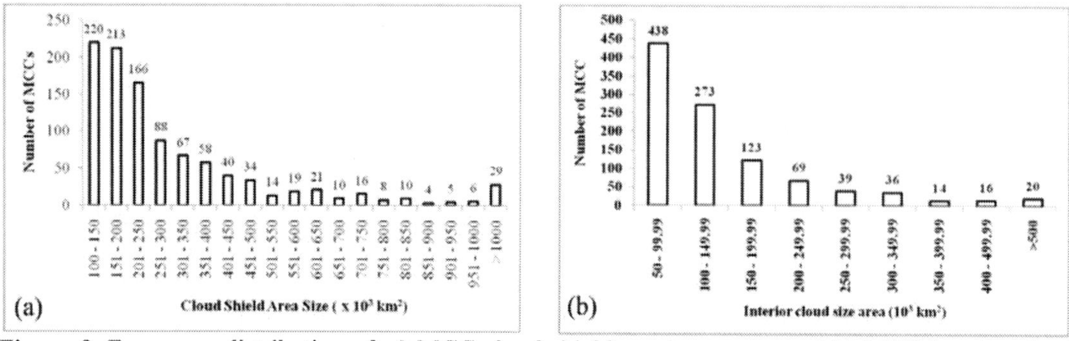

Figure 9. Frequency distribution of; a) MCC cloud shield maximum area and b) MCC interior cloud maximum area during 15-years.

3.6 MCC duration

In this research, we define the MCC duration as the duration between the time of initial stage and the time of dissipation stage. The frequency of distribution of the duration of all MCC systems is shown in **Figure 10(a).** The average duration is approximately ~9 .5 hours, and the maximum duration is 10 hours. The distribution and average duration are very similar to the distribution and average for MCCs in the southern Africa (~9.5 h; [32]). However, it is slightly shorter than the global average (~10 h; [2]) and much less than those found in subtropical South America (14 h; [21]). Most of the MCCs around 686 events occurred during 8 - 12 hours as shown in **Figure 10(a).** Only 20 events are the long-lived MCCs that occurred during more than 20 hours. A total of 57 events indicate that short-lived MCCs has duration only 6 - 7 hours. The long-lived MCCs occurred in March, May, July, September and November as shown in **Figure 10(b).** Most of the short-lived MCCs occurred in March-May, July, September, and December.

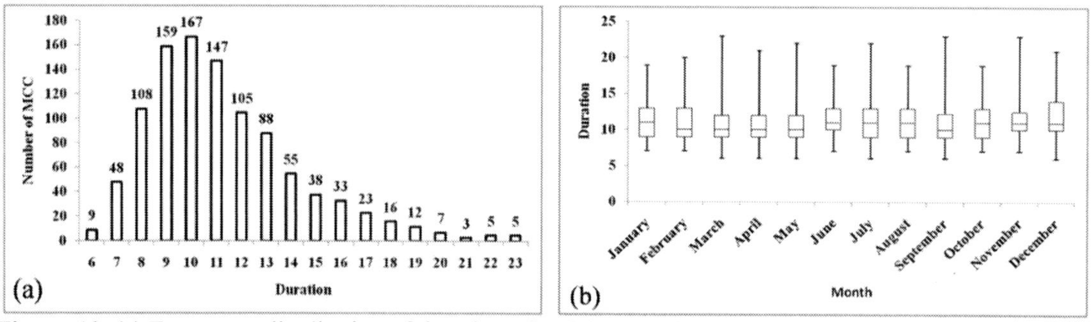

Figure 10. (a) Frequency distribution of duration of MCCs over the IMC during 15-years. (b) Box and Whisker plot of the duration of MCC frequency during 15-years.

Figure 11(a) shows that there is no relation between the duration and MCC locations because the correlation is slight (r = 0.01). **Figure 11(c)** shows that the MCC locations are not related to a maximum area of MCCs (correlation coefficient r = 0.06). **Figure 11(b)** indicates that there is a tendency for relation of duration and maximum area of MCCs over the IMC that is positively correlated (correlation coefficient r = 0.58). It indicates that systems which generate large cold-cloud shields tend to persist longer than those which produce small shields. This result is similar to global MCC population by Laing and Fritsch [2] who stated there are relation between duration and maximum area in the global population of MCCs that has correlation coefficient r = 0.385. Tollerud et al. [30] also found a similar relationship, although the relationship is weak. This result is also very similar to Durkee and Mote [21] who stated MCC maximum size and duration of subtropical South America exhibited a significant relationship with correlation coefficient r = 0.56.

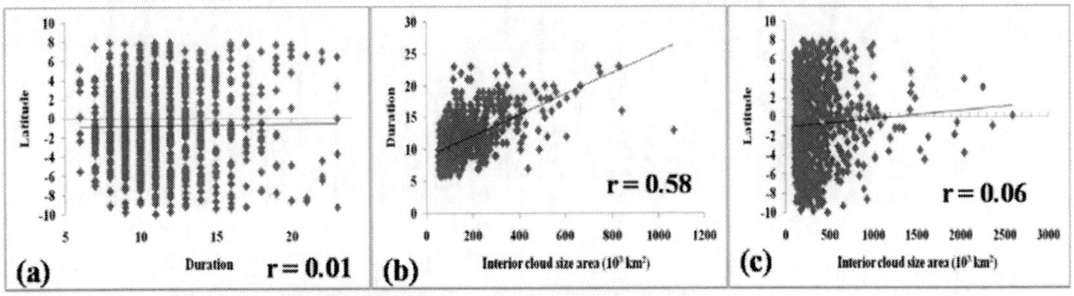

Figure 11. Scatter plots show relationship between; (a) latitude and duration, (b) duration and interior cloud size and (c) latitude and interior cloud size. Latitude, duration and interior cloud size at the time of maximum extent or mature stage. Correlation coefficient ***r*** is shown for each case.

3.7 MCC life-cycle

Figure 12 shows that in general, the MCCs in IMC usually develop in the late afternoon and by around sunset in a few times before the initial stage. The greatest frequency of occurrences for initiation occurred between 1900 – 2200 LT. The systems reached the maximum size extent after midnight or early morning in predominantly between 0100 - 0500 LT and then MCC started to decay in the morning a few hours after sunrise, predominantly around 0700 - 0900 LT. The systems dissipated from the daytime until the afternoon, around 1000 - 1500 LT. It indicates that characteristics of MCCs life-cycle tend to nocturnal characteristics. This result is similar to the previous studies that documented regional populations of MCCs were found to be nocturnal, among others; MCC in the Americas and the Western Pacific region, southern Africa [1, 20, 29, 21, 32]. It follows that global population is also predominantly nocturnal [2]. It is also strengthened by Sui et al [28] who stated that convective activity in the tropics has three variations. Using the results of observations, COARE TOGA (Tropical Ocean Global Atmosphere Coupled Ocean- Atmosphere Response Experiment) acquired three diurnal variations of atmospheric convective activity in the tropics. The three variations are warm cumulus morning, afternoon convective showers, and nocturnal convective systems. Rainfall afternoons are particularly true for convective cells, but nocturnal rainfall caused by cells deeper convective and stratiform clouds in a wide area. Further stated that convective precipitation afternoon (afternoon convective showers) is more apparent in the period of large-scale undisturbed than when the diurnal cycle of sea surface temperature occurs strongly. However, the nocturnal convective system (nocturnal convective systems) and cumulus morning (warm morning cumulus) are higher in the period disrupted when more moisture is available [28].

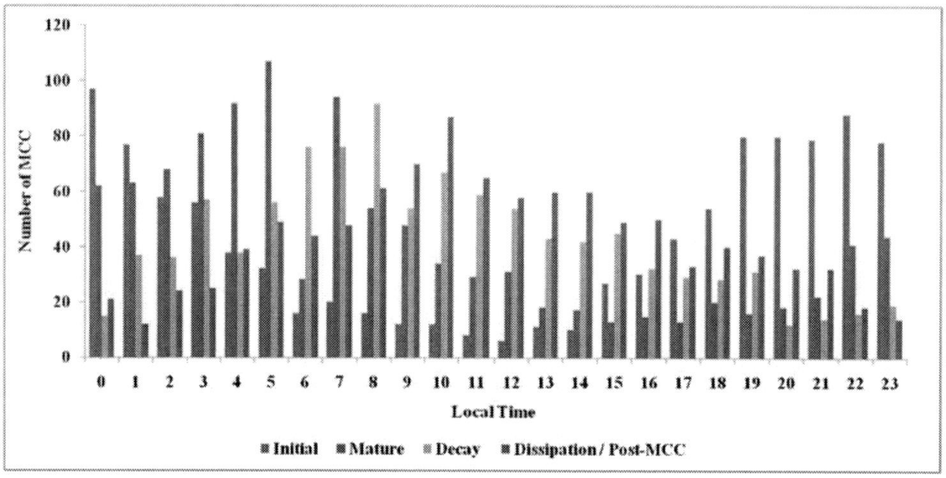

Figure 12. Frequency distribution of all MCCs during critical stage (initial, mature, decay and dissipation or post-MCC stage).

4. Conclusion

A total of 1028 MCCs were identified and tracked during the 15-year period from 2001 to 2015. Most of these MCCs were over the continental area, mainly near the mountains and the high elevation areas which have similar characteristic with the global population of MCCs in the world. There are nine favored regions of MCC occurrences, they are the Indian Ocean near Sumatra Island, along the western coastal of Sumatra Island, the South China Sea near Kalimantan Island, the Central Kalimantan around 3°S-1°N; 111° -115°E, the East Kalimantan around 1° - 4°N; 116° - 120°E, Makassar Strait near coasts of Kalimantan Island, the Central Sulawesi around 4° - 2°S; 120° - 123°E, Merauke over Papua Island in location around of 8° - 5°S and 135.5° - 140°E. The peak of MCC occurrence over the IMC is during MAM with around 33.56%. The season with fewer events is JJA, around of 14.79%. The average of cloud shield area size of MCCs is around 315,000 km^2 which slightly less than with global MCC in the world around of 354,000 km^2. However, the MCCs in the IMC are more than from MCC in the western Pacific region and subtropical South America. The average duration of the MCCs over IMC was approximate ~9 .5 hours, and the maximum duration was 10 hours that very similar to the distribution and average for MCCs in the southern Africa (~9.5 h), but it is slightly shorter than the global average (~10 h) and much less than those found in subtropical South America (14 h). The long-lived MCCs with duration more than 20 hours occurred in March, May, July, September, and November and most of the short-lived MCCs with duration 6 - 7 hours occurred in March - May, July, September, and December. In general, the greatest frequency of occurrences for initiation occurred between 1900 – 2200 LT. The systems reached the maximum size extent after midnight or early morning in predominantly between 0100 - 0500 LT and then MCC started to decay in the morning a few hours after sunrise, predominantly around 0700 - 0900 LT. The systems dissipated from the daytime until the afternoon, around 1000 - 1500 LT. It indicates that the majority of MCCs over IMC exhibit a nocturnal life cycle that begins in the mid-to-late afternoon, peaks after midnight and ends shortly after daybreak. Nocturnal growth in convective storms has been hypothetically related to the differential radiative heating between the MCC and the environment, which leads to peripheral environmental subsidence and low-level convergence into the system.

5. Reference

[1] Maddox R A 1980 Mesoscale Convective Complexes *Bull. Amer. Meteor. Soc.* **61** 1374 - 1387.

[2] Laing A G and Fritsch J M 1997 The Global Population of Mesoscale Convective Complexes *Q. J. R. Meteorol. Soc.* **123** 389-405.

[3] Trismidianto 2012 A study of mesoscale convective complexes (MCCs) activities in the western Indian Ocean and their effects on convection over Sumatra Island *Magister Thesis FITB ITB*.

[4] Trismidianto, Tri W H, Sachinobu I, Qoosaku M, Atsuyoshi M, and Satoshi I 2016 Development Processes of Oceanic Convective Systems Inducing the Heavy Rainfall over the Western Coast of Sumatra on 28 October 2007 *Scientific Online Letters on the Atmosphere (SOLA)*. **Vol. 12** 6–11 doi:10.2151/sola.2016-002.

[5] Ismanto H 2011 Characteristics of Mesoscale Convective Complexes over Martitime Continent *Magister Thesis FITB ITB*.

[6] Anderson C J and Arrit R W 1998 Mesoscale convective complexes and persistent elongated convective systems over the United States during 1992 and 1993 *Mon. weather Rev.* **126** 578 - 599.

[7] Durkee J D, Mote T L and Shepherd M J 2009 The Contribution of Mesoscale Convective Complexes to Rainfall across, Subtropical South America *Int J. Clim.* **vol 22**.

[8] Laing A G and Fritsch J M 1993a Mesoscale Convective Complexes Over the Indian Monsoon Region. *American Meteorology Society,* **121** 2254 – 2263.

[9] Laing A G and Fritsch J M 1993b Mesoscale Convective Complexes in Africa *Monthly Weather Review,* **121** 2254-2263.

[10] Morel C and Senesi S 2002 A climatology of mesoscale convective systems over Europe using satellite infrared imagery II: Characteristics of European mesoscale convective systems *Q. J. R. Meteorol. Soc.* **128** 1973-1995.

[11] Carvalho L M V and Jones C 2001 A Satellite method to identify structural properties of mesoscale convective systems based on the Maximum Spatial Correlation Tracking Technique (MASCOTTE) *J. of Applied Meteorology. American Meteorology Society,* **40** 1683-1701.

[12] Strachan J 2007 Understanding and Modelling the Climate of the Maritime Continent *Doctoral Thesis University of Reading*. **227 pp**.

[13] Lindzen R S, Hou A Y 1988 Hadley circulations for zonally averaged heating centered off the equator *J. Atmos. Sci.* **45** 2416-2427.

[14] Ashley W S 2003 A distribution of mesoscale convective complexe rainfall in the United State *American Meteorology Society,* **131** 3003 – 301.

[15] Laing A G and Fritsch J M 2000 The Large Scale environment of the Global Population of Mesoscale Convective Complexes *Monthly Weather Review,* **128** 2756 - 2776.

[16] Kawashima M, Fujiyoshi Y, Ohi M, Honda T, Mori S, Sakurai N, Abe Y, Harjupa W, Syamsudin F, Yamanaka M D 2011 Case study of an intense wind event associated with a mesoscale convective system in west Sumatera during the HARIMAU2006 campaign *J. Meteorol. Soc.* **89A** 239–257.

[17] Garstang M, Kelbe B E, Emmitt G D, and London W B 1987 Generation of convective storms over the escarpment of northeastern south africa *Monthly weather review.* **115** 429–443.

[18] Curtis S 2008 The El Nin˜ o–Southern Oscillation and global precipitation *Geogr. Compass.* **2** 600–619.

[19] Silva V B S and Ambrizzi T 2006 Inter-El Nin˜ o variability and its impact on the South American low-level jet east of the Andes during austral summer—Two case studies *Adv. Geosci.* **6** 283–287.

[20] Velasco I and Fritsch J M 1987 Mesoscale Convective Complexes in the America *J. Geophysic. Res.* **92** 9591-9613.

[21] Durkee J D and Mote T L 2009 A Climatology of Warm Season Mesoscale Convective Complexes in Subtropical South America *Int J. Clim.* **12 pp**.

[22] Mori S, Jun-Ichi H, Tauhid Y I, Yamanaka M D, Okamoto N, Murata F, Sakurai N, Hashiguchi H and Sribimawati T 2004 Diurnal Land-Sea Rainfall Peak Migration over Sumatera Island, Indonesian Maritime Continent, Observed by TRMM Satellite and Intensive Rawinsonde Soundings *American Meteorology Society.* **132** 2021 – 2039.

[23] Tan J, Jakob C, Rossow W B and Tselioudis G 2015 Increases in tropical rainfall driven by changes in frequency of organized deep convection *Nature.* **vol 519** doi:10.1038/nature14339.

[24] Tyson P D and Preston-Whyte R A 2000 The Atmosphere and Weather of Southern Africa *Oxford University Press.* **396 pp**.

[25] Mason S J and Jury M R 1997 Climate variability and change over southern Africa: A reflection on underlying processes *Prog. Phys. Geogr.* **21** 23–50.

[26] Murakami M 1983 Analysis of the deep convective activity over the western Pacific and Southeast Asia Part I: Diurnal variation *J. Meteor. Soc.* **61** 60–75.

[27] Nitta T and Sekine S 1994 Diurnal variation of convective activity over tropical western Pacific. *J. Meteor Soc. Japan.* **72** 627-641.

[28] Sui C H, Lau K M, Takayabu Y N and Short D A 1997 Diurnal variations in tropical oceanic cumulus convection during TOGA COARE. *J. Atmos. Sci.* **54** 639–655.

[29] Miller D and Fritsch J M 1991 Mesoscale Convective Complexe in Western Pasific Region *Amer. Meteor. Soc.* **119** 2978 – 2992.

[30] Tollerud E I, Augustine J A and Jamison B D 1992 Cloud top characteristics of mesoscale convective systems in 1986 *Am. Meteorol. Soc. Atlanta, USA.*

[31] Laurent H 1996 Tracking of convective cloud clusters from Meteosat data *Tenth Meteosat Scientific Users' Conference.*

[32] Blamey R C and Reason C J C 2012 The role of mesoscale convective complexes in southern Africa Summer Rainfall *J. Clim.* **26.**

IOP Conf. Series: Earth and Environmental Science **166** (2018) 012041 doi:10.1088/1755-1315/166/1/012041

Carbon Emission from Peat Fire in 2015

W Setyawati[1*] and Suwarsono[2]

[1]Center of Atmospheric Science and Technology
Indonesian National Institute of Aeronautics and Space (LAPAN)
Jl. Dr. Djundjunan 133, Bandung 4017, Indonesia
[2]Remote Sensing Applications Center
Indonesian National Institute of Aeronautics and Space (LAPAN)
Jl. Kalisari No. 8, Pekayon, Pasar Rebo, Jakarta 13710, Indonesia

E-mail: wiwieksetyawati21@gmail.com

Abstract. Peat fire emits a large amount of carbon to the atmosphere. This emission can affect atmospheric composition and also the climate system. This paper discussed the carbon emission from the 2015 peat fire, the largest after 1997 one. Indonesian National Carbon Accounting System (INCAS) published by The Indonesian Ministry of Environment and Forestry was used to calculate the carbon emission from the peat fire. Peat fire emission factors used in the calculation for region of Sumatera and Kalimantan were selected from previous study, while for Papua's region extrapolation method was applied. Peat area burned was estimated by using Normalized Burn Ratio (NBR) based on hotspot data enquired from MODIS. The cumulative peatland area burned from July – October 2015 in Indonesia was estimated to be about 623,304 ha, where about 270,691 ha (43%), 320,756 ha (51%) and 31,857 ha (5%) were found in Sumatera, Kalimantan and Papua, respectively. By considering only the three biggest gaseous carbon compounds released by biomass burning (CO_2, CO, and CH_4), therefore total carbon emitted to the atmosphere during these four months peat fire was estimated to be about 0.002 Gtonnes, of which 81% was in the form of CO_2; 16% CO and 2.3% CH_4.

1. Introduction

Peatland is defined as a land composes of organic soils. The land is an area with cumulated half decomposed organic material, ash contents equal or less than 35%, peat depth equal or more than 50 cm and organic carbon contents at least 12% [1]. Indonesia has the largest tropical peatland in the world [2] where about 69.61% of its total area is found in Sumatra and Kalimantan [3]. Tropical peatland is estimated to preserve about 40% of terrestrial carbon [4], but due to traditional slash and burn practices for agricultural and plantation purposes especially in Sumatra and Kalimantan, it has led to large peat fires that released a large amount of carbon to the atmosphere [5], that could affect atmospheric composition and also the climate system. This traditional practice of land clearing is still the cheapest and most efficient way of getting rid of rats, wood debris and grass; the traditional farmers also believe that ash residue from burning can improve the soil pH and fertility [6][7].

In 2015 large peat fires stroked regions of Sumatera, Kalimantan and Papua. Human activities relating to agricultural or plantation purposes were behind the catastrophe [5], while the prolonged drought induced by a strong El Nino enhanced the fires [8]. This 2015 peat fire was the largest after the 1997 one. The carbon emissions from peat fire have been investigated by several researchers [5][8][9]. But the results are still inaccurate because the emission was calculated by assuming that peat fire in Sumatra, Kalimantan and Papua emitted the same amount of carbon per-kg of dry peat burned, called emission factor. This study is trying to improve the carbon estimation for the 2015 peat fire by using corresponding emission factor for Sumatra, Kalimantan and Papua peat fires.

2. Materials and Methods

2.1. FIRMS: Hotspot

Daily cumulative hotspots data as indicator of fire used in this analysis were obtained from Fire Information for Resource Management System (FIRMS) website (https://earthdata.nasa.gov/earth-observation-data/near-real-time/firms) retrieved from Moderate Resolution Imaging Spectro radiometer (MODIS) on board of Aqua and Terra satellites. The hotspot from July-October 2015 with confidence level bigger than or equal to 95% was selected for further analysis. Daily cumulative hotspots are used to indicate fires. Only hotspot with level of confidence equal to or bigger than 95% are selected.

2.2. TRMM: precipitation rate (mm/hr)

Precipitation rate is a measure of the rainfall intensity by calculating the amount of rain that would fall over a given interval of time if the rainfall intensity were constant over that time period (American Meteorological Society, 2017). Area averaged daily precipitation rate in unit of mm/month with spatial resolution of $0.25^0 \times 0.25^0$ from July – October 2015 were acquired from Tropical Rainfall Measuring Mission (TRMM) and downloaded from GES-DISC Interactive Online Analysis Infrastructure (Giovanni) NASA website (https://giovanni.gsfc.nasa.gov/giovanni/) as part of the NASA Goddard Earth Sciences Information Services Center (DISC).

2.3. Methods

Indonesian National Carbon Accounting System (INCAS) [1] has followed the model used by IPCC [10] to estimate carbon emission from the forest and peat fire as follows:

$$E = a \times F_l \times C_f \times EF \times 10^{-3} \quad (1)$$

where E is amount of CO_2 or non-CO_2 emission (ton), a is total area burned annually (ha), F_l is dry fuel mass available for combustion (ton/ha), C_f is a dimensionless combustion factor and EF refers to emission factor for each gas (g/kg dry mass burned). Dry fuel mass available for combustion (ton/ha) is calculated by multiplying depth of burned peat (m) with bulk density (ton/m^3).

Carbon emission from forest and peat fire was limited to peat fire in Sumatra, Kalimantan and Papua. It was calculated for CO_2, CO, and CH_4. These three were the largest gaseous carbon compounds emitted by peat fire and contributed more than 95% of total carbon emitted [11-13]. The other trace gaseous and particulate carbons were neglected.

Emission factor values used to calculate carbon emission from peat fires are listed in **Table 1**. Emission factor for Kalimantan peat fire was the average of emission factors from three previous studies [12-14]. Because there were no previous studies for Papua and West Papua peat fires, therefore we used emission factor for CO_2 of 1,111 g/kg by extrapolating peat carbon mass fraction of 0.3053 for hemic peat [15] to the regression linear equation of emission factor for smoldering peat fire as shown in **Figure 1** [16]. The corresponding emission for CO and CH_4 can be calculated by multiplying their emission ratio, by using CO_2 as a reference species, with the calculated emission of CO_2 by applying the equation below [17]:

$$E_x = E_{CO_2} \times ER_{x/CO_2} \quad (2)$$

where E_x is amount of emission of x (CO or CH_4) (ton), E_{CO2} is amount of emission of CO_2 and $ER_{x/CO2}$ refers to emission ratio of x with respect to CO_2 (mol/mol). For the purpose of this study, therefore $ER_{CO/CO2}$ is 0.153 mol/mol and $ER_{CH4/CO2}$ is 0.029 mol/mol [16].

Table 1. Emission factors (g/kg dry peat burned) used for the calculation of carbon emission

	Sumatra peat [11]	Kalimantan peat [12, 13, 14]
CO_2	1,703	1,677
CO	210.3	221
CH_4	20.80	13.1

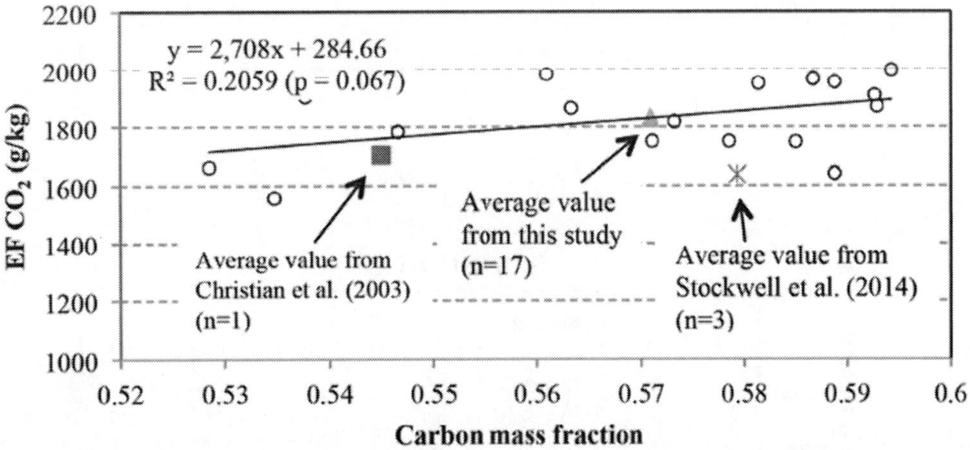

Figure 1. Simple linier regression between peat carbon content and emission factor (EF) of CO_2 for smoldering combustion [16]

Total peatland area burned was calculated for four months peat fire from July – October 2015 based on normalized burn ratio (NBR) method. This method used hotspot data as an indicator of peat fire from MODIS.

Dry fuel mass available for combustion is a product of depth of burned peat and bulk density [1]. For depth of burned peat, INCAS recommends using 0.18 m for newly burned peat and 0.11 m and 0.04 m for peat burned two and more than two times, respectively [1]. Bulk density used for Sumatra [18] [19], Kalimantan [16] [19] [20] and Papua [15] peats are 0.1716, 0.23 and 0.23, respectively.

Although INCAS recommends using 1 as the combustion factor, which means peat fire is a complete combustion, but in this study, we used 0.8 for Sumatra and 0.7 for Kalimantan and Papua peat fires [11] [13]. It was because peat fire mostly was dominated by smoldering combustion, a type of combustion when no flame is visually observed but apparent thin or thick smokes [16].

3. Results and Discussion

Analysis of hotspot and precipitation rate in Sumatera, Kalimantan and Papua based on satellite data was presented in **Figure 2.** In 2015, a strong El Nino hit Indonesia and the region suffered with prolonged drought [9]. The minimum precipitation rate in Sumatera, Kalimantan and Papua were found in July, September and August, respectively. Lower precipitation rate in Sumatera and Kalimantan had triggered higher number of hotspots in both regions (**Figure 2a** and **b**). On the contrary, low precipitation rate did not increase number of hotpots dramatically in Papua (**Figure 2c**). Land clearing for palm oil and pulpwood plantations was reported to be responsible for mostly peat fire incidents in Indonesia [21]. In 2015 Sumatera has the largest palm plantation in Indonesia with area of about 7,139,060 ha (63%), Kalimantan had about 3,639,737 ha (32%) and Papua had about

106,402 ha (0.9%) [22]. It is very interesting fact that proved human activities were the main factors that led to peat fires, since Papua and West Papua both has the smallest palm plantation area [22] and also the smallest population density in Indonesia [23] during that year.

a)

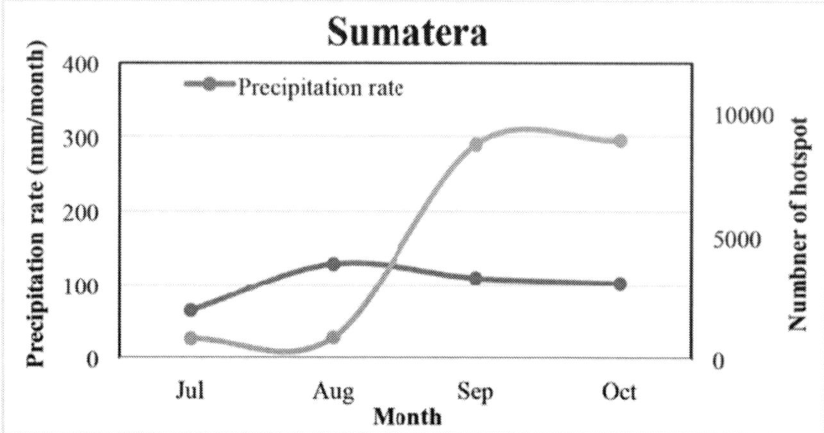

b)

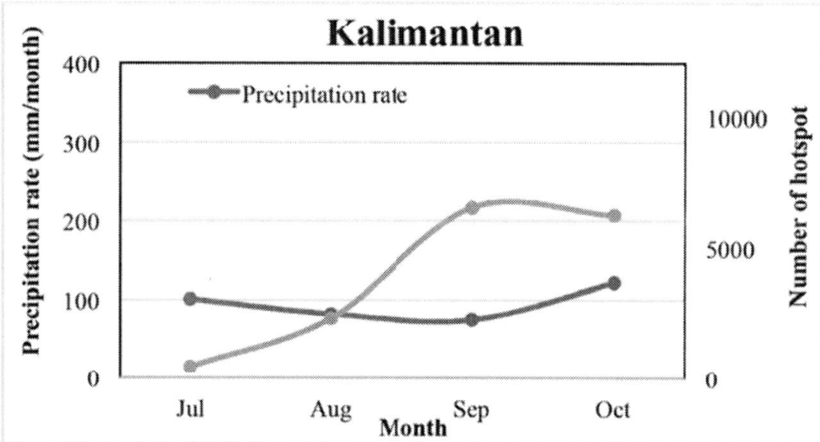

c)

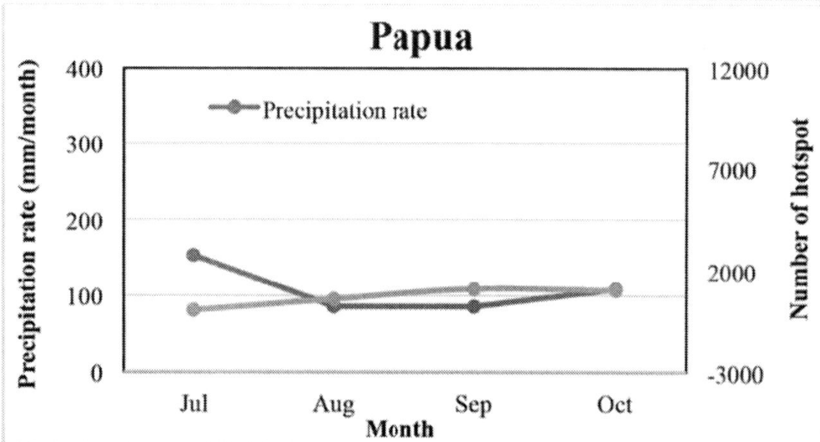

Figure 2. Precipitation rate and number of hotspot during July – October 2015 in a) Sumatera, b) Kalimantan and c) Papua

Figure 3 shows the estimation of cumulative peatland area burned in Indonesia during peat fire in July – December 2015 and their related number of hotspot. The three largest peatland areas burned were found in Central Kalimantan (197,486 ha), South Sumatra (146,986 ha) and West Kalimantan (74,858 ha) provinces. Total peat areas burned in Sumatra, Kalimantan and Papua were 270,691 ha, 320,756 ha and 31,857 ha, respectively. Based on The Ministry of Environment and Forestry statistic data in 2015 [24], it was reported that since 2010 – 2015 the government has allowed the conversion of forest to plantation/agricultural purposes in Sumatra, Kalimantan and Papua with total area of 3.063 million ha, 1.910 million ha and 1.279 million ha, respectively. Unfortunately, there was no detailed information regarding the proportion of peatland and non-peatland conversion in the data. By assuming that all of these conversions and also the agricultural/farming practices were conducted by slash and burn method, it meant that most fires were found in the existing plantation/agricultural areas. Because of this assumption, therefore for this study we used 0.04 m as a depth of peat burned by assuming that most of peat areas were burned more than two times already and they were located in the existing plantation/agricultural areas.

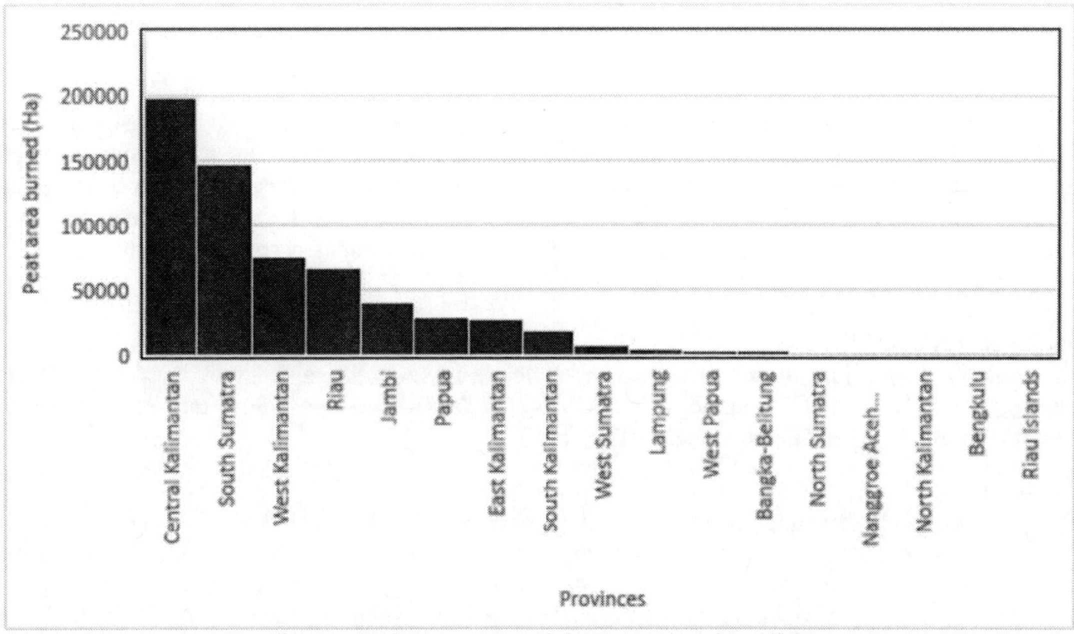

Figure 3. Peat area burned in July – October 2015

Table 2 presents the carbon emission calculated for Sumatra, Kalimantan and Papua and their corresponding emission variables. During four months peat fire in Indonesia from July – October 2015, total carbon released to the atmosphere was estimated to be about 1.848 million tonnes or 0.002 Gtonnes C, of which 81% is in the form of CO_2 (0.0062 Gtonnes CO_2); 16% CO (0.0008 Gtonnes CO) and 2.3% CH_4 (0.000065 Gtonnes CH_4). Huijnen et al. [9] reported their overestimated carbon emission calculation result during two months peat fire (September – October 2015) to be about 0.227 Gtonnes of which 83% was in the form of CO_2 (0.692 Gtonnes CO_2), 16% CO (0.084 Gtonnes CO) and 1% CH_4 (0.0032 Gtonnes CH_4). The reason was probably because the area of forest and peatland burned in Huijnen et al.'s estimation was much larger, covering all the South-East Asia region. Secondly, they also neglected the possibility of peat with different carbon content that would emit different carbon concentration. In this case, they used the same emission factor for Sumatra, Kalimantan and Papua peat fires. Page et al. [2002] reported that during four months (July – October), the 1997 peat fire in Indonesia emitted about 0.81 – 2.57 Gtonnes carbon to the atmosphere which was much higher than the one we calculated for the 2015 peat fire. The reason was because the 1997 peat fire burned much larger peatland areas (about 6.8 million ha), and also, they used the same emission factor for Sumatra, Kalimantan and Papua peat fire.

Table 2. Emission and their corresponding variables for peat fire in 2015

No	Descriptions		Sumatra	Kalimantan	Papua	Total
1.	Total peat area burned (ha)		270,691	320,756	31,857	623,304
2.	Dry fuel mass available for combustion (ton/ha)		0.0069	0.0092	0.0092	
3.	Combustion factor (dimensionless)		0.8 [11]	0.7 [13]	0.7	
4.	Emission factor (g/kg dry fuel burned)	CO_2	1,703 [11]	1,677 [13]	1,111	
		CO	210.3 [11]	221 [13]		
		CH_4	20.80 [11]	13.1 [13]		
5.	Emission CO_2 (million ton)		2.545	3.464	0.228	6.237
6.	Emission CO (million ton)		0.314	0.457	0.035	0.806
7.	Emission CH_4 (million ton)		0.031	0.027	0.007	0.065
8.	Emission CO_2 eq. (million ton)		2.756	3.765	0.253	6.774
9.	Emission C (million ton)		0.752	1.027	0.069	1.848

4. Conclusion

During the 2015 peat fire, total peat areas burned in Sumatra, Kalimantan and Papua were about 270,691 ha, 320,756 ha and 31,857 ha, respectively. By considering that peat with different carbon content would emit different carbon concentration, in this case specific emission factors were applied for Sumatera, Kalimantan and Papua peat fires, therefore this study revealed that during July – October 2015 peat fire in Indonesia had released about 0.002 Gtonnes carbon to the atmosphere, of which 81% was in the form of CO_2; 16% CO and 2.3% CH_4.

Acknowledgement

We thank to Tuti Budiwati for her great valuable discussion. We also thank to the Center for Atmospheric Science and Technology – LAPAN for facilitating this study and also NASA for providing satellite data in Giovanni and FIRMS webs.

5. References

[1] Krisnawati H, Imanuddin R and Adinugroho W C 2015 *Standard method to estimate greenhouse gases emission from forest and peatland in Indonesia, version 2* Bogor: The Agency for Research, Development and Innovation, The Ministry of Environment and Forestry.

[2] The Ministry of Agriculture 2011 *Peta Lahan Gambut Indonesia skala 1:250.000* Bogor: Balai Besar Penelitian dan Pengembangan Sumberdaya Lahan Pertanian.

[3] The Ministry of Environment and Forestry 2017 *Map of Indonesia Hydrology Unit. DKI Jakarta, Indonesia.* Retrieved from Map of Indonesia Peat Hydrology Unit: http://appgis.dephut.go.id/appgis/KHG/INDONESIA_KHGAMBUT.jpg.

[4] Cochrane M A 2009 *Tropical fire ecology: Climate change, land use and ecosystem dynamics* Chichester (The UK: Springer).

[5] Page S E, Siegert F, Rieley J O, Boehm H D, Jaya A and Limin S 2002 The amount of carbon released from peat and forest fires in Indonesia during 1997 *Nature* **420** p 61- 65.

[6] Shiv S, Boer R and Conrad E 2017 *World Resources Report Case Study. Managing Peatland Fire Risk in Central Kalimantan, Indonesia* World Resources Report, Washington DC. Available online at http://www.worldresourcesreport.org.

[7] Matt D G, Gray A, Rein G and Legg C J 2013 Peat consumption and carbon loss due to smoldering wildfire in a temperate peatland *Forest Ecology and Management* **308** p 169 – 177.

[8] Huijnen V, Wooster M, Kaiser J, Gaveau D, Flemming J, Parrington M,Inness A, Murdiyarso D, Main B, and van Weele, M. (2016). Fire carbon emission over maritime southeast Asia in 2015 largest since 1997. Scientific report, 1-8.

[9] Gaveau D L A, Salim M A, Hergouac'h K, Locatelli B, Sloan S, Wooster M, Marlier M E, Molidena E, Yoen H, DeFries R, Verchat L, Murdiyarso D, Nasi R, Holmgren P and Sheil D

2014 Major atmospheric emission from peat fires in Southeast Asia during non-drought years: evidence from the 2013 Sumatran fires Scientific reports 4:6112 DOI: 10.1038/srep06112.

[10] IPCC 2013 *Supplement to the 2006 IPPC guidelines for national greenhouse gas inventories: Wetlands. Methodological guidance on lands with wet and drained soils and constructed wetlands for wastewater treatment* Gyldenkaerne S and Lin E (eds). Japan: Institute for Global Environmental Strategies [IGES].

[11] Christian, T J, Kleiss B, Yokelson R J, Holzinger R, Crutzen P J, Hao W M, Saharjo B H and Ward D E, Comprehensive laboratory measurements of biomass-burning emissions: 1. Emissions from Indonesian, African, and other fuels *J. Geophys. Res.* **108** (D23) 4719 DOI:10.1029/2003JD003704, 2003.

[12] Stockwell C E, Yokelson R J, Kreindeweis S M, Robinson A L, DeMott, P J, Sullivan R C, Reardon J, Ryan K C, Griffith D W T, and Steven L 2014 Trace gas emission from combustion of peat, crop residue, domestic biofuel, grasses and other fuels: configuration and fourire transform infrared (FTIR) component of the fourth fire lab at Missoula Experiment (FLAME-4) *Atmospheric chemistry and physics* **14** p 9727-9754.

[13] Stockwell C E, Jayarathne T, Cochrane M A, Ryan K C, Putra E I, Saharjo B H, Nurhayati A D, Albar I, Black D R, Simpson I J, Stone E A and Yokelson R J 2016 Field measurements of trace gases and aerosol emitted by peat fires in Central Kalimantan, Indonesia during the 2015 El Nino *Atmospheric Chemistry and Physics Discussion* DOI: 10.5194/acp-2016-411.

[14] Setyawati W, Damanhuri E, Lestari P, Dewi K 2017 Emission factor from small scale tropical peat combustion *IOP Conf. Series: Materials Science and Engineering* **180** p 012113 doi:10.1088/1757-899X/180/1/012113.

[15] Wahyunto, Heryanto B, Bekti H and Widiastuti F 2006 *Peta-Peta Sebaran Lahan Gambut, Luas dan Kandungan Karbon di Papua / Maps of Peatland Distribution, Area and Carbon Content in Papua, 2000 – 2001* Wetlands International – Indonesia Programme & Wildlife Habitat Canada (WHC).

[16] Setyawati W 2017 *Development of peat fire emission factor in Indonesia to support greenhouse gas inventory emission, case study: Kalimantan* Doctoral Dissertation Institut Teknologi Bandung.

[17] Penmann J, Gytarsky M, Hiraishi T, Krug T, Kruger D, Pipatti R, Buendia L, Miwa K, Ngara T, Tanabe K and Wagner F 2003 *Good practise guidance for land use, land use-change and forestry, IPCC National Green House Gas Inventories Programmes, IPCC Guidelines* the Institute for Global Environmental Strategies (IGES) Japan.

[18] Wahyunto, Ritung S and Subagjo H 2003 *Peta Luas Sebaran Lahan Gambut dan Kandungan Karbon di Pulau Sumatra / Maps of Area of Peatland Distribution and Carbon Content in Sumatra, 1990 – 2002* Wetlands International - Indonesia Programme & Wildlife Habitat Canada (WHC).

[19] Hikmatullah and Sukarman 2014 Physical and chemical properties of cultivated peat soils in four trial sites of ICCTF in Kalimantan and Sumatra, Indonesia *Journal of Tropical Soils* **19** No 3 pp 131-141 http://journal.unila.ac.id/index.php/tropicalsoil DOI: 10.5400/jts.2014.19.3.131.

[20] Wahyunto, Ritung S and Subagjo H 2004 *Peta Sebaran Lahan Gambut, Luas dan Kandungan Karbon di Kalimantan / Map of Peatland Distribution Area and Carbon Content in Kalimantan, 2000 – 2002* Wetlands International - Indonesia Programme & Wildlife Habitat Canada (WHC).

[21] The Guardian 2015 Indonesia's forest fires: everything you need to know.

[22] Statistics Indonesia 2015 *Indonesian oil palm statistics 2015*

[23] Statistics Indonesia 2015 *Population of Indonesia: Result of the 2015 Intercensal Population Census.*

[24] The Ministry of Environment and Forestry 2016 *Statistik Kementerian Lingkungan Hidup dan Kehutanan Tahun 2015* Data and Information Center of The Ministry of Environment and Forestry.

IOP Conf. Series: Earth and Environmental Science **166** (2018) 012042 doi:10.1088/1755-1315/166/1/012042

Agricultural Production and Access to In-kind Finance from Government: A Study of Banana Farms in Cianjur District, West Java

E Wulandari[1], Ernah[1] and D Supyandi[1]

[1]Faculty of Agriculture, Padjadjaran University (UNPAD)
Jl. Raya Bandung – Sumedang KM. 21, Jatinangor 45363, Indonesia

E-mail: eliana.wulandari@unpad.ac.id

Abstract. Finance can be accessed from many sources of finance, for instance from government. One of the forms of finance provided by government in Indonesia is in-kind finance such as seeds, fertilizer and harvesting tools. The objective of this study is to analyse the agricultural production related to access to in-kind finance from government. Survey method was used to collect data especially from fifty banana farmers in Cianjur District in 2017. The collected data were analysed using independent t-test. The results showed that access to in-kind finance from the government had helped farmers in growing bananas. Having access to in-kind finance from the government, farmers have higher production of bananas compared to the farmers who did not have access to similar in-kind financing. This study suggests the need to enhance access to in-kind finance from Indonesian government for farmers. To increase the opportunities for farmers to obtain the in-kind finance from the government, this study further suggests among others, to encourage farmers take part in the membership of farmers' group.

1. Introduction

In general, farmers financing has been provided from several financial sources such as from banks and cooperatives [1], micro finance institutions (MFIs) [2], and from government [3-6]. Finance from government has been distributed through farmers' association in the form of subsidy and in-kind finance such as seed, fertilizer and harvesting equipment [6].

Farmers can access sources of funds from different financial sources, including banana farmers. Banana has been determined by the agricultural ministry of Indonesia as one of the important products for the development of horticulture in Indonesia. Banana production in Indonesia from 2012 to 2016 is presented in **Table 1**.

Table 1. Banana production in Indonesia from 2012 to 2016

Province	Production (tonnes)					Average	Growth 2016 over 2012 (%)
	2012	2013	2014	2015	2016		
West Java	1192860	1095325	1237171	1306287	1204083	1207145	0.94
East Java	1362881	1527375	1336685	1629437	1865772	1544430	36.90

Central Java	617456	560985	519628	581782	591649	574300	-4.18
Lampung	817606	938280	1481692	1937348	1517004	1338386	85.54
North Sumatera	363061	342297	298910	139541	137886	256339	-62.02
Others	1835179	1815017	1988472	1704871	1690723	1806852	-7.87
Indonesia	6189043	6279279	6862558	7299266	7007117	6727453	13.22

Source: Indonesian statistics (2013, 2014, 2015, 2016, 2017).

Data from **Table 1** shows that banana production in Indonesia has a positive growth from 2012 to 2016 as much as 13.22%. On average, the production of banana in Indonesia has reached 6,727,453 tonnes during the last five years. West Java is one of the centre of banana production in Indonesia that also has a positive growth as much as 0.94%. The production of banana in West Java from 2012 to 2016 is presented in **Table 2**.

Table 2. The production of banana in West Java from 2012 to 2016

District	Production (tonnes)					Average
	2012	2013	2014	2015	2016	
Cianjur	341162	200238	201317	201317	191561	227119
Sukabumi	128818	78595	192639	192639	165319	151602
Tasikmalaya	107817	115589	91022	91022	121255	105341
Ciamis	102963	132405	204862	204862	213971	171813
Garut	96980	116761	110290	110290	120323	110929
Others	415120	451737	437041	506158	391654	440342
West Java	1192860	1095325	1237171	1306288	1204083	1207145

Source: Indonesian statistics (2013, 2014, 2015, 2016, 2017)

The **Table 2** shows an increase of banana production in West Java from 2012 to 2016. During the last five years, on average, Cianjur had the highest production of banana, which produced 227,119 tonnes. The average banana production in Cianjur contributed 19 percent of the total production of banana during the last five years.

Previous studies have found that finance from government had contributed to better innovation performances [5] and to an increase of agricultural production by motivating farmers in technology application [7]. Furthermore, finance provided by government has positively associated with productivity improvement [8] and horticultural technical efficiency [6].

Government has provided financial support for farmers distributed in different forms such as subsidised credit, agricultural equipments, agricultural inputs aid such as seed and fertizer [3]. Existing literatures mostly explore the importance of subsidised credit from government. However, only few studies assessed the role of in-kind finance from government. Therefore, the objective of this study is to analyse the agricultural production related to farmers access to in-kind financing from government. Results of this study provide insights to the government about the importance of in-kind finance from the government to farmers in supporting the production of agriculture.

2. Materials and methods

The Indonesian Ministry of Agriculture has identified banana as one of the important crops for the development of horticulture in Indonesia [9]. Based on the identification, this study chose to focus on banana farming and financing. This study used survey technique to collect data from May to September 2017.

IOP Conf. Series: Earth and Environmental Science **166** (2018) 012042 doi:10.1088/1755-1315/166/1/012042

The data were collected from 50 banana farmers in Cianjur District, West Java, Indonesia. Cianjur district was chosen for study area as it has contributed a significant number of banana production in Indonesia. The data needed for this study included the experiences from farmers in obtaining in-kind finance from the government, the production of banana produced by each farmers, the amount of in-kind financing from government in the region of Cianjur that farmers had received, and the number of farmers' group that received in-kind finance from government.

The collected data were analysed using descriptive statistics and independent t-test. The independent t-test was used to investigate the difference of banana production between two groups of farmers, i.e. farmers who had access to in-kind finance from government and those who did not have any access to in-kind financing from the government. The results of the analysis were interpreted and discussed involving relevant literatures that supports the findings of the study.

3. Results and discussion

Cianjur district is known to become the largest banana crop produces in West Java. The average production of banana in Cianjur has reached 227119 tonnes from 2012 to 2016. Among the subdistricts within Cianjur, Sukaresmi has the highest production of banana during the last five years, in which it contributed 35% of the whole banana production in Cianjur. Following Sukaresmi is Cibeber, that has the second highest production of banana crops during the years, in which Cibeber contributed 13% of banana production in Cianjur. The production of banana in Cianjur from 2012 to 2016 is presented in **Table 3**.

Table 3. The production of banana (tonnes) in Cianjur

Sub-districts	2012	2013	2014	2015	2016	Average
Agrabinta	1123	2105	1096	4104	2685	2223
Leles	2133	2097	2714	2876	3582	2680
Sindangbarang	15831	14294	7646	1929	1325	8205
Cidaun	17136	15187	10428	16817	23205	16555
Naringgul	4160	1661	2582	1679	1876	2392
Cibinong	480	683	486	594	1227	694
Cikadu	7998	14467	10351	10990	11196	11000
Tanggeung	12509	4835	5861	10115	7247	8113
Pasirkuda	268	349	182	84	86	194
Kadupandak	12	52	110	217	215	121
Cijati	22	55	45	64	40	45
Takokak	249	224	200	206	144	205
Sukanagara	246	3576	173	1805	453	1250
Pagelaran	191	213	2264	670	798	827
Campaka	8921	8934	8982	9053	9074	8993
Campaka mulya	115	214	120	68	71	117
Cibeber	28282	28243	33849	41018	31183	32515
Warungkondang	3025	90	89	29	66	660
Gekbrong	2439	3241	2842	1772	2115	2482
Cilaku	483	551	334	181	201	350
Sukaluyu	234	127	90	69	202	144
Bojongpicung	4999	6026	1875	937	2689	3305
Haurwangi	837	859	1909	6755	6278	3327

Sub-districts	2012	2013	2014	2015	2016	Average
Ciranjang	3533	4645	1099	3897	2573	3149
Mande	300	888	330	546	456	504
Karangtengah	7031	19880	24465	15618	13040	16007
Cianjur	29	48	97	401	135	142
Cugenang	5242	8019	3761	14199	11193	8483
Pacet	6	7	12	19	11	11
Cipanas	58	7	9	5	8	17
Sukaresmi	183961	54925	45206	50400	53634	77625
Cikalongkulon	29308	3737	3277	4200	4457	8996

Source: Agricultural Office of Cianjur

To support banana production in Cianjur, Indonesian government has provided finance for farmers. These agricultural finance subsidies by government, or other types of incentive are stimulants for farmers to improve their agricultural production [4]. Finance distributed for banana farmers in Cianjur was in form of in-kind finance including seed, fertilizer and harvesting equipment. The distribution of in-kind finance for Cianjur farmers in the period of 2015 and 2016 is presented in **Table 4**.

Table 4. In-kind finance distributed by Cianjur Government.

Description	2015	2016
Number of farmers' groups	5	6
Number of sub-districts	4	4

The **Table 4** shows that the government distributed in-kind finance through 5 farmers' groups in 2015 especially for banana farmers. The government increased the number of farmers' groups receiving in-kind financing to 6 farmers' groups in 2016. This implies that the government has paid more attention to farmers' group to improve farmers' performance in agricultural production. The study of Indraningsih [10] concluded that farmers' group has important role in improving the performance of farmers especially in communicating farming innovation. Zakaria [11] further showed the benefit of government aid, which would be more appropriate if distributed through farmers' association. The Cianjur Government distributed the in-kind finance to similar number of sub-districts in 2015 and 2016.

In distributing in-kind financial support, the government required farmers to join with a farmer group. This policy is congruent with previous studies that found the role of farmers' association in financial access [12]. Farmers' association has important role in farming development [13]. Joining farmers' association benefits to farmers. Through farmers' association, farmers can improve their attitude, knowledge and skills in farming [14]. Being a membership of farmers' association, farmers can share experiences and information in agriculture [15]. Furthermore, farmers who are members of farmers' association have positively associated with the technical efficiency of farm [16].

In-kind finance program provided by government implies that the government supports farmers to develop their crop production to increase their income [11]. The descriptive statistics of banana produced by the respondents in 2016 is presented in **Table 5**. The **Table 5** shows that, overall, banana was produced by the respondents as much as 3.27 tonnes in 2016. **Table 5** further shows that the banana production of the respondents who obtained in-kind finance from the government is about twice higher than the banana produced by the respondents who didnot obtain in-kind finance from the government.

Table 5. Descriptive statistics of banana (tonnes) produced by the respondents in 2016

Description	Access to in-kind finance from government		Overall
	Have access	No access	
Mean	4.39	1.99	3.27
Minimum	0.60	0.10	0.10
Maximum	18.00	12.00	18.00
Variance	18.21	7.67	14.17
Std. deviation	4.27	2.77	3.77

The mean of banana production in Cianjur from two groups of farmer, i.e. farmers with and without in-kind finance from the government is presented in Table 6. Based on the results of the different test, the test shows that there is a significant difference of banana production between the two groups of farmers implying that the farmers who obtained in-kind finance from the government had a higher banana production compared to the farmers who did not receive in-kind finance from the government. This may happen because the government has distributed high quality of banana seeds, which lead to higher banana production. Muchtar et al. [7] revealed that in-kind finance from government increased the production of agriculture.

Table 6. The mean banana production for the different test

Access to in-kind finance from government	Banana production (tonnes)
Have access	4.36*
No access	1.99*

* significantly different at 5% level

4. Conclusion

The results showed that access to in-kind finance from the government had helped farmers in increasing their banana crop production, compared to the farmers who did not have access to the in-kind finance. This study suggests the need to enhance access to in-kind finance from Indonesian government for farmers. To increase the opportunity of farmers to obtain the in-kind finance from the government, therefore, this study also suggests farmers to be member of farmers' group.

Acknowledgments

The authors would like to thank all the farmers who agreed to participate in this study. The authors would also like to gratefully acknowledge data support from the agricultural office of Cianjur District. Furthermore, the authors would like to thank the directorate of research, community service and innovation of Universitas Padjadjaran for the financial support of this study.

5. References

[1] Armendariz B and Labie M 2011 *The Handbook of Microfinance* ISBN 139789814295659 World Scientific Publishing Co. Pte. Ltd, Singapore

[2] Bastin A and Matteucci N 2007 Financing coffee farmers in Ethiopia: Challenges and opportunities *Savings and Development* **31**(3) 251-282

[3] Swastika DK 2012 Harvest and post-harvest technologies: Adoption constraints and development strategy *Analisis Kebijakan Pertanian* **10**(4) 331-346

[4] Elizabeth R 2017 The restructuritation of implementation and effectivity of financial program to the improvemnent of rice capacity and productivity *UNES Journal of Agricultural Scientie*s **1**(1) 88-104

IOP Conf. Series: Earth and Environmental Science **166** (2018) 012042 doi:10.1088/1755-1315/166/1/012042

[5] Szczygielskia K, Grabowskib W, Pamukcuc MT and Tandogan V S 2017 Does government support for private innovation matter? Firm-level evidence from two catching-up countries *Research Policy* **46** 219–237
[6] Wulandari E, Meuwissen M, Karmana MH and Oude Lansink A 2017 Performance and access to finance in Indonesian horticulture *British Food Journal* **119**(3) 625-638
[7] Muchtar K, Susanto D and Purnaningsih N 2015 Adoption of innovation of technology by farmer in field school of comprehensive agriculture land-use management (SL-PTT) *Jurnal Penyuluhan* **11**(2)176-185
[8] Dorward A and Chirwa E 2011 The Malawi agricultural input subsidy programme: 2005/06 to 2008/09 *International Journal of Agricultural Sustainability* **9**(1) 232-24
[9] Agricultural Ministry of Indonesia 2011 Technical guidelines for operationalization of horticultural development 2012 Directorate General of Horticulture Agricultural Ministry of Indonesia Jakarta
[10] Indraningsih KS 2011 The influence of extension on farmer's decision in innovation adopting of integrated farming technology *Jurnal Agro Ekonomi* **29**(1) 1-24
[11] Zakaria AK 2011 Anticipatory policy and farmers consolidating strategy toward national corn self-sufficiency *Analisis Kebijakan Pertanian* **9**(3) 261-274
[12] Machethe, CL 2004 Landbouw en armoede in Zuid-Afrika: Can agriculture reduce poverty. In *Paper presented at the Overcoming Underdevelopment Conference held in Pretoria*, 28, 29
[13] Wahyuni S 2003 The performance of farmers' group in paddy farming system and its empowerment methods *Jurnal Litbang Pertanian* **22**(1) 1-8
[14] Novia R A 2011 The response of farmers on the field school of integrated farming in Ajibarang Sub-district, Banyumas District *Mediagro* **7**(2) 48-60
[15] Moobi MN and Oladele OI 2012 Factors influencing small-scale farmers' participation in informal financial markets in Mafikeng Municipality *Journal of Food, Agriculture & Environment* **10**(2) 1133-1137
[16] Waryanto B, Chozin MA, Dadang and Putri EIK 2014 Environmental efficiency analysis of shallot farming: A stochastic frontier translog regression approach *Journal of Biology, Agriculture and Healthcare* **4**(19) 87-100

IOP Conf. Series: Earth and Environmental Science **166** (2018) 012043 doi:10.1088/1755-1315/166/1/012043

The Supporting and Impeding Factors of Java Preanger Coffee Agribusiness on Margamulya of Pangalengan Sub District of Bandung Regency

E Djuwendah[1], T Karyani[1], E Rasmikayati[1], S Fatimah[1] and Deliana[1]

[1] Agribusiness Study Program, Faculty of Agriculture, Padjadjaran University (UNPAD)
Jl. Raya Bandung – Sumedang KM. 21, Jatinangor 45363, Indonesia

E-mail: endah_djuwendah@yahoo.com

Abstract. Java preanger coffee is internationally recognized as superior arabica coffee from West Java, particularly in coffee trading. The high demand and selling price of coffee has no fully supported the coffee agribusiness because there are other factors influencing it. This research aims to determine the factors supporting and inhibiting the agribusiness of sustainable Java preanger coffee in the Margamulya village the District of Pangalengan. The research design used is qualitative descriptive with case study method. The results showed that in subsystem agro-input, the main supporting factors are coffee plant seedlings and labor, while the main constraint factors are land, fertilizers, weather, and capital. In the agro-processing subsystem, the main supporting factors are the knowledge of farmers in the cultivation of sustainable coffee and selling prices of coffee, while the constraining factors are among others climate changes, and capital. In the agro-industry subsystem, the main supporting factor is processing technology, while the inhibiting factor is capital and processing equipment. On the agriculture marketing subsystem, the supporting factors are market certainty, while the inhibiting factor is capital. On agri-service subsystem, the main supporting factor is the availability of cooperatives, extension agencies, farmer groups, mass media, while the inhibiting factors are less coordinations between supporting institutions with farmers and the absence of financial institutions that provide credit by the characteristics of coffee farming.

1. Introduction

Coffee is one of the plantation commodities that has high economic value, as it plays a role in the acquisition of foreign exchange, source of farmers' income, labour absorption, encourage agribusiness and regional development in Indonesia. Therefore, coffee is included in the seven main plantation commodities of Indonesia, which receives priority development from the government.

Java Preanger coffee is one type of coffee Arabicas, especially originated from West Java. According to coffee lovers, Java Preanger coffee has a distinct in the strong floral nuances (floral) with flowery-fruity-nutty-dark chocolate aromas [1]. With the issuance of Government Regulation No. 51 of 2007 on Geographical Indication, based on the suggestions from the community of geographical indication protection (MPIG), Java Preanger coffee and West Java government in 2013 Java Preanger Arabica coffee (JPAC) is certified by the geographical indication from the Indonesian Ministry of Law and Human Rights.

Since 2009 JPAC has penetrated international markets such as Southern Korea, Morocco, Taiwan, Hongkong, China, Germany, Japan, England, Netherlands and Germany [2]. The specialty coffee market is currently growing in major consumer countries such as the United States, the European Union, Japan, Southern Korea, Brazil and Indonesia. The countries, which become the main market of coffee demand the quality of coffee to be in accordance with the demands of consumers such as food security, environmental conservation and welfare of farmers.

In order to increase the competitiveness of Java Preanger Arabica coffee in regional and international markets, the development of this commodity is done through a sustainable agribusiness approach. Sustainable agribusiness is a functionally integrated ecological agribusiness based on upstream subsystem, farming, yield processing, marketing, and supporting subsystems aligned with the surrounding environment and realizing sustainable agricultural development.

Based on the statistic data of West Java from Provincial Plantation Office, the Bandung Regency is a potential coffee development center with an area of 8,656 Ha (31.39%) of coffee plantation area in West Java. Pangalengan Sub district as one of the areas in Bandung with the largest production of coffee. There are 5 villages in Pangalengan Sub district as coffee farming center namely Losari, Margamulya, Wanasari, Sukaluyu and Lamajang.

The development of crops and coffee production in Margamulya Village every year tend to increase. Based on interviews with 4 coffee producers in Margahayu village namely Margamulya Coffee Producer Cooperative (CPCM), Malabar mountain, Malabar civet coffee and Coffee Prisca, it was found that the area of coffee farming in Margamulya village in 2013 is 225 Ha, in 2014 is 400 Ha and in 2015 it reached 580 Ha with production respectively of 669 Ton, 1,200 tons and 1740 tons.

Nevertheless Potential coffee commodities in Margamulya Village, still faced with some challenges are still found farmers who implement coffee cultivation that does not fit with the technical guidelines of sustainable farming so that the productivity is not optimal i.e 930 kilograms ha^{-1} or about 62-93% of the standard of national coffee productivity 1300 - 1400 kg ha^{-1} according to Pusat Penelitian dan Pengembangan Perkebunan. In the stage of agro-industry is still limited the farmers who do the processing of coffee until the stage of horn skin (HS) and 70% of coffee produced by agro-industry in the form of green bean.

On the other hand, Java Preanger Arabica coffee from Margamulya village has succeeded in penetrating international market through processing and marketing by CPCM in partnership with PT Taman Delta Indonesia, an exporter, for obtaining UTZ certificate which essentially in the management of its business to pay attention of environment. Therefore it is interesting to study what factors are the driving forces and obstacles in the development of sustainable Java Preanger coffee agribusiness in Margamulya Village.

2. Research Method

The research area was determined purposively in Margamulya Village Pangalengan Sub district of Bandung regency because Pangalengan is one of Java Preanger coffee production center in Bandung regency. This research used qualitative method. Qualitative research method emphasizes its analysis on the deductive and inductive processes and on the analysis of the dynamics of relationships between the phenomena [3]. The research technique is case study. The data sources are primary and secondary data. Technical data collection is done by observation, interview and literature study. Interviews are conducted with 34 coffee farmers,

IOP Conf. Series: Earth and Environmental Science 166 (2018) 012043 doi:10.1088/1755-1315/166/1/012043

village officials, the cooperative's member and management. Data were analyzed descriptively and the technique of data presentation was done by tabulation.

3. Results and Discussion

3.1. Overview of Java Preanger Arabica Coffee Agribusiness in Margamulya Village

According to Hanafie [4] agribusiness as an integrated system is defined as all activities ranging from procurement and distribution of production facilities to the marketing of products produced from a farm or agroindustry that are associated with each other. Agribussiness consists of subsystems procurement of production facilities, farming subsystem, agro-industry subsystem, marketing subsystem and sub system supporting facilities. Each subsystems has a backward and forward linkage.

Margamulya Village is one of the villages in Pangalengan sub-district that develops sustainable of arabica coffee agribusiness. This is evident through the implementation of UTZ certification by the Margamulya farmers group. UTZ certification certifies the production process traceability which can indicate that the product with this certification has been through a series of good plantation practices, paying attention to social and environmental aspects, food safety and quality improvement.

Coffee farming area in Margamulya village ranges from 0.2 to 1.0 H.a.making them known as smallholder coffee plantations. The arabica coffee plantations in Margamulya village are mostly (73.50%) consist of of coffee farming practices on narrow fields (land area < 0.5 ha),14.7% was done on moderate (0.5-1, 0 ha) and 11.80% land area (> 1 Ha). Land tenure is generally owned by state forest company of Indonesia (*Perhutani*), which managed by farmers through community forest management program. There are 51% of farmers who plant coffee in *Perhutani's* land with a profit sharing of 10-20%. There are 40.20% of farmers who plant coffee on their own land and 8.80% plant coffee trees simultaneously on their own land and *Perhutani 's* land.

Currently cultivation of Arabica Java Preanger coffee in Margamulya Village has been increasing because many farmers had switched from planting vegetable commodities to coffee plants both on their own land and *Perhutani 's* land. The transition of practices of these planting commodities is partly due to the motivation of farmers, environmental awareness, the suggestion of *Perhutani* to grow coffee in forest areas through community forestry program, as they provided assistances for coffee seeds. Furthermore, coffee farming risk is lower than vegetable farming. According to leaders of the Rahayu forest village community institution and the head of the Margamulya farmer group, the concept of forest management with the community provides two benefits: increased economic income and practices on sustainable forest conservation.

Varieties of Arabica coffee are grown by farmers in Margamulya Village in the early days of 2001-2003 were brought from Central Aceh coffee seedlings (Ateng) and Kartika I. However, since 2010 farmers started to plant Linne S 795 coffee seeds, and in 2013 planted Sigarar Utang which has the peculiarity to bear fruit following the pattern of local rainfall distribution. Plant densities used are varied from 1 X 1 meter and 1.5 X 1.5 meters.

Arabica coffee plants start flowering at the age of 2 years and then it can be harvested from 2.5 to 3 years old and continued until the plants are 20 years. Arabica coffee takes 6- 8 months from flower bud to harvest. Harvesting are done by picking ripe coffee fruit. The harvest season takes place between March and August.

Generally produce processing is done by large plant traders and cooperatives. The Cooperative of Coffee Margamulya Producer (CPCM) has applied the standard operating procedure of coffee processing which refers to UTZ certification. Java Preanger coffee processing is done immediately after the farmers implement the harvest of red coffee (cherry) by wet processing method. Wet processing process begins with sorting of cherry coffee, peeling the outer coffee skin with pulper to produce Horned Skin (HS) coffee. Further proccesses are

12-36 hours fermentation and continues with washing. Drying uses partition or drying floors until moisture content around 12.5%. Deep skin peeling are done by huller machine, followed by drying processes to reach 12% moisture content. Finally, the coffees are sorted by grader machines and then packed using sacks and stored in storage facility.

There are some farmers who chose to process cherry coffee into horn skin (HS), because they think HS processing can increase their income. However, farmers who process HS are still very small, or at least 10% of farmers population. this is due to the limited capital that they have. This is in line with research done by Kirana S and Karyani [5] showing the price differences between HS Rp which amounts to Rp 23.000 per kilogram and cherry coffee which is Rp 8,500 per kg. Processing of cherry coffee to HS gives added value to farmers, equal to Rp 528,74 per kilogram of cherry coffee. Large traders which has a processing equipment and The Cooperative of Coffee Margamulya Producer to process coffee into green beans. Cooperative gets the added value of green beans processing ranging from Rp 2,100.68 - 6,1015.68 kg^{-1}, depending on the destination market. While the add value roasted bean and gounded bean processing are Rp 15,772.95 and 23,462.71 kg^{-1} respectively.

Production capacity of Margamulya coffee is about 350 ton cherry coffee per- harvest season. Production capacity of cooperative producers of margamulya coffee are about 70 tons of green beans. Good quality of green beans (grade 1 and 2) are sold to the exporter company, PT Taman Delta Indonesia, which located in Semarang for further export to Japan by labeling the product as UTZ certified. Low quality of green beans are sold to local market in Medan, Bandung and other cities. The cooperative also processes and markets roasted coffee and ground bean coffee with the Java preanger brand of Gunung Tilu coffee sold at souvenir shops, exhibition and e-commerce such as through blibli.com. Some collectors who process coffee sell the processed products to PT Indocom Citra Persada in Surabaya and PT Sari Makmur Tunggal in Medan.

Java preanger coffee produced by Margamulya Village is favored by consumers because it has a distinctive taste. The result of Java arabica preabook from Margamulya Village which was conducted by Indonesian coffee and cocoa research center (*Puslitkoka*) at Jember in 2012, concluded that their coffee have the characteristic of spicy, floral and flowery flavors, scored 84,08 and in 2014 with score 84,67 [6]. Based on the Indonesian National Standard (SNI) 01-2907-2008, the coffee beans of Margamulya village are categorized as good quality (grade 1) which means the value of coffee defects is less than 11 [7].

3.2. Supporting and inhibiting factors of sustainable Java preanger coffee agribusiness

Java preanger coffee agribusiness system includes 5 subsystems in which the implementation should be integrated and synergised.The driving factor is something that helped support the omset of Java Preanger Arabica Coffee Agribussiness. The inhibiting factor is a condition that can complicate the development of Java prenger coffee agribusiness. Internal factors are factors that come from within the agribusiness actors or organizations and external factors are factors that come from outside the actors or agribusiness organizations.

3.2.1 Agro-input Subsystem

The agro-input subsystem is all the activities of procurement and distribution of inputs of production inputs for the implementation of a farming technology, and the optimum utilization of agricultural resources. Coffee farmers in the village of Margamulya obtain coffee plant seeds easily, either for free from cooperatives or buy from seedlings in the village of Margamulya. Based on the Decree of the Minister of Agriculture No. 65 / Kpts / SR.120 / 2/2014, there are two h.a. coffee nurseries managed by forest village community institutions Rahayu Tani in Margahayu village. The provincial government also provides free coffee seedlings to farmer [2]. Since the year 2013 CPCM also become a coffee seed cultivator of West Java provincial government aid assistance with a target of 20,000 coffee seeds per year.

IOP Conf. Series: Earth and Environmental Science **166** (2018) 012043 doi:10.1088/1755-1315/166/1/012043

Table 1. Supporting and inhibiting factors on agro-input subsystem

Factor	Internal	External
Supporting	Seed, feertilizer, land, labor	Cooperation with *Perhutani* in the provision of land
Inhibiting	The scale of land tenure, capital, quality of human resources	Traditional technology of Agricultural

Production inputs are available at agricultural stores which are located in the subdistrict, and labors are available from within and outside of farmer's family. Farm lands are provided by *Perhutani* through community forest management program, as it became one of the supporting factor for the agro-input subsystem.

The inhibiting factors of sustainable coffee farming development are the narrow scales of land management, the low level of farmer education and the traditional cultivation techniques. Most (51%) of coffee farmers had primary school education, 2.9% were out of school, 43.6% went to middle school (junior and senior high school) and 2.5% participated in higher education. Banking institutions that provides credit to coffee farmers are still limited, hence many farmers had to borrow capital to cooperatives or to collecting traders.

3.2.2 Farming Subsystems

Sustainable coffee cultivation practices are culturally cultivated and utilize green gardening practices. Among the practices are for farmers not to open steep forest areas and should take care of the garden including the maintenance of tree protection and pruning; farmers should use organic liquid fertilizer and organic compost fertilizer; and use litter as ground cover; and finally using pest control for Coffee Fruit using traps and biological control. Such practice require control of pests and other diseases using vegetable pesticides; and finally, the use and processing of coffee waste into compost.

Table 2. Supporting and inhibiting factors on farming subsystem

Factor	Internal	External
Supporting	Suitability of agroclimate, high motivation of farmers, SOP and UTZ certified, integration of coffee plants with livestock	The provision of certified coffee seeds, market certainty
Inhibiting	Cultivation is not in accordance with the technical guidance of sustainable coffee cultivation and limited capital	The climate change

Sustainable coffee cultivation is in line with the semi organic coffee cultivation required in UTZ Certification. In principle, farmers should do not use non-organic fertilizers except at the time of initial planting. Therefore, the fertilizers and pesticides used are organic. UTZ Certification is a traceability system to ensure cultivation activities to proces and distribute production in environmentally friendly activities and positive impact for socio-economic community.

The altitudes of suitable places for Arabica coffee plants range from 1,000 - 2,000 m asl, annual rainfall 1250 - 2,500 mm dry months (rainfall <60 mm / month) 1-3 months per year with an average air temperature of 15- 25° C and climate types A through B (Schmidt and Ferguson). Andosol is a type of soil that is relatively fertile and suitable for coffee plants. The content of organic matter is high, that is above 3% with pH pH ranging from 6 to 7 [8]. Margamulya village is a mountainous area located at an altitude of 1416 m asl with rainfall of 2350 mm per year with 6 months rainy month, daily average temperature 18-23° C, ph ranges from 4.8 to 5.6, air humidity 60, 7% with type B climate. Based on appropriate environmental conditions for arabica coffee farming which described by the Director General of Plantation and Sari NP, et al (2013) [8], this area is geographically suitable for Arabica coffee farming.

The other supporting factor of sustainable coffee agribusiness is the strong motivation of the farmers who are driven by concerns about the increasingly destructive condition of the forest area, resulting in the desire to preserve the forest area by growing coffee as revealed by Pak Enjang and Pak Aleh one of Margamulya group members. Coffee cultivation follows the method of conservation coffee program.

3.2.3 Agro-industry Subsystem

The quality assurance, supply availability, timely and sustainable according to consumer demand are some requirements needed for sustainable agribusiness of Java preanger coffee. As consumer demand for safe and environmentally friendly coffee product, the standard reference of quality for the coffee beans must accommodate the principle of good harvest post-treatment and true Good Handling Practices (GHP) [9].

There is a processing SOP at CPCM referring to SNI and UTZ certified to ensure that the activities of cultivation and processing of coffee are environmentally friendly and have a positive impact on the economy and people's welfare. One of the activities which is environmentally friendly is the use of waste coffee skin as organic fertilizer.

Table 3. Supporting and inhibiting factors on agro-industry subsystem

Factor	Internal	External
Supporting	Availability of labor, there is SOP of coffee processing according to SNI and UTZ certified procedures, specialty coffee products	The provision for building and coffee processing equipment to cooperatives, existence of demand for instant coffee, roasted coffee and extract coffee
Inhibiting	Capital limitations, fluctuating coffee beans supply, inconsistent quality of coffee	The lack of coffee processing technology to produce instant coffee, roasted coffee and extract coffee

Internal constraints faced by agro-industry coffee actors such as raw material quality that are not uniformed, fluctuating raw material supply, and limited capital. Limited capital causes only about 10% of farmers who can access process cherry coffee into horn skin coffee. There are still many farmers who pick up coffee that are still green or coffee beans that are attacked by pests and diseases. Coffee beans that are still green or after being treated for disease will produce low quality green beans because it was already damaged during processing. There is a mismatch of coffee processing in Margamulya coffee cooperative with its SOP when processing green bean during fruit harvesting activities; sorting and mining as well as stripping of the outer skin. Capital limitations cause agro-industry actors to choose traditional equipment and can not immediately buy and process coffee harvested by farmers. In fact, if coffee is left one night it will experience fermentation and will no longer be perfect so it will reduce the quality of processed coffee.

The external constraints faced by agroindustry actors to respon demand of coffee variants from the niche of modern market had kept growing the powder coffee processing technology to its limits. Most of coffee export in the form of green been. whereas if the exports are in the form of roasted coffee and grounded coffee it will increase the added value for agro-industry actors in Margamulya village.

3.2.4 Agricultural marketing sub-system

The supporting factors in agriculture subsystem of Java Preanger Coffee are the special product quality and easy marketing. The types of Java preanger coffee products marketed through CPCM are horn skin, green beans, roasted and grounded coffee. The Java Preanger Coffee that produced by CPCM has a taste test value of 84.67 in 2014, thus can be classified as specialty coffee, because the taste has a score above 80. According to SCAA (2009) [10], the value of coffee has total value flavor ≥ 80 on a scale of 100, it can be categorized as spescialty.

There is a diversity of marketing institutions in the village of Margamulya including collector and cooperative. As many as 74 percent of farmers sell in the form of cherry coffee to the cooperative the remaining 26 percent sell to the middlemen. The middlemen sell to cooperatives in the form of green beans. A total of 14.70 percent of farmers said that guarantee of price certainty is a factor that encourages them to farm coffee. The selling price of cherry coffee ranges from Rp 7000 to 8,500 per kilogram while in the form of horn skin coffee, it can be sold for 20,000 s.d 23.00 per kilogram. Payment of coffee to farmers are in cash.

Table 4. Supporting and inhibiting factors of agricultural marketing subsystem

Factor	Internal	External
Supporting	Available coffee marketing agencies, as Java preanger coffee is specialty coffee	The high price of coffee in world markets, clear of market target, increasing the consumption of world coffee
Inhibiting	The quality and quantity of coffee supply are not fixed, has not penetrated the consumer market directly	low levels of consumption and purchasing power of domestic consumers, the absence of coffee exporters from west Java, coffee competitor from vietnam with cheaper price

Java Preanger Arabica Coffee Margamulya has a clear target market with the number of continuous demand. 70% of coffee products are sold to PT Taman Delta Indonesia (TDI), and 30% are sold to Aroma stores, PT Indocom, Glory morning, cafes and gift shops located in Pangalengan Subdistrict, Bandung and surrounding areas. The selling price of green beans to PT Delta is Rp 62.000 / kg, Indocom Rp 75.000 / kg and cafe Rp 95.000 / kg. The selling price of Roated beans and goundaed beans is Rp 175,000 and 250,000 per kilogram respectively. Payment of coffee from exporter (TDI) and local cafe to cooperative is made 3-5 days after delivery of goods.

Internal inhibiting factors of Java preanger coffee are inconsistent quality and quantity of supply of coffee products to local and export markets and have not touched the consumer market directly. While external barriers include low levels of coffee consumption and consumer purchasing power, the absence of coffee exporters from western Java and the cheaper competitor of Vietnamese coffee. This is in line with the results of Kustiari's research (2007) [11], which states that the consumption of Indonesian coffee are by 0.6 kg per capita per year, which is lower than the coffee consuming countries, yet the price of coffee in Indonesia is influenced by world coffee prices which are exogenous variable and unstable rupiah exchange rate .

3.2.5 Agriservice Subsystem

The agriservice sub-system is a supporting institutional subsystem that plays an important role for the development of the agribusiness system as a whole. Including the extension component, information, education and training, credit, etc.

Table 5. Supporting and Inhibiting Factors on Agriservices Subsystem

Factor	Internal	External
Supporting	The presence of farmer groups, coffee producers cooperatives and banking institutions	The presence of coffee business association (AEKI), research center for coffee and chocolate in Jember, the opening of West Java coffee exports (Law No. 12 of 2012)
Inhibiting	The low access of agribusiness actors to banking	Lack of coordination among agribu-siness support institutions

The supporting institutions in the form of associations of coffee entrepreneurs (AEKI), governments, research center for coffee and chocolate, farmer groups and cooperative coffee producers, have an influence on the mindset and working patterns of farmers and coffee entrepreneurs. Therefore farmers and entrepreneurs who have joined these institutions improved the quality and quantity of coffee they produced. The support of West Java provincial government in coffee development in Margamulya Village are coffee seed certification services, giving coffee seeds, strengthening of farmer groups and cooperatives. In 2010 provincial government give coffee processing equipment and coffee processing factory in 2012. While the Geographical Indicator observer community (MPIG) and AEKI helped provide market information to the agribusiness actors in Margamulya Village. The results of Kustiari's research [11], show that macro government which gather with AEKI proactively monitor the development of world coffee demand and price, implement coffee market development to countries that still import coffee from Indonesia such as Eastern Europe, China and Russia.

Inhibiting factors in agriservice are the low access of agribusiness actors to banking, lack of coordination between farmer groups, so there are still members of the farmer group who have not implemented standard operating procedures in accordance with the UTZ certified.

4. Conclusion

Agribusiness of Java Preanger coffee in Margamulya village develop from time to time. This is because the Margamulya Village has potential natural resources for the development of Java Preanger Coffee. In order to improve the competitiveness of products in the global market, the implementation of local sustainable coffee based on local resources is required.

The supporting factors for the development of sustainable agribusiness of Java Preanger coffee are the availability of input production, the suitability of agro-climates, the high farmers motivation and the experience of farming, the technical guidance of sustainable coffee farming and operational standard of procedur on coffee processing according to UTZ certified, integration of coffee plants with livestock, and market targets, specialty coffee, high coffee prices in global markets, instant coffee, coffee opportunities from modern markets and international markets, seedlings support, building and government coffee processing equipment, farmer group institutions, cooperatives producers coffee, financial institutions, association of Indonesian coffee exporter (AEKI), coffee and cocoa research centre (Puslitkoka) and the opening of West Java coffee exporting faucet (Law No. 12 of 2012).

Inhibiting factors for the development of Java Preanger coffee agribusiness include limited capital and quality of human resources, the implementation of cultivation is not in accordance with the technical guidance of sustainable coffee farming, climate change, quantity and quality of coffee is not maintained, the limitations of the final processing technology and has not touched the consumer market directly. On the other hand the level of consumption and purchasing power of national consumers is relatively low, added with the fluctuations in the price of coffee in the international market, the majority of coffee exporters which are not from outside West Java, and Vietnam coffee competitor that has 30% cheaper price from the national coffee price.

Acknowledgments
Thanks to Padjadjaran University (UNPAD) the directorate of community service research and innovation which has funded this research through internal grant (HIU) in 2017.

5. References

[1] Dinas Perkebunan Jawa Barat 2016 *Mengenal Kopi Arabika Java Preanger (KAJP) yang Telah Mendapat Perlindungan Indikasi Geografis* http://disbun.jabarprov.go.id/ index.php/artikel/detailartikel/119
[2] Sunjaya P and Ferry Y 2015 Keragaan kopi arabika Java Preanger di Jawa Barat *Jurnal Sirkuler Inovasi Tanaman Industri dan Penyegar (SIRINOV)* **3**(3) 113-126
[3] Wirartha IM 2006 *Metode Penelitian Sosial Ekonomi* Andi Offset Yogyakarta 134
[4] Hanafie R 2010 *Pengantar Ekonomi Pertanian* Andi Offset Yogyakarta 31-34
[5] Kirana S and Karyani T 2017 Nilai tambah rantai pasok kopi pada Koperasi Produsen Kopi Margamulya di Kecamatan Pangalengan Kabupaten Bandung: Komparasi antara petani dan pengolah kopi *AGRISEP* **16** 165-176
[6] Ima M and Djuwendah E 2015 Analisis kesiapan dan strategi pengembangan bisnis Koperasi Produsen Kopi Margamulya Pangalengan Bandung *Proceeding Semnas Kristalisasi Paradigma Agribisnis dalam Pembangunan Ekonomi dan Pendidikan Tinggi* IPB Bogor 257-266
[7] Xena LG and Trimo L 2014 Potensi pengembangan Preanger coffee *Proceeding Seminar Nasional Pembangunan Inklusif di Sektor Pertanian* Departemen Sosial Ekonomi Faperta UNPAD Bandung 205-212
[8] Sari NP, Santoso TI and Mawardi S 2013 Sebaran tingkat kesuburan tanah pada Perkebunan Rakyat Kopi Arabika di dataran tinggi Ijen-Raung menurut ketinggian tempat dan tanaman penaung *Pelita Perkebunan* **29** 93–107
[9] Henny M 2013 Kebijakan penyediaan teknologi pascapanen kopi dan masalah pengembangannya *Forum Penelitian Agro Ekonomi* **31** 31-49 http://ejurnal.litbang.pertanian.go.id/index.php/fae/article/view/3842
[10] Speciality Coffee Association of America (SCAA) 2009 What is specialty coffee? http://www.scaa.org/18 Juni 2014
[11] Kuastiari 2007 Perkembangan pasar kopi dunia dan implikasinya bagi Indonesia *Forum Penelitian Agro Ekonomi* **25** 43-55

IOP Conf. Series: Earth and Environmental Science **166** (2018) 012044 doi:10.1088/1755-1315/166/1/012044

ENSO and PDO Influence to Climate Variability in Monsoon Region of Indonesia

E Yulihastin[1,2], N Cholianawati[1], G. A Nugroho[1], T Sinatra[1] and H Satyawardhana[1]

[1]Atmospheric Science Research Group, Faculty of Earth Sciences and Technology, Institut Teknologi Bandung, Indonesia
[2]Center of Atmospheric Science and Technology, National Institute of Aeronautics and Space, Indonesia

E-mail: erma.yulihastin@gmail.com

Abstract. Climate variability can be identified based on two climate parameters: Outgoing Longwave Radiation (OLR) and Precipitable Water (PW). South of Indonesia (2-10°S, 90-15°E) is a rainfall monsoon region with relatively homogeneous characteristics in seasonal variations. However, the region has some certain sensitivities in responding variability from intraseasonal to interannual. Using the method of Empirical Mode Decomposition (EMD), S-Transform and Wavelet for 30 years monthly data of OLR and PW (1980-2011), it was found that both of data indicate signal strengthening after 2010 which is related to Tropical Multi-decadal Mode. Regional response from Intraseasonal to interannual variations confirm that both parameters are sensitive to 1-3 monthly (Intraseasonal Variation) related to MJO, 6 month and 1 year related to monsoon, the 2-7 year variability associated with El Niño Southern Oscillation (ENSO) on short term and medium scale period, whereas 10 yearly due to ENSO in longterm scale period and probably represent the phenomenon of the solar cycle. However it may be suggested that the 15 and 30 year variability would probably corresponded with ENSO and the Pacific Decadal Oscillation (PDO).

1. Introduction

Knowledge of climate variability is more important than the average climate [1]. This is because the value of statistical variability can indicate a variety of important information outside normal climatic conditions such as anomalies, trends and extreme events. The pattern of climate variability in the climate scenario of the changing world is becoming increasingly important to note because the events of weather and climate extremes often acts as a catalyst which is sensitive to climate variability [1].

Climate variability can be identified by two climate parameters, namely Outgoing Longwave Radiation (OLR) [1-7] and Precipitable Water (PW) [8-12] which is a non-stationary data. OLR is associated with wavelengths of the sun (with a wavelength around the wavelength infrared) reflected /emitted by the earth's surface, or reflected by clouds. OLR is the longwave radiation of the sun that is absorbed by the earth's surface and reflected back into space. Some of this long-wave radiation is either retained or trapped in the atmosphere. Thus, the high (low) value of OLR parameters indicates few (many) clouds present in the atmosphere.

Another climate parameter is the PW defined as the amount of water vapor collected in a precipitated column of the atmospheric atmosphere as well as decreased as rainfall when the moisture content in the column has been condensed. The PW value is derived from the specific moisture data, which is a function of pressure and temperature.

The relationship between OLR and PW is negatively correlated, meaning that if a region has a large PW value it means a small OLR value is in the region, and vice versa. This can occur due to high PW value that may indicate more and more clouds are formed as a result of convection activity. When clouds are formed, the atmosphere will absorb more radiation, hence the longwave radiation that is reflected back to space are small so that OLR values will be small or low.

Annual variation of OLR in southern Indonesia is showed by high OLR values indicating the dry season, and vice versa. Meanwhile, the relationship between PW and OLR is negatively correlated with a very high correlation coefficient value of -0.95 [13]. In addition, the annual cycle is clearly identified based on PW and OLR data [13]. However, the opposite relationship between PW and OLR over a longer period of time and variations from year to year in southern Indonesia have not been studied. Yet research on the variation of OLR and PW needs to be done to determine the consistency, sensitivity and change, which is useful as the initial identification of climate change in Indonesia. It due to OLR represented Earth's radiation budget which is a critical component to understand the climate change [14] and precipitable water reflected water vapors as mainly parameter which is contributes at about 50% of the present-day global greenhouse effect [15].

This paper aims to present the results of climate variability studies in the Indonesian monsoon region from the highest (intraseasonal) to the lowest (antartahunan) frequency. In addition, this study also aims to determine the consistency and sensitivity of both parameters in southern Indonesia, in relation to the monsoon phenomenon; i.e. the El Niño Southern Oscillation (ENSO), the 11-yr solar cycle, and the Pacific Decadal Oscillation (PDO).

2. Data and Method

All data in this study has a monthly period ranging from 1980-2011. To examine the periodicity and variance of data used, two methods based on the analysis of the frequency spectrum are deployed: wavelet and S-Transform. Both methods are used so that the resulting frequency spectrum can be interconnected. In addition, to know the variability of the data in more detail by performing the separation of the type of frequency from highest to lowest, the EMD method is performed. The EMD method can also generate data trends indicated by recent IMF results and can provide information on weakening and signaling through wave envelope over the data period. The EMD method is suitable for long-term non-stationary data as is the characteristic of OLR and PW data. The data and methods are described in more detail as follows.

2.1. Data

The main data used are PW data with monthly temporal resolution and 2.5-degree spatial resolution that can be downloaded freely via the following sites: http://www.esrl.noaa.gov/psd/data/gridded/data. ncep.reanalysis.derived.surface. html and OLR data with temporal resolution and 0.05 degree spatial resolution http://www.esrl.noaa.gov/psd/data/gridded/data.interp_OLR.html. PW data is available from 1974 to 2011 while OLR data is available from 1948 to 2011. As supporting data, this study utilize sun spot data obtained from Solar Influences Data Analysis Center (SIDC) http://sidc.oma.be/sunspot-data/ and SOI data Southern Oscillation Index (SOI) http://www.cpc.ncep.noaa.gov/data/indices/. The period of data used is data from January 1980 to December 2011 (30 years).

2.2. Methods

2.2.1 Morlet Wavelet

Wavelet morlet is a wavelet type of complex exponential signal that is modulated with Gaussian window [14].

$$\psi_0(\eta) = \pi^{-1/4} e^{-i\omega_0\eta} e^{-\eta^2/2}$$

Where ψ is the wavelet value at non-dimensional time, while ψ_0 is the wave number.

2.2.2 Stockwell Transform Method

The Stockwell transform (S-Transform) produces a representation of the relationship between time-frequency in the time domain. This method is a combination of resolution settings based on frequency and spectral arrangement of real and imaginary values simultaneously [17].

$$S(\tau, f, p) = \int_{-\infty}^{\infty} h(t)\omega(\tau - t, f, p)e^{-j2\pi ft}dt$$

Where p is the parameter that determines the shape and value of f, whereas ψ is the window function of S-transform (Gaussian function with window width depends on its frequency).

$$\omega(t, f, p) = \frac{|f|}{\sqrt{2\pi p}} e^{\frac{t^2 f^2}{2p^2}}$$

2.2.3 EMD Method

The Empirical Mode Decomposition (EMD) principle is to decompose signals based on the direct extraction of signal energy associated with various intrinsic time scales [19]. This technique adaptively decomposes the non-stationary signal into a set of intrinsic oscillatory modes. Components called intrinsic mode functions (IMFs), allow instant component frequency calculations based on the Hilbert transform. Thus, it can potentially localize events in time and frequency.

The starting point of EMD is by reviewing the signal oscillations at a very local level. Given the signal $x(t)$, the effective EMD algorithm can be summarized as follows: (1) identifies all extracts $x(t)$; (2) interpolate between minima and maxima so as to produce some envelope $e_{min}(t)$ and $e_{max}(t)$; (3) calculate the average $m(t) = (e_{min}(t) + e_{max}(t))/2$; (4) extracting detail $d(t) = x(t) - m(t)$; (5) iteration on residual $m(t)$. The sifting process includes the first iteration of steps 1 through 4 to obtain detailed $d(t)$ signals, then reviewed as zero-mean according to the termination criteria.

Once obtained, the detail is referred to as Intrisic Mode Function (IMF), the corresponding residual is calculated and followed by step 5. The extreme amount will decrease as it moves from one residual to the next, and the overall decomposition is fitted with a finite number of modes. A mode extraction is obtained when the shifting process is complete. Two conditions must be met: (1) the number of extremes and the number of zero-crossings must vary by a maximum of 1; (2) the average between the upper and lower limits of the envelope should be close to zero. Or with sifting process termination criteria: SD value = 0.2 or 0.3.

$$SD = \sum_{t=0}^{T} \left[\frac{|(h_{1(k-1)}(t) - h_{1k}(t))|^2}{h_{1(k-1)}^2(t)} \right]$$

3. Result and Discussion

Figure 1 shows areas in southern Indonesia that have seasonally different contrasts in precipitation and OLR patterns especially between the DJF (rainy season) and the JJA (dry season). This distinction in precipitation intensity and the seasonal wind direction is what causes the region to be referred to as the region with monsoonal rain pattern in Indonesia. Periodicity of OLR and PW data in the region will be further investigated using wavelet analysis.

The wavelet results in **Figure 2a** show the amplitude of OLR anomaly data varying from month to month with values between +2 and -2. **Figures 2b-c** show that the one-year signal is the dominant period of OLR in southern Indonesia. Other emerging signals are 2.5 yr, 5.3 yr and 10.6 yr.

The annual signal is closely related to the monsoon phenomenon. Meanwhile, the 2.5 yr and 5.3 yr signals are influenced by the irregular cycles of ENSO [18, 21-23]. Signal strengths of 2-4 yr and 11 yr on OLR data have the same power spectrum. In addition, for the 2.5 yr, signal strengthening occurred in 1994-2003 while for the 5.3 yr lasted between 1997-2000. For the 10.6 yr, no significant signal strengthening occurs in OLR data. For OLR data variance for 30 years there appears to be no threshold value (**Figure 2d**), which shows that OLR data has a high degree of regularity. It also proved that wavelet method has limitation to decompose OLR signal, so that we need other method which can decompose the non-stationary and non-linear signal data in more details and consistent, i.e. EMD (Empirical Mode Decomposition). It is because wavelets concept considering both low frequency as well as high frequency events simultaneously [19]. In the other hand, the limitation of wavelet morlet which is applied in the current research, depends on the characteristics of the data, also the data border distortion can affect the decomposition process at the two limits of the signal. However, wavelet morlet are the most commonly used for periodic data such as climatology and hydro-meteorology [20].

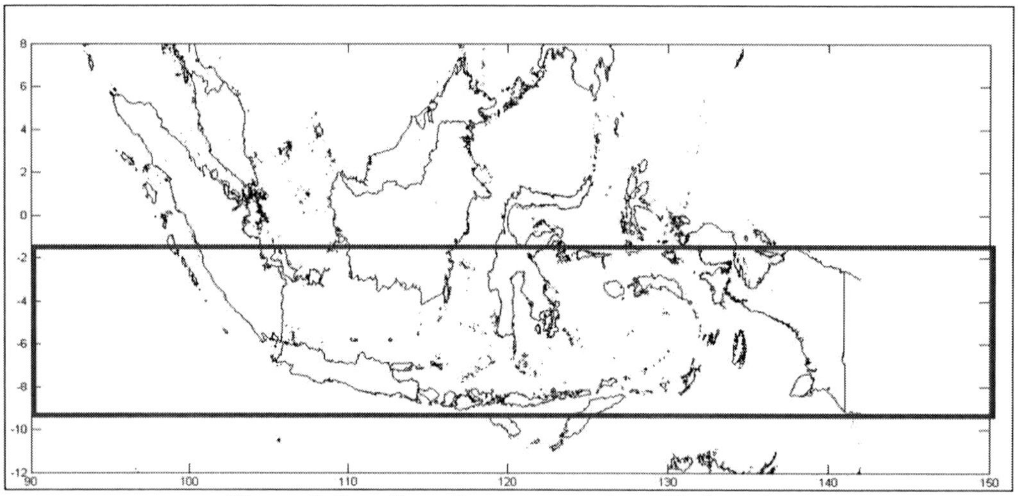

Figure 1. Location of monsoon region of Indonesia

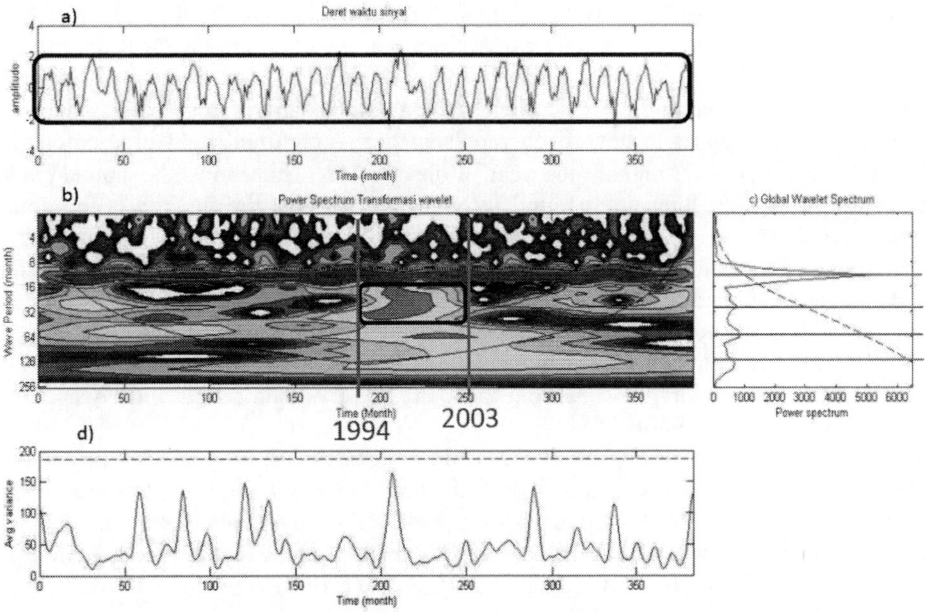

Figure 2. Time series of (a), frequency spectrum (b), and variance (c) based on wavelet method of monthly climatology of OLR (1980-2011)

Meanwhile, in the PW data (Figures 3 (b) and (c)), the dominant periods that emerged were annual, 10.6 yr, 5.3 yr. The 10.6 yr signal strengthened in 1996-2011. For a 2.5 yr, the strongest signal was shown in 1994-2003. The wavelet results for PW data also show that the maximum value of PW after 2010 shows a variance that exceeds its threshold value (Fig. 3 (d)). These findings show different realities to the OLR data.

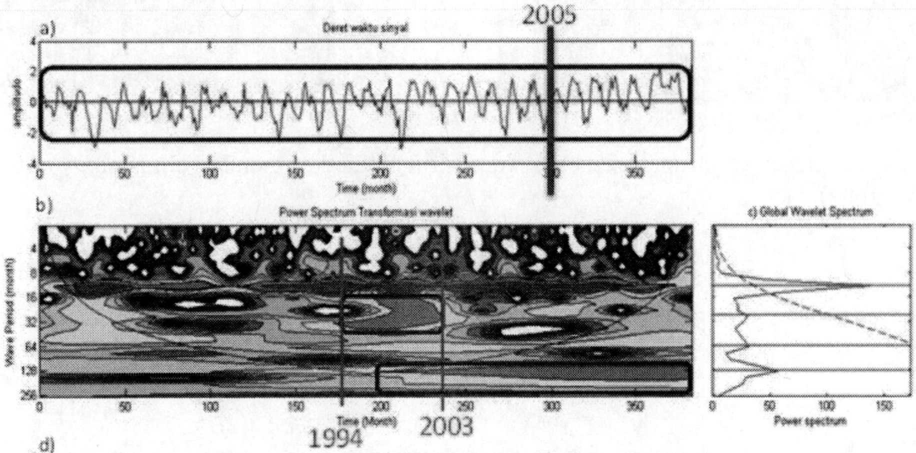

Figure 3. Time series of (a), frequency spectrum (b), and variance (c) based on wavelet method of monthly climatology of PW data (1980-2011)

Different results are exhibited by the S-Transform method, where OLR and PW both show the strongest signal about 5-6 years. This is evident from the five cycles in the OLR data and six cycles in the PW data for 30 years of observation shown by red (**Figure 4**). The frequencies of these five to six times were probably related to the ENSO phenomenon that has an irregular cycle between 2-5 years.

Other signals that appear to have been represented by green and light blue colours, are seen as cycles that always appear throughout the year of observation. This shows the annual cycle associated with monsoon. The interesting point found through the S-Transform method is the dominant signal that appears in 2010 to 2011. Seen in **Figure 4**, both OLR and PW data show the same signal.

Furthermore, to test the consistency of OLR and PW data, cross-correlation between OLR and PW period data were generated from wavelet and S-Transform method. The cross-correlation results describe that both OLR and PW have lag data period of 5 months. This shows the difference in phase calculation between S-Transform and wavelet against OLR and PW data is 5 months. In addition, for the same lag phase (5 months) it seems that the OLR and PW data consistently over 30 years have a correlation that is opposite (**Figure 5**).

Figure 6 explains that by using the EMD method, more detailed signals can be seen in both OLR and PW data. The EMD results in the OLR data show 1-3 months (IMF 1), 6 months (IMF 2), 1.3 years (IMF 3), 2.9 years (IMF 4), 4 years (IMF 5 and IMF 6), 6.4 years (IMF 7), 10.6 years (IMF 8), 15 years (IMF 9) and 30 years (IMF 10). In the PW data, periodic signals are 1-3 (IMF 1), 6 months (IMF 2), 1.3 years (IMF 3), 2.6 years (IMF 4), 3.75 years (IMF 5), 5 years (IMF 6), 7.5 years (IMF 7), 10 years (IMF 8), 15 and 30 years (IMF 9 and 10).

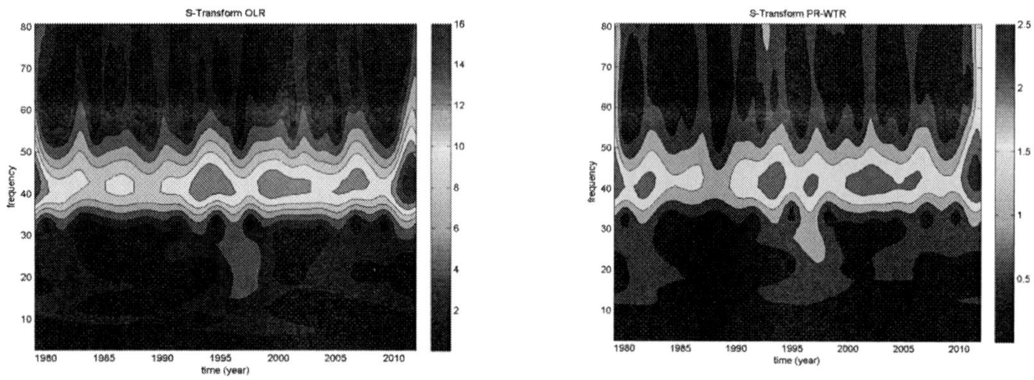

Figure 4. Frequency spectrum based on S-Transform method of monthly climatology of OLR (left) and PW (right) data.

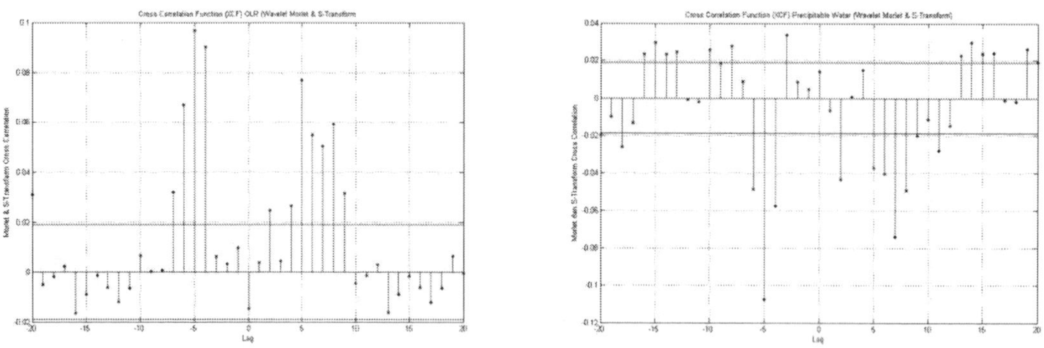

Figure 5. Cross-correlation of OLR (left) and PW (hand right) between frequancy spectrum of wavelet and S-Transform.

IMF 2 and 3 present the 6 monthly and annual signals associated with monsoon activity in southern Indonesia so as to impact the formation of Semi Annual Oscillation and Annual Oscillation. It also appears that signal strengthening for OLR data does not correspond consistently with PW data. Reinforcement of seasonal signals on OLR data generally occurs over a span of two consecutive years (1984-86, 1987-89, 1999-01) except in the period 1992-96 and 2004-08. Similarly, PW data, signal strengthening with the longest duration (8 years) only occurred in 1998-2006. In addition, strong signals took place in 1983-86, 1994-96, 2009-11. For the period 2009-2011, the facts show that there has been a wet in dry season in the Indonesian monsoon region even though in 2009 El Niño occurred in the Pacific Ocean. Some of the events that allegedly can be the cause of the a wet in dry season is because the MJO and tropical cyclone activity that has a higher frequency that is 30-96 days during that period [6].

The Intraseasonal Cycle can be identified through both OLR and PW data through IMF 1 (Figure 6). In fact, the increasing trend of convective activity and the increase in total humidity in atmospheric columns has occurred since 2000 to 2011. This can be demonstrated through the trend pattern of OLR and PW data through IMF 11 (**Figure 7**).

While IMF 4 and IMF 5 are related to the occurrence of ENSO in the Pacific Ocean with a period of 2-4 years, similarly, IMF 6 represents ENSO which has a longer period of 6 years. The relationship between IMF 4 (2 years) and ENSO events is shown in **Figures 8** and **9**. Linear relationships can be identified in the case of strong El Niño in 1997. PW shows a very large value reduction yet at the same time the value of OLR has increased enormously.

The IMF 8 (10 years) is suggested to be related to the 11th annual solar cycle. The 11-yr solar cycle has had an impact on climate change and weather in Europe by altering atmospheric circulation in the North Atlantic Oscillation and Arctic Oscillation indirectly by extending the hot anomalous memory that occurs in the ocean so as to generate a mutual response in the atmosphere [24-25]. It is also possible to explain how the 11yr solar cycle can affect the Asian winter monsoon [22]. In the Pacific Ocean, solar cycles can affect the formation of temperature hiatus conditions due to changes in ocean temperatures in the tropical Pacific [28].

Figure 7 showed IMF 8 (10 years), 9 (15 years), and 10 (30 years) can also be related to the ENSO phenomenon which has low frequency. It proves that ENSO also has lower frequency signals of 13 and 32 years, hereinafter referred to as Multivariate ENSO Index (MEI) as a phenomenon that has a global impact on various climate anomalies, ENSO itself has multi-decadal evolution and dynamics [29-30]. However, there is another possibility of PDO occurrence that were shown by IMF 9 (15 years) and IMF 10 (30 years) through OLR and PW data regarding to previous research that PDO has a period of 16.7 and 32.8 years [29]. Moreover, ENSO and PDO have mutual relations and can modulate each other [31].

This is reinforced by the findings in this study which show that IMF 5 (3-4 years), 6 (5-6 years), and 9 (15 years) are mutually correspondent. Although the same is not shown by IMF 4 (2 years) and IMF 10 (30 years). Thus, the monsoon region in southern Indonesia is merely sensitive to MEI that has 15 years oscillations. Another opinion was expressed by Chelliah which suggests that the 15-20 years relate to another phenomenon called Tropical Multi-Decadal Mode which is evident in precipitation data during the DJF period in the western Pacific region (including Indonesia) [4].

Figure 9 shows that over a 12 year period, two PW wave cycles with peak cycles occurred in 1989-1990 and 2003-2004. Compared with the 11-yr solar cycle data, it appears that at the time of maximum PW during that period, the sun reached the maximum of cycles that occurred in 1989-1992 and 2000-2004.

IOP Conf. Series: Earth and Environmental Science **166** (2018) 012044 doi:10.1088/1755-1315/166/1/012044

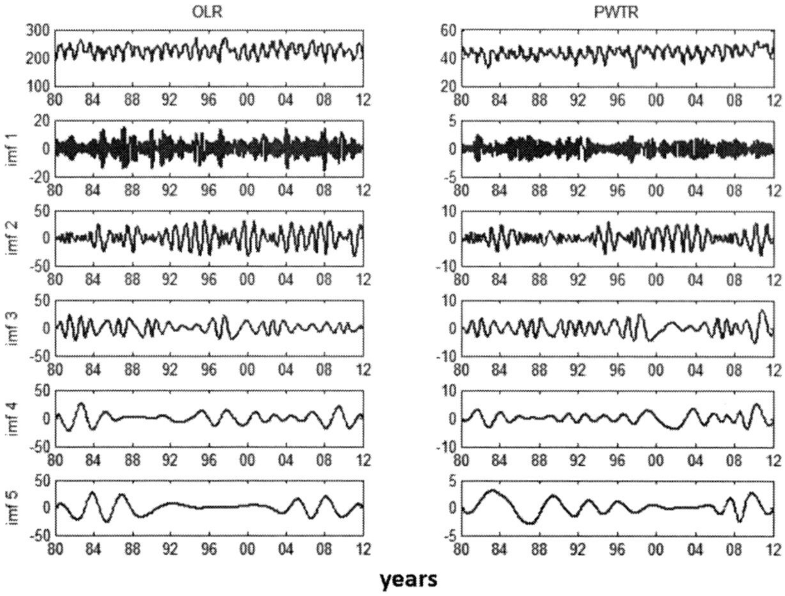

Figure 6. IMF 1-5 based on EMD method for OLR (left) and PW (right) data

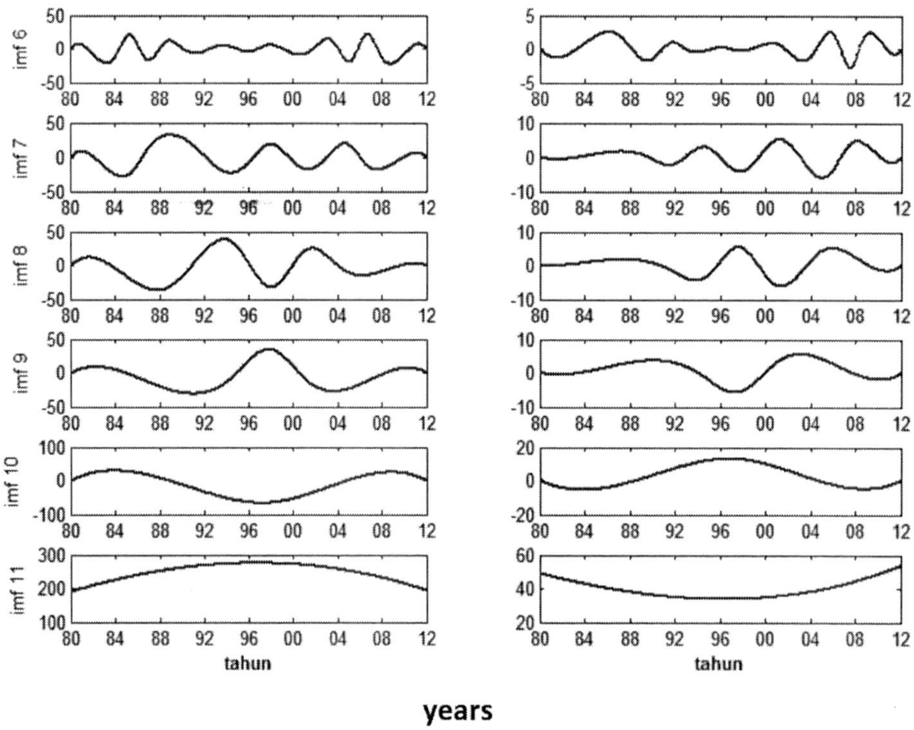

Figure 7. IMF 6-11 based on EMD method for OLR (left) and PW (right) data

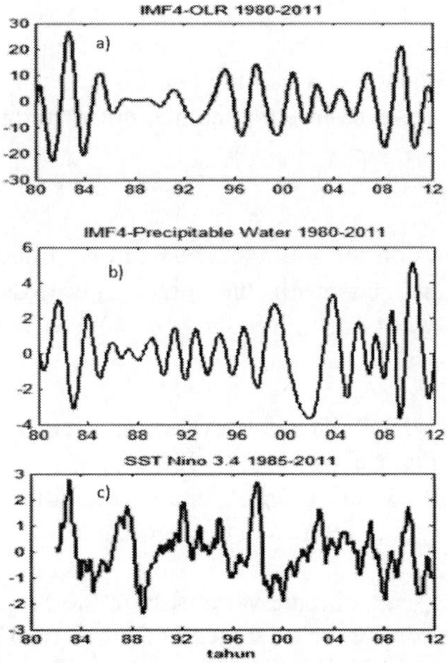

Figure 8. IMF 4 OLR (a), PW (b) compared with SST Nino 3.4 (c)

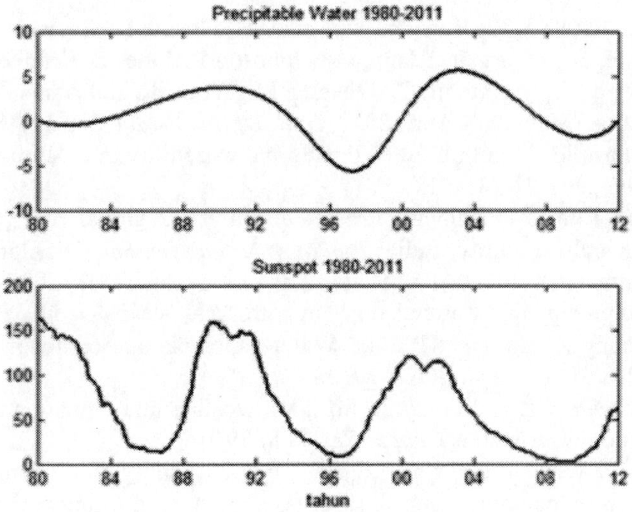

Figure 9. PW and sunspot monthly data of 1980-2011.

4. Conclusion

The wavelet and S-Transform results show that the one-year signal is the dominant period of OLR in southern Indonesia. Other signals that appear are periods of 2, 5, and 10 years. In the wavelets, the 2, 5, and 10 years signals in the OLR data have the same power spectrum. However, in the PW data, the dominant periods that emerged were 1, 10, 2 and 5 years, respectively. The wavelet results for the PW data also show that the maximum value of PW after 2010 shows an anomaly that exceeds the climatological average of the data for 30 years. The S-Transform results of OLR and PW data also clearly show signal strengthening after the 2010 period. While the EMD results show in detail the 1-3 monthly cycles associated with MJO, 6 monthly and 1 yearly associated with monsoon activity, 2-3 yearly related with ENSO short-term, 5-7 years ENSO long periods, 10 years that can be attributed ENSO (EMI) or the 11-yr solar cycle. Signals with lower frequencies of 15 and 30 years can be associated with EMI and PDO phenomena as they can affect each other. This study also found that OLR and PW have consistently for 30 years maintained correlations of the opposite shown from the results of the cross-correlation.

Acknowledgments
The author would like to thank Indonesia Endowment Fund for Education, Ministry of Finance of the Republic of Indonesia for their scholarship in the completion of doctoral studies in Earth Sciences, Faculty of Science and Technology of Earth, Bandung Institute of Technology.

5. References

[1] Katz RW and Brown BG 1992 Extreme events in a changing climate: Variability is more important than averages *Clim. Chang.* **21** 289-302
[2] Lau K-M and Chan PH 1983 Short-term climate variability and atmospheric teleconnections from satellite observed outgoing longwave radiation, part I: Simultaneus relationship *J. Of Atm. Sci.* **40** 2735-2750
[3] Graham NE 1994 Decadal-scale climate variability in the Tropical and North Pacific during the 1970s and 1980s: Observations and model results *Clim. Dyn.* **10** 135-162
[4] Chelliah M and Bell GD 2004 Tropical multidecadal and interannual climate variability in the ncep-ncar reanalysis *J. of Clim.* **17** 1777-1803
[5] England MH, Ummenhofer CC and Santoso 2006 Interannual rainfall extremes over southwest western australia to indian ocean climate variability *J. of Clim* **19** 1948-1969
[6] Waliser DE, Moncrieff MW, Burridge D, Fink AH, Gochis D, Goswami BN, Guan B, Harr P, Heming J, Hsu H.-H., J Christian, Janiga M, Johnson R, Jones S, Knippertz P, M Jose, Nguyen H, Pope M, Serra Y, Thorncroft C, Wheeler M, Wood R and Yuter S 2012 The "year" of tropical convection (May 2008-April 2010) *Bull. Of Amer. Met. Soc.*1189-1218
[7] Trenberth KE, Fasullo J, Smith L 2005 Trends and variability in column-integrated atmospheric water vapor *Clim. Dyn.* **24** 741-748
[8] Amenu GG and Kumar P 2005 NVAP and reanalysis-2 global precipitable water products: Intercomparison and variability studies *Bull of America Met. Soc.* 245-256
[9] Mears CA, Santer BD, Wentz FJ, Taylor KE and Wehner MF 2007 Relationship between temperature and precipitable water changes over tropical oceans *Geoph. Res. Lett.* **34** 1-5
[10] Dessler AE, Zhang Z and Yang P 2008 Water-vapor climate feedback inferred from climate fluctuations, 2003-2008 *Geoph. Res. Lett.* **35** 1-4
[11] Haar THV, Bytheway JL and Forsythe JM 2012 weather and climate analyses using improved global water vapor observations *Geoph. Res. Lett.* **39** 1-6
[12] Bock O, Willis P, Wang J and Mears C 2013 A high-quality, homogenized, global, long-term (1993-2008) doris precipitable water data set for climate monitoring and model verification *J. of Geop. Res.* 7209-72031
[13] Yulihastin E and Hermawan E 2012 Annual migration of monsoon over Indonesia maritime continent based on olr data *J. Tek. Indo.* **35** 1-12
[14] Susskind J, Gyula M, Lena I and Norman GL 2012 Interannual variability of outgoing longwave radiation as observed by AIRS and CERES *J. of Geop. Res.* **117**, 1-8
[15] Schmidt GA, Ruedy RA, Miller RL and Lacis AA 2010 Attribution of the present-day total greenhouse effect *J. of Geop. Res.* **115**, 1-6
[16] Tang YY, Yang LH, Liu J and Ma H 2000 Wavelet theory and its application *World Scient* **36** p 53
[17] Stockwell RG 2007 A basis for efficient of the s-transform *Dig Sig Proc* **17** 371-393
[18] Coughlin K, and Tung KK 2004 Eleven-year solar cycle signal throughout the lower atmosphere *Geophysics Research Letter* **109** DOI:10.1029/2004

IOP Conf. Series: Earth and Environmental Science **166** (2018) 012044 doi:10.1088/1755-1315/166/1/012044

[19] Karthikeyan L and Kumar DN 2013 Predictability of nonstationary time series using wavelet and EMD based ARMA models *J. of Hyd.* **502** 103-119
[20] Araghi AA, Mousavi M, Baygi A, Adamowski JB, Malard JB , Nalley DB, and Hasheminia SMA 2015 Using wavelet transforms to estimate surface temperature trends and dominant periodicities in Iran based on gridded reanalysis data *Atm. Res* **155** 52-72
[21] Ren H-L, Jin F-F, Stuecker MF and Xie R 2013 ENSO Regime change since the late 1970s as manifested by two types of ENSO *J. of the Met. Soc. Japan* **91** 835-842
[22] An S-I and Jin F-F 2000 An eigen analysis of the interdecadal changes in the structure and frequency of ENSO Mode *Geophys. Res. Lett.* **27** 2573-2576
[23] An S-I and Wang B 2000 Interdecadal change of the structure of the ENSO mode and its impact on the ENSO frequency *J. Clim.* **13** 2044-2055
[24] Fedorov AV and Philander SG 2000 Is El Niño changing? *Sci.* **288** 1997-2002
[25] Chen W and Zhou Q 2012 Modulation of the arctic oscillation and the East Asian winter climate relationships by the 11-year solar cycle *Adv. In Atm. Sci.* **29** 217-226
[26] Gray LJ, Scaife AA, Mitchell DM, Osprey S, Ineson S, Hardiman S, Butchart N, Knight J, Sutton R, Kodera K 2013 A lagged response to the 11 year solar cycle in observed winter Atlantic/European weather patterns *J. of Geoph. Res. Atm.* **118** 13405-13420
[27] Scaife AA, Ineson S, Knight JR, Gray L, Kunihiko K and Smith DM 2014 A mechanism for lagged North Atlantic response to solar variability *Geop. Res. Lett.* **40** 434-439
[28] Kosaka Y and Xie S-P Recent global-warming hiatus tied to Equatorial Pacific surface cooling nature **501** 403-407
[29] Lee HS, Yamashita T and Mishima T 2012 Multi-decadal variations of ENSO, the Pacific decadal oscillation and tropical cyclones in the Western North Pacific *Prog. in Ocean* **105**, 67-80
[30] Wang S, Huang J, He Y and Guan Y 2014 Combined effects of the Pacific decadal oscillation and El Niño Southern oscillation on global land dry-wet changes *Sci Rep* **4** 1-8
[31] Capotondi A and Sardeshmukh PD 2017 Is El Niño really changing? *Amer. Geop. Uni.* 10.1002/2017GL074515 1-15

IOP Conf. Series: Earth and Environmental Science **166** (2018) 012045 doi:10.1088/1755-1315/166/1/012045

Economic Valuation of Nipa Palm (*Nypa fruticans* Wurmb.) Sap as Bioethanol Material

Imawan Wahyu Hidayat[1]

[1]Cibodas Botanical Garden, Indonesian Institute of Sciences (LIPI), P.O. Box 19 SDL, Sindanglaya-Cipanas, Cianjur 43253, West Java, Indonesia

Email: imawan.wahyu.hidayat@lipi.go.id

Abstract. Nipa palm (*Nypa fruticans* Wurmb.) widely recognized as one of the biological resources with various aspect of utilization. Nipa palm sap can be processed into bioethanol. Bioethanol also has a quite high selling price, especially for the fulfilment of energy and for fossil fuels substitution. The main purpose of the study was to evaluate the economic value of nipa sap for further processed into bioethanol, in order to improve the income and job opportunity of coastal households. This research was conducted at Sungsang, South Sumatera with focusing on nipa palm sap further processing into bioethanol and its impact on increasing household income. This research was statistically testing the relationship between potential production of bioethanol, average revenue and labour recruitment. Based on the 2013 data, potential production of bioethanol was capable to contribute of nett income about IDR 507,940 to 3,503,280 ha^{-1} $month^{-1}$ or IDR 6,095,274 to 42,039,360 ha^{-1} $year^{-1}$. Therefore each hectare of nipa palm can generate labour recruitment up to 3.7 person ha^{-1}. This assessment was expected to provide a better description of the importance of local biological resources utilization of the nipa palm, in order to improve community welfare and the efforts of biodiversity conservation on a local scale.

1. Introduction

The total primary energy share of around 78.9% was produced from fossil fuel that precludes nuclear, hydro, biofuel and other energy sources [1]. For comparison, in 1973, fossil fuel percentage of the total primary energy share was 86.7%. At the global level, 57.7% of the energy is accounted for transportation system [2]. Petroleum-based fuels lead to environmental pollution, which results in global warming, health hazards, and ecological imbalances [3]. The shift towards sustainable and environmentally friendly energy sources has generated significant interest in developing biofuel production from plant biomass [4].

In 2013, Indonesian oil consumption reached 1.61 million barrels per day, yet the country is "only" capable to produce as much as 874.79 thousand barrels of crude oil per day [1]. The dependence of the energy needs on fossil fuels of Indonesia is also seen in the proportion of the total fossil fuel consumption by 73% compared to other sources, such as biomass and other renewable sources, by 27% in 2012. This certainly implies concerns for policymakers to encourage explorations of alternatives that would be viable and regenerative to attain sustainability. Renewable energy opened up prospects for appropriate resource conservation and an eco-friendly solution directed to energy security [5].

Indonesia with its high biodiversity resources has high potentials for developing new and renewable energy derived from plants. Bioethanol derived from nipa palm (*Nypa fruticans* Wurmb.) sap is one of them. Ethanol which is produced by nipa sap is better than sugarcane, even other sugar sources such as cassava, coconut and potato [6]. Nipa sap is a potential material to be processed into bioethanol [7][8]. Bioethanol is one of the sources of renewable energy which can replace or a mixture of fossil fuels, by its widely used in beverages, cosmetics, in the health field as an antiseptic substance, a solvent, as well as an industrial raw material. The total nipa sap chemical composition is 19.5 wt%, mainly consisting of sucrose, glucose and fructose [7]. The potential of nipa sap that can be generated is ranged from 0.4 to 1.2 L d^{-1} per palm [8]. *N. fruticans*, an underutilized palm, was assessed recently for bioethanol yield by several authors. Tamunaidu et al. [8] conducted a detailed study for ethanol production from *N. fruticans*. If a sap yield of 0.5–1.0 L $palm^{-1}$ day^{-1} is assumed, with a tapping period of 100 day $year^{-1}$, and with a population of 1,000 palm ha^{-1}, then an annual sap yield of 50,000-100,000 L ha^{-1} $year^{-1}$ could be achieved from this palm. In alcoholic fermentation, sap with a minimum sugar content of 15 wt.% is converted to ethanol with a yield of ~0.48 gr-ethanol $gr\text{-}sugars^{-1}$. Consequently, the annual ethanol yield is estimated to be 4,550–9,100 L ha^{-1} $year^{-1}$. Using this result for 1.23 Mha of *N. fruticans* plantations in Indonesia, Papua New Guinea, Malaysia and the Philippines [7], the annual ethanol production may reach 5,590–11,180 ML $year^{-1}$ [9]. Because palms do not require much fertilizer or particular care, and since they are able to live long and can be tapped for up to 50 years, their utilization would be more environmentally-friendlier compared with corn and sugarcane. Sustainable bioethanol production from palm sap could be achieved with the proper integration of sap exploitation alongside existing palm industries and plantations.

Nevertheless, exploitation of mangrove areas as nipa habitat which continuously occurred will be potentially reducing the diversity of plant species including its benefits. Mangrove as a major habitat for nipa continuously harmed, as happened in most of the northern coast of Sumatra. Mangrove ranges in the Banyuasin (including Sungsang) were decreased by 20,546.5 ha from 1992 to 2003 [10]. The data in 2006 were also discovered that 55.4% (158,989.39 ha) mangrove region in Banyuasin II area (including in Sungsang) categorized as moderate to severe harmed and only 44.6% (127,983.57 ha) categorized in great condition [11].

Therefore, it needs a habitat conservation strategy to be able to build collaborative ecological interests and socioeconomic benefits for coastal households. This research is conducted in order to conserve the existence of nipa in its habitat as well as to assess the economic value as ethanol producer in Sungsang-South Sumatera, as one of wide nipa habitat. This research calculated the potential income for households (i.e. average revenue) and labour recruitments. This research is expected to advance the description of *N. fruticans* benefits as a non-timber commodity, which at the end deforestation and land degradation that occurs in the mangrove forest can be reduced.

2. Methods

The study was conducted in Sungsang, which administratively located in Banyuasin II District, Banyuasin Regency, South Sumatra Province. The area of study is located at 2° 16' 33" S to 2° 30' 31" S and 104° 45' 10" E to 104° 55' 4" E. Topographically, the site is a swampy area with a moderate tidal influence.

2.1. Data acquisition

The number of fruit stalks was collected from Hidayat [12]. Sampling acquisition of nipa was conducted on eight-plots with three sub-plots repetition in each plot [12]. Each sub-plot samplings were measured in 10 x 10 square meters. The Plot I was adjacent to agricultural land and plantations, Plot II was adjacent to plantations. Plot III and VII were located contiguous with the settlement; Plot IV, V and VI were as undisturbed sites type; and, Plot VIII was located adjacent to agricultural land, plantations and the settlement. Data were collected from February to March 2013 [12]. If a sap yield of 0.5–2.0 L $palm^{-1}$ day^{-1} is assumed, with a tapping period of 100 day $year^{-1}$ [8, 13], then, further processing of nipa sap would be capable to generate 8.98% [14] to15% [8] per volume of bioethanol, then the potential bioethanol which can be produced is = [8.98% x nipa sap production ≤ potential bioethanol ≤ 15% x nipa sap production] L $ha^{-1}day^{-1}$. Therefore, the potential bioethanol, which can be produced in Sungsang, is shown in **Table 1**.

2.2. Data analysis

Based on 2013 data [15], the price of bioethanol was about IDR 9,200 to 9,400 L^{-1}. Therefore, potential total revenue (*TR*) would be 0.12 x IDR 9,200 ≤ *TR* ≤ 136.65 x IDR 9,400 or IDR 1,104 to 1,284,510 ha^{-1} day^{-1}. If total production cost (*TC*) to produce bioethanol is approximately IDR 3,000 [15], then *TC* would be around IDR 360 to 409,950. Therefore, nett income (*NI*) received would potentially be *TR* – *TC* or about IDR 744 to 874,650 ha^{-1} day^{-1}.

In order to evaluate the expediency of labour recruitment, the previous values are compared with South Sumatera Province minimum regional wage. Based on 2013 data [15], minimum regional wage was IDR 1,974,000 $month^{-1}$. Therefore, labour recruitment can be estimated as = *NI* in a year /12/ 1,974,000.

Table 1. Potential bioethanol production in the field

Plot Number[a]	Ave. number of fruit stalks[a] (ha^{-1})	Nipa sap[b] (L ha^{-1} day^{-1})		Bioethanol production[c] (L ha^{-1} day^{-1})	
		Min.	Max.	Min.	Max.
I	354.22	177.11	708.44	15.90	106.27
II	198.05	99.03	396.10	8.89	59.42
III	392.32	196.16	784.64	17.62	117.70
IV	269.83	134.92	539.66	12.12	80.95
V	455.49	227.75	910.98	20.45	136.65
VI	63.73	31.87	127.46	2.86	19.12
VII	15.33	7.67	30.66	0.69	4.60
VIII	2.67	1.34	5.34	0.12	0.80

[a] Primary data generate from Hidayat [12].
[b] Secondary data generate from Tamunaidu et al. [8] that 0.5 L $palm^{-1}$ day^{-1} and based on Matsui et al. [13] that 2.0 L $palm^{-1}$ day^{-1}.
[c] Secondary data generate from Trisasiwi et al. [14] that nipa sap would be capable to generate 8.98% per volume of bioethanol and based on Tamunaidu et al. [8] generate up to 15% per volume of bioethanol.

In order to evaluate the correlation between the average number of fruit stalks (x) and *NI* (y), and the average number of fruit stalks (x) and the labour recruitment (y'), a Pearson correlation coefficient test was conducted using the following formula:

$$r = \frac{n\sum x_i y_i - \sum x_i \sum y_i}{[(n\sum x_i{}^2(\sum x_i)^2]^{1/2}\ [(n\sum y_i{}^2(\sum y_i)^2]^{1/2}} \quad (1)$$

Because both of the results were different in value and unit, the number will be transferred in natural logarithm, $\boldsymbol{ln(x)}$ and $\boldsymbol{ln(y)}$, and $\boldsymbol{ln(x)}$ and $\boldsymbol{ln(y')}$ before this test. In order to evaluate the correlation coefficient (r) between the average number of fruit stalks (x) and the labour recruitment (y') use the above formula with y replaced by y'. To statistically test, then arranged hypothetic that H_0: $r = 0$; H_1: $r \neq 0$, $\alpha = 0.01$. The value of $r = 0$ means that there isn't any correlation between the number of fruit stalks (x) and NI (y), and the number of fruit stalks (x) and the labour recruitment (y'), in reverse, the value of $r \neq 0$ means that there is a correlation. The r value will not exceed from $|1|$ when approaching to the value of -1 means that there is a negative linear relationship, and when approaching to the value of 1 means that there is a positive linear relationship. In order to assess the degree of significance, a t-test in *ANOVA* with $\alpha = 0.01$ was also conducted. If $r \neq 0$, p-value $< \alpha$ means that the correlation between two factors is significant, and if p-value $< \alpha$ means that the correlation between two factors is not significant.

The results were expected to be a general description of advantages of the nipa palm sap utilization in order to increase the welfare of the community. If the result significant shows the benefits, from the economic viewpoint, even though tapping of nipa saps could be labour-intensive, it will create a considerable number of jobs and help generate sustainable livelihood [8]. As the result, deforestation and land degradation that occurs in the mangrove forest can be reduced.

3. Results

Nipa thrives well in the mangrove environment, favouring brackish water environments such as estuaries or shallow lagoons. Nipa also colonizes the upper tidal reaches of rivers, semi-liquid mud of estuaries. At the study area, the abundant number of nipa community were found in the area that has not been disturbed by human activities (such as plantation, settlements and other massive infrastructures). Based on **Table 1**, the data can generate the potential economic value of bioethanol derived from nipa palm sap in a year as described in **Table 2**. The values were various, but generally, the abundant number of nipa communities showed high value potential economic advantages.

For example, in plot samples I to V potential nett income can be generated about IDR 5,513,288 to 87,454,080 ha^{-1} $year^{-1}$ or about IDR 459,441 to 7,287,840 ha^{-1} $month^{-1}$. In addition, the number of sap production days and fruit stalk length was found to be highly correlated with sap yield [16]. This value also corresponds with potential value in other sites. The estimated annual ethanol yield from nipa sap in the east coast of Southern Thailand of 4,550-9,100 L ha^{-1} $year^{-1}$ [7] was as competitive as 5,300-6,500 L ha^{-1} $year^{-1}$ estimated in sugarcane and 3,100-3,900 L ha^{-1} $year^{-1}$ for corn as reported by Marris [17]. This demonstrates the wide ecological amplitude of the nipa palm [18,19]. Nipa palm populations collected from China, Thailand, Japan and Vietnam showed low genetic diversity [20]. Most of the nipa palm populations were found in the more brackish mangrove forest strips, situated further inland and away from the direct exposure to pure seawater [21].

Table 2. Potential economic value of bioethanol from nipa sap in a year

Plot Number[a]	Bioethanol production[b] (L ha^{-1} $year^{-1}$)		*TR*[c] (IDR ha^{-1} $year^{-1}$)		*TC*[d] (IDR ha^{-1} $year^{-1}$)		*NI*[e] (IDR ha^{-1} $year^{-1}$)	
	Min.	Max.	Min.	Max.	Min.	Max.	Min.	Max.
I	1,590.45	10,626.6	14,632,140	99,890,040	4,771,350	31,879,800	9,860,790	68,010,240
II	889.24	5,941.5	8,181,008	55,850,100	2,667,720	17,824,500	5,513,288	38,025,600
III	1,761.52	11,769.6	16,205,984	110,634,240	5,284,560	35,308,800	10,921,424	75,325,440
IV	1,211.54	8,094.9	11,146,168	76,092,060	3,634,620	24,284,700	7,511,548	51,807,360
V	2,045.15	13,664.7	18,815,380	128,448,180	6,135,450	40,994,100	12,679,930	87,454,080
VI	286.15	1,911.9	2,632,580	17,971,860	858,450	5,735,700	1,774,130	12,236,160
VII	68.83	459.9	633,236	4,323,060	206,490	1,379,700	426,746	2,943,360
VIII	11.99	80.1	110,308	752,940	35,970	240,300	74,338	512,640

[a] Primary data generate from Hidayat [12].
[b] Secondary data generate from Tamunaidu et al. [8] that tapping period of 100 day $year^{-1}$.
[c, d, e] Secondary data modified from *Bappeda Kabupaten Banyuasin* [15] and self-data measurement.

Tapping palm for bioethanol production can generate work and income for local farmers and encourage them to maintain the local palm population rather than replacing it with other industrial crops. In order to evaluate labour recruitment, the value can be generated from South Sumatera Province minimum regional wage compared to the value of *NI*. To compare the nett income and the number of labour recruitment with the minimum regional wage was IDR 1,974,000 month^{-1}based on 2013 data [15], **Table 3** is presented below.

Table 3. Potential monthly labour recruitment based on nett income compared to minimum regional wage.

Plot Number[a]	Labour recruitment[b] (person ha^{-1})	
	Min.	Max.
I	0.42	2.87
II	0.23	1.61
III	0.46	3.18
IV	0.32	2.19
V	0.54	3.69
VI	0.07	0.52
VII	0.02	0.12
VIII	0.003	0.022

[a] Primary data generate from Hidayat [12].
[b] Secondary data modified from *Bappeda Kabupaten Banyuasin* [15] and self-data measurement.

The evaluations showed that the potential labour recruitment on the utilization of nipa palm sap into bioethanol can generate up to 3.69 person ha^{-1} month^{-1}. This number is significant to increase the welfare of the households when compared to available or existing job field. In addition to growing nipa for its bioethanol, there are other advantages of growing nipa. The continuous productivity of nipa means no displaced labour, which is one of a major problem in sugarcane ecosystem [6]. Furthermore, replanting and rotation do not interrupt production of nipa. Other advantages are no bagasse disposal problem and nipa does not compete with other crops for agricultural land except where total reclamation is undertaken on mangrove land [6]. Tapping of nipa saps could be labour-intensive and it will create a considerable number of jobs and help to generate sustainable livelihood for the coastal communities [8, 22].

4. Discussions

These advantages were also observable through correlation between the average number of fruit stalks (**Table 1**) and *NI* (**Table 2**). The test using Pearson correlation coefficient, the value of r = 1 (reject H_0), with p-value = 0.003 < α = 0.01 shows high significance. This can conclude that the number of fruit stalks was linearly a significant influence on the nett income of the households at a high level. The same results were also achieved through correlation between the average number of fruit stalks (**Table 1**) and labour recruitment (**Table 3**). Based on the test using Pearson correlation coefficient, the value of r = 0.99 (reject H_0), with p-value = 0.003 < α = 0.01 also shows high significance. This also can be described that the number of fruit stalks was linearly a significant influence on labour recruitment in a high level. These results indicate that the importance of preservation of nipa palm fruit stalks to produce benefits to the communities. Therefore, mangrove ecosystem as *N. fruticans* habitat must be immediately conserved in order to sustain these resources. Since nipa palm sap is still underutilized for its further potential uses such as bioethanol, its survival and sustainability are threatened by other human needs.

Palms can yield sap for many years: 30 years for *B. flabellifer*, 20 years for *C. nucifera*, 50 years for *N. fruticans*, and 25 years for *P. sylvestris*. Logging and replanting, which are necessary for the harvesting sugar and starch crops can, therefore, be reduced for tapped palms [9]. Harvesting of traditional energy crops often produces large volumes of biomass wastes such as straw, leaves and bagasse [6, 23]. Sugar extraction from sugarcane involves extra investment in equipment and energy for compression of the stem [6]. Thus, direct tapping of palms for bioethanol production may become more convenient and economical than conventional crops. Tapping palm sap for bioethanol production may compete with other palm products. For example, tapping female flowers sacrifices fruit formation. Tapping may also reduce fruit yield from the palm. Sap and other palm products should be considered to achieve sustainable bioethanol production. Palms that yield sap as the main product are suitable for ethanol production.

Free sugars in palm sap can be converted directly into ethanol without complex pre-treatment and hydrolysis that are required for starch or lignocellulosic feedstock [4]. Moreover, fermentation of lignocellulosic materials requires an external supply of nutrients to maintain microorganism activity [24]. In contrast, palm saps contain inorganics, amino acids, and vitamins, which can play their role as nutrients during fermentations [25]. Nipa sap and oil palm trunk sap were shown to produce ethanol without nutrient supplements [25]. Thus, using palm sap instead of lignocellulosics could reduce nutrient costs, although additional fermentation time may be required to achieve a high ethanol yield [8].

Generally, the level of income and education of the coastal communities which relatively low are some of the weakness factors and an internal constraint in managing sustainable mangrove ecosystems. Where large-scale removal and degradation of mangroves have already taken place [26], then the recognition of the role of mangrove in the present policy scenario is very important. Institutions (i.e. government and non-government organizations) as one component of mangrove ecosystem management play an important role in triggering and enhancing the roles as well as the community. Legislations, good governance and role of stakeholders are important to avoid conflict of interest between institutions. Proper protections are needed as considering this mangrove ecosystem as its habitat are vulnerable to human interest, for example for ponds. The habitat needs to be protected as an integral part of mangrove forests. Such studies are useful for better understanding of goods and services provided by mangroves. This may influence the awareness and perception of the local people as well as the policymakers to go along way in eliciting support for the conservation of mangrove ecosystems [26].

These indicative results of the study suggest that contribution of nipa sap is of significant economic importance, as it increases the resilience and sustainability of the local economy. This study can be an initial assessment of the usefulness of nipa palm to be further processed into a potential economic resource, which can increase income and labour recruitment to the coastal communities. Furthermore, this must be complemented with the further study of the impact of incremental changes in the mangrove ecosystems and their productivity.

5. Conclusions

The research has shown the advantages of the usefulness of nipa palm sap processed into bioethanol that from potential economic value can increase the income of the households. The highest potential annual nett income received by community up to IDR 87,454,080 or equally to IDR 12,679,930 month^{-1}. These benefits can also recruit the labour up to 3,69 person ha^{-1} month^{-1}. The result has also described that the more nipa communities can be found then significantly the potential income of the households could potentially be increased and more labour can be recruited. Based on the research, it is important to conserve *N. fruticans* habitat to continue providing benefits to coastal households, especially utilization of nipa sap as a non-timber product.

6. References

[1] U.S. EIA [United States of Energy Information Administration] 2014 http://www.eia.gov/

countries/cab.cfm?fips=ID [16.10.2017].
[2] Kumar A, Kumar K, Kaushik N, Sharma S and Mishra S 2010 Renewable energy in India: Current status and future potentials *Renew. Sustain. Energy Rev.* **14** 2434-42
[3] Gupta N and Kushwaha H 2011 Date palm as a source of bioethanol producing microorganisms In *Date Palm Biotechnology*, eds Jain SM, Al-Khayri JM and Johnson DV, Springer, Dordrecht 711-27
[4] Zabed H, Faruq G, Sahu JN, Azirun MS, Hashim R and Nasrulhaq Boyce A 2014 Bioethanol production from fermentable sugar juice *Sci. World J.* **2014** 957102
[5] Everett B, Boyle G, Peake S and Ramage J (Eds.) 2012 *Energy Systems and Sustainability: Power for a Sustainable Future*, 2nd ed, University Press, Oxford
[6] Hamilton LS and Murphy DH 1988 Use and management of nipa palm (*Nypa fruticans*, Arecaceae): A review *Econ. Bot.* **42** 206-13
[7] Tamunaidu P and Saka S 2013 Comparative study of nutrient supplements and natural inorganic components in ethanolic fermentation of nipa sap *J Jpn. Inst. Energy* **92** 181-6
[8] Tamunaidu P, Matsui N, Okimori Y and Saka S 2013 Nipa (*Nypa fruticans*) sap as a potential feedstock for ethanol production *Biom. Bioe* **52** 96-102
[9] Van Nguyen D, Harifara, Rabemalonontsoa and Saka S 2016 Sap from various palms as a renewable energy source for bioethanol production *Chem. Ind. Chem. Eng. Q.* **22** 355-73
[10] Ridho RM, Sundoko A and Ulqodry TZ 2006 Analysis of changes of mangrove area in Estuary of Banyuasin River, Sungsang and Upang South Sumatra Province using Landsat-TM Satellite Imagery *J. Pengelolaan Lingk. SDA* **4**(2) 11-18 [Indonesian]
[11] Indonesian Ministry of Forestry 2006 *Inventory and Identification of Mangrove Province of South Sumatra and Bangka Belitung Islands. Book I (Main)* Multima Krida Cipta, Jakarta [Indonesian]
[12] Hidayat IW 2015 Natural production potency of nipa (*Nypa fruticans*) sap as production commodity for bioethanol *Pros. Sem. Nas. Masy. Biodiv. Indon.* **1**(1) 109-13
[13] Matsui N, Okimori Y, Takahashi F, Matsumura K, Bamroongrugsa N 2014 Nipa (*Nypa fruticans* Wurmb.) sap collection in Southern Thailand: I. Sap production and farm management *Environ. Nat. Resour. Res.* **4**(4) 75-88
[14] Trisasiwi W, Asnani A and Setyawati R 2011 Optimization of bacterial doses and incubation time on bio-ethanol fermentation of nipah (*Nypa fruticans*) for biofuel energy *J. Life Sci.* **5** 1022-9
[15] Regional Development Planning Board and Capital Investment of Banyuasin Regency [*Bappeda Kabupaten Banyuasin*] 2014 *Banyuasin in Number 2013* (Banyuasin: Regional Development Planning Board and Capital Investment of Banyuasin Regency) [Indonesian]
[16] Rasco Jr ET, Ragas RG and Junio RG 2012 Morphological and sap yield variation in Nipa (*Nypa fruticans* Wurmb.) *Asia Life Sci.* **21** 123-32
[17] Marris E 2006 Sugar cane and ethanol: Drink the best and drive the rest *Nature* **444** 670-2
[18] Giesen W, Wulffraat S, Zieren M and Scholten L 2006 *Mangrove Guidebook for Southeast Asia*, Part VIII: Palms, Cycads & Pandans (Bangkok: FAO and Wetlands International)
[19] Teo S, Ang WF, Lok AFSL, Kurukulasuriya BR and Tan HTW 2010 The status and distribution of the Nipah palm, *Nypa fruticans* Wurmb (Arecaceae), In Singapore *Nat. Singapore* **3** 45-52
[20] Jian S, Ban J, Ren H and Yan H 2010 Low genetic variation detected within the widespread mangrove species *Nypa fruticans* (Palmae) from Southeast Asia *Aquat. Bot.* **92** 23-7
[21] Theerawitaya C, Samphumphaung T, Cha-um S, Yamada N and Takabe T 2014 Responses of Nipa palm (*Nypa fruticans*) seedlings, a mangrove species, to salt stress in pot culture *Flora* **209** 597-603
[22] Ame RB, Ame EC and Ayson JP 2011 Management of the nypa mangrove as a mitigating measure against resource over-utilization in Pamplona, Cagayan *Kuroshio Sci.* **5** 77-85
[23] Thatoi H, Dash PK, Mohapatra S and Swain MR 2016 Bioethanol production from tuber crops using fermentation technology: A review *Int. J. Sustain. Energy* **35** 443-68

IOP Conf. Series: Earth and Environmental Science **166** (2018) 012045 doi:10.1088/1755-1315/166/1/012045

[24] Ingram LO, Gomez PF, Lai X, Moniruzzaman M, Wood BE, Yomano LP and York SW 1998 Metabolic engineering of bacteria for ethanol production *Biotechnol. Bioeng.* **58** 204-14

[25] Kosugi A, Tanaka R, Magara K, Murata Y, Arai T, Sulaiman O, Hashim R, Hamid ZAA, Yahya MKA, Yusof MNM, Ibrahim WA and Mori Y 2010 Ethanol and lactic acid production using sap squeezed from old oil palm trunks felled for replanting *J. Biosci. Bioeng.* **110** 322-5

[26] Hussain SA and Badola R 2010 Valuing mangrove benefits: Contribution of mangrove forests to local livelihoods in Bhitarkanika Conservation Area, East Coast of India *Wetlands Ecol. Manage.* **18** 321-31

IOP Conf. Series: Earth and Environmental Science **166** (2018) 012046 doi:10.1088/1755-1315/166/1/012046

Agricultural Local Knowledge System: Lesson Learn from Rural Development In Mekarjaya Village, Kabupaten Purwakarta

M. A Heryanto[1], D Supyandi[1], Y Sukayat[1]

[1]Department of Agricultural Socio-Economic, Faculty of Agriculture, University of Padjadjaran (UNPAD) Campus Jatinangor, Sumedang-West Java 45363, Indonesia

E-mail: mahra.arari@unpad.ac.id

Abstract. Local knowledge termed traditional knowledge contributes to the quality of human life improvement in rural area. As the local knowledge which held by Mekar Jaya villagers, traditional agricuture practices emphasizes continuity and long-term practices. This article discussed about how local knowledge contribute to rural development as a whole of agricultural system. How external modern technology meets the accumulated traditional knowledge are elaborated in a social-ecological system using system archetype. The value of traditional agriculture practices have shifted gradually from food security purpose to food commercialization in line with the modernization technology. Social element in local agriculture practices looks more respect to environment than modern agriculture practices, while social element in modern agriculture tend much more to economic aspect. Finally, personal curiosity and willingness to learn, together with social networking among farmers and community, and supportive formal knowledge and governance structures, are learning point for successful sustainable agriculture development

1. Introduction

Traditional ecological knowledge represent accumulated experience in a long period of direct human contact with their environment. Local people developed their practice of resource use on their perspective based on experience and knowledge. Shortly, traditional knowledge is a complex interaction knowledge-practice-belief [1]. The result of many Traditional Ecological Knowledge systems shows that there are many components built the system: a) local observational knowledge of species and environmental phenomena, b) practice in the way people use their resources in activities, c) belief about how people fit into relationship to their ecosystem.

Traditional knowledge can be describe as informal knowledge because it is not a modern research product. Traditional and local knowledge systems are dynamic expressions in perceiving and understanding the world [2]. Traditional knowledge in the type of informal knowledge can be as local context, practice-based, traditional and indigenous and endogenous with self-education, learning by doing and social interaction. Informal farmers knowledge come from their experimentations and practical experiences on the farm, holistic approach, own by local community and locally specific solutions [3].

Traditional knowledge or informal knowledge as a holistic approach in agriculture practices and concern with the ecosystem in many cases, have strong relationship between peoples and the environment. Traditional ecological knowledge is the term used to describe the knowledge and beliefs that local peoples hold of their environments that is handed down through the generations include indigenous peoples. With or without ecological knowledge, traditional knowledge or informal knowledge should have a good correlation to environment and human. Traditional knowledge is adaptive by nature and holistic in outlook, practically it accumulates incrementally, tested by trial-and-error and transmitted to future generations orally or by shared practical experiences directly [4-6].

Since Green Revolution replicated in Asia, farmers in Indonesia have become dependent on synthetic fertilizers and agricultural chemicals input such as pesticide to sustain the food productivity. Farmer independency have been decrease slightly in the last four decades during the introduction and implementation of modern knowledge which came from the outside of rural area. Relation between human and technology as socio-technique relationship in the farmer society growing faster and effective as economy accelerator in the 80's until 90's era. At the other side, foreign pesticide companies have benefitted as the demand for high imported pesticide product has ben increased [7].

This paper analyse the agricultural local knowledge on farmer communities through a study of rural agricultural system in Mekarjaya village, Sub Dictrict of Kiarapedes, District of Purwakarta, West Java. Our analysis focus on the adaptation mechanism of traditional knowledge facing modern knowledge in agriculture activity, mainly rice or paddy as food and also discuss changes in economic income in rural development.

2. Material and Methodology

Ecological agriculture is a complex system involving ecology, economics, industry, human behavior, policy and society. Each element interrelated one to another built an agricultural rice system include human-ecosystem relationship. Recognizing these complexities, we do an interdisciplinary approach in data collection and analysis, combining with community-based ethnographic studies.

Focusing on adaptation mechanism of traditional knowledge facing modern knowledge in agriculture activity, we chose rural community who applied traditional knowledge in agriculture practices of Mekarjaya village Sub Dictrict of Kiarapedes, District of Purwakarta, West Java. Most of people who live there are wetland paddy farmers in the medium land about 500 meters above sea level. Mekarjaya village is one of the village in Purwakarta district where the traditional culture still upheld by their peoples. Nowadays, no one of the farmers in Mekarjaya applied the traditional knowledge in paddy farming completely except chief of traditional village. All the farmers have adapted modern technology since 1977 after a terrible pest disaster one year before.

The first stage of data collection involved the use of three key informants who were local experts and praticioners in paddy farming, they have adapted from traditional to modern technology. Another key informants are government in extension services and head of farmer group association in Mekarjaya.

The next stage was 15 farmers of the communities who were chosen purposively by the key informant, involved into a focus group interviews during September to October 2017. Comprehensive information to answer the research question in qualitative research sampling is the main consideration, because of the external validity of case studies is based on the logic of replication rather than on sampling logic. Individually, this study also identify technology used in farming by surveying farmer experience of farmer group members that divided into two group of modern or traditional technology.

After conducted several number of interviews, all data (mental and written database) on traditional knowledge and its adaptation were analysed. We looked for patterns, definition, stories and lessons related to the topic from all informants [8]. The data was analysed with system thinking approach. Complexities and interdependencies in knowledge management were then drawn clearly in a causal system diagram, it is powerful to sort the problem. Traditional knowledge as an object accumulating over time is a massive and complex interdependent system. Traditional knowledge is a cyclic learning process on how to learn continuously and effectively and how to manage knowledge of people in a favourable environment. Traditional knowledge helpful to guide an effective decisions, cultivate trust and mapping the relationship of knowledge flow and information within the organization and its environment [9,10].

Validation is confidence building of the research model [11], we used a triangulation method to collect information from different person or informant, then compared to the existing theory and the relationship of causality as information control. Feedback loops in system thinking are research control to identify logic causality of empiric evidence. Charting feedback loop in this article created by Vensim DSS Software to obtain the cause of the feedback and the leverage point that occurs in the model structured.

3. Result and Discussion

Based on field research, traditional knowledge application in paddy farming in Mekarjaya is characterized the variety of cultivated paddy by farmers. The vilagers known this practice with the term "buhun" (Sundanese : old or traditional). The old cultivated paddy was a long term growing variety (between 5-6 months), so that only cultivate once a year. Nowadays, farmers cultivate paddy's "buhun" very rarely because of long term growing, these has begun since 1977 and become the change milestone of traditional farming practices. Great pest attack to their local paddy plantation forced farmer to switch their paddy to the short term growing and pest resistant variety. Since the disaster many modern technologies arrived to Mekarjaya and adopted progressively in crop agriculture system, such as synthetical fertilizer and chemical pesticides until now.

Food Commercialization and Food Security

Rice (*Oryza sativa* L.) is the main food of Mekarjaya villagers. Before 1977, people cultivated paddy only for food availability reason. They stocked up paddy as dry grains, then processed them to rice for daily consumption. Most of cultivated wetland in Mekarjaya are forest area managed by Indonesian Forest Enterprises (Perhutani). The forest conversion to agriculture land are allowed through annual rent payment once a year.

According to information from chief of farmers group association, deforestation in Mekarjaya have been started since before of modern agriculture technology was officially introduced, the speed run slowly at that moment. However, agriculture land availability in the 1970-an era were insufficient to meet food requirement of Mekarjaya people, therefore forest conversion for agriculture land was unnecessary.

The situation changed since modern technology applied in Mekarjaya, the practices of modern agriculture through numerous components manufactured externally to the farm (fertilizers, pesticides, and technology) manipulates the land to make it amenable to industrial processes [12]. Socially, the value of agriculture was shifted gradually from food security purpose to food commercialization. The use of short growing period paddy variety resulted in a more frequence paddy cultivation (2-3 times a year) that lead to a higher yield and increase accumulated excess production of paddy grains. Farmer household have much more rice stock in their storage so they can sale their rice beside for own consumption in order to increase household income. Today, paddy is a commercial plant commodity for farmers as a result of surplus grains from their household consumption. These has changed people's value to paddy, they not only respect to paddy as food fulfilment but also as commercial goods, even worse the next generation people only know paddy as commercial goods as farmers statement in a small focused group discussion.

These changing behaviour to the commercialization agriculture forced more deforestation converted to wider agriculture land production. Farmers have become more aggressive converting the forest area to production field for their better income through agriculture (**Figure 1**).

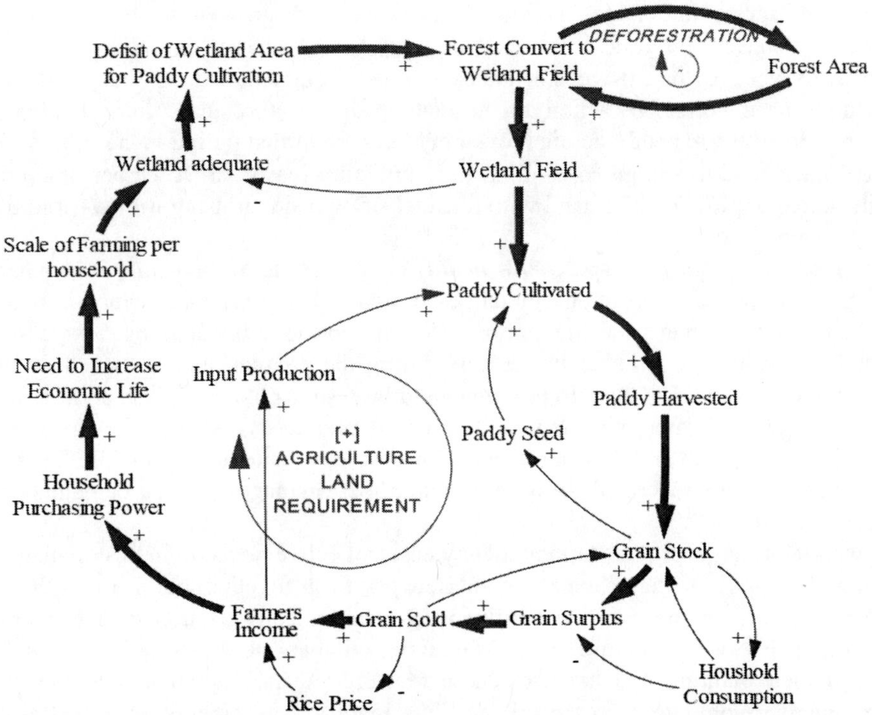

Figure 1. Causal Diagram of Deforestation for Agriculture Land in Mekarjaya

Integration Traditional Knowledge Practices and Modern Practices

Traditional knowledge practices and modern practices are two different systems viewed from its origin and approach. Traditional knowledge originated from experienced local peoples through many practices while modern practices come from research and scientific studies. Traditional or informal knowledge and modern or formal knowledge competition can lead to conflict with each other. Informal knowledge influences in the grass root level while formal knowledge have a strong network in the researcher and policy maker. Informal knowledge operates in a fuzzier networks, relying on community or group ties, community and personal relations, neighbourhood associations, peer groups, and territorial communication structures and traditions. These networks are more local, but not exclusively mainly for community. Knowledge integration and dissemination between traditional and modern practices often occurs through farmers networking both in formal organisations and in informal structures [3,13].

The growth period is an important phase in paddy farming system. Interrelated of social and ecological aspect in traditional practices shows the principles of relationships between humans and their environment. It derived from the principles of relationships among humans, consist of the core moral and ethical values: honesty, fairness, responsibility, compassion, and respect [13]. There are at least two values including in traditional practices during paddy cultivation in Mekarjaya: responsibility and respect. Pest management is an activity which responsibility and respect value were applied in traditional knowledge practices wisely. These practices did not kill pest during paddy's growth period alternatively farmer drove away the pest to move out from their paddy plantation or shifted insect food to another media such as little crab carcass which was clapped using bamboo and put it in the middle of paddy's area. Traditional practices show us how local knowledge respect and responsible one to another, between human with their ecosystem.

Pest control and cropping pattern management in traditional practices considered pest life cycle. For example, food shifting of *lembing batu* insect or *Scotinophara coarctata* into residual straw was done in the post harvest time. Paddy's residual straws were left growing for the next 3-5 months until reach generative phase. The grain from residual straw become insect's food around September until November every year, so that it will not attack the paddy plantation in the next cultivation session. These methods need collective action to cultivate paddy at the same time. Traditional practices forbade farmer to cultivate paddy before November because of that period is the time for lembing batu insect sought their food in a large population. Entering rainy season in December, traditional practices farmer will started to plowing their land with residual straw inside it using *wuluku* (traditional plowing tool) pulled by a buffalo.

Controlling *Walang Sangit* (*Leptocorisa acuta*) or in local term *kungkang* is another example of respecting pest life found in traditional practices It attacked in April-May every year or in pregnant phase of their paddy. Farmer used fumigation technique to move out *Walang Sangit* from plant then burned out fragrant leafage together in the land during the pregnant phase. Based on these, farmer respect to another organism as they do to human and be responsible not to kill another organism even though it is harmful. Farmer who applied traditional practices said that they responsible and be civilized to their ecosystem and human ethical. Those have become main principle for them because it will preserve food chain in wetland ecosystem (natural pest management and damage of food chain in **Figure 2**).

The powers of social capital of farmer society ease collective decision in pest management such as fumigation and cropping pattern. Social capital network form togetherness in ritual thanksgiving and traditional practices in cultivation have influenced their collective action in pest management and cropping pattern. Every year before 1977, Mekarjaya villagers periodically held thanksgiving ritual after finished their planting and harvest. These reinforce social capital among the villagers, both farmers and another community in the village. Systematically its process described as the power of social capital in pest management (**Figure 2**).

Any research institutions have tried to introduce ecological agriculture practices such as organic farming system at least for two years pilot project. Even though its impact improves their rice production almost two times higher than synthetic fertilizer and chemical pesticide application, but these practices discontinued. Lack of collective actions among the young farmers generate the benefeciary of organic programme stopped their experiment and decided to use synthetic fertilizer and chemical pesticide as others farmer. The organic beneficiary said that organic system requires a big effort in resources management and social collective to ensure the availability of natural input especially fertilizer and pesticide.

Another local knowledge in traditional practices is about frequency of paddy cultivation in the same land. Local paddy variety with 6 months cultivation period was planted almost only once a year. Naturally, the land have 3 months for recovery until next season of cultivation. Today with the short term growing and pest resistant variety (IR series, Ciherang, Mekonga), there are no recovery time for the land because of more frequent cultivation (2-3 times)in one year. Intensive cultivation tends to degrade soil quality in a variety of ways, so that land depletion cannot be avoided in industrial and commercialization agriculture. Application of synthetic fertilizer and intensive cultivation on annual crops dramatically reduce the ability of soils to sequester and store carbon because overuse chemicals use reduce the soil's biological activity and expose its organic matter to depletion by erosion, chemical degradation, and bacterial respiration. In addition, the mineral components of synthetic fertilizers are easily leached out of the soil. In wetland irrigated systems, the leaching problem may be particularly acute because of a large amount of the fertilizer applied to fields actually ends up in streams, lakes, and rivers [14].

Based on farmer experiences, April-May is the time for kungkang pest (Leptocorisa acuta) and every September-November for Lembing Batu (Scotinophara coarctata) to attack their paddy crop. Since cropping pattern has become uninterrupted because of intensive cultivation, chemical pesticide and synthetical fertilizer become an effective choice to secure the food production.

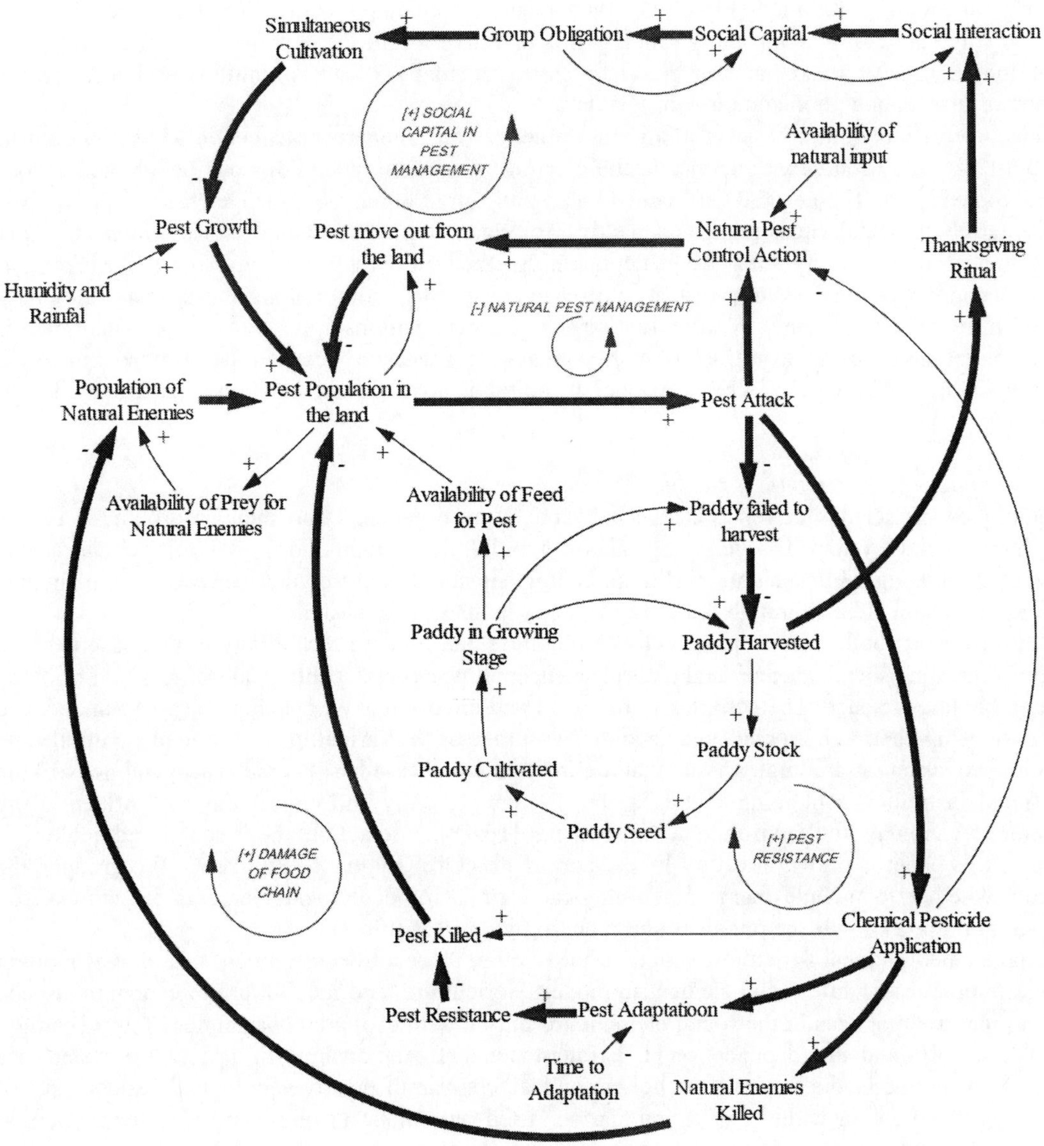

Figure 2. Causal Loop Diagram of Local and Modern Agriculture Practices Integration in Fertilization and Pest Control Activity

Nowadays, natural pest management has been abandoned because of resources allocation and duration for its process turned out not as simple and fast as chemical input. Farmer prefers to use chemical pesticide for pest and weed control because they can dramatically lower pest populations in the short term and at the same time increase pest resistance gradually. When pest resistance to pesticide is happening, farmers are forced to apply larger amounts of pesticide or using another pesticides, that might actually accelerate even greater pest resistance [14]. Actually, farmer realize about chemical application will kill pests' natural enemies, but they do not realize about its impact which is pest populations can often quickly rebound and reach a greater numbers than before as Gliessman [14] explained in agroecology concept . This situation can be captured from their confusion in pest control especially for insect population which has doubled than a few years ago and attacked randomly during the cultivation (**Figure 2**). Increasing pesticide application dosage has become their choice to solve the pest attack usually. Shortly it is effective to kill the pest using chemical pesticide, but in long period it make another problem environmentally (water pollution) and also greater numbers of insect population and randomly attack.

Trajectory of integration between traditional practices and modern practices in Mekarjaya during 1977 until now has been very dynamic. Technically all of modern technology can be adopted as long as economically profitable and efficient, meanwhile traditional knowledge practices are still maintained about social ritual related to paddy farming system. Eventhough social ritual still exist among the villagers, slowly it has been depreciating since modern technology came to Mekarjaya. Modern technology with less labors utility decrease social connection among the farmers, they have changed from collective farmer to individual farmer. Farmer relationship today is more distant than in the past, as relationships grow in trust, now they have grown weaker because trust in social capital has been depleted. Unfortunately, when the social capital is lost, farmer relationships are no longer sustainable [13].

Socio-Ecological Interaction in Agriculture Practices

Ecology is the science of relationships between living organisms and their environment, on the other hand social system is everything about people, their population, psychology and social organization shaping their behaviour. Human ecology is about analyses the consequences of human activities as a chain effects through the ecosystem and human social system [15].

Point of contact both traditional knowledge practices and modern technology in this case is about relationship among social sciences and natural sciences. It was social–ecological systems that bridging social and natural science. The complexity of social and environmental changes trigger to understand the relationship between social and ecological processes. Agriculture as social, natural and technological system is a complex issue that are interrelated each other. People as a social networking have social institution containing social capital that conceptually and empirically are influenced by economic, technology and environment components [16-18]. Interaction about social and ecological aspect on this paper is drawn in causal loop diagram structure (**Figure 1**, **Figure 2**). We grouped the elements whether to include them in sociological element or ecological elements for comparison between traditional knowledge practices and modern practices **Table 2**).

Social element in local agriculture practices looks more respect to environment than that of modern agriculture practices, while social element in modern agriculture tend much more to economic aspect. Local agriculture practices in the social element are in line with ecological principles of agroecology: holism, diversity, and interdependence [13]. Interdependent relationships in agriculture system are necessary to transform the potential of holism and diversity into positive ecological reality, in this case to avoid failure in paddy harvesting, farmer used natural pest control such as fumigation or shifting pest feed to another media around paddy crop. These methods do not kill natural enemies or another microorganism that may lead to disturb the food chain in wetland ecosystem. In the other hand when farmer used modern agriculture practices, pesticide application kill almost microorganism in the ecosystem include pests' natural enemies.

Sustainable development can be achieved in practice most importantly with preserving the ecosystem, secondly to do things nature. Sustainable development operationally can be reached through social institution to protect common property resources from tragedy of the commons as impact of unsustainable human-ecosystem interaction. Enviromentally, we have to preserve the ecosystem such an extent to provide essential services and carefully to avoid negative impacts for environmental or social when using new technologies. Agriculture as economy sector need to be sustain for the next generation and as a whole-systems approach to production, sustainable agriculture have to balances environmental soundness, social equity, and economic viability [14,15].

Table 1. Social and Ecological Elements in Local and Modern Agriculture Practices in Mekarjaya Villages

Comparison	Social Elements	Ecological Elements
Local Agriculture Practices	- Paddy harvested, - Paddy stock, - Natural pest control action, - Thanksgiving ceremony, - Social interaction, - Social capital, - Group obligation, - Simultaneous cultivation	- Availability of feed for pest, - Pest population in the land, - Pest attack, - Paddy failed to harvest, - Paddy in Growing Stage, - Pest move out from the land, - Pest growth, - Population of Natural Enemies
Modern Agriculture Practices	- Chemical pesticide application - Synthetical fertilizer application - Intensive cultivation - Paddy Grain Sales	- Pest adaptation, - Pest resistance, - Natural predators killed, - Population of Natural predators, - Pest population in the land, - Pest attack

The practices of industrial or commercial agriculture should be tending to compromise between present and future productivity. The ways in industrial agriculture puts future productivity at many risk [14]. Agricultural resources such as soil, water, and genetic diversity are overdrawn and degraded, human health suffers, and the social conditions are weakened and dismantled to resource conservation

Personal curiosity and willingness to learn, together with social networking among farmers and community, and supportive formal knowledge and governance structures, are all important learning points for success in social and ecological aspect [3]. Human as decision maker have to manage their ecological aspect in order to improve economy and welfare.

4. Conclusion

The values of traditional agriculture practices in Mekarjaya village were shifted gradually from food security purpose to food commercialization in line with the modernization technology in rural area. Trajectory of integration between traditional practices and modern practices in the last four decades is very dynamic. Technically all of modern technology can be adopted as long as economically profitable and efficient, meanwhile traditional knowledge practices which are still maintained about social ritual related to paddy farming system. Farmer relationship now is more distant than in the past, as relationships grow in trust, now social institution grows weaker because trust in social capital has been depleted.

Social element in local agriculture practices looks more respect to environment than that of modern agriculture practices, while social element in modern agriculture tend much more to economic aspect. These force commercial agriculture to compromise between present and future productivity for sustainability reason. Personal curiosity and willingness to learn, together with social networking among farmers and community, and supportive formal knowledge and governance structures, are all important learning points for successful in social and ecological aspect for sustainable agriculture development.

Acknowledgement

We would like to thank Universitas Padjadjaran that founded the Fundamental Research Universitas Padjadajran with the title "Study of Traditional Ecological Knowledge in Agricultural Pratices Using

Social-Ecological Studies (SES) Approach (Case Study in Mekarjaya Village, Subdistrict Kiarapedes, Purwakarta)".

References

[1] Berkes F, Colding J, and Folke C 2000 Rediscovery of traditional ecological knowledge as adaptive management *Ecological applications*, 10(5) 1251-1262

[2] Nakashima D and Elias D eds., 2002 Science, traditional knowledge and sustainable development. ICSU

[3] Šūmane S, Kunda I, Knickel K, Strauss A, Tisenkopfs T, des Ios Rios I, Rivera M, Chebach T and Ashkenazy A 2017 Local and farmers' knowledge matters! How integrating informal and formal knowledge enhances sustainable and resilient agriculture. *Journal of Rural Studies*

[4] Menzies CR ed. 2006 *Traditional ecological knowledge and natural resource management.* U of Nebraska Press

[5] Berkes F and Folke C 1998 Linking social and ecological systems for resilience and sustainability. Linking social and ecological systems: management practices and social mechanisms for building resilience 1(4)

[6] Ohmagari K and Berkes F 1997 Transmission of indigenous knowledge and bush skills among the Western James Bay Cree women of subarctic Canada *Human Ecology 25*(2) 197-222

[7] Gliessman SR 2014 *Agroecology: the ecology of sustainable food systems* CRC press

[8] Luna Reyes LF and Andersen DL 2003 Collecting and analyzing qualitative data for system dynamics: methods and models *System Dynamics Review 19*(4) 271-296

[9] Gao F, Li M and Nakamori Y 2002 Systems thinking on knowledge and its management: systems methodology for knowledge management. *Journal of Knowledge Management 6*(1) 7-17.

[10] Kim DH 2000 *Systems Archetypes I: Diagnosing System Issues and Designing High-Leverage Interventions* Waltham: Pegasus Communications Inc

[11] Sterman JD 2000 *Business dynamics: systems thinking and modeling for a complex world* (No. HD30.2S7835 2000)

[12] Kirschenmann F 2010 Alternative agriculture in an energy-and resource-depleting future *Renewable Agriculture and Food Systems 25*(2)85-89

[13] Lamine C 2015 Sustainability and resilience in agrifood systems: reconnecting agriculture, food and the environment *Sociologia ruralis 55*(1)41-61

[14] Gliessman SR 2014 *Agroecology: the ecology of sustainable food systems.*CRC press

[15] Marten GG 2001 *Human ecology: Basic concepts for sustainable development* Earthscan

[16] Folke C, Colding J and Berkes F 2003 Synthesis: building resilience and adaptive capacity in social-ecological systems. *Navigating social-ecological systems: Building resilience for complexity and change 9*(1)352-387

[17] Cote M and Nightingale AJ 2012 Resilience thinking meets social theory: situating social change in socio-ecological systems (SES) research *Progress in Human Geography 36*(4) 475-489.

[18] Supyandi D, Pardian P and Heryanto MA 2017 March. Organic Rice Development: How Social-Ecological System Perspective Could Contribute In *Proceeding of International Symposium for Sustainable Humanosphere* 118-124

Comparison of *Value-Added* Analysis of Coffee Processing with Different Market Orientations: (Case Study on Cooperative of Producer Coffee Margamulya)

T Karyani[1], E Djuwendah[1] and E Supriyadi[2]

[1]Faculty of Agriculture, Universitas Padjadjaran, Jl. Raya Bandung Sumedang km.21 West Java 45363, Indonesia
[2]Institute of Koperasi Indonesia (IKOPIN), Jatinangor, Indonesia

E-mail: tutikaryani@unpad.ac.id

Abstract. Cooperative of Producer Coffee Margamulya (CPCM) consists of 140 members who produced Arabica java preanger coffee with an average of 70 tons green bean per year. The cherry coffee is first processed into green bean using wet processing method before being sold. Eighty percent of green beans are marketed to exporters (PT. Taman Delta Indonesia/ PT. TDI) while 10% to Indocom and the other 10% to various markets with various products. This study was more focused on the analysis of difference of added value obtained by cooperatives from the dominant market for green bean, i.e., to exporters and Indocom. Also, this study aimed to determine the factors that influence market selection. The method used was a case study, with Hayami value-added analysis technique. The result showed that the added value of cherry processing became green bean for the export market was 19% while the value added for Indocom was 34%. The reason for choosing the market to exporters is that PT. TDI receives a large number of green beans, and provides UTZ (utilize) certification assistance. On the contrary, Indocom only bought smaller volumes than PT. TDI with non-fixed frequencies.

1. Introduction

West Java has great potential in coffee development as it is supported by its coffee plantation area of 12,000 ha in 2016. The West Java Provincial Government seeks to develop West Java specialty coffee which since 2012 has obtained geographical indication by the name of java preanger. Java preanger coffee is categorized as a specialty coffee with a good taste and distinctive.

The widest area and the biggest coffee production in West Java are in Bandung Regency with the type of arabica coffee. This is because Bandung Regency has some plains area with altitude over than 1,000 m a.s.l. which is suitable for arabica coffee [1]. One of the largest coffee plantations in Bandung Regency is located in Pangalengan District It is 3% of West Java [2] and 10.24% of Bandung Regency coffee plantations [3]. Initially, through the village community forest institutions program, farmers and Perhutani then cooperate and plant coffee in the land of Perhutani. After realized that coffee is profitable, the farmers plant the coffee in their own land. The existence and increased enthusiasm of farmers to grow this coffee strongly support the conservation forest protection program, thereby reducing and diverting farmers to grab the forest. This land grabbing is an indication of the carrying capacity of land to meet the needs of farmers who do not live a decent life [4]. The fact that coffee is profitable and can alleviate farmers around the forest from poverty. In Pangalengan District, a cooperative under the name of cooperative of producer coffee margamulya (CPCM) was formed in 2014. CPCM currently has 140 members of coffee farmers. This cooperative holds and distributes coffee beans to PT. Taman Delta Indonesia (PT. TDI) as much as 80%, and for other consumers around Bandung as much as 20% of total production. Demand for PT. TDI 70 tons per year in the form of green beans. Besides, green beans are also marketed to Indocom and the cafes around Bandung. Processing coffee fruit (cherry) to green bean certainly can increase its added value. Therefore, this study tried to compare the added value of coffee after processing in cooperative from 2 different markets, and analyze the factors influencing the added value.

2. Methodology

2.1 Literature Review

2.1.1 Arabica Coffee

Arabica coffee grows at an altitude of 600-2,000 m a.s.l. This plant can grow up to 3 meters at the optimum environment condition. The optimum growing temperature is 18-26^0 C. The coffee beans are small and have green to dark-red color. West Java has a typical type of arabica coffee called java preanger. Arabica coffee java preanger is a typical arabica coffee in West Java with Geographical Indication (IG) of java preanger. It is cultivated by people in some mountainous areas of West Java such as Cikuray Mountain, Papandayan Mountain, Malabar Mountain, Caringin Mountain, Tilu Mountain, Patuha Mountain, Halu Mountain, Beser Mountain, Burangrang Mountain, Tangkuban Perahu Mountain and Manglayang Mountain [5]. All those mountains have altitude over than 1,000 m a.s.l. IG is one of the most important components of intellectual property rights in trading activities, particularly on protecting trade commodities that are firmly related to the name of the region or place of origin of the products. The existence of the IG certificate can be used at least for promotion, competitiveness and selling value of a product in the broader market. So that it will be more economically beneficial to the welfare of the local community. A label is attached to a product that has received an IG certificate. It could be names of places, regions, words, images, letters, or combinations of those elements.

Coffee plantation initially is only in Perhutani and is a form of agroforestry. Nowadays, after realizing that coffee is beneficial, coffee is not only planted in the land of Perhutani but also in private area as intercropping plant among vegetables or as the main plant. However, farmers still produce coffee and sell it directly in the form of fruits or called cherry. The process of handling coffee is still quite long, including conversion of cherry into grain, and wet or dry processing of grain into green beans. Processing from cherry to the green bean is done at cooperative level. So far, the market accepts in the form of green beans because of its durability. However, to increase the added value of coffee, processing into coffee powder is essential to get the business opportunities. While Pangalengan is one of the potential tourist destinations, and CPCM is located between the Bandung-Pangalengan routes, it fits the needs of cafes or coffee shops that are planned to be built.

2.1.2 Value-added

The purpose of the value-added analysis is to measure the rewards received by business actors and employment opportunities from derivatization of commodity. Value added is the result of reduced selling prices minus the cost of purchasing raw materials and is often used as a focal point for cost analysis [6].

According to [7], the added value created in each value chain actors due to the activity or different stages of each actor. The means of added value by [7] relates to high quality, lower cost, delivery time, innovation, and so on. According to [7], one important factor that affects the added value of agricultural products is quality. The quality of agricultural products is divided into two, namely intrinsic characteristics (such as color, physical characteristics, and taste) and extrinsic characteristics (such as fair trade and organic farming). Value added in the value chain is influenced by market objectives and the number of actors in the value chain. The common method used in measuring added value is the Hayami method. According to [8], the added value is the difference between the selling price and the cost during the process. The sources of added value are the utilization of factors such as labor, capital, human resources, and management. Term of added value means a value that occurs on a commodity because the commodity has undergone processing, transport, and storage in a production process. The Hayami method can be used in analyzing the added value of the sub-processing system or secondary production. The results obtained from the analysis of value-added Hayami is in the form of productivity, output value, added value, labor remuneration, and the benefits obtained from the processing [9]

2.2 Data Collection and Analysis

This research used qualitative research design with case study research technique and was done in CPCM. The data used in this research were primary and secondary data. Primary data were obtained through direct interviews with farmers and cooperative management. Secondary data, as complementary and supporting for primary data, were obtained through literature study, internet search, Central Bureau of Statistics and data from other related institutions and relevant to this research. The Hayami value-added analysis was used to analyze added value at sub-processing system or secondary production. Descriptive analysis was used to answer the identification of problem.

3. Results and Discussion

The first marketing channel is marketing green bean coffee to PT. TDI, which is a coffee exporter partner of CPCM. The first cooperation between the two parties is started when the Chairman of CPCM meets with the representative of PT. TDI at the coffee exhibition in Batam. PT. TDI has an obligation from the Association of Indonesian Coffee Exporters (AEKI) to conduct coaching to Small and Medium Enterprises (SMEs) coffee. Thus, PT. TDI conducts the financial training for UTZ (utilize) certification to the coffee processing of CPCM and also all farmers who are registered as CPCM members. UTZ certification assures the physical quality of sustainable products and coffee businesses [7]. UTZ certified is a requirement for exporting coffee to Japan. In the agreement of both parties, it is stated that the CPCM can only include UTZ certified on products sold to PT. TDI.

Another form of partnership with PT. TDI is member accompaniment to get UTZ certified and UTZ quality control every year. In addition, PT. TDI also provides certified coffee seeds and lends some coffee processing machines to CPCM. Delivery of coffee to PT. TDI is conducted every week with a minimum quantity of 3 tons of green beans.

To meet the qualifications set by PT. TDI, CPCM established Standard Operational Procedure (SOP) together with PT. TDI. The CPCM also implements the standard for sorting determined by AEKI.CPCM also sells green beans to local coffee companies. Green beans purchased by local coffee companies are green bean *reject* and also grade II and III of green beans. Local coffee companies such as Indocom buy the grade II and III of green beans. Currently, CPCM has developed its business in addition to selling green bean products. However, since the green bean product is the most substantial part of the output generated by CPCM, this paper only discussed the value-added analysis for green bean.

3.1 Value-Added Analysis on Processing of Chery Coffee to Green Bean Coffee

Processing cherry into green bean is the dominant activity of CPCM in order to fulfill the request of its business partner. The results of green bean value-added analysis for PT. TDI market can be seen in **Table 1**.

Table 1. Green Beans Value-Added Analysis for PT. TDI

Output, Input, Price		Value
Output (Kg)	A	500.00
Input (Kg)	B	3.000.00
Labor (Man work days)	C	17.14
Conversion Factor (A/B)	D	0.18
The coefficient of labor (C/B)	E	0.01
Price of Output per Kg (IDR)	F	62.000.00
Wage of Labor (IDR)	G	29.248.77
Receipts and Profit (IDR/Raw Material)		**Value**
Raw material per kg (IDR)	H	8.500.00
Price of other Input per Kg (IDR)	I	559.32
Output Value (DxF) (IDR/kg)	J	11.160.00
a. Added Value (IDR/Kg)	K=J-H-I	2.100.68
b. Added Value Ratio (%)	L=K/J	0.19
c. Labor Income (IDR/mwd)	M=ExG	167.14
d. The Share of Labor (%)	N=M/K	0,08
e. Profit (IDR/Kg)	P=K-M	1.933.55
f. Profit Rate (%)	Q=P/J	0.17
Remuneration Service Owner Factor Production		**Value**
Margin (IDR/kg)	S=J-H	2.660.00
a. Labor Income (%)	T=M/S	0.06
b. Share of other input (%)	U=I/S	0.21
c. Profit (%)	V=P/S	0.73

PT. TDI gives the price according to green bean's world price. When the research was done, price determined by PT. TDI was IDR. 62.000,00 per kilogram green bean. The delivery of green bean to PT. TDI is conducted once a week with quantity reaching 3 tons per shipment. In addition to processing cost of cherry, there were also other input costs. Additional input costs other than purchasing cherry by CPCM include fuel for pulping and hulling, electricity, packaging, warehousing, promotion, depreciation of equipment and machinery, and distribution costs. Processing undertaken by the CPCM adds value to the cherry of IDR 2,100.68 per kilogram of cherry. The profit gained from processing is IDR 1,933.55 per kilogram of processed cherry. It is accounted for 73% of all margins obtained from green bean processing for PT. TDI. The share of labor from horn skin processing is 8%. This indicates that 8% of the added value obtained in the processing is for labor cost. CPCM also sells green beans specialties to the local coffee company, Indocom. Green beans purchased by Indocom are specialty coffee with grade II and III. The selling price for this grade is IDR 75.000,00 per kilogram. Other input costs are same as the PT. TDI with cost of IDR 459.32 per kilogram. The green beans sold to Indocom adds value to cherry ofIDR 4,540.68 per kilogram. The profit gained from processing is IDR 4,373.55 per kilogram of processed cherry. The profit gained is 87% of all margins earned from the processing of green beans for Indocom. The share of labor from this green bean processing is 4%. This indicates that 4% of the added value obtained in the processing is labor cost.

When compared to the increase of value-added per kg of cherry processing into the green bean with the aim of the market to Indocom higher (IDR 4,540.68) compared to PT TDI (IDR 2,100,68) because the price received by CPCM from Indocom is higher than that of PT TDI. This is because the price of green bean per kg received from Indocom is higher (IDR 75,000), while the green bean price per kg received from PT TDI is IDR 62,000. This is consistent with [8] said that factors affecting the value added of processing could be categorized into two factors, which are, technical and market factors.

The price of PT TDI follows the green bean price of the world which in turn will also affect the price of cherry paid by the cooperative to the farmers. Price to Indocom is higher because that determines it is the price of green bean competitors. However, the volume of purchases from Indocom is not as much as PT TDI.

Table 2. Green Beans Value-Added Analysis for Indocom

Output, Input, Price		Value
Output (Kg)	A	500,00
Input (Kg)	B	3.000,00
Labor (Man work days)	C	17.14
Conversion Factor (A/B)	D	0.18
The coefficient of labor (C/B)	E	0.01
Price of Output per Kg (Rp)	F	75.000,00
Wage of Labor (IDR)	G	29.248.77
Receipts and Profit (IDR/Raw Material)		**Value**
Raw material per kg (IDR)	H	
Price of other Input per Kg (IDR)	I	8.500,00
Output Value (DxF) (IDR/kg)	J	459.32
a. Added Value (IDR/Kg)	K=J-H-I	13.500,00
b. Added Value Ratio (%)	L=K/J	4.540.68
c. Labor Income (IDR/mwd)	M=ExG	0.34
d. The Share of Labor (%)	N=M/K	167.14
e. Profit (IDR/Kg)	P=K-M	0.04
f. Profit Rate (%)	Q=P/J	4.373.55
Remuneration Service Owner Factor Production		**Value**
Margin (IDR/kg)	S=J-H	5.000,00
a. Labor Income (%)	T=M/S	0.03
b. Share of other input (%)	U=I/S	0.09
c. Profit (%)	V=P/S	0.87

3.2 Factors Affecting the Selection of Marketing Channels

From the results of interviews with the chairman of the board of CPCM, it was revealed that almost 80% of coffee produced in cooperative was marketed to PT TDI. There are factors that become the reason for the occurrence of partnership with PT. TDI.

- Market guarantees and prices
  With the partnership with PT. TDI, the green beans produced by the cooperative have definite market and price. It results in high demand volume of 3 tons of green beans per week or about 17 tons in the form of red cherry.
- Risk reduction
  The existence of certainty of market, quantity, quality and price reduces the risk of unsold products.
- Transfer of knowledge
  In partnership with PT. TDI, there is a process of transfer of knowledge in the form of SOP for cultivation, harvesting, post-harvesting, and processing results. The factor of quality is very important for exporter. So by fulfilling the SOP, the quality of green beans could be maximized.The partnership of this model according to [10] is called a productive partnership.
- Improving access to financial institutions.
  Cooperatives are more trusted by financial institutions because there is market certainty from PT.TDI, so that loans given to cooperatives will be paid.
  This condition occurs in the case of mango in Cirebon where Samimulya Farmers Group get a loan from the bank after a partnership with the exporter [11]. Nevertheless, because the need of cooperatives on capital is enormous, then the investors help to loan the money.

Reasons marketing to non-exporters:

- Looking for an alternative market for unsold green beans of export market. Indocom and café are the market for those unsold green beans. Although the quantity is small, the price is higher than to the exporter.
- Providing an opportunity to expand partnerships with other companies or parties.
- Diversification of products such as producing green beans into roasted beans or grounded beans (coffee powder) with different packaging for business development.

4. Conclusions

The added value of coffee production process for Indocom market is higher than for market to PT. TDI. However, because the volume of green beans marketed to PT. TDI is larger than Indocom, it makes PT. TDI more profitable. The factors that cause the cooperative choosing the market are the certainty of market, risk reduction, transfer of technology, and increasing access to the financial institution.

The choosing to other markets such as Indocom, is more likely as an alternative market to the remaining coffee. In addition, it provides an opportunity to expand partnerships with other companies, and for business development. The amount of green beans marketed to Indicom and different market is further enhanced, but the marketing and relationship to PT. TDI is maintained steadily.

Product and market development still continue until the powdered coffee products are ready in brew form or even ready to drink. Cooperative further increases its role not only as an exporter supplier, but also to be a direct exporter.

Acknowledgments

This research supported by Padjadjaran University (UNPAD) Internal Grant in RKDU scheme in 2017. The research team would like to thank to all the support of DRPM UNPAD. Likewise, we would like to thank the management of CPCM for their cooperation.

References

[1] West Java Provincial Agriculture, Plantation, and Forestry Office. 2013. Laporan Tahunan Dinas Pertanian, Perkebunan Dan Kehutanan Tahun 2013.

[2] Badan Pusat Statistik Jawa Barat "Provinsi Jawa Barat Dalam Angka 2017" Melalui https://jabar.bps.go.id/ (Accessed 8 Januari 2018 on 18.20 WIB)
[3] Badan Pusat Statistik Kabupaten Bandung "Kabupaten Bandung Dalam Angka 2017". Melalui https://bandungkab.bps.go.id/ (Accessed 8 Januari on 18.08 WIB)
[4] Soemarwoto 1985 A Qualitative of population Pressure and It's Potential Use in Development Planning Majalah Demografi Indonesia **Vol** 12 (24) Jakarta
[5] Purnama Siti 2016 Mengenal Kopi Arabika Java Preanger (KAJP) Yang Telah Mendapat Perlindungan Indikasi Geografis Dinas Perkebunan Provinsi Jawa Barat http://disbun.jabarprov.go.id/Accessed 20 Januari 2017)
[6] Porter M 1994 Keunggulan Bersaing: Menciptakan dan Mempertahankan Kinerja Unggul Terj Binarupa Jakarta
[7] Trienekens JH 2011 Agricultural Value Chains in Developing Countries: A Framework To Analysis *International Food and Agribusiness Management Review* **Vol 14** Issue 2 Belanda:Wageningen University
[8] Hayami Y Toshihiko K Yoshinori M and Masdjidin S 1987 Agricultural Marketing and Processing in Upland Java: A Perspective From A Sunda Village Bogor CGPRT Centre
[9] Kuswanti Heny 2015 Analisis Nilai Tambah Usaha Tempe di Kabupaten Bogor. Yogyakarta: Prosiding Seminar Nasional
[10] Wibisono Y 2007 Membedah Konsep dan Aplikasi CSR (Corporate Social Responsibility) Fascho Publishing Gresik
[11] Karyani Tuti 2016 Mango Agricultural Supply Chain: Actors, Business Process, And Financing Scheme *International Journal of Applied Business Economics Research* **Vol 14** No 11 7751-7764

Business Model for Promoting Organic Vegetables

Y Deliana[1], D Supyandi[2] and E Djuwendah[3]

[1] Padjadjaran University, Jl. Raya Bandung- Sumedang KM 21, Jatinangor, Indonesia

E-mail: y.deliana@gmail.com

Abstract. Consumers' interest in organic vegetables increases every year, because they aware on healthy life and environment. However, these organic vegetables have not been marketed everywhere. Only high-income people often eat organic vegetables. Producers of organic vegetables intend to enlarge their earnings by expanding the market. One of the organic vegetable producers began to apply business model by training, and th e other producers established organic community and used "*instagram*" as a media to communicate amongst the member. The objective of this paper is to explain type of factor that motivates consumers to join in organic community, constraint faced by consumers and producers, and expectation of consumers and producers. This research was conducted in Bandung. The respondents of this study were entirely members who joined together in this community. This research expects to provide inputs to policymakers, stakeholders, producers and related institutions. Also, this research is aimed to provide a prototype of the business model in developing organic products and educating other societies for a healthy life. The research found that trust is the main factor of respondents in the business model using"*instagram*", while word of mouth is the main factor of respondents in business model by training. The main constraint of producers is waste from environment that hampers organic vegetables cultivation. For the customers, the high price of organic vegetables is the main constraint. The main expectation for producers is the low cost of certification, while for costumers is the presence of a trusted Indonesian organic emblem.

1. Introduction

Organic vegetables farmers earn better yield (40-60%) than the conventional agricultural vegetables farmers [1]. This is due to more guaranteed quality of organic vegetable products than those of non-organic vegetables. This quality standard can be seen from the existence of certificates and organic label to differentiate with those of non-organic vegetables. The use of packaging and brands will result in premium vegetable product prices [1]. Moreover, organic vegetables have a better nutrient content than that of non-organic vegetables. Therefore, people who have at least once consumed organic vegetables and know the information that organic vegetables are good for health. Having potential and opportunities from existing markets, several producers are currently working on organic vegetable products in Bandung. They are Kurnia Hijau Farm, Ina Green Farm, Organic Garden, Semai Organic, Farm Organic, and Warung Langit. These six producers are regular suppliers for supermarket in Bandung. Kurnia Hijau Farm, Ina Green Farm, Organic Garden, and Semai Organic run the business without community. Farm organic provides training and build organic communities, while Warung Langit use "*instagram*" and build organic community. They use Instagram because it is considered a new way in in social media to promote a product and not have to pay.The purpose of this study is to analyze a business model in promoting organic vegetables.

2. Methodology

2.1 Data Collection

This research applied survey method in collecting data. The object of this research was a business model conducted by two companies. Farm Organic conducts training to promote organic products, while Warung Langit uses social media, notably "*instagram*" in promoting organic products.

The data used were primary data and secondary data. Primary data were collected from interviews with a producer who applied a business training model, and a producer who used "Instagram" for promoting business." In addition, 30 respondents were selected from each community. Community in this right is a group. There are two groups of people who often consume organic vegetables. The two groups are (1) groups of respondents who have participated in the organic training. This training explains how organic vegetable cultivation and organic vegetable benefits for health; (2) group of respondents incorporated in the service of organic vegetable delivery through Instagram. The population of respondents from group one about 150 people, while group two about 100 people. From each group taken 30 people by random sampling. Respondents were selected because they knew about the benefits of organic vegetables, the reasons for buying organic vegetables and bias to provide information about all organic to neighbors, family, friends and other words of mouth or to provide information related to organic vegetables in Instagram.The observed variables were the products, effectiveness, words of mouth (WOM), lifestyle, purpose, and belief. Each variable has five indicators, and each indicator was measured using a Likert scale from 1 to 5. Secondary data were collected from related agencies. Place of the study was in Bandung from March to May 2016. Organic vegetables have long been known, especially since there is a government program 2010 about "Go Green", although there are still a few people who regularly buy organic vegetables because the price is expensive and currently only available in supermarkets. From the field revealed there are two ways to educate the public about the importance of organic vegetables for health, with the training and through Instagram. Both ways are called business models because the business initiator of this model wants his business to run smoothly and sustainably generate profit, but also educate other people will live healthy.

2.2 Data Analysis

The business model was analyzed descriptively, while dominant consumer factor in choosing business model was estimated using Partial Least Square (PLS). The method was developed as an alternative to structural equation modeling (SEM) with a weak theoretical basis. PLS can be used as an analytical testing and generates a relationship with no theoretical basis (exploratory). The steps in the PLS are design the structural model (inner model), the assessment model (outer model), the flowchart, the conversion of the flowchart to the system of equations, the estimation of the path coefficient, loading and weight, the estimation of goodness of fit, and hypothesis testing [2]

2.3. Literature Review

2.3.1 Organic Vegetables. Organic vegetables are vegetables that are naturally grown without any chemical substances. An essential concept of organic vegetables is the processing and cultivation techniques without the use of chemicals. The advantages of organic vegetables according to [3] are:

1. Organic vegetable products are healthy to be consumed because they do not contain pesticide residues and toxic chemicals that are harmful to health.
2. Organic vegetable products have crispier texture and sweeter taste.
3. Products of organic agricultural inputs (manure, bio-pesticides) do not cause environmental pollution, safe for the health of the user and readily biodegradable.
4. Improve and conserve soil fertility and biodiversity.
5. Suppressing production costs, which is economically profitable in the long term.

However, organic vegetables also have weaknesses, such as:

1. Organic vegetable products have poor physical appearance compared to those of conventionally grown crops.

2. Organic vegetable needs more labor than conventional, especially for fertilization and pest control activities.
3. In terms of the process of nutrient absorption from organic fertilizers and the effectiveness of plant botanical pesticides, the effect is slower than those of synthetic chemicals.
4. Plant maintenance activities are more intensive than that of conventional.

In addition, the color of organic vegetables is more concentrated, and it has a distinctive and strong vegetable aroma. Moreover, after cooked or processed, organic vegetables still have the smell of fresh vegetables. Organic vegetables are not heat resistant and easy to yellowing than those of non-organic vegetables. On the other hand, non-organic vegetables tend to be more durable although stored for months. If rotted, non-organic vegetables usually have an unusual smell of rottenness like a strong acidic smell due to the use of preservative.

2.3.2 Business Model Definition. The business model is defined as a model that describes how a company operates to gain profit. According to [4], there are three main characteristics of a business model. First, the business model must be following company goals. Second, the business model must be able to strengthen itself. Lastly, the business model must be tough and reliable. The business model that can be implemented to farmers is a business model that produces profit for farmers but there is a social business. Social business is done between the fabric of the introduction of all about organic activities that come to his garden for free. The hope after understanding the importance of health, then there is a tendency to consume organic vegetables increased and business become sustainable

Stages to create a business model according to [5] are to determine customer segment; determine the products, service or value proposition; deliver products to customers and how relationships will be established between customers and the business; determine where the source of income or revenue streams; identify the source of resources needed or key resources; determine the main activities in the business or key activities; and determine who the main partner is needed in the business or key partners and from where the source of expenditure or cost structure.

3. Results and Discussion

3.1 Characteristics of Producers

Organic vegetable producers have 10-15 years' experience, various vegetable products, with target of organic community market and supermarket of 32 and 68%, respectively. Techniques to produce organic vegetables are based on experience, self-learning, and training, as well as cooperation with research institutes (Indonesian Research Institute), international institutions (Hivos) and non-governmental organizations (NGOs). Therefore, the help of Hivos is complete, the farmer must finance his own business. With the sale of vegetables through Instagram, farmers seem to get the capital first to do business. Capital is obtained by farmers from members who participated in the Instagram group. Farmers reduce the risk in their business because the results are definitely purchased and distributed to members who are members of Instagram

3.2 Characteristics of Consumers

Organic community respondents who used training and "Instagram" are mostly women, between 20 - 40 years of age, bachelor degree, entrepreneurs, having income between IDR 5 million and IDR 10 million per month. Commonly, the consumers buy organic vegetables because of a healthier diet, healthy life, maintaining the environment, motivating others to go green and to build a business. Fanatic consumers of organic vegetables can distinguish between organic and non-organic vegetables.

3.3 Business Models of Organic Vegetables

3.3.1. Using Social Media "Instagram" to Promote Organic Vegetables. The use of social media "*Instagram*" in promoting organic vegetables has been started since 2015 by producers. These producers created a small market program, where the community uses "*Instagram*" to communicate. The program was suspended in mid-2016 and can be used again in 2017. Small markets bring farmers and consumers together. To grow the organic vegetables, farmers are assisted by funding from consumer membership fee of IDR 200,000 per person. For 1 kilogram of vegetables with eight deliveries, consumers pay IDR 160,000 per month or IDR 20,000 per delivery.

There are three places for organic vegetable delivery, namely Arcamanik, Martadinata, and Dago. Delivery is usually done using bamboo baskets. Each basket contains four to five vegetables. Every consumer should take the vegetables at predetermined point, usually on Monday and Thursday. Consumers directly receive the organic vegetable products, but consumers cannot choose the type of vegetables. However, producers have prepared the latest recipes to process a food using the vegetables. These producers develop their business by making a café which specially provides organic foods with the principle of 4 P [without "pengawet" (preservative), "pewarna" (coloring), "penyedap" (flavoring) and "pemanis buatan" (artificial sweeteners)]. Business model training and Instagram also include elements of the element. Its customer segment is clear that the respondents who follow the training and who follow Instagram. Value proportions are the benefits customers get. Channels in this case directly to customers. Revenue streams are a stream of profits, both business models promote organic vegetables, in order for the business to be sustainable. Customer relationship is a company relationship with the customer. The hope after the customer understands all about organic vegetables it will spread it to family, neighbors, and friends. Key activities created to create value proportion. Key resources are to manage the business resources as much as possible, this has been done both business model. The key partnership is maintaining a good relationship with all parties. The cost structure is managing the business efficiently and minimizing risk. This has been done business training model by offering its products by subscribing to trainees. Similarly, the Instagram business model minimizes business by obtaining capital from members of Instagram. Monthly contributions paid in advance by members can be used as capital for farmers to manage their business. The business model is as follows:

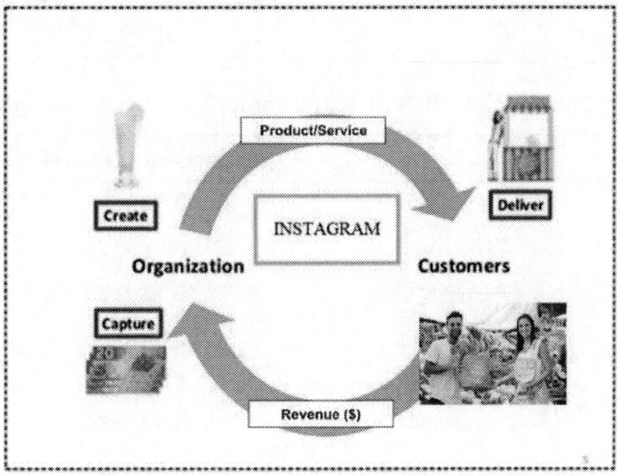

Figure 1. Business model with "*instagram*"

3.3.2. Conducting Training to Promote Organic Vegetables. Producers who do training in promoting organic vegetables already have a website in promoting organic vegetables. From this website, people know what programs are offered. From this training, it also forms an organic community. Training is conducted every month on Saturday, but the date is adjustable. This depends on the number of trainees. If there are at least 20 participants, then this training is carried out. The cost of group training is IDR 300,000 per person, and an additional cost of IDR 50,000 per person is needed to see the garden. The cost of private training is IDR 400,000 per person. The business model is as follows:

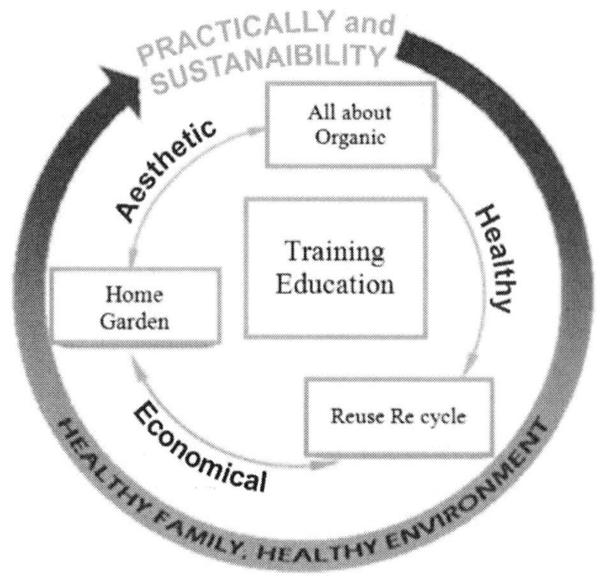

Figure 2. Business model with training

3.4 Dominant Factor of Consumers to Join in Organic Community

3.4.1. Business Model with Social Media "Instagram". **Table 1** shows the values of sixth latent composite reliabilities. The values are above 0.7. Hence, it can be said that the variable has good reliability as a measuring tool.

Table 1. Results of Composite Reliability of "Instagram"

	Composite Reliability
Aim	0,739128
Effectiveness	0,789574
Life Style	0,754109
Product	0,800164
Producer Factor	0,926160
Trust	0,778476
WOM	0,755370

Source: Data Processing Results, 2017

From the Confirmatory Factor Analysis (CFA) model with the second order, information of the dominant factors and indicator are available. From the value of loading factor, the *trust* factor has the highest loading factor among others, which is equal to 0.959. This shows that the reason of respondents follows the community with the "*instagram*" model business is *trust*, followed by effectiveness (with a loading factor of 0.904), lifestyle (with a loading factor of 0.903), and the other three are considered lower, namely the product, destination, and WOM.

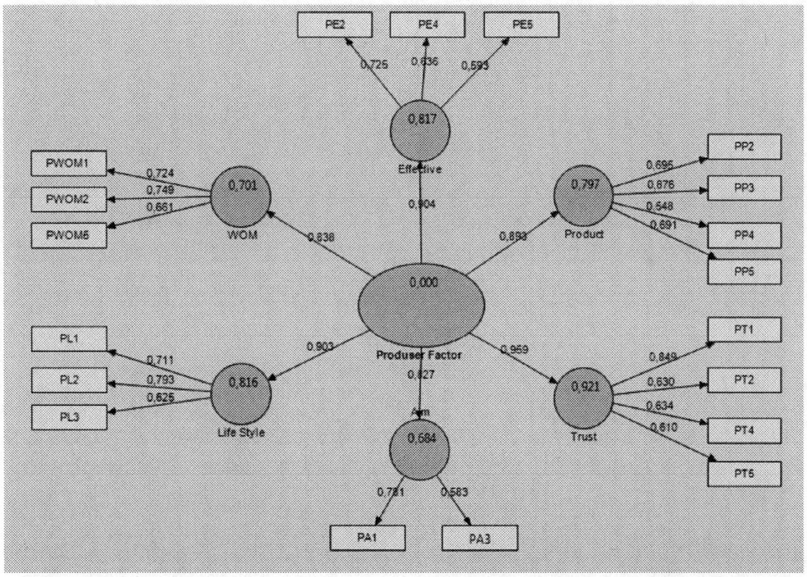

Figure 3. *Standardized Loading Factor* Inner dan Outer Model Business with Instagram

Trust is the dominant factor of respondents follow in organic vegetables community by using "*instagram*". It is because "*instagram*" spreads information quickly and can be trusted. This trust also correlates with the reputation of organic producers.

3.4.2. Business Model with Training. **Table 2** shows the values of sixth latent composite reliabilities of training. All values are above 0.7. Thus, it can be said that the variable has good reliability as a measuring tool.

Table 2. Results of Composite Reliability of Training

	Composite Reliability
Aim	0,815825
Effectiveness	0,6717417
Life Style	0,774438
Producer Factor	0,845010
Product	0,798011
Trust	0,728889
WOM	0,771241

From the value of loading factorthe WOM factor has the highest loading factor among others, which is equal to 0.919. This shows the reason of respondents follow the business model of training is the existence WOM, followed by lifestyle (with loading factor of 0.881); trust (with loading factor of 0.878), and three other factors are considered lower, namely the product, objectives, and effectiveness.

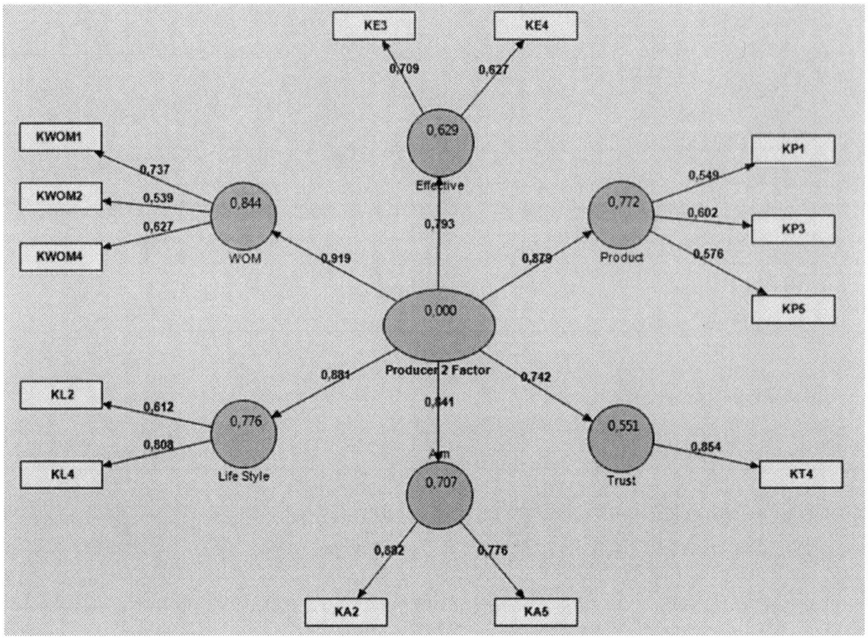

Figure 4. *Standardized Loading Factor* Model Business Training

The WOM is the dominant factor of respondents follow in the organic vegetables community by using training as media promotion. Training alumnus usually spread the information indirectly to friends, relatives, peer groups and people around them. WOM is the most effective medium in promoting a product or service [6], [7].

3.5. Constraints Faced by Producers and Consumers for Organic Vegetables

For producers, the main constraint in organic cultivation is the waste from the surrounding environment that hampers the cultivation of organic vegetables. While for consumers, the main constraint is the price of organic vegetables, which is expensive. The high price of organic vegetables can be understood because organic vegetable producers are still small. Moreover, it requires persistence and patience in the process of farming. Constraints of producers and consumers can be seen in **Table 3**.

Table 3. Constraints Faced by Producers and Consumersof Organic Vegetable

Producers	%	Consumers	%
- Long cultivation time	34.00	- High price of organic vegetables	50.00
- Poor physical appearance	20.00		
- The existence of waste from the surrounding environment	46.00	- Place of sale in supermarket	26.67
		- Availability of desired vegetables	23.33

3.6. Expectation of Producers and Consumers of Organic Vegetables

The main expectation of producers of organic vegetables is the low cost for certification, while the main expectation of consumers is the presence of a reliable Indonesian organic emblem. Expectations of producers and consumers can be seen in **Table 4**.

Table 4. Expectation of Producers and Consumers of Organic Vegetables

Producers	%	Consumers	%
- Low cost of certification	60.00	- The existence of quality control institutions	13.33
- The existence of facilities from the government	20.00	- The presence of symbol of organic vegetables of Indonesia	26.67
- Easiness to get investors	7.00	- Easiness to get	16.67
- Easiness to get the loan	13.00	- An affordable price	20.00
		- Green packaging	6.66
		- Honesty of producers	16.67

4. Conclusions

There are two business models in promoting organic vegetables, using "*instagram*" and training. Trust is the main factor of respondents in the business model using "*instagram*," while WOM is the main factor of respondents in business model by training. The main constraint of producers in organic vegetable is the waste from the environment that hampers organic vegetables cultivation. For the consumer, the high price of organic vegetables is the main constraint. The main expectation of producers in organic vegetables is the low cost of certification, while for consumers is the presence of a trusted Indonesian organic emblem. The main constraint of producers in organic vegetable is the waste from the environment that hampers organic vegetables cultivation. For the consumer, the high price of organic vegetables is the main constraint. The main expectation of producers in organic vegetables is the low cost of certification, while for consumers is the presence of a trusted Indonesian organic emblem.

References

[1] Hubeis, Musa. 2014. Prospek Cerah Produksi Sayuran Organik Bernilai Tambah Tinggi Berbasis Petani. *Jurnal Risalah Kebijakan Pertanian dan Lingkungan*. **Vol. 1 No. 2**. pp. 112.

[2] Monecke, Armin and Friedrich Leisch. 2012. SEM PLS: Structural Equation Modeling Using Partial Least Square. *Journal of Statistics Software*. **Vol 48. Issue 8** pp. 1-32

[3] Samsudin dan Satrio, T. 2004. *Kiat Bercocok Tanam Sayuran Organik*. Jakarta: Lembaga Pertanian Sehat Dompet Dhuafa Republika.

[4] Ramon Casadesus-Masanell and Joan Enric Ricart. 2010. *From Strategy to Business Models and onto Tactics*. Long Range Planning 43 (2010). http://www.elsevier.com/locate/lIDR

[5] Osterwalder, Alexander, and Pigneur, Yves. 2012. Designing Business Models and Similar Strategic Objects: The Contribution of IS," *Journal of the Association for Information Systems*: **Vol. 14: Iss. 5** https://aisel.aisnet.org/jais/vol14/iss5/3

[6] Murtiasih, Sri., Sucherly., Hotniar Siringoringo. 2013. How Words of Mouth Influence Brand Equity for Automotive Products in Indonesia. *Procedia- Social and Behavioral Sciences* **81**. Pp 40-44

[7] Deliana, Yosini, Lucyana Trimo, Sulistyodewi and Agriani Hermita Sadeli. 2016. Magic of Words of WOM in Promoting Agriculture Products. *International Journal in Economics Research*. **Vol. 13. Number 5**. pp 2312 - 2332

Attachment

Goodness Of Fit Business Model with Instagram

To validate the overall model, goodness of fit (GoF) is used. The GoF index is a single measure used to validate the combined performance of the measurement model and the structural model. This GoF value is obtained from the average communal index multiplied by the value of R^2 model. Here are the results of the calculation of goodness of fit model:

Tabel 1. Results of *Average Communalities Index*

	AVE	R Square
Aim	0,574824	0,683645
Effective	0,527189	0,816787
Life Style	0,507838	0,815939
Product	0,506993	0,796672
Produser Factor 1	0,510426	
Trust	0,572767	0,920527
WOM	0,507886	0,701474
Rata-rata	0,529703	0,789174
GOF	0,646551	

Goodness Of Fit Model Business Training

Table 2. Hasil *Average Communalities Index*

	AVE	R Square
Aim	0,689820	0,706781
Effective	0,547494	0,628562
Life Style	0,513481	0,776211
Product	0,593329	
Produser Factor 2	0,531802	0,772289
Trust	0,728889	0,550682
WOM	0,508844	0,844409
Rata-rata	0,587666	0,713156
GOF	0,647377	

Based on **Table 2**, the average result of communalities is 0, 529703. This value is then multiplied by R^2 and is rooted. The calculation results show that the GoF value of 0.646551 is more than 0.36. Therefore, it is categorized as a big GoF, meaning that the model is very good (has the high ability) in explaining empirical data.

IOP Conf. Series: Earth and Environmental Science **166** (2018) 012049 doi:10.1088/1755-1315/166/1/012049

The Role of an Interdisciplinary Approach to Improving Farmers' Resilience to Climate Change: Its Potentials and Challenges

Y. T Winarto[1]

[1] Department of Anthropology, Faculty of Social and Political Sciences, Universitas Indonesia Kampus Baru UI, Depok, Jawa Barat, 16424 Indonesia

E-mail: yunita.winarto@gmail.com

Abstract. Farmers are now experiencing hardships resulting from the increasing variability of climate, global warming, and the more frequent (and severe) extreme events. Those consequences of climate change on agriculture produced increasing vulnerabilities of food crops ecosystems and threats to farmers' resilience in the changing environment. Without any timely and appropriate services from the government, the farmers would be "trapped in their cages" by not knowing the phenomena beyond their ability to observe. Since farmers are the decision makers of their own fields, providing new scientific knowledge in agrometeorology is urgent. The absence of effective extension services, however, challenged the scientists in their efforts to assist farmers. Based on the experience of institutionalizing an educational commitment to groups of farmers in Indonesia, an inter-disciplinary approach between the natural scientist (agrometeorology) and social-cultural scientist (anthropology) has proved to be an effective means in providing climate services to farmers. The paper examines the ways the inter-disciplinary approach has been developed in the past decade in two regencies in Indonesia (Indramayu Regency and East Lombok Regency). An ongoing reflexivity and inter- subjectivity between the scientists from different disciplines and between scientists and farmers has been a significant means in making the collaboration works effectively.

1. Facing Climate Change, Failing to Understand

"Why has the rain stopped all of a sudden? Why does the dry spell last a very long time (up to 30 days)?" asked farmers curiously after experiencing a long dry spell after very intensive rainfall in early October 2007. These are only two questions among many others that have been raised by farmers over the past decade. Since farmers thought that the rainy season had already started, they began to start planting. During the long drought in the dry rain-fed fields, maize and rice were damaged (see Winarto and Stigter, 2011; Anantasari, Winarto, and Stigter, 2011). Unfortunately, such a risk was also experienced by many farmers in various other places in Java under the same unusual consequences of climate change. The phenomenon of a "false start of rain" grew, yet farmers were not well-equipped with the appropriate anticipation capabilities to avoid repeatedly creating wet nursery beds.

Our observation was validated by some rainfall observers in the regency of Indramayu in West Java who explained that the risks farmers experience were due to their ignorance of future climate conditions. This ignorance has persisted given the absence of an early warning system and climate services provided to the farmers by those in authority. Farmers are left on their own to struggle against various kinds of "puzzling phenomena" emerging in their habitat.

Farmers are meticulous observers of their own environment. Their rich empirical observations are a good means to improve their knowledge. For them, "seeing is believing" and "seeing is also knowing". Without seeing, they do not gain any belief, and without any belief, they would not consider what they see as relevant to their farming. Thus, their knowing is significantly based on their empirical senses, and not on any other tools to help them know things beyond their ability to observe, see, taste and experience. Another factor enriching their knowledge is the extent to which the things they observe have cultural meaning (see Bentley, 1989, 1992; Winarto, 2004).

Brown planthopper (BPH), a very devastating pest for rice nowadays, was previously a non-pest for farmers when they were planting their own traditional rice. BPH became a pest after the introduction of the Green Revolution technological package, which involved the planting of high yield rice varieties. These varieties were susceptible to pests/diseases and thus chemical pesticides were introduced to farmers as a way to control these pests/diseases (Fox, 1991). Whilst spraying chemical pesticides has gradually become part of their farming culture, knowledge of the existence of "small insect-natural enemies" and their role in preying on insect-pests have been kept outside their empirical reach. Only after the introduction of the Integrated Pest Management Farmer Field Schools (IPM FFSs) in early 1990 has the function and role of insect-natural enemies gradually become part of the IPM-farmers' schema of controlling pests/diseases. Thus, natural enemies had a new cultural meaning for the IPM farmers (see Pontius, Dilts and Bartlett, 2002; Winarto, 2004). This is an example of how observable phenomena, which are also culturally important, can become part of farmers' schema (Bentley 1989, 1992; Winarto, 2004). Within this way of knowing and learning, farmers cannot foresee what is really going to happen in the future. As this is the case, how do they make decisions on what should be taken into account to sustain their farming in the context of ongoing climate change?

Farmers' empirical knowledge, their experience as a result of the accumulated precedents in their life, and the inherited local cosmology from previous generations have become their points of reference in decision making. A combination of elements - stored in their schema – that is recalled by farmers at a particular time depends on their past experiences. These experiences result in lessons learned regarding the best or otherwise failed strategies for a rainy, dry, or second dry-season planting. However, if the forthcoming season produces different conditions than the past, their conventional adaptation strategies could yield a maladaptive result for both the environment and their own welfare (see Bartlett, 1980). Without any provision of knowledge (from scientists, governments, or other parties) about foreseeable climate conditions, farmers remain trapped in the "cages" of their past and present (see Winarto, Stigter, and Ariefiansyah, 2015, 2017; Winarto, Walker, and Ariefiansyah, 2017). While farmers experience increasing unusual risks and/or opportunities as a result of weather conditions, they do not understand the relationship between these conditions and climate change. They have only heard about "climate change" and "global warming" through the media, and they do not know of the significant consequences of climate change for agriculture derived from the increasing variability of the climate, global warming, and the more frequent (and more severe) extreme events (see Stigter and Winarto, 2015).

In the case of a "false start of a rainy season", once farmers know the probability of this situation, and understand "dry spells" and the increasing number of dry days in the rainy season, they can decide to wait until the true rainy season starts. "Without understanding these factors, farmers would hastily start to plant. As explained by a rainfall observer who joined the Science Field Shops (SFSs) in 2009, "once the dry-spell comes, the nursery would dry up, and the seeds would fail to grow. Therefore, in these instances, farmers have to redo the nursery beds several times which is very costly."

SFSs are arenas where farmers, scientists, other experts and extension workers meet to discuss and share any vulnerabilities found by farmers and how to solve them. In the SFSs, farmers also learn new knowledge on agrometeorology to enrich their schema of farming in the context of climate change (see Stigter and Winarto, 2012, 2013, 2015; Winarto and Stigter, 2013, 2016, 2017). In Bahasa Indonesia, we call the SFSs Warung Ilmiah Lapangan.

IOP Conf. Series: Earth and Environmental Science **166** (2018) 012049 doi:10.1088/1755-1315/166/1/012049

2. Transferring Scientific Knowledge to the Local Domain of Knowing

Farmers were indeed grateful to have some help in order to improve their anticipation capabilities (see Winarto, Stigter and Ariefiansyah, 2015, 2017). Once they received a "climate seasonal scenario" from the agrometeorologist (the late Stigter and later, Walker), which was then translated into Indonesian by an anthropologist and sent to farmers via mobile phone, the farmers could imagine that the forthcoming season would have near normal, or either above or below normal rainfall. Sometimes, they also received the warning of a weak or strong El-Niño or a weak or strong La-Niña. Gradually, they gained the skills to interpret these "alien" terms and relate them to their existing knowledge of planting seasons. They confessed that without practicing agrometeorology in their own fields and having regular monthly discussions, they would not have been able to understand these alien terms.

Seasonal climate scenarios are only one among seven climate services provided by scientists for farmers in the SFSs. The following are the seven climate services:

1. Guidance for daily rainfall measurements by all rainfall observers in their own plots;
2. Guidance about daily agroecological observations;
3. Measurement of yields and the understanding of differences between fields, seasons and years;
4. The organization of the SFS;
5. The development and exchange of monthly updated seasonal climate predictions in the form of seasonal rainfall scenarios;
6. The exchange of new knowledge related to the above, and
7. Guidance on the establishment of field experiments to develop best practices and obtain on-farm answers to urgent local questions.

See Stigter (2016); Winarto and Stigter (2016, 2017); Walker (2017).

Measuring daily rainfall in their own fields by using farmer-made rain gauges was one of the prominent services introduced by the scientists. Observing their fields' agroecosystems in detail, supplemented with written documentation, was another service provided in the SFSs. In combination with measuring daily rainfall, the agroecosystem documentation and analysis become a strong foundation for advancing farmers' understanding of the impacts of a particular rainfall pattern on the growth of crops and the condition of fields. Gradually, they became able to relate the rainfall data and the qualitative nature of rain with their own lexicons and taxonomy (see Winarto *et al.*, 2011, 2013). By experiencing this exercise daily for several years in a row, they have developed their senses and ability to classify "near normal, above normal, or below normal" rainfall and their impact on plants and fields.

Together with the new forms of knowledge the farmers received through dialogues and discussions in the SFS monthly meetings, the farmers have gradually been able to enrich their schema of farming by incorporating previously "unseen and unforeseen" phenomena. For them, such a wider horizon in interpreting their "world of farming" has been significant indeed to improve their anticipation capability towards the ongoing and the forthcoming planting season. Some farmers expressed their gratitude of knowing the probable "wet-dry season planting following the first dry season in 2016", as they could avoid planting "watermelons" in fallow periods. Whilst some farmers experienced harvest failure and spent their capital without earning anything, some rainfall observers did not lose their capital as they did not plant this commodity. Knowing the forthcoming "wet dry season", a group of rainfall observers in a village in East Lombok, who used to plant tobacco in the dry season, decided to diversify their cropping strategies. Those who did not decide to plant tobacco and made deeper drainage, preferred to plant rice and tobacco to spread risks, or cultivated rice or maize instead of tobacco. These farmers could save their

IOP Conf. Series: Earth and Environmental Science 166 (2018) 012049 doi:10.1088/1755-1315/166/1/012049

harvests in the midst of the hardships faced by those who kept planting tobacco as usual. In the previous long drought of 2015 under a strong El-Niño, some rainfall observers - followed by their neighbouring fields - gained benefits by planting mung-beans which could withstand drought (see Winarto, Stigter, Ariefiansyah, 2015, 2017; Winarto *et al.*, in preparation).

These stories are exemplars of the improvement in farmers' anticipation capabilities once they are able to "foresee" the forthcoming unusual risks of El-Niño, La-Niña, or normal seasons with increasing variability of rainfall. The provision of new agrometeorological knowledge as an important component of climate services has proved to be beneficial for farmers who have to rely only on their empirical knowledge and precedents from past and present experiences. Face-to-face dialogue with scientists and an enduring commitment between farmers and scientists in advancing farmers' learning capabilities are evidence of the effective means of improving farmers' resilience to climate change. How can this be further developed?

3. Developing an Inter-Disciplinary Approach: Practicing Anthropology in Applying Agrometeorology

"Where are the IPM facilitators? Why don't they return to us? We have so many questions to ask in this ongoing pest infestation," complained some IPM farmers at the time when I was the only one left in the field following their practices in the post-period of IPM FFS training in the early 1990s. Interestingly, yet not surprisingly, I experienced the same situation in 2007 in the post-period of the Climate Field School (CFS) in another place (in Gunungkidul Regency in Yogyakarta). Farmers were "puzzled" after following the introduction of the "water harvesting method" which involved making dikes inside their small fields to retain water in the soil in their karst-dry rain-fed ecosystem. That was the introduced technology in the CFS at the end of their training. Without any warning of the incoming La-Niña in the following rainy season, farmers kept their additional dikes in the field. Accordingly, their young plants were damaged in flooded fields due to the excessive rains and the existing dikes. None of the CFS facilitators returned to the farmers and no climate services were provided in the post-period of the training. Instead, the facilitator asked us, the anthropologists, to assist farmers during our stay in the field. As anthropologists, we did stay to continue our ethnographic fieldwork. But, why, the CFS facilitator asked us as anthropologists to assist as we did not have any expertise in meteorological climatology, or agrometeorology, stay (see Winarto and Stigter, 2011, 2017)?

The reality was that any new knowledge and technology provided by the state was packaged as a "one-season-long training program" by adopting the method of Farmer Field Schools, which were developed in the early 1990s to introduce IPM principles and strategies. Yet no follow-up activities were institutionalized with the farmers to help them internalize this new knowledge and introduced technology. Even though a number of multi-disciplinary scientists had been working together to prepare the IPM FFSs curriculum and learning methods (see Pontius, Dilts and Bartlett, 2002), the implementation of the "schools" were entirely in the hands of agricultural officials. If there were some anthropologists observing the program and the post-period of the training, they were positioned more as "observers". Based on such an experience and the intention to help farmers in their adaptation to the consequences of climate change, the SFSs were not designed as "short-period training programs" with a pre-fixed curriculum prepared in advance.

From the beginning, an inter-disciplinary approach was initiated by accident with the presence of anthropologists in the field. A Dutch agrometeorologist - who passed away in May 2016 during his visit to farmers in Indramayu - was willingly assisting the farmers to learn agrometeorology. He designed the method of measuring rainfall and observing the agroecosystem in detail. Instead of doing this in an "observation or demonstration plot", as practiced in the IPM FFS and CFS, he asked each farmer to collect the rainfall and agroecosystem data in their own field. During his visits, he answered farmers' questions on various agronomic problems and not the problems of climate change per se. Yet enabling the farmers to carry out the nitty-gritty of agrometeorological observation and documentation was beyond his capacity. As anthropologists left in the field, we developed our own innovative ideas to facilitate the farmers *in situ* by referring to the rules set up by the agrometeorologist. By following a group of farmers and observing how they innterpreted the new ideas, how they dealt with the problems they faced in mounting the rain gauges, the difficulties they had in documenting the data, and their way of evaluating the data and discussing the problems, we learned - together with the farmers - how to improve their learning day by day (see Prahara, Winarto, and Kristiyanto, 2011; Winarto and Stigter, 2011). Once we discovered the core problems related to the agrometeorological methods or content, we consulted the agrometeorologist via email. As a professor at several universities in Africa, he could not stay in the field for long periods of time.

Over time, the collaboration grew. From season to season, we both discovered the complexity of farmers' vulnerabilities, the constraints in implementing and continuing the learning, the kinds of help the farmers needed to improve their understanding, as well as the finances we needed to keep the educational commitment going, and the challenges from other parties, including state authorities. Gradually, the agrometeorologist developed an understanding of the problems which led to the "garbage in, garbage out" style of data collection. "Why were the farmers sloppy in documenting the data?" questioned the agrometeorologist when he saw the incomplete data the farmers had collected. The anthropologists knew how difficult it was for the farmers to develop a new kind of *habitus*, writing the complexity of their fields' ecosystem into numbers and words. Only through time and persistent efforts by both parties did the tradition of carrying out more detailed observations, more precise documentation, and better ways of interpreting and analyzing their own data become incrementally established (see Winarto and Stigter, 2013, 2016). The establishment of new learning institutions could only be accomplished through intimate interactions between the farmers and the anthropologists (Winarto and Stigter, 2013, 2016; Taqiuddin, 2016), and the building of "trust" between them (see Prihandiani, 2016). Such is the strength of anthropological methodology which involves immersing oneself into the community they are studying.

The anthropologists pursued their traditions of learning from the local communities by listening to their voices, following their activities, and observing their circumstances to understand better why some ideas were internalized by farmers and some were not. Ongoing reflexivity was the main method of improving the ways of establishing the farmers' learning. Such an accomplishment would not have been successful without the persistent curiosity and interests of the agrometeorologist to make the agrometeorological science simpler so that it could be applied by farmers *in situ*. The main aim of applying agrometeorology was to help as many farmers as possible. Providing climate services to farmers in as simple way as possible was his great motivation. The seven climate services developed were also the product of gradual thinking by the agrometeorologist, who incorporated lessons learnt from the reality on the ground. On many occasions, as exemplified in his publications, he told me and my colleagues that he would not have been able to do this work alone without our help. "I used to talk to my students and agricultural extension staff in the roving seminars I developed. But reaching farmers in their own fields would not have been possible without your help. Only in Indonesia am I able to do this," claimed Stigter. For us, it was also the first time we had developed an inter-disciplinary project.

Similar to the agrometeorologist's claims, the anthropologists would have never been able to assist farmers to improve their resilience to the consequences of climate change without collaboration with the agrometeorologist. Facing the increasing uncertainty and complexity of today's world, there was no other effective way to help local people but to draw on the willingness of each mono-disciplinary scientist to cross the borders of their own discipline. Figure 1 presents a visualization of our collaboration, which incorporated both interdisciplinary and trans-disciplinary approaches (Winarto *et al.*, in preparation; also see Stigter and Winarto, 2016).

IOP Conf. Series: Earth and Environmental Science 166 (2018) 012049 doi:10.1088/1755-1315/166/1/012049

4. Developing Why it is Not Easy to Develop an Interdisciplinary Approach: A Reflection

In the past decade, I have been trying to approach scientists of two different disciplines: agricultural scientists and anthropologists from several other universities throughout Indonesia. I introduced the SFS and its benefits for farmers who are facing increasing uncertainty due to climate change. Unfortunately, these efforts have not produced any significant results. Why? Firstly, crossing the border of each disciplinary boundary and even the faculty's boundary to establish an interdisciplinary network has not been a tradition in Indonesian universities. Reflecting on my own work in the past decade, I can honestly say that I have been able to develop such collaborations with scientists from abroad, since there is no Faculty of Agriculture at my university (Universitas Indonesia). However, my efforts to establish a similar network with agrometeorologists from another university, Bogor Agricultural Institute, failed to produce an agreement on working collaboratively. It is likely that it would be possible to introduce applied agrometeorology in an arena such as SFSs, yet anthropological methods of working with the local community have not been a "tradition" of Indonesian natural scientists.

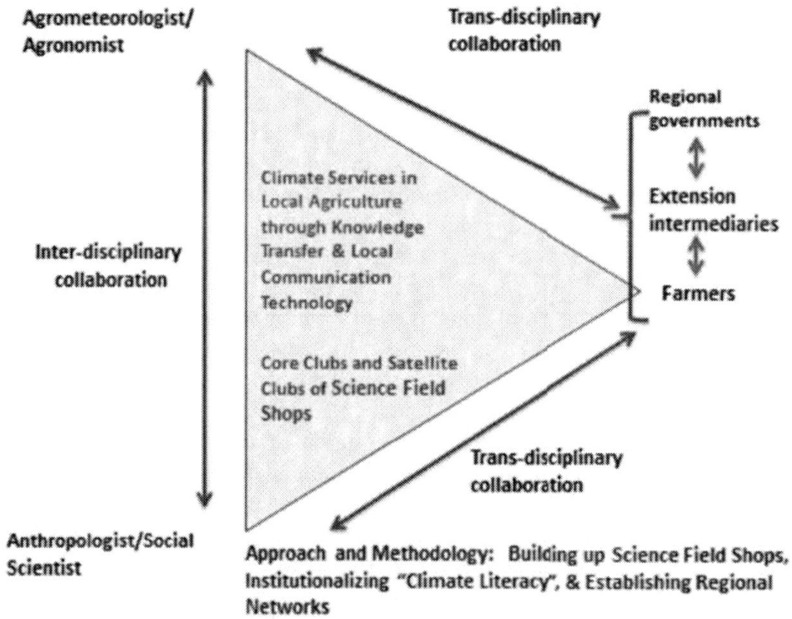

Figure 1. The inter- and trans-disciplinary collaboration in "Science Field Shops"
Source: Center for Anthropological Studies, FISIP Universitas Indonesia for FAO Webinar on FFS and Climate Change, 3rd November 2016

Secondly, attempts to collaborate with three other universities which have both agricultural scientists and anthropologists also failed. Not all anthropologists in Indonesian universities have an interest in developing their field of studies and expertise in the area of environmental anthropology, and in particular in the area of climate change. Therefore, without any interest in the same issues, it was difficult to initiate any collaboration with scientists from other faculties. If there was some interest from the faculty of agriculture, who are the lecturers specialized in agrometeorology or other related disciplines who could take the lead as the initiators?

Thirdly, the technology-driven approach has been the basis of various kinds of state development projects, including community engagement programs at universities. It is much easier to propose a project with a "short-period of training to introduce new technology or practices" rather than a project that has an educational commitment for a long-term period involving longitudinal fieldwork. Related to this is the constraint of securing financial support for long-term projects, or projects that need a thorough design incorporating entry and exit strategies.

The nature of the educational commitment itself, which necessitates ongoing inter-subjectivity and reflexivity between all parties involved, is indeed very demanding and challenging. However, as scientists we have to seriously develop our commitment to helping millions of farmers in Indonesia and throughout world in their effort to survive amidst increasing uncertainties as a result of climate change.

5. References

[1] Anantasari E Winarto YT and Stigter K 2011Learning and Understanding Some Scientific Knowledge of Climate in Winarto YT and Stigter K (Eds.). *Agrometeorological Leraning: Coping Better with Climate Change with a Foreword by Niels Röling.* Saarbrűcken: LAP Lambert Academic Publishing GmbH & Co. KG. Pp.44—85.

[2] Bentley JW 1989What Farmers Don't Know Can't Help Them: The Strength and Weaknesses of Indigenous Technical Knowledge in Honduras *Agriculture and Human Values* 6(3):25-31.

[3] Bentley JW 1992 Alternatives to Pesticides in Central America: Applied Studies of Local Knowledge *Culture and Agriculture* 44:10-13.

[4] Fox JJ 1991 Managing the Ecology of Rice Production in Indonesia, in J. Hardjono (Ed.) *Indonesia: Resources, Ecology and Environment* Kuala Lumpur: Oxford University Press. Pp.61—84.

[5] Pontius JR Dilts and Bartlett A 2002 *From Farmer Field School to Community IPM: Ten years of IPM training in Asia.* Bangkok: FAO Community IPM Programme, Food and Agricultural Organization of the United Nations Regional Office for Asia and the Pacific.

[6] Prahara H Winarto YT and Kristiyanto 2011The Joint Production of Knowledge: Its Dynamics, in Y.T. Winarto and K. Stigter (Eds.). *Agrometeorological Leraning: Coping Better with Climate Change with a Foreword by Niels Röling.* Saarbrűcken: LAP Lambert Academic Publishing GmbH & Co. KG 145-180.

[7] Prihandiani AF 2017 Menyebarluaskan Pembelajaran Agrometeorologi: Variasi Luaran dari Keagenan Petani Pengukur Curah Hujan Indramayu (Disseminating the Agromeeteorological Learning: Variation of Outputs based on the Agency of Indramayu Rainfall Observers)', Undergraduate Thesis. Depok: Universitas Indonesia.

[8] Stigter K 2016 Science Field Shops with Farmer Extension Intermediaries for Climate Services in Agriculture, *Training of Trainers (ToT) Roving Seminar Training Guidance, Selong and Indramayu, April—March 2016.* Translated into Indonesian by Y. T. Winarto. Bruchem/Bondowoso and Depok: Agromet Vision and the Centre for Anthropological Studies, Faculty of Social and Political Sciences, Universitas Indonesia.

[9] Stigter C(K)J and Winarto YT 2012 Coping with Climate Change: An Active Agrometeorological Learning Approach to Response Farming. Invited opening keynote presentation of the *APEC Climate Symposium 2012: Harnessing and Using Climate Information for Decision Making: An In-Depth Look at the Agriculture Sector, St. Petersburg, Russia.* Extended Abstract in Proceedings.

[10] http://www.apcc21.org/eng/acts/pastsym/japcc0202_viw.jsp

[11] Stigter C(K)J and Winarto YT 2013Science Field Shops in Indonesia. A Start of Improved Agricultural Extension that Fits a Rural Response to Climate Change *J Agric Sc Appl* 2(2) 112-123.

[12] Stigter C(K)J and Winarto YT 2015 Agrometeorological Learning: Exchanges of Traditional, More Recent Empirical and Scientific Knowledge between Farmers, Extension Intermediaries and Scientists', *Learning Agrometeorology with Farmers: Why is it Necessary to be Institutionalized?, Regional Workshop, Indramayu, 29 October 2015 and Selong, 5 November 2015.* Knowledge Sector Initiative (Bilateral Agreement between Australian and Indonesian Government) and the Centre for Anthropological Studies, Faculty of Social and Political Sciences, Universitas Indonesia. Represented in the *Environmental Education in Indonesia Workshop, Depok, 13—14 December 2015* University of Western Australia (Australia Research Council) and the Centre for Anthropological Studies, Faculty of Social and Political Sciences, Universitas Indonesia.

[13] Stigter C(K.)J. and Winarto YT 2016 Science Field Shops Approach, in S. Sala, F. Rossi and S. David (Eds.) *Supporting Agricultural Extension for Climate Smart Agriculture: An Overview of Existing Tools.* Compendium Climate-smart Agriculture and Extension. Rome: Global Alliance for Climate-smart Agriculture. 22-24.

[14] Taqiuddin M 2017 Struktur Sosial dalam Perubahan: Implikasi Pembelajaran Agrometeorologi dalam "Warung Ilmiah Lapangan" pada Dinamika Struktur Sosial Petani Tadah Hujan di Pulau Lombok (Social Structure in its Changing: The Implication of Agrometeorological Learning in the "Science Field Shops" on the Dynamics of Social Structure among Dry Rain-fed Farmers in Lombok Island)', Doctoral Thesis. Depok: Universitas Indonesia.

[15] Walker S 2017 Hand-out for Extension Trainers and Farmer-Facilitators in facilitating 'Science Field Shops adapted from K. Stigter (2016c) and translated by Y.T. Winarto, R. Ariefiansyah and A.F. Prihandiani. Depok: Center for Anthropological Studies, Faculty of Social and Political Sciences, Universitas Indonesia.

[16] Winarto YT 2004 *Seeds of Knowledge: The Beginning of Integrated Pest Management in Java.* Yale Southeast Asia Studies Series, Monograph 53. New Haven: Yale Southeast Asia Studies.

[17] Winarto YT and Stigter K(Eds.) 2011 *Agrometeorological Learning: Coping Better with Climate Change with a Foreword by Niels Röling.* Saarbrűcken, Germany: LAP Lambert Academic Publishing.

[18] Winarto YT and Stigter K 2013Science Field Shops to Reduce Climate Vulnerabilities: An Inter- and Trans-disciplinary Educational Commitment, *Collaborative Anthropologies* 6(1) 420-441.

[19] Winarto YT and Stigter K 2016 Incremental Learning and Gradual Changes: Science Field Shops as an Educational Approach to Coping Better with Climate Change in Agriculture", in L. Wilson and C. Stevenson (Eds.) *Promoting Climate Change Awareness through Environmental Education.* Hershey PA: Information Science Reference (An imprint of IGI GLOBAL) 60-96.

[20] Winarto YT and C(Kees).J. Stigter 2017 Anthropology Teaming Up with Agrometeorology: Getting University Science to Assist Farmers with Climate Resilience *Asian Politics & Policy* 9(3):479—492.

[21] Winarto YT Stigter K and Ariefiansyah R 2015 Interpreting the Present, Anticipating the Future: Continuous Learning in Ongoing Climate Change Paper & ppt.file presentation in the workshop on *Environmental Education in Indonesia.* Asia Research Council Grant: University of Western Australia and Center for Anthropological Studies, Faculty of Social & Political Sciences, Universitas Indonesia, Depok: 13th-14h of December.

[22] Winarto YT Stigter K and Arifiansyah R 2017 Anticipating the Future, *Inside Indonesia* 127 Jan--March 2017, http://www.insideindonesia.org/anticipating-the-future

[23] Winarto YT Walker S and Ariefiansyah R 2017 People, Clouds and Roots: Farmers Learning to Adapt Paper to be presented in the conference on *Agriculture in the Anthropocene* Cambridge: The Centre for Research in the Arts, Social Science and Humanities, Cambridge University, October 27, 2017.

IOP Conf. Series: Earth and Environmental Science **166** (2018) 012049 doi:10.1088/1755-1315/166/1/012049

[24] Winarto YT Stigter K Prahara HAnantasari E and Kristiyanto 2011 Collaborating on Establishing Agro-meteorological Learning Situation among Farmers in Java', *Anthropological Forum* 21(2) 175-198.
[25] Winarto YT Stigter K Dwisatrio B Nurhaga M and Bowolaksono A 2013 Agrometeorological Learning Increasing Farmers' Knowledge in Coping with Climate Change and Unusual Risks *Southeast Asian Studies* 2(2) 323-349.
[26] Winarto YT Walker S Ariefiansyah R Prihandiani AF Taqiuddin M and Nugroho ZC. *In preparation*. "Institutionalizing the Science Field Shops: Developing Response Farming to Climate Change" Manuscript submitted to Global Alliance for Climate Smart Agriculture (GACSA). Rome: FAO.

IOP Conf. Series: Earth and Environmental Science **166** (2018) 012050 doi:10.1088/1755-1315/166/1/012050

Phytase production by fungi based on palm oil mill effluent

A Sugiharto[1]

[1] Research Center for Biology, Indonesian Institute of Sciences
Jl. Raya Bogor Km 46, Cibinong Science Center, Cibinong 16911, Indonesia.

E-mail: one.myconano@yahoo.co.id

Abstract. The utilization of phytase enzyme containing feed to monogastric and digastric livestock could increase the efficiency of nutrients uptake and livestock resistance to disease attacks. Palm oil mill effluent (POME) is one type of waste that has not been widely used in enzyme production. Some fungi that grow on POME indicate capable of producing phytase. The aims of this study is to utilize POME as a media production of phytase through fermentation by fungi with the eddition of C and N as sources. The experiment was arranged by complete randomized design with 5 replications. The result showed that isolated fungi from palm oil mill effluent (POME) was able to grow with pH value of 4 to 8. Furthermore, the two isolated fungi could optimally grow in pH value of 5 to 7. Through spores microscopic of the cross section of two fungi, the fungi were identified as *Aspergillus niger* and *Neurospora crassa.* Further study showed different results regarding on phytase enzyme activity of these two fungi in a modified media consists of 1% and 5% of POME with the addition of sucrose as the C source and 1% of peptone as the N source. The highest phytase production by *Aspergillus niger* was 0.393 U/mL substrate in media contain with 5% of POME and addition of 1% of peptone, for 96 hours and at 30^{0}C. While for *Neurospora crassa,* the highest phytase production was 0.115 U/mL substrate, in media contain 1% of POME, 1% of sucrose, for 48 hours and at 30^{0}C.

1. Introduction

The palm oil sludge and palm oil mill effluent (POME) has received considerable attention in recent years [1]. Most of this material is just disposed away and polluted the environment [1-8]. POME also is most expensive and difficult waste to manage [6], becausethe factory requires a relative large amount of money to dispose the waste. POME have not been efficiently utilized by the factory except for fertilizer. Beside that, POME as the one of wastes still contains large concentrations of carbohydrate, proteins, nitrogen compouds, lipid and minerals [4,6]. So POME is an excellent raw material for bioconversion by biotechnological technique [9]. Utilization of POME as a raw material to produce phytase enzyme has never been report. Through preliminary research, it is known that some of fungi are able to grow in POME and have ability to produce phytase enzyme.

Phytase enzyme plays an important role to help breaking the phytate bond in animal feed derived from nuts. Phytase compounds are classified as anti-nutritional substances. The existence of this bond in animal feed can result in the feed not being easily digested. This bond will bind the essential elements for animal in the feed, such as protein and minerals [10]. The absence of the phytase enzyme in the gastrointestinal tract in monogastric animals results in low feed digestibility value [8,11-14].

The use of phytase as a mixture of monogastric and digastric animal feed has been done in several countries, but it has not been done in Indonesia. It might be caused by the expensive production costs. Nowadays, some innovation to utilize cheap materials for phytase production must be continuously pursued. Phytase production is available on fungal sources [15], but there is no report on phytase production by fungi based on POME as a media.

The aims of this study is to utilize POME as a media production of phytase through fermentation by fungi with the eddition of C and N as sources. The experiment was arranged by complete randomized design with 5 replications. The results of this study are expected to be able to provide better alternative on POME utilization and provide a new method to produce phytase enzyme using a cheap material.

2. Materials and Methods

2.1. Materials

POME, *Aspergillus niger, Neurospora crassa.,* potato dextrose agar (PDA)(Oxoid), buffer solution, mixed reagent, Glucose, Sucrose, Peptone, (Merck) Yeast (Difco), Ca-phytate (Sigma), distillated water, $MnSO_4$, $CaCl_2$, $MgSO_4$, $FeSO_4$, KCl, and NaCl (Merck) were used as materials in this study. *Aspergillus niger* and *Neurospora crassa* used in this study were obtained from our own culture collection, through isolation from POME derived from Palm Oil Processing, PT Perkebunan 4, Medan, North Sumatera.

2.2. Methods

2.2.1. Fungi inoculants

The *Aspergillus niger* and *Neurospora crassa* were cultivated in PDA and incubated for 96 hours, at pH 7, and 30^0C. After incubation, 20 ml of distillated water was added. This suspension was used as inoculants. Biomass inoculants were measured using a method from Garraway and Robert [16].

2.2.2. Phytase production in POME media

The mixture of 1% and 5% POME, 1% of sucrose and peptone solutions, and 200 ml of solution consists of 6 g of $C_6H_{12}0_6$; 0.1 g of $MgSO_4.7H_2O$; 0.1 g of KCl; 0.02 g of $FeSO_4$; and 0.02 g of NaOH were added to erlenmeyer flask. The solution was then homogenized and sterilized for 20 minutes in temperature of 121^0C and pressure of 1 atm. This liquid medium was conditioned for 24 hours in temperature of 30^0C. The prepared inoculant was then added to the solution and fermented in room temperature, and under stirring of 150 rpm.

2.2.3. Measurement of phytase

The test of phytase production capability by *A. niger* and *N. crassa* was done in 1% and 5% of POME medium and with addition of 1% of sucrose as carbon source and 1% of peptone as nitrogen source. The phytase analysis was performed at 30^0C for 48 hours, 96 hours and 144 hours after fermentation.

Phytase activity was measured using Engelen, *et al.*[17]. The centrifugation supernatant was taken for 50 μL and transferred into test tube. After that 50 μL of calcium phytate 1% was added to the extract enzyme and incubated for 30 minutes. Then 160 μL of mixed reagent prepared by mixing 10 mL of H_2SO_4 5N, 6 mL of ammonium molybdate 0.0032 M, 3 mL of ascorbic acid 0.1 M, and 1 mL of potassium antimonyl 0.0086 M was added into the solution. After that the sample was incubated for 3 hours, and then the phytase activity was measured using UV-spectrophotometer at wavelength of 880 nm.

One unit of phytase activity (U) is expressed as the amount of enzyme released for 1 μM phosphorus/min. The phytase activity unit is calculated using formula below:

$$\text{Enzyme Activity (U/g)} = \frac{\frac{[\text{phosphate}]\text{mg}}{1} x 1000\mu \frac{\text{g}}{\text{mg}} x \frac{1\text{L}}{1000\text{mL}} x 5\ \text{mL} \frac{\text{enzyme}}{1} g}{\text{BM}\ x\ \text{v}\ x\ \text{t}} x\ \text{fp}$$

Note:

v : Phytase volume

t : Incubation time

BM : Molecular weight

Fp : Dilution factor

3. Results and Discussion

Aspergillus niger and *Neurospora crassa* were grown on PDA for 4 days at 30^{0}C and pH 7 (**Figure 1**). The biomass used from both isolates *Aspergillus niger* and *Neurospora crassa* was about 0,0345 g and 0.0337 g (**Figure 2A**). This biomass was obtained by growing both isolates on PDA slank at 30^{0}C, pH 7 for 5 days using in reaction tube with 2.5 cm diameter.

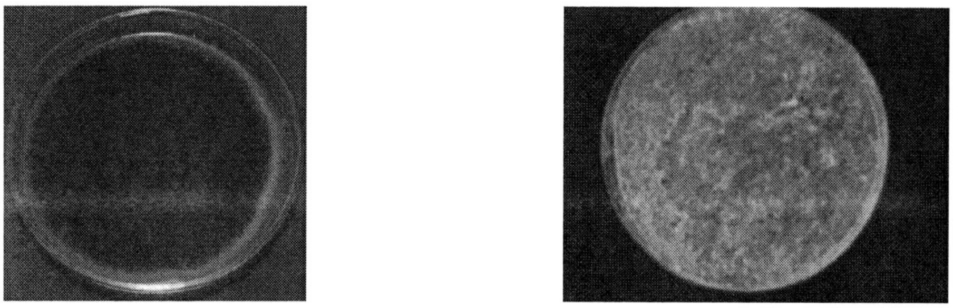

A B

Figure 1. Morphology of *A. niger* (A) and *N. crassa* (B) grown in PDA after 4 days incubation at pH 7 and 30^{0}C.

The results of pH and POME concentration treatment on the phytase activities showed that both isolates, *Aspergillus niger* and *Neurospora crassa,* had quite different results (**Figure 2B**). In general, *N. crassa* had higher phytase activities than that of *A. niger* on both treatments of 1% or 5% of POME and different pH levels. However, the addition of POME may not always increase the phytase activities. The highest phytase activity for *A. niger* was 0.0389 U/mL substrate, at pH 5 and 5% of POME. While for *N. crassa,* the highest phytase activity was 0.0653 U/mL substrate at pH 6 and 5% of POME.

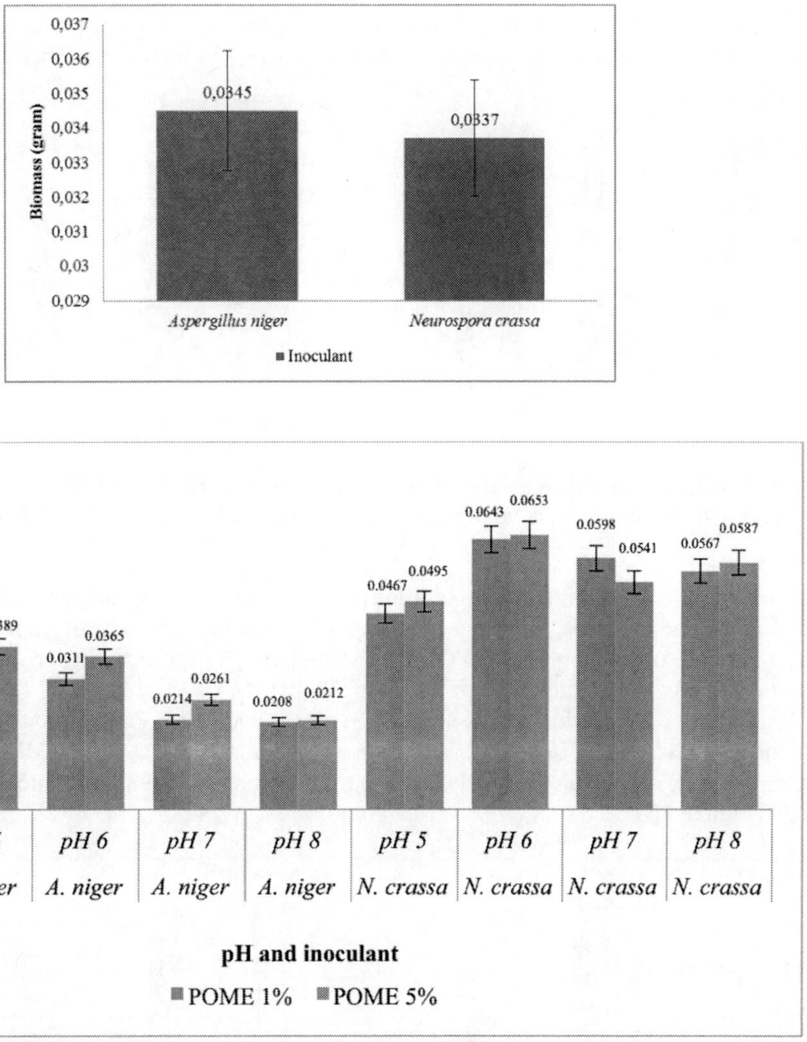

Figure 2. Biomass for *A. niger* and *N. crassa* grown in PDA slank after 96 hours (above), Phytase actvity for *A. niger* and *N. crassa* in different pH values with addition 1% and 5% POME (below)

The effect of adding 1% of sucrose to the production of phytase from *Aspergillus niger* and *Neurospora crassa* on media containing 1% and 5% of POME showed different results. In general, the addition of 1% of sucrose can increase the production of phytase compared to the control (**Figures 2B** and **3A**). In *A. niger* , the maximum results were obtained on 1% of POME for 48 and 144 hours incubation, 0,145 U/mL substrate and 0.144 U/mL substrate. While in 5% of POME the maximum result was 0.142 U/mL substrate, and wasobtained after 96 hours incubation (**Figure 3A**). This condition is different from the isolate of *N. crassa*, where the addition of 1% of sucrose, in 1% and 5% of POME will decrease the phytase production as increases in incubation time. The maximum result was 0.115 U/mL substrate, for 48 hours incubation and 1% of POME (**Figure 3B**). This condition indicates that the carbon source of sucrose is better used for *A. niger*. The carbon source used will greatly affect the phytase production [18,19] and sucrose is the best carbon source for phytase production by *A. niger* [19]. This condition is possible because the addition of carbon source will increase cell production after 24 hours [20]

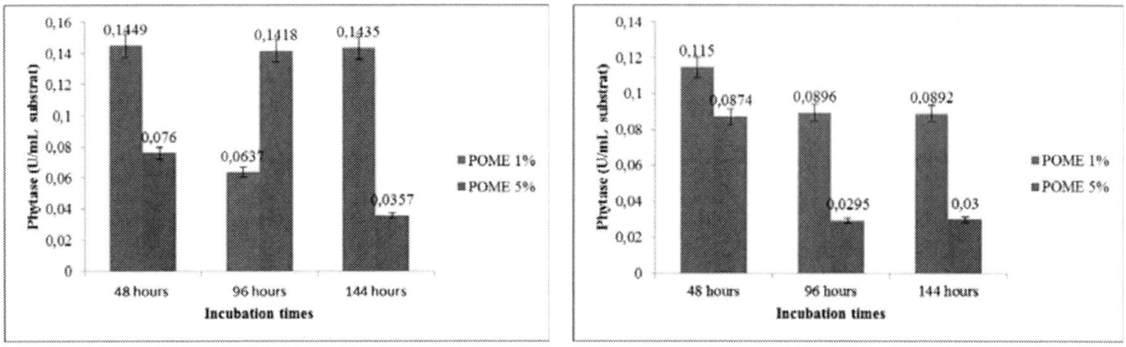

A B

Figure 3. Phytase activity of A. niger at pH 5, and at 30^{0}C with 1% and 5% of POME and 1% of sucrose (A) Phytase activity of N. crassa at pH 6, and at 30^{0}C with 1% and 5% of POME and 1% of sucrose (B)

Different results were obtained in addition of 1% of peptone in medium containing 1% and 5% of POME for the same isolate. In *Aspergillus niger*, the addition of 1% of peptone gave better phytase activity on media containing 5% of POME. The highest phytase activity for *A. niger* was 0,393 U/mL substrate obtained at incubation time of 96 hours (**Figure 4A**)

In *Neurospora crassa*, addition 1% of peptone on 1% and 5% of POME produced different results. The maximum of phytase of 0.109 U/mL substrate was obtained at 1% of POME medium after 48 hours incubation period, meanwhile 5% of POME had the best result at 98 hours incubation with 0.105 U/mL substrate (**Figure 4B**).

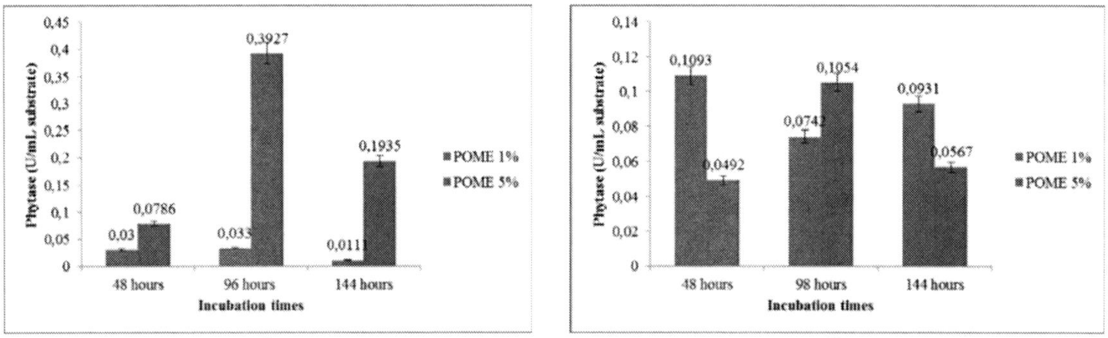

A B

Figure 4. Phytase activity of A. niger at pH 5, and at 30^{0}C with 1% and 5% of POME and 1% of peptone (A), Phytase activity of N. crassa at pH 5, and at 30^{0}C with 1% and 5% of POME and 1% of peptone (B)

Several bacteria, fungi, yeasts and actinomycetes were identified having capability to produce phytase in several different environmental and media conditions [13,18-19, 21-24]. Some fungi such as *Aspergillus niger* and *Neurospora crassa* have the ability to produce phytase [21,22]. Purwadaria *et al* [21] has reported that *Aspergillus oryzae* has abilities to produce phytase with POME and rice bran media.

The results indicated that the pH and type of inoculant used were important factors in controlling the production of phytase enzymes (**Figure 2B**). *Aspergillus niger* and *Neurospora crassa* have different abilities in the production of phytase in POME media. As shown in **Figure 3**, the phytase production by *A. niger* was optimum at pH 5 (0.0389 U/mL substrate) and *N. crassa* was at pH 6 (0,0653 U/mL substrate). The productivity of some fungi in producing phytases is very different and one of the factors to be considered is pH [8,18,19]. Sucrose became one of the best sources of carbon to produce phytase by *A. niger* [19]. *A. niger* has the ability to produce phytase on POME [21], and it is commonly used for production of phytase [25]. Addition of small amounts of nitrogen to the fermentation medium can trigger the production of phytase [18]. Phytase production are available on fungal sources [26] but there is no report on phytase production by fungi base on POME as a media. The content of the material, pH and the eddition of C or N in the media is known to affect the microbes activity in producing phytase enzymes [18-20].

4. Conclusion

Both *Aspergillus niger* and *Neurospora crassa* are types of fungi which are able to grow well in POME. The optimal phytase production of *A. niger* was 0.3927 U/mL substrate, and was obtained in the addition of 1% of peptone with 5% of POME, and incubated for 96 hours, at 30^0C. The optimal phytase production of *N. crassa* was 0.115 U/mL substrate, and was obtained in the treatment of 1% of POME, after 48 hours of incubation time with the addition of 1% of sucrose at 30^0C.

Acknowledgements
This research was funded by Research Center for Biology 2016. My appreciation and thanks are dedicated to Mr Maman Rahmansyah and Devi Ayu for their helpful insights and stimulating comments.

References

[1] Nwuche CO, Doris CE, Chijioke NE, Hideki A and James CO 2014 Use of Palm Oil Mill Effluent as media for cultivation of *chlorella sorokinia British Biotechnology Journal* **4**(3) 305-16

[2] Okwute, Loretta O and Isu NR 2007 The environmental impact of palm oil mill effluent (POME) on some physico-chemical parameters and total aerobic bioload of soil at a dump site in Anyigba, Kogi State, Nigeria *African Journal of Agricultural Research* **2**(12) 656-62.

[3] Rupani PT, Rajeev PS, Hakimi I and Norizan E 2010 Review of current palm oil mill effluent (POME) treatment methods: vermicomposting as a sustainable practice *World Applied Sciences Journal* **10**(10) 1190-1201

[4] Yejian Z, Hairen Y, Xiangyong Z, Zhenjia Z and Li Y 2011 High-rate mesophilic anaerobic digestion of palm oil mill effluent (POME) in expanded granular sludge bed (EGSB) reactor *Int.Confer.Agricul, Natural Res. Engine. Adva. Biomed. Engine* **3**(5) 214-19

[5] Hassan S, Kee LS and Hussain HA 2013 Experimental study of palm oil mill effluent and oil palm frond waste mixture as an alternative biomass fuel *Journal of Engineering Science and Technology* **8**(6) 703-12

[6] Madaki YS and Lau S 2013 Palm Oil Mill Effluent (POME) from Malaysia palm oil *International Journal of Science, Environmentand Technology* **2**(6) 1138-55

[7] Bala JD, Japareng L and Norli I 2014 Palm oil mill effluent (POME) treatment "Microbial communities in an anaerobic digester": A Review *Interntional Journal of Scientific and Research Publications* **4**(6) 1-24

[8] Sandhya A, Sridevi A, Suvarnalatha D and Narasimha G 2015 Production and optimization of phytase by *Aspergillus niger Der Pharmacia Lettre.* **7**(12) 148-53

[9] Habib MAB, Yusoff FM, Phang SM, Ang K and Mohammed S 1997 Nutritional values of chironomid larvae grown in palm oil mill effluent and algae culture *Aquaculture* **158** 95-105

[10] Hurrell RF, Lynch S, Bothwell T, Cori H, Glahn R, Hertrampf E, Kratky Z, Miller D, Rodenstein M, Streekstra H, Teucher B, Turner E, Yeung CK and Zimmermann MB 2004 Enhancing the absorption of fortification iron A sustain task force report 2014 *International Journal of Vitamin and Nutrition Research* **74** 387-401

[11] Mullaney Ej, Daly C and Ullah AJH 2000 Advances in phytase research *Adv. Appl. Microbiol* **47** 157-99

[12] Joshi JB 2014 Phytase – A Key to unlock phytate complex *International Journal of Pure and Applied Bioscience* **2**(6) 304-13.

[13] Selvamohan T, Ramadas V and Rejibeula M 2012 Optimization of phytase production by *Pseudomonas sp.* isolated from poultry faces *International Journal of Modern Enggineering Research* **2**(3) 1326-30

[14] Shianna GB and Govindarajulu V 2014 Phytase production by *Aspergillus niger* CFR 335 and *Aspergillus ficuum* SGA 01 through submerged and solid state fermentation *The Scientific World Journal* **Article ID 392615** 1-6

[15] Awad GEA, Mohamed MIH, Enas ND and Mona AE 2014 Optimization of phytase production by *Penicillium purpurogenum* GE1 under solid state fermentation by using Box-Bahnker design *Saudi Journal Sciences* **21** 81-88

[16] Garraway MO and Robert CE 1991 *Fungal nutrition and physiology* Kreger Publishing Company, Malabar, Florida. P

[17] Engelen AJ, Van der Heeft FC, Randsdorp PH and Smit EL 1994 Simple and rapid determination of phytase activity *J.AOAC. Int* **77**(3) 760-64

[18] Moreira, KA, Polyanna NH, Marilia HCM, Porto TS, Micheler RS, Cristina MS, Anna LFP and Carlos RS 2014 Optimation of phytase production by *Aspergillus japonicus* Saito URM 5633 using cassava bast as substrate in solid state fermentation *African Journal of microbiology Research* **8**(9) 929-38

[19] Suleimenova Z, Nurlan A, Aigul K, Kairat M and Zhazira S 2016 Effect of different cultural conditions for phytase production by *Aspergillus niger* in submerged fermentation *Advance in Enzyme Research* **4** 62-67

[20] Subagiyo, Sebastian M and Triyanto 2015 Pengaruh penambahan berbagai jenis sumber karbon, nitrogen dan fosfat pada medium deMan Rogosa and Sharpe (MRS) terhadap pertumbuhan bakteri asam laktat terpilih yang diisolasi dari intestinum udang penaeid *Jurnal Kelautan Tropis* **18**(3) 127-132

[21] Purwadaria T, Roswita I, Arnold P S and Susana IWR 2003 The activity of phytase and phosphorus content of fermented dry palm oil mill effluent (POME) and rice bran with *Aspergillus oryzae* GS-66 *Jurnal Bioteknologi Pertanian* **8**(2) 46 – 51

[22] Kanti A and I. Made S 2016 Comparison of *Neurospora crassa* and *Neurospora sitophila* for phytase production at various temperatures *Biodiversitas* 7(2) 769-77

[23] Lee J, Park I and Cho J 2014 Extracellular phytase production by *Bacillus sp* T4 using solid state fermentation *The Journal of Animal and Plant Sciences* **24**(4) 1116 – 22

[24] Aly MM, Sanaa T, Soleh MA and Saleh AK 2015 Production and characterization of phytase from *Streptomyces luteogriseus* R10 isolated from decaying wood sampels *International Journal of Agriculture and Biology* **17**(3) 515-22

[25] Saithi S and Anan T 2016 Phytase production of *Aspergillus niger* on soybean meal by solid state fermentation using a rotating drum bioreactor *Agriculture and Agricultural Science Procedia* **11** 25-30

[26] Awad GEA, Mohamed MIH, Enas ND and Mona AE 2014 Optimazation of phytase production by *Penicillium purpurogenum* GE1 under solid state fermentation by using Box-Behnken disign *Saudi Journal of Biological Sciences* **21** 81-88